Compendium of Plant Genomes

Series editor

Chittaranjan Kole, Raja Ramanna Fellow, Department of Atomic Energy, Government of India, Kalyani, Nadia, West Bengal, India

Whole-genome sequencing is at the cutting edge of life sciences in the new millennium. Since the first genome sequencing of the model plant Arabidopsis thaliana in 2000, whole genomes of about 70 plant species have been sequenced and genome sequences of several other plants are in the pipeline. Research publications on these genome initiatives are scattered on dedicated web sites and in journals with all too brief descriptions. The individual volumes elucidate the background history of the national and international genome initiatives; public and private partners involved; strategies and genomic resources and tools utilized; enumeration on the sequences and their assembly; repetitive sequences; gene annotation and genome duplication. In addition, synteny with other sequences, comparison of gene families and most importantly potential of the genome sequence information for gene pool characterization and genetic improvement of crop plants are described.

Interested in editing a volume on a crop or model plant? Please contact Dr. Kole, Series Editor, at ckole2012@gmail.com

More information about this series at http://www.springer.com/series/11805

Shengyi Liu · Rod Snowdon
Boulos Chalhoub
Editors

The Brassica napus Genome

Springer

Editors
Shengyi Liu
Key Laboratory of Biology and Genetic Improvement of Oil Crops, the Ministry of Agriculture
Oil Crops Research Institute of the Chinese Academy of Agricultural Sciences
Wuhan, China

Rod Snowdon
Department of Plant Breeding, IFZ Research Centre for Biosystems, Land Use and Nutrition
Justus Liebig University
Giessen, Germany

Boulos Chalhoub
Institute of Crop Science
Zhejiang University
Hangzhou, China

and

Unité de Recherche en Génomique Végétale, Institut National de Recherche Agronomique (INRA), UMR1165, Organization and Evolution of Plant Genomes
Université d'Evry Val d'Essone
Evry, France

ISSN 2199-4781 ISSN 2199-479X (electronic)
Compendium of Plant Genomes
ISBN 978-3-030-09583-3 ISBN 978-3-319-43694-4 (eBook)
https://doi.org/10.1007/978-3-319-43694-4

This Springer imprint is published by the registered company Springer Nature Switzerland AG
The registered company address is: Gewerbestrasse 11, 6330 Cham, Switzerland

This book series is dedicated to
my wife Phullara, and our children
Sourav, and Devleena

Chittaranjan Kole

This book is dedicated to Prof. Tingdong Fu on the occasion of his 80th birthday

Preface to the Series

Genome sequencing has emerged as the leading discipline in the plant sciences coinciding with the start of the new century. For much of the twentieth century, plant geneticists were only successful in delineating putative chromosomal location, function, and changes in genes indirectly through the use of a number of 'markers' physically linked to them. These included visible or morphological, cytological, protein, and molecular or DNA markers. Among them, the first DNA marker, the RFLPs, introduced a revolutionary change in plant genetics and breeding in the mid-1980s, mainly because of their infinite number and thus potential to cover maximum chromosomal regions, phenotypic neutrality, absence of epistasis, and codominant nature. An array of other hybridization-based markers, PCR-based markers, and markers based on both facilitated construction of genetic linkage maps, mapping of genes controlling simply inherited traits, and even gene clusters (QTLs) controlling polygenic traits in a large number of model and crop plants. During this period, a number of new mapping populations beyond F2 were utilized and a number of computer programs were developed for map construction, mapping of genes, and for mapping of polygenic clusters or QTLs. Molecular markers were also used in studies of evolution and phylogenetic relationship, genetic diversity, DNA-fingerprinting, and map-based cloning. Markers tightly linked to the genes were used in crop improvement employing the so-called marker-assisted selection. These strategies of molecular genetic mapping and molecular breeding made a spectacular impact during the last one and a half decades of the twentieth century. But still they remained 'indirect' approaches for elucidation and utilization of plant genomes since much of the chromosomes remained unknown and the complete chemical depiction of them was yet to be unraveled.

Physical mapping of genomes was the obvious consequence that facilitated development of the 'genomic resources' including BAC and YAC libraries to develop physical maps in some plant genomes. Subsequently, integrated genetic-physical maps were also developed in many plants. This led to the concept of structural genomics. Later on, emphasis was laid on EST and transcriptome analysis to decipher the function of the active gene sequences leading to another concept defined as functional genomics. The advent of techniques of bacteriophage gene and DNA sequencing in the 1970s was extended to facilitate sequencing of these genomic resources in the last decade of the twentieth century.

As expected, sequencing of chromosomal regions would have led to too much data to store, characterize, and utilize with the-then available computer software could handle. But development of information technology made the life of biologists easier by leading to a swift and sweet marriage of biology and informatics, and a new subject was born—bioinformatics.

Thus, evolution of the concepts, strategies, and tools of sequencing and bioinformatics reinforced the subject of genomics—structural and functional. Today, genome sequencing has traveled much beyond biology and involves biophysics, biochemistry, and bioinformatics!

Thanks to the efforts of both public and private agencies, genome sequencing strategies are evolving very fast, leading to cheaper, quicker, and automated techniques right from clone-by-clone and whole-genome shotgun approaches to a succession of second-generation sequencing methods. Development of software of different generations facilitated this genome sequencing. At the same time, newer concepts and strategies were emerging to handle sequencing of the complex genomes, particularly the polyploids.

It became a reality to chemically—and so directly—define plant genomes, popularly called whole-genome sequencing or simply genome sequencing.

The history of plant genome sequencing will always cite the sequencing of the genome of the model plant Arabidopsis thaliana in 2000 that was followed by sequencing the genome of the crop and model plant rice in 2002. Since then, the number of sequenced genomes of higher plants has been increasing exponentially, mainly due to the development of cheaper and quicker genomic techniques and, most importantly, development of collaborative platforms such as national and international consortia involving partners from public and/or private agencies.

As I write this preface for the first volume of the new series 'Compendium of Plant Genomes,' a net search tells me that complete or nearly complete whole-genome sequencing of 45 crop plants, eight crop and model plants, eight model plants, 15 crop progenitors and relatives, and three basal plants is accomplished, the majority of which are in the public domain. This means that we nowadays know many of our model and crop plants chemically, i.e., directly, and we may depict them and utilize them precisely better than ever. Genome sequencing has covered all groups of crop plants. Hence, information on the precise depiction of plant genomes and the scope of their utilization is growing rapidly every day. However, the information is scattered in research articles and review papers in journals and dedicated Web pages of the consortia and databases. There is no compilation of plant genomes and the opportunity of using the information in sequence-assisted breeding or further genomic studies. This is the underlying rationale for starting this book series, with each volume dedicated to a particular plant.

Plant genome science has emerged as an important subject in academia, and the present compendium of plant genomes will be highly useful both to students and teaching faculties. Most importantly, research scientists involved in genomics research will have access to systematic deliberations on the plant genomes of their interest. Elucidation of plant genomes is of interest not only for the geneticists and breeders, but also for practitioners of an array of plant science disciplines, such as taxonomy, evolution, cytology, physiology, pathology, entomology, nematology, crop production, biochemistry, and obviously bioinformatics. It must be mentioned that information regarding each plant genome is ever-growing. The contents of the volumes of this compendium are therefore focusing on the basic aspects of the genomes and their utility. They include information on the academic and/ or economic importance of the plants, description of their genomes from a molecular genetic and cytogenetic point of view, and the genomic resources developed. Detailed deliberations focus on the background history of the national and international genome initiatives, public and private partners involved, strategies and genomic resources and tools utilized, enumeration on the sequences and their assembly, repetitive sequences, gene annotation, and genome duplication. In addition, synteny with other sequences, comparison of gene families, and, most importantly, potential of the genome sequence information for gene pool characterization through genotyping by sequencing (GBS) and genetic improvement of crop plants have been described. As expected, there is a lot of variation of these topics in the volumes based on the information available on the crop, model, or reference plants.

I must confess that as the series editor, it has been a daunting task for me to work on such a huge and broad knowledge base that spans so many diverse plant species. However, pioneering scientists with lifetime experience and expertise on the particular crops did excellent jobs editing the respective volumes. I myself have been a small science worker on plant genomes since the mid-1980s and that provided me the opportunity to personally know several stalwarts of plant genomics from all over the globe. Most, if not all, of the volume editors are my longtime friends and colleagues. It has been highly comfortable and enriching for me to work with them on this book series. To be honest, while working on this series I have been and will remain a student first, a science worker second, and a series editor last. And I must express my gratitude to the volume editors and the chapter authors for providing me the opportunity to work with them on this compendium.

I also wish to mention here my thanks and gratitude to the Springer staff, Dr. Christina Eckey and Dr. Jutta Lindenborn in particular, for all their constant and cordial support right from the inception of the idea.

I always had to set aside additional hours to edit books besides my professional and personal commitments—hours I could and should have given to my wife, Phullara, and our kids, Sourav, and Devleena. I must mention that they not only allowed me the freedom to take away those hours from them but also offered their support in the editing job itself. I am really not sure whether my dedication of this compendium to them will suffice to do justice to their sacrifices for the interest of science and the science community.

Kalyani, India | Chittaranjan Kole

Preface

The major global crop *Brassica napus* L. (rapeseed, oilseed rape, canola, kale, swede, rutabaga; genome AACC, 2n = 4x = 38) is a recent allopolyploid species, formed during the past ~7500 years by interspecific hybridization between *B. rapa* (AA, 2n = 2x = 20) and *B. oleracea* (CC, 2n = 2x = 18). This hybridization is believed to have occurred spontaneously, with at least three independent origins, but most likely under human cultivation as no wild forms of *B. napus* are known. The diploid progenitors each derived via ancestral hexaploidy from a common origin, but despite relatively close genome homology, evolved into separate species with distinct karyotypes and genome structures. Their hybrid, *B. napus*, represents the collision of two related, highly duplicated genomes in a single nucleus. As such, *B. napus* has become an important model for investigation of the consequences of polyploidy on duplicated selective forces during crop plant evolution. The availability of assembled *B. napus* genomes thus provides intriguing insight into the genome restructuring and selection processes associated with polyploidization and human selection from agricultural traits.

At the time of completion, the *B. napus* genome was the most highly duplicated plant genome yet sequenced and also the genome with the highest content of annotated genes (more than 100,000). The genome sequence therefore provided a unique opportunity to examine the consequences of large-scale gene duplication, structural and functional crosstalk within and among highly duplicated gene pathways and epigenetic regulation of gene expression and modification. The ability to readily generate and resequence synthetic *B. napus* forms, derived by embryo rescue from new interspecific crosses between different A-subgenome and C-subgenome diploid progenitors, provided an unprecedented view of widespread homeologous exchanges during the allopolyploidization process. Large-scale and small-scale genome restructuring through homeologous exchanges, which was also found to be widespread and prevalent in natural *B. napus*, appears to have shaped the modern genomes of different *B. napus* accessions, creating a basis for quantitative trait variation and leading to human selection of ecogeographically and morphologically divergent crop types. As an example, breeding selection for specific genome rearrangements led to loss of glucosinolate genes but expansion of oil biosynthesis genes, providing a genetic basis for a globally important oilseed crop. The availability of high-quality *B. napus* genome sequence assemblies thus enables novel insights into recent allopolyploid genome evolution and its impact on plant domestication and crop improvement.

In contrast to other concerted international plant genome sequencing efforts, many of which have been initiated within the framework of coordinated international sequencing consortia, the ultimate completion of the *B. napus* genome was enabled by informal cooperation between independent genome assembly efforts on different reference genotypes in Europe (winter-type oilseed rape), China (semi-winter-type rapeseed), and Canada (spring-type canola), respectively. The exchange among these programs was facilitated and encouraged by the steering committee of the 'Multi-national *Brassica* Genome Project,' which promotes and coordinates international cooperation in the area of *Brassica* genomics. The published reference assembly of the European winter-type oilseed rape genotype Darmor-*bzh* (Chalhoub et al. 2014) represents the result of a highly successful international research community effort to exchange and share data from competing research programs in the interests of scientific progress. The result was a unique genome assembly, at the time the most complex plant genome to be successfully assembled into a high-quality reference, which provided a hugely valuable resource for research into allopolyploid crop evolution and for breeding and genetics in *B. napus* and related crops.

In this volume, authors from the thriving international *B. napus* research community present deep insight into genetic and genomic analysis and applications enabled by the *B. napus* genome. Introductory chapters outlined the importance of *B. napus* as a crop and as a cytogenetic model for the consequences and importance of polyploidy and introduced the state of the art with regard to mapping of genes and quantitative trait loci for agronomic traits; many of mapping researches were based on the assembled reference genome resource. Five chapters broadly cover genome organization, one of the most interesting and complex features of the *B. napus* genome, with detailed contributions on genome and gene duplication, organization and evolution of repeat sequences, homeologous exchanges and the influence of these factors on gene expression and epigenetic regulation. Insight into the mitochondrial and chloroplast genomes of *B. napus* is presented in the context of *Brassica* evolution and crop differentiation, a topic which is also at the core of gene family differentiation among different *Brassica* species and forms. The impact of allopolyploidization on selection for important agronomic traits is underlined by three chapters which describe the complexities of trait-related gene evolution in relation to oil biosynthesis pathway genes, glucosinolate pathway genes, and resistance genes, respectively. The book closes with an overview of valuable *B. napus* genomic resources and outlooks on future applications of the B. napus genome for genome-facilitated breeding of oilseed rape and for research on structural, evolutionary, and functional genomics in *B. napus*.

As sequencing technologies and genome assembly strategies become increasingly cost-effective, efficient, and accurate, the first reference genome assembly of *B. napus* was likely one of the last complex crop genomes to be assembled on a backbone of Titanium Roche 454 and Sanger sequences. Ultra-cheap, ultra-high throughput next-generation sequencing, the ever-increasing accuracy and cost-effectiveness of long-read sequencing technologies, and new assembly procedures including scaffolding and phasing-based chromatin conformation technologies present completely new opportunities to accurately sequence complex crop genomes. As this volume is published, a multitude of new *B. napus* genomes has already been assembled using new-generation strategies, and many will almost certainly be published in the near future. This will give rise to a new era of crop genome analysis, moving far beyond single reference genome sequences and toward an association of pan-genome variation with agronomic and biological trait information. Implementing this great magnitude of new information to advance breeding will be one of the great challenges for coming generations of *Brassica* geneticists and breeders. Even with new possibilities offered by genome editing in association with genomic knowledge, considerable challenges still lie ahead: The complex genetics underlying quantitative disease resistances, nutrient and water use efficiency and heterosis must be better understood in order to make targeted use of genome diversity in agriculture. A better understanding of chromosome structure, homeologous pairing, recombination, and genome stability will be essential to make best use of available (and *de novo*) diversity for *B. napus* improvement, for example, by better control of new interspecific hybridization to exploit the vast diversity present in the diploid progenitors of *B. napus*. Finally, maximizing the value of genome data in *B. napus* and other crops will rely in future on coordinated, integrated data management and analysis systems as a basis to navigate between diverse, multidimensional omics datasets from the international research community and implement them to draw biological insight into the complex relationship between genotype and environment. The first *B. napus* genome has laid an excellent foundation for this quest, and we look forward to working together with future *Brassica napus* researchers to continue this momentum.

Wuhan, China — Shengyi Liu
Giessen, Germany — Rod Snowdon
Evry, France — Boulos Chalhoub

Contents

Contributors

Abdul Baten Southern Cross Plant Science, Southern Cross University, Lismore, NSW, Australia

Jacqueline Batley School of Biological Sciences and Institute of Agriculture, University of Western Australia, Crawley, WA, Australia

Anne-Sophie Bouchet IGEPP, AGROCAMPUS OUEST, INRA, University Rennes, Le Rheu, France

Boulos Chalhoub Institute of Crop Science, Zhejiang University, Hangzhou, China; Institut National de Recherche Agronomique (INRA)/Université d'Evry Val d'Essone, Unité de Recherche en Génomique Végétale, UMR1165, Organization and Evolution of Plant Genomes, Evry, France

Jianmei Chen National Navel Orange Engineering Research Center, College of Navel Orange, Gannan Normal University, Ganzhou, China

Xiaohui Cheng Key Laboratory of Biology and Genetic Improvement of Oil Crops, the Ministry of Agriculture, Oil Crops Research Institute, Chinese Academy of Agricultural Sciences, Wuhan, China; Hubei Collaborative Innovation Center for Green Transformation of Bio-Resources, Hubei University, Wuhan, China

Anne-Marie Chèvre INRA, UMR1349 IGEPP, Le Rheu, France

Pu Chu State Key Laboratory of Crop Genetics and Germplasm Enhancement, College of Agriculture, Nanjing Agricultural University, Nanjing, China

Olivier Coriton INRA, UMR1349 IGEPP, Le Rheu, France

Régine Delourme IGEPP, AGROCAMPUS OUEST, INRA, University Rennes, Le Rheu, France

Aria Dolatabadian School of Biological Sciences, University of Western Australia, Crawley, WA, Australia

Wolfgang Friedt Department of Plant Breeding, University of Giessen, Giessen, Germany

Tingdong Fu College of Plant Science and Technology, Huazhong Agricultural University, Wuhan, China

Rafaqat Ali Gill Key Laboratory of Biology and Genetic Improvement of Oil Crops, the Ministry of Agriculture, Oil Crops Research Institute, Chinese Academy of Agricultural Sciences, Wuhan, China

Laurie Grandont Institut Jean-Pierre Bourgin, INRA, AgroParisTech, CNRS, Université Paris-Saclay, Versailles Cedex, France

Rongzhan Guan State Key Laboratory of Crop Genetics and Germplasm Enhancement, College of Agriculture, Nanjing Agricultural University, Nanjing, China

Zhiyong Hu Key Laboratory of Biology and Genetic Improvement of Oil Crops, the Ministry of Agriculture, Oil Crops Research Institute, Chinese Academy of Agricultural Sciences, Wuhan, China

Wei Hua Key Laboratory of Biology and Genetic Improvement of Oil Crops, the Ministry of Agriculture, Oil Crops Research Institute, Chinese Academy of Agricultural Sciences, Wuhan, China

Junyan Huang Key Laboratory of Biology and Genetic Improvement of Oil Crops, the Ministry of Agriculture, Oil Crops Research Institute, Chinese Academy of Agricultural Sciences, Wuhan, China; Hubei Collaborative Innovation Center for Green Transformation of Bio-Resources, Hubei University, Wuhan, China

Eric Jenczewski Institut Jean-Pierre Bourgin, INRA, AgroParisTech, CNRS, Université Paris-Saclay, Versailles Cedex, France

Mélanie Jubault IGEPP, AGROCAMPUS OUEST, INRA, University Rennes, Le Rheu, France

Hyun Hee Kim Department of Life Science, Chromosome Research Institute, Sahmyook University, Seoul, Republic of Korea

Graham J. King Southern Cross Plant Science, Southern Cross University, Lismore, NSW, Australia

Anne Laperche IGEPP, AGROCAMPUS OUEST, INRA, University Rennes, Le Rheu, France

Jonghoon Lee Department of Plant Science, Plant Genomics and Breeding Institute, and Research Institute of Agriculture and Life Sciences, College of Agriculture and Life Sciences, Seoul National University, Seoul, Republic of Korea

Jiana Li Chongqing Rapeseed Engineering Research Center, College of Agronomy and Biotechnology, Academy of Agricultural Sciences, Southwest University, Beibei, Chongqing, PR China

Shengyi Liu Key Laboratory of Biology and Genetic Improvement of Oil Crops, the Ministry of Agriculture, Oil Crops Research Institute of the Chinese Academy of Agricultural Sciences, Wuhan, China; Hubei Collaborative Innovation Center for Green Transformation of Bio-Resources, Hubei University, Wuhan, China

Kun Lu Chongqing Rapeseed Engineering Research Center, College of Agronomy and Biotechnology, Academy of Agricultural Sciences, Southwest University, Beibei, Chongqing, PR China

Eric Lyons School of Plant Sciences, BIO5 Institute, iPlant Collaborative, University of Arizona, Tucson, AZ, USA

Martin A. Lysak Central European Institute of Technology (CEITEC), Masaryk University, Brno, Czech Republic

Lixin Ma Hubei Collaborative Innovation Center for Green Transformation of Bio-Resources, Hubei University, Wuhan, China

Maria-J. Manzanares-Dauleux IGEPP, AGROCAMPUS OUEST, INRA, University Rennes, Le Rheu, France

Annaliese S. Mason Department of Plant Breeding, IFZ Research Centre for Biosystems, Land Use and Nutrition, Justus-Liebig University, Giessen, Germany

Nathalie Nesi IGEPP, AGROCAMPUS OUEST, INRA, University Rennes, Le Rheu, France

Sophie Paillard IGEPP, AGROCAMPUS OUEST, INRA, University Rennes, Le Rheu, France

Sampath Perumal Department of Plant Science, Plant Genomics and Breeding Institute, and Research Institute of Agriculture and Life Sciences, College of Agriculture and Life Sciences, Seoul National University, Seoul, Republic of Korea; Agriculture and Agri-Food Canada, Saskatoon, SK, Canada

Birgit Samans Plant Breeding Department, IFZ Research Centre for Biosystems, Land Use and Nutrition, Justus Liebig University, Giessen, Germany

Rod Snowdon Plant Breeding Department, IFZ Research Centre for Biosystems, Land Use and Nutrition, Justus Liebig University, Giessen, Germany

Fengming Sun Key Laboratory of Biology and Genetic Improvement of Oil Crops, the Ministry of Agriculture, Oil Crops Research Institute, Chinese Academy of Agricultural Sciences, Wuhan, China

Haibao Tang Center for Genomics and Biotechnology, Fujian Agriculture and Forestry University, Fuzhou, Fujian Province, China; School of Plant Sciences, BIO5 Institute, iPlant Collaborative, University of Arizona, Tucson, AZ, USA

Chaobo Tong Key Laboratory of Biology and Genetic Improvement of Oil Crops, the Ministry of Agriculture, Oil Crops Research Institute, Chinese Academy of Agricultural Sciences, Wuhan, China; Hubei Collaborative Innovation Center for Green Transformation of Bio-Resources, Hubei University, Wuhan, China

Jingxing Tu College of Plant Science and Technology, Huazhong Agricultural University, Wuhan, China

Nomar Espinosa Waminal Department of Plant Science, Plant Genomics and Breeding Institute, and Research Institute of Agriculture and Life Sciences, College of Agriculture and Life Sciences, Seoul National University, Seoul, Republic of Korea; Department of Life Science, Plant Biotechnology Institute, Sahmyook University, Seoul, Republic of Korea; Department of Life Science, Chromosome Research Institute, Sahmyook University, Seoul, Republic of Korea

Christian Werner Department of Plant Breeding, Justus Liebig University, Giessen, Germany

Yang Xiang Guizhou Rapeseed Institute, Guizhou Academy of Agricultural Sciences, Guiyang, China

Hua Yang School of Agriculture and Food Science, The University of Queensland, St Lucia, QLD, Australia

Tae-Jin Yang Department of Plant Science, Plant Genomics and Breeding Institute, and Research Institute of Agriculture and Life Sciences, College of Agriculture and Life Sciences, Seoul National University, Seoul, Republic of Korea

Xingtan Zhang Center for Genomics and Biotechnology, Fujian Agriculture and Forestry University, Fuzhou, Fujian Province, China

Academic and Economic Importance of *Brassica napus* Rapeseed

1

Wolfgang Friedt, Jingxing Tu and Tingdong Fu

Abstract

Rapeseed or canola (B. napus) is the second most important oilseed crop of the world. It is also a favourite plant for basic and breeding research. Due to its origin and evolution, rapeseed has a complex polyploid genome. Recent sequencing of the corresponding genomes provides the basis for a better understanding and exploitation of the genetic diversity involved in major rapeseed traits. However, directed selection for major quality characteristics, i.e. minimal erucic acid content and low glucosinolate level, has caused genetic bottlenecks limiting genetic variation in the current gene pools of cultivated oilseed rape (OSR). Therefore, broadening genetic diversity is an important aim of research and a necessary prerequisite for further progress by OSR breeding. In agricultural production, rapeseed is nowadays an indispensable component of crop rotations in major growing areas such as Australia, Western Canada, Central China and many countries of the European Union. In many cases, OSR is the only leaf crop among dominating cereal species. Therefore, OSR as a component of crop rotations helps to maintain soil fertility and contributes to sustainable production therefore. As a major cash crop OSR substantially contributes to farmers' incomes and therefore helps to stabilize rural populations. Beyond that, as major globally traded agricultural commodities rapeseed/canola and rapeseed/canola oil and meal significantly input the national products of a number of countries e.g. Canada. Rapeseed/canola is a raw material for vegetable oil and extraction meal as feed, food and fuel. The oil is mainly used as a high-value salad oil for dressings etc. due to its high contents of oleic acid (ca. 60%) and poly-unsaturated linolenic acid (omega-3, ca. 10%). Nevertheless, a large part is also used as a mobility fuel for diesel cars and tractors, particularly in Germany and Europe. The extraction meal (and protein) from oil processing is now recognized as a highly valuable animal feed, particularly for ruminants (cattle) but also for monogastric farm animals (pigs, poultry). Furthermore, the interest in rapeseed protein for the purpose of human nutrition is increasing. Optimal contents of the major compounds mentioned before represent the main requirements for rapeseed/canola varieties today. Consequently, quality characteristics are major criteria for variety testing and registration therefore. Other major requirements for modern rapeseed varieties are yield

W. Friedt (✉)
Department of Plant Breeding, University of Giessen, 35392 Giessen, Germany
e-mail: Wolfgang.Friedt@agrar.uni-giessen.de

J. Tu · T. Fu
College of Plant Science and Technology, Huazhong Agricultural University, Wuhan 430070, China

S. Liu et al. (eds.), *The Brassica napus Genome*, Compendium of Plant Genomes,
https://doi.org/10.1007/978-3-319-43694-4_1

and various agronomic traits securing seed yield, i.e. resistance against fungal diseases and insect pests. Because of the environmental concerns and the rejection of agrochemicals such as insecticides (e.g. ban of neonicotinoids in the EU), genetic approaches of establishing resistant crop cultivars constantly gain importance. OSR breeding has long been a relatively ordinary process of repeated selection for resistance, quality and yield, the propagation of improved populations and their release as new open pollinated (OP) varieties. Since this approach is not very effective, breeders have been interested in breeding hybrids instead. Today, F1 hybrids represent the major variety type. They are produced with genetic male sterility systems, most of which are based on cytoplasmic mutations causing male sterility (cms). The higher performance of hybrids is caused by "heterotic effect" which largely depends on the genetic distance between the parents. Therefore, distant genetic pools need to be generated for the development of female and male parents. The future potential of variety design is consequently determined by the usefulness of the genetic pools and the performance of hybrid parents extracted from them. Modern breeding tools based on biotechnology and genomics can substantially contribute to a better exploitation of useful genetic diversity, i.e. specific genes and genetic networks. Better varieties are a precondition for further crop improvement. Future quality OSR cultivars deserve high yield potential, combined with good stability due to disease and pest resistance. This will be the basis for exploiting the great agronomical and industrial advantages of the rapeseed plant.

1.1 General Relevance of *Brassica napus*

Scientific research on *Brassica napus* is particularly relevant because of the following reasons:

(i) It is a major crop plant and member of a large plant family (*Brassicaceae*) including many crops and weeds with many unique features.
(ii) It also represents one of the few dicot plants which are particularly important today to break plant production systems (rotations) often dominated by cereals and maize.
(iii) Rapeseed is a very valuable plant, used for various purposes: as a forage for cattle, as an oil plant ("oilseed rape") for nutritional and industrial purposes and as a protein crop for producing compound animal feed.
(iv) As a major seed component, rapeseed oil can have very different fatty acid profiles and corresponding uses, i.e. high erucic, canola (hi-oleic, hi-linolenic), high oleic (HOLLI).
(v) Other than many other species, it is a allopolyploid organism derived from an ancient hybridization between two (crop) species which again derived from prehistoric hybridizations; for this reason, genome analysis of *B. napus* can give deep and unique insights into the structure, function and regulation of a complex genome showing variation on the sub-genomic, chromosomal, sub-chromosomal and molecular levels.

1.2 Origin and Evolution of Rapeseed—A Plant with a Complex Polyploid Genome

The mustard family (*Brassicaceae*, *Cruciferae*) comprises about 300 genera with around 4000 species. These include a number of agriculturally important plants, either as crops (vegetables, spices and oil plants) or weeds (e.g. *Raphanus raphanistrum*, *Sinapis arvensis*). Major crop plants are domesticated forms of wild cabbage (*Brassica oleracea*), e.g. white and red cabbage, kohlrabi (turnip cabbage), Brussels sprouts, and broccoli. Further major crop members of the genus *Brassica* are pak choi, Chinese cabbage (*B. rapa* subsp. *chinensis*), white beet (*B. rapa* subsp. *rapa*), turnip rape (*B. rapa* subsp. *oleifera*), swede (*B. napus* subsp. *rapifera*) and rapeseed (*B. napus* subsp. *napus*). Other nameable members are *B. nigra* (black mustard) and *Sinapis alba* (white mustard). Representatives of the genus *Raphanus* are radish types (*R. sativus*), horseradish (*Armoracia rusticana*) and cresse (*Lepidium*). Another very important member of the family is the model plant *Arabidopsis thaliana* (common wallcress) which has been very useful for basic scientific research in the family and far beyond.

The taxonomic relationships within the genus *Brassica* were basically revealed by cytological studies in the 1930s. There are obviously two ploidy levels represented (i) by three primary species: *B. rapa* (A, $n = 10$), *B. oleracea* (C, $n = 9$) and *B. nigra* (B, $n = 8$) and (ii) three secondary species: *B. napus* (AC, $n = 19$), *B. juncea* (AB, $n = 18$) and *B. carinata* (BC, $n = 17$). The latter are amphidiploids (allotetraploids) derived from the primary species as has been confirmed by the re-synthesis of *B. napus* and other amphiploids ("*The triangle of U*"; Fig. 1.1; see U 1935).

More recent genetic and molecular studies give closer insights into the taxonomic relationships between the species and related genera (e.g. Mei et al. 2011; Liu et al. 2014; Chalhoub et al. 2014). According to the latter authors, rapeseed was formed presumably approximately 7500 years ago or later by hybridization between *B. rapa* and *B. oleracea*, followed by chromosome doubling, resulting in an allopolyploid progeny. Together with more ancient polyploidizations, this conferred an aggregate 72 × genome multiplication that we could detect by comparative genomic and bioinformatic analyses since the origin of angiosperms. It has been speculated that rapeseed originated in the Mediterranean where the distribution areas of the parental species overlap.

It is still debated from what time on rapeseed has been cultivated in Europe and used instead of animal fat. It seems that this was first the case in Western Europe from where the plant migrated to northern and later to southern Germany (sixteenth century). In the eighteenth and nineteenth centuries, rapeseed became widely used in central Europe where rapeseed oil (from *B. napus* or *B. rapa*) was mainly used as a lamp oil or for technical purposes. As oil plants, mainly the winter forms of rapeseed (*B. napus* ssp. *oleifera* L.) and turnip rape (*B. campestris* L. ssp. *oleifera*) became widely distributed due to the existing diversity. Summer forms are relatively unimportant in Germany and would

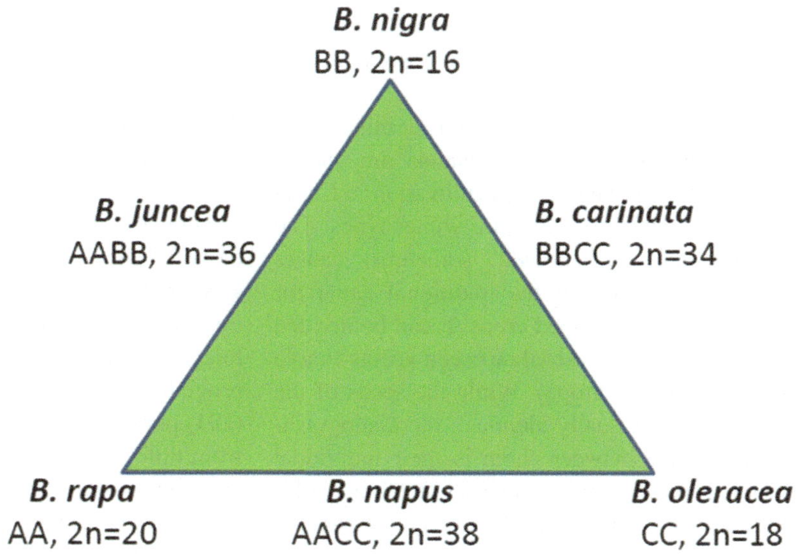

Fig. 1.1 U's triangle illustrates the origin of the polyploid crop plants *B. carinata* (Ethiop. mustard), *B. juncea* (Indian mustard) and *B. napus* (rapeseed) from crosses between the diploid parents indicated

mainly be grown in northern latitudes, e.g. as a substitute after winterkill of a previous winter (rapeseed) crop.

The genetic basis of winter rapeseed adapted to central Europe is thought to be rather limited: only a few landraces are known as basic germplasm formed under different climatic conditions and deviating in their vegetative growth and winter hardiness, "Lembkes Winterraps", the first German variety, selected from a landrace from Mecklenburg has obviously been used frequently as a cross parent in France, Poland and Sweden. The introduction of 0 (low erucic) from the summer forage variety, "Liho" and 00 quality (plus low glucosinolates from the polish summer variety, "Bronowski") caused a bottle-neck effect due to the fixation of several genomic regions involved in these traits. For example, Körber et al. (2012) evaluated the patterns of phenotypic diversity in a species-wide *B. napus* germplasm set of more than 500 inbreds with respect to various seedling development, agronomic and seed quality traits in greenhouse and field trials. They observed differences in phenotypic diversity among the examined eight germplasm types. The reduction of phenotypic diversity was on average more pronounced for the seedling development traits than for the agronomic and seed quality traits (Körber et al. 2012). These results along with other studies suggest that new genetic variation has to be introduced into current rapeseed elite material as a basis of future breeding progress.

Whereas winter 00 rapeseed dominates in Europe and spring types of canola are widely used in Canada, the rapeseed cultivation in China is based on alternative or semi-winter types. The use of the term "rapeseed" sometimes causes confusion: rapeseed is the traditional name for the *Brassicaceae* oilseed crops. It can be divided into two types—industrial rapeseed versus canola (00-rapeseed in Europe). While the seeds of the two types are visually identical, the distinguishing difference between them is their individual chemical profiles regarding fatty acids and glucosinolates. Generally, "industrial rapeseed" refers to any rapeseed with a high content (45% or more) of erucic acid in the oil. The name "Canola" was registered in 1979 in Canada and refers to the edible oil crop that is characterized by low erucic acid (less than 2%) and low levels of glucosinolates. In this chapter, the term rapeseed is used in a more general sense for any type of *B. napus*. Types especially developed for the use of seed oil (and protein) are also known as "oilseed rape" (OSR).

Crucifers including the brassicas are basically outcrossing species. Self-fertilization (selfing) is often impeded by genetic self-incompatibility systems preventing sexual crosses between parents with identical S alleles. In the cultivated types, self-fertile varieties have been selected, allowing the generation of inbred pure lines. Therefore, species like *B. juncea* or *B. napus* are in fact facultative inbreeding/outcrossing species, where either open-pollinated (OP) varieties or hybrids can be developed by breeding. Current commercial varieties of oilseed rape (*B. napus*) are predominantly hybrids all over the world. For example, in the European Union including Germany WOSR crops today are predominantly single-cross hybrids based on the application of male sterility systems such as the "Ogu-INRA" cytoplasmic male sterility (CMS) or the "Male Sterility Lembke" (MSL) system (Fig. 1.2).

1.3 Broadening Genetic Diversity for Oilseed Rape Research and Breeding

1.3.1 Using Existing Intraspecific Variation

Genetically diverse populations are necessary tools for phenotypic and genotypic analyses of a species. Such populations may represent collections of random inbred lines (e.g. RIL or SSD lines), introgression lines (IL), segregating F2 progenies, fixed inbreds or doubled haploid (DH) populations. Numerous collections derived from different parents and developed for various traits have been propagated and are used for genetic approaches such as mapping and gene cloning. The combined use of different resources like this enables the establishment of consensus

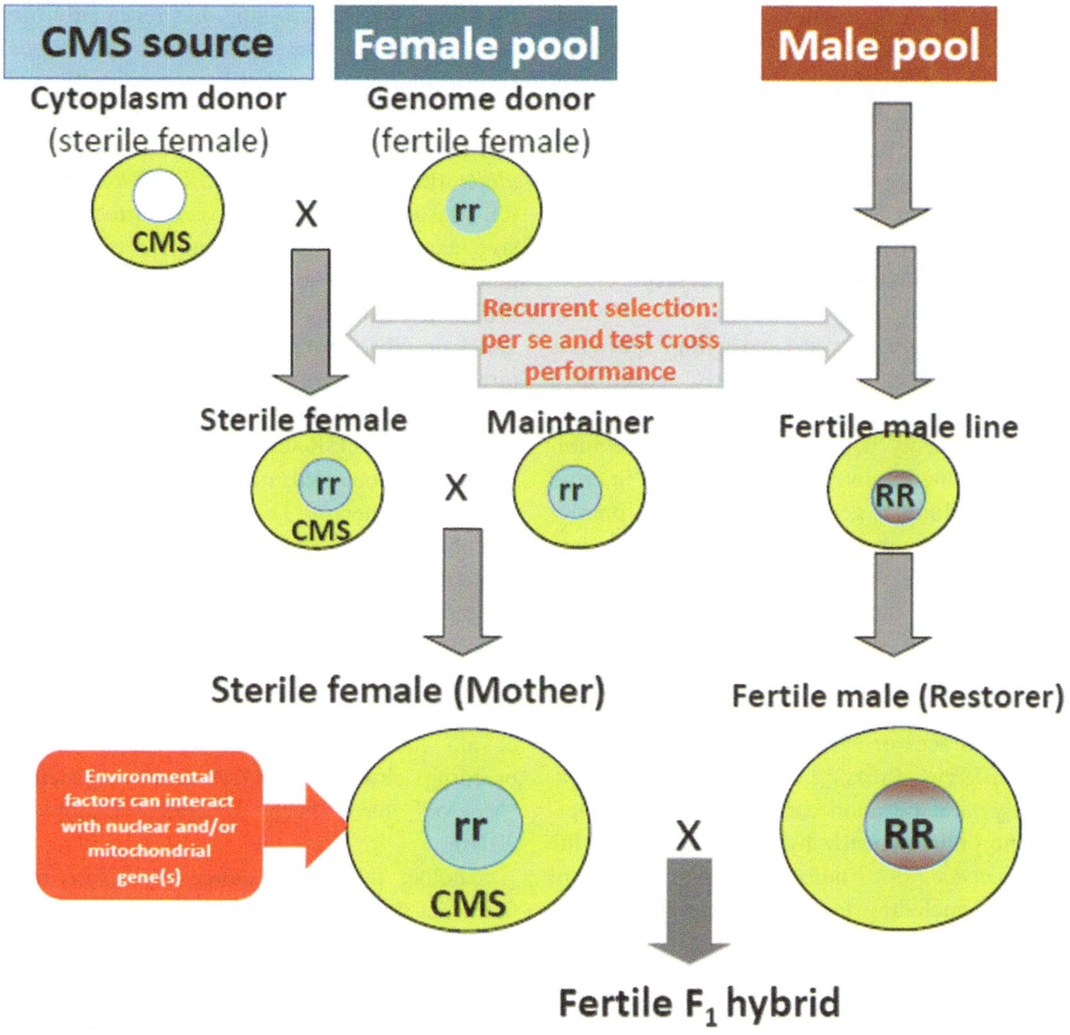

Fig. 1.2 Simplified scheme of rapeseed breeding based on a cytoplasmic male sterile (CMS) female (reproduced by a complementary fertile "maintainer" line) The CMS line is used as a mother (seed parent) for seed multiplication via pollination by a suitable male parent (restorer, pollen parent), restoring fertility by the action of a nuclear R allele corresponding to the rr allele of the maintainer

maps for the deep genomic analysis of complex traits. For example, Raman et al. (2013) have applied nine DH populations developed by various authors for different traits and purposes and developed a high-density consensus map:

- Ag-Castle/Topas (segregating for resistance to *Leptosphaeria maculans*)
- BLN2762/Surpass400 (*L. maculans* and shatter resistance, water-soluble carbohydrates)
- Lynx-037DH/Monty028DH (flowering time, seed oil quality)
- Maxol1/Westar (resistance to *L. maculans*)
- Skipton/Ag-Spectrum (various components of flowering time, race-specific and non-specific *L. maculans* resistance, C isotope discrimination, water-soluble carbohydrates)
- Tapidor/Ningyou7 (various components of flowering time, oil content, erucic acid, α-tocopherol content, glucosinolate concentration).

The markers represented on their array were sequenced and aligned with the *B. rapa* and *B. oleracea* genomes, and they provide insight into the Brassica A and C genomes (Raman et al. 2013).

1.3.2 Exploiting Interspecific and Intergeneric Diversity

Genetic diversity between and within species is generally valuable and particularly needed as a basis for crop breeding and enhancement. Genetic diversity is a precondition for recombination leading to novel genetic variation in cross progeny. Minimal genetic distance is a prerequisite for achieving heterotic effects, i.e. the yield advantage of F1 versus its parents, and therefore necessary for breeding hybrid cultivars (see below) with a high yield potential. Genetic variance can be induced by sexual hybridization (crossing) of released cultivars, adapted breeding lines, landraces or even alien exotic germplasm (i.e. different *Brassica* species). In addition, entirely new variation can be created by interspecific or intergeneric hybridization within the *Brassicaceae*. As indicated, novel forms of *B. napus* including forage and oil types can be created by interspecific hybridization between selected genotypes of the two parental species (Fig. 1.1). The progeny of such hybrids ("resynthetic [RS] rapeseed") can give rise to completely new progenies differing from commercial OSR cultivars and elite material, as demonstrated, e.g. by Seyis et al. (2003, 2006). At the same time, different interspecific families show obvious genetic distance and form separate sub-clusters within the progenies. The development of inbred lines derived from such wide crosses can lead to new test hybrids (F1) with elevated seed and oil yield potential as compared to former variety types (Friedt et al. 2004; Gehringer et al. 2007; Seyis et al. 2003, 2006).

Such introgressive breeding, whereby a novel phenotypic or allelic variant in a wild or related species is crossed with a crop species followed by successive backcrossing, has repeatedly been carried out in rapeseed/canola (*B. napus*). Hybridization with related species such as *B. rapa*, *B. juncea* and *B. carinata* has resulted in the transfer of resistance to *Leptosphaeria maculans* (blackleg) and *Verticillium longisporum* (cf. Mason and Batley 2015). A further impressive example is the development of breeding material resistant to clubroot (*Plasmodiophora brassicae*). Already in 1987, the clubroot resistant kale (*B. oleracea* ECD-15) was crossed to turnip rape (*B. rapa* ECD-04) to create new *B. napus* RS 15/04 ($2n = 38$). This progeny was crossed to WOSR cv. "Falcon". Haploid F1 pollen was cultured in vitro, and DH plants regenerated thereof. The selected DH line 47/19 was then backcrossed two times to cv. Falcon to generate a BC2F1 population segregating for three dominant, race-specific clubroot resistance (CR) genes. After further crossing and haploidization steps, the resistant WOSR cv. "Mendel" could be released in 2001 (Diederichsen et al. 2009). Using this cultivar in further breeding new CR WOSR cultivars such as "Mentor" have been generated (Anonymus 2016).

Another interesting source of disease resistance is seen in the B genome of *B. nigra* (BB) and *B. carinata* (BBCC). In this regard, the work of Fredua-Agyeman et al. (2014) using *B. napus* × *B. carinata* interspecific hybrids provides evidence that the *Brassica* B genome chromosome B3 carries blackleg resistance gene (s). The authors showed that the B genome chromosomes were inherited several generations along with the *B. napus* chromosomes.

Re-synthesis of rapeseed has also been used to create new genetic variation for quality traits. In such cases, it is usually necessary to backcross the RS material with elite rapeseed (*B. napus*) lines or varieties. An example is represented by the winter rapeseed progeny with the pedigree {[(*B. rapa* subsp. *oleifera* (DC.) Metzg. × Quinta) × Quinta] × Belinda} × rapeseed line, which has been used as a valuable source of the yellow seed trait (J. Koch, pers. Mitt.)

accompanied by a low dietary fibre content of the seed and associated with better digestibility, particularly for monogastric farm animals.

Even wider hybridizations across the genus level are achievable. Prominent examples are fusions of isolated cells (protoplasts) of rapeseed and radish (*R. sativus*). Such protoplast fusions have, for example, been the basis for the development of the Ogu-INRA CMS system for hybrid breeding (e.g. Primard-Brisset et al. 2005; see below).

1.4 Agronomy, Environment and Sustainability

1.4.1 Rapeseed Production and Crop Rotation

Environmentally friendly and sustainable land use and agricultural production are based on many factors, one of the main being diversity of crops and agricultural production system(s). Nowadays, the production systems in many countries are characterized by a small number of plant species which are grown in short rotations on relatively large areas as compared to former times. Consequently, few (cash) crops dominate the land use; for example, only four crop plants, i.e. wheat, maize, barley and oilseed rape (OSR), cover about 80% of the arable land in Germany; OSR (ca. 12% of acreage) has been the major non-cereal crop recently. For ecological but also economic reasons, it is necessary to maintain a minimum acreage of leaf crops such as OSR in cereal-dominated agricultural systems in order to avoid or reduce negative effects of short rotations or monoculture, e.g. increased incidence of diseases and pests. On the other hand, OSR itself is also subject to many pests and diseases often causing damage to the crop: insect pests such as crucifer flea beetles (*Phyllotreta cruciferae*), pollen beetles (*Brassicogethes aeneus*) or cabbage seedpod weevils (*Ceutorhynchus obstrictus*) can induce the loss of flower buds, and fungal pathogens such as blackleg, cancer and clubroot affect the productivity and seed yield of OSR crops (e.g. Hwang et al. 2012). Whereas the resistance of rapeseed cultivars against diseases such as blackleg or clubroot has been substantially improved during the last decades (e.g. Diederichsen et al. 2009; Obermeier et al. 2013; Rahman et al. 2014), resistance against insect pests has not been achieved yet. Basic studies have been initiated however to develop resistant plants via genetic engineering, e.g. by RNAi technology (P. Krause, pers. comm.).

Altogether, agricultural practices have changed dramatically in the last millennium due to the farmers' needs and huge technical progress. For example, during earlier centuries in Europe, a 3-year crop rotation was practiced by farmers rotating winter rye or wheat in year 1, followed by spring oats or barley in year 2 and followed by a 3rd year of no crop (fallow year). With the introduction of turnips, this all changed: the farmer did not have to lay the arable land fallow to get rid of the established weeds. Since a crop of turnips grown in rows could be hoed to remove weeds, the area of fallow land could be greatly reduced or even avoided.

Farmers in those days were essentially self-supporters. Today, the agricultural production—particularly in Australia, America and Europe—is strongly depending on (international) markets, and the production is directly influenced by the demand and consumption. This led to an increasing expansion and dominance of cereals such as wheat, rice and corn in plant production. From an agronomical point of view, it is therefore very important (i) to diversify crop production as much as possible and (ii) to extend the share of dicot crops in crop rotations. For temperate regions of the world, rapeseed (*B. napus*) has long been known as a very suitable complement for rotations with cereals like wheat and barley. For example, wheat and other cereals cover more than 50% of the arable land in Germany today. The share of OSR is only 12%, and typical rotations would be dominated by winter cereals (cf. Table 1.1).

According to recent field studies in Germany, the extension of "leaf crops" (dicots) such as faba beans or field peas in rotations can lead to positive effects regarding field sanitation and the

Table 1.1 Examples of typical rotations for crop production with oilseed rape as a main component practiced in Germany (W. Sauermann, pers. comm.)

Sequence of crops	Frequency of rapeseed (%)
WOSR–WW–WW	33
WOSR–WW–WB	33
WOSR–WW–WW–WW	25
WOSR–WW–FB–WW–WB	20

WOSR winter oilseed rape, *WB* winter barley, *WW* winter wheat, *FB* faba beans

protection of beneficial organisms. In addition, it allows yield increases of 10–20% in comparison with wheat as a previous crop. In direct-seeded field experiments at five locations in Western Canada from 2008 to 2013, involving continuous canola and all rotation phases of wheat and canola or field pea, barley and canola were conducted. It was found that for each annual increase in the number of crops between canola, the yield of canola increased by 0.20–0.36 t ha^{-1} (Harker et al. 2015). In addition to the direct yield advantage, a leaf crop makes the soil preparation for the following (cereal) crop easier (and cheaper), saves plant protection costs (for herbicides, fungicides, a.o.) and breaks work peaks. In addition, the extensive root system of OSR allows the exploitation of deeper soil layers and saves nutrients for the following crop(s).

1.4.2 Impact of Rapeseed Cultivation on the Environment

One of the major environmental impacts of modern agriculture is increased greenhouse gas (GHG) emission leading to global warming and climate change. The Renewable Energy Directive (RED) of the European Union from 2009 requests that until 2017, greenhouse gas emissions from the production and use of biofuels for transport must be at least 35% lower than those from fossil fuels, thereafter, 50% lower. According to the EU commission's studies, rapeseed oil meets the RED requirement, delivering greenhouse gas cuts of at least 38% compared with conventional fuels. Regarding the sustainability of OSR and RME production and use, a detailed life-cycle assessment (LCA) study of biodiesel has been carried out (Herrmann et al. 2013). Based on today's climate change potential from the production and use of biodiesel, the authors have assessed the specific environmental impacts from the production and use of biodiesel as it is today, based on rapeseed oil and different types of alcohols, and using technologies that are currently available or will be available shortly. Based on their analysis, the authors recommend investigating additional options and incentives regarding the better use of OSR straw, particularly considering carbon sequestration issues of using bio-alcohol instead of petrochemical alcohol for the transesterification process in order to further reduce greenhouse gas emissions.

1.5 Impact on Agriculture and the Rural Population

1.5.1 The Value of Rapeseed and Its Products

Rapeseed is not only an important oil crop, but also a source of high-quality vegetable protein determining its value as an excellent feedstuff for farm animals, mainly ruminants (cattle, sheep, etc.). As shown in Table 1.2, almost three-quarters of the world's rapeseed production is harvested in Canada, China and the EU (71%). These figures demonstrate the outstanding importance of this crop for agriculture and the whole economy of these countries.

The major "rapeseed" growing countries besides China are India (mainly mustard), Canada, France, Germany and Australia. Further countries with a significant production are the Ukraine, Poland, the UK, Russia, the USA and Pakistan (FAOSTAT 2016). In Europe, oilseed rape (OSR) is mostly grown as a winter crop

Table 1.2 Annual rapeseed yield and total production in major countries or regions and in the world, 2004, 2009 and 2014 (FAOSTAT 2016)

Country or region	2004		2009		2014		2014 versus 2004
	Yield t/ha	Prod. Mio t	Yield t/ha	Prod. Mio t	Yield t/ha	Prod. Mio t	(%)[a]
Australia	1.12	1.54	1.12	1.92	1.41	3.83	126/249
Canada	1.58	7.67	1.95	12.89	1.93	15.56	122/203
China	1.81	13.18	1.88	13.66	1.77	11.60	98/88
EU	3.39	15.49	3.30	21.48	3.61	24.29	106/157
World	1.84	46.54	1.97	62.59	1.98	70.95	108/152

[a]Yield in tons per hectare and total production 2014 versus 2004 (%)
Source http://faostat.fao.org

(WOSR). The highest seed and oil yields are harvested in Central and Western Europe, where the annual average yield of WOSR crops varies between 3.3 and 4.3 t/ha (Table 1.2).

1.5.2 Special Relevance of OSR for China

The Chinese harvest corresponds to an annual production of 5 mio tons of edible oil (accounting for 55% of total vegetable oil production) and more than 6 mio tons of high-quality protein feedstuff. Together, this provides an agricultural income of nearly 100 billion yuan for more than 100 million peasants. Therefore, the Chinese rapeseed producers have a great importance to effectively supply the processing industry with vegetable oil and forage protein. In addition to established feeding purposes, the use of rapeseed extraction meal (REM) or cake in aquaculture is also promoted today in China and south-east Asia but also in other parts of the world (e.g. Scandinavia, South America).

Major rapeseed production regions in China are: (i) the spring rapeseed area (Qinghai–Tibet, Xinjiang and Northeastern China); (ii) the upper region of Yangtze River (Yunnan–Guizhou Plateau, Sichuan Basin and Chongqing); (iii) the middle region of Yangtze River (Hubei, Hunan and Jiangxi provinces); (iv) the lower region of Yangtze River (Anhui, Jiangsu, Zhejiang Provinces, Shanghai City) and (v) the Huanghe-Huaihe River plain (Henan, Shanxi). Overall, the Yangtze River Basin is the main growing area of oilseed rape in China, and OSR is the only oil crop grown there during winter.

As is obvious from Table 1.2, in the year 2004 the total rapeseed production of China was comparable to the EU. Due to yield improvement based on the introduction of high-yielding hybrid cultivars and an extension of the OSR cultivation area, the production of the EU is now two times higher than that of China, where at the same time yield and acreage have significantly decreased. This is obviously due to a lower competitiveness of OSR in comparison with wheat, maize and other crops in China.

However, it is important to mention that the rapeseed industry does not influence the cereal food supply for the Chinese people, because rapeseed does not compete for land with major summer food crops such as rice and maize, and in this region, most areas suitable for winter OSR are not suited for winter wheat because of quality restrictions. At present, there is a limited potential of arable land to develop other summer oil crops like soybean and peanut in China. But the potential for OSR development is very large as there are still 4 million ha of winter fallow farmland in the Yangtze River Basin which can be used. According to estimations, the rapeseed production could be extended by 9 million tons (t) per year if the necessary investments and technological progresses would be put into place.

1.6 Economy and Global Trade

1.6.1 The Global and Regional Production of OSR

As indicated, rapeseed is one of the major sources of vegetable oil in the world. Regarding oil production, it ranks third after oil palm and soybean. Rapeseed and soybean are the world's major oil and protein seed crops with different regional importance due to (i) the plants' requirements regarding soil quality and ambient temperature and (ii) the specific national or local traditions regarding human nutrition and feeding practices of farm animals. As indicated, the term "rapeseed" as a commodity represents different species of the *Brassicaceae*: Its major, economically most important representative is oilseed rape (*B. napus*), grown widely in Australia, East Asia, Europe and America (mainly North America). Other species grown for seed oil production include turnip rape (*B. rapa*) in higher northern latitudes (Canada, Scandinavia) and brown mustard (*B. juncea*), mainly on the Indian subcontinent.

The world's total acreage of *B. napus* OSR/canola along with related "rapeseed" brassicas is about 35 million (mio) hectares. The total and regional cultivation of rapeseed is as follows: Australia 2.78 mio ha, Canada 8.15 mio ha, China 7.17 mio ha, Europe 9.31 mio ha (averages of 3-years, 2012–2014); this corresponds to 75% of the world's rapeseed cultivation area (36.37 mio ha). According to the FAO, about 72.5 mio t of rapeseed have been harvested globally in the year 2013, in comparison with a mass of 276.4 mio t of soybeans (Table 1.3). More than one-third of the total OSR has been harvested in Europe, where the production was more than duplicated in the last decade. The increase in China has been comparatively small (+26%). Whereas sunflowers play a major role in (Eastern) Europe, they actually represent a minor crop in China (Table 1.3). The current cultivation of soybean in China, its country of origin, is also relatively restricted and has even declined during the last decade. Rapeseed is appreciated not only as an important oil crop, but also as a high-quality protein feed crop. In China, it is number 5 of the main crops following rice, maize, wheat and soybean. Besides soybean and rapeseed, groundnut is an important source of vegetable oil in China but not in temperate regions such as Europe.

A great potential for extending OSR production is seen in Eastern Europe, e.g. Poland, Romania, Ukraine, Western Russia, depending on growing conditions and the economic situation. The yield potential in other major rapeseed growing regions of the world is much lower. This is either due to the harsh environment and short growing season of spring canola in Western Canada, to the small-scale production of alternative rapeseed types (semi-winter types) as a second crop (after rice) in Central China. The possibility of enhancing the average yield level under such production conditions seems to be rather limited to favourable locations and optimum cultivation practices based on new hybrids with a genetically high yield potential. For example, the rapeseed production has doubled in Canada in the last decade, based on a strong improvement of average yield (Table 1.2). The increase of production in the EU and the whole world was primarily due to an extension of oilseed rape (canola) cultivation area by about 50%.

Table 1.3 Comparison of annual production of major oilseeds in the world, in China and whole Europe, 2003, 2008 and 2013 (million tons)

Year	Rapeseed			Sunflower			Soybean		
	World	China	Europe	World	China	Europe	World	China	Europe
2003	36.78	11.42	11.45	27.56	1.74	16.32	190.65	15.39	1.85
2008	57.93	12.10	23.36	36.33	1.79	21.95	231.27	15.54	2.74
2013[a]	72.53	14.40	25.59	44.75	2.38	31.89	276.41	12.50	5.94
Rel. (%)[b]	197	126	223	162	137	195	145	81	321

Source FAOSTAT (http://faostat3.fao.org)
[a]Estimations for 2013; [b]2013 versus 2003 (%)

1.7 Food, Feed and Biofuel Uses of Rapeseed

1.7.1 Rapeseed Oil Quality and Its Improvement

The major useful compounds of rape seeds are oil and protein. Whereas the average protein content is approximately 23% (19–26%), the oil content varies around 44% (38–49%) (Wittkop and Friedt unpubl.). The seed oil of 00-rapeseed/canola is widely used both for human nutrition and as an important renewable resource for non-food purposes (e.g. biodiesel). Almost all of the OSR production in Europe is of 00-quality. A relatively small amount of high erucic acid rapeseed (HEAR, also called "traditional rapeseed") is still grown in Canada and Europe for industrial end-users. Due to its favourable fatty acid composition, the oil extracted from current 00-rapeseed varieties is very suitable for the production of biodiesel (rapeseed methyl ester, RME). Its relatively high mono-unsaturated fatty acid (oleic acid, 18:1) content together with comparatively low contents of poly-unsaturated fatty acids (linoleic acid, 18:2; linolenic acid, 18:3) are determinants of high RME quality. High-oleic types (HOLL or HOLLI) with elevated contents of C18:1 would be even better suited. They are also very useful for frying purposes ("hot kitchen"); see Table 1.4.

The achievable oil and methyl ester yield from a given amount of seed depends primarily on its oil content. Further progress in oil content and yield can be expected from breeding in the future. However, a large number of genes or quantitative trait loci (QTL) are involved in seed oil biosynthesis, and a strong environmental modification of seed oil content has been observed (Nesi et al. 2008). Rahman et al. (2013) reported that the QTL associated with oil quality was distributed among 17 of the 19 chromosomes of *B. napus* which on the one hand complicates recombination breeding but on the other hand allows recombinations between unlinked loci. Further investigations are essential to obtain deeper insight into this complex trait in order to further improve oil content in high-yielding cultivars. In this context, the identification and utilization of genes contributing to oil content via genetic analysis in OSR, e.g. comparative QTL mapping in different genetic backgrounds, will help to identify gene loci with a key function for this complex trait by knowledge-based breeding. For example, in a comparative study homoeologous genomic regions involved in oil content in different genetic backgrounds could be identified (Delourme et al. 2006), and novel alleles were found in individual genotypes. Zhu et al. (2012) have identified dozens of genes which were differentially expressed in rapeseed lines differing in a QTL influencing oil content. Among these genes, six were differentially expressed regardless of temperature, indicating the major relevance for oil content. Such work aims at the marker-assisted combination of favourable alleles at different genetic loci to increase seed oil content in OSR by breeding.

Table 1.4 Typical fatty acid compositions of selected genotypes of three major oilseed crops (Wittkop et al. 2009)

Type	Fatty acid composition (% of total fatty acids)[a]									
	12:0	14:0	16:0	18:0	18:1	18:2	18:3	20:1	22:1	Rest
High erucic	–	–	3	1	11	12	9	8	52	4
Canola/00	–	–	4	2	60	21	10	1	1	1
Low linolenic	–	–	4	2	61	28	3	1	–	1
Lauric	37	4	3	1	33	12	7	–	–	3
High myristic	–	18	23	2	34	15	4	–	–	4
HOLLI			4	1	84	5	3	1	–	2

[a]12:0 lauric, 14:0 myristic, 16:0 palmitic, 18:0 stearic, 18:1 oleic, 18:2 linoleic, 18:2 linolenic, 20:1 eicosenic, 22:1 erucic acid

1.7.2 Oil Consumption Versus Need of a Growing Population

The consumption of vegetable oils continues to grow globally and particularly in China with the largest population of the world. Whereas domestic vegetable oil consumption in 2001/02 was only 14.54 mio tons, it reached 28 mio tons in 2011/12. On the other hand, domestic vegetable oil production is only about 11 mio tons so that the self-sufficiency rate is 35.4%. Consequently, China has become the largest importer of edible oil and oil products in the world. OSR has contributed most in domestic edible oil, which is considered the most important oil crop to secure the edible oil supply of China. Although the total production of soybean is much higher, due to its low oil content (20–23%), the usage of soybean oil is only 30% since the majority of home-grown soybeans is for direct consumption. Peanut production is high, but it is mainly for direct eating and is largely exported, so that the annual consumption of peanut oil is only about 2 Mio tons. Regarding oil production, OSR ranks first and contributes most to the self-sufficiency of oil in China. In 2012, China's consumption of edible vegetable oil was 28 mio tons, including imports (palm oil and soybean oil) of 17 mio tons and domestic oils of 11 mio tons (55% rapeseed oil). Considering the quality, consumption, yield and other factors regarding edible oil supply, rapeseed oil plays an important role in China. Therefore, the development of rapeseed production is of great significance to ensure the supply of edible oil and to meet the consumers' needs.

1.7.3 Protein-Rich By-Products of Rapeseed Oil Extraction as Animal Feed or Human Food

As indicated above, rapeseed is also a valuable source of vegetable protein. Today, rapeseed extraction meal (REM) is widely used as a protein component in compound feeds for farm animals. Due to the relatively high fibre content of REM, it is particularly useful for ruminant feeding, and sophisticated diets have been elaborated for feeding cattle (mainly cows) and sheep. Based on the improvements of the digestibility of REM, it can now even be used for feeding monogastric animals such as pigs and poultry (broilers, laying hens and turkeys). By further improvement of OSR varieties aiming at lower fibre (lignin) content and high protein with better amino acid composition, larger quantities of oilseed rape meal may be added to the compound feeds to replace soybean meal in the diets. Respective mutants with differential fibre composition are available for future breeding (e.g. Liu et al. 2012).

A recent study with canola cultivars released in Australia between 1978 and 2012 and tested in 2008 and 2014 at different field sites has shown that oil and protein concentrations have increased in this period by 0.09% $year^{-1}$ and 0.05% $year^{-1}$ respectively (Potter et al. 2016). This demonstrates how lengthy quality improvements can be if the focus of breeding has to be on yield and disease (blackleg) resistance (for more details on quality, see, e.g. Friedt and Snowdon 2010, Wittkop et al. 2009).

1.8 Major Requirements for Modern Rapeseed Varieties

1.8.1 Technical Innovations Changing Rapeseed Production Schemes

With social changes, labour prices and other aspects, China's rapeseed planting area has shrunk year by year. The main reason is the low production potential and high production costs (mainly high labour input). Meeting the demands of rapeseed industry, one of the current rapeseed breeding goals is to increase the yield per unit area and improve traits necessary for mechanized planting, such as herbicide tolerance. On the farm level, the average rapeseed yield of 1.8 t/ha in China is 20–50% lower than in advanced European countries. The potential yield of new varieties in regional trials in Canada is close to

4.5 t/ha, while the yield in the Yangtze River Basin is about 3.0 t/ha, indicating that there may be space to improve rapeseed yield. Heterosis utilization could be one of the most effective approaches for yield enhancement. It depends on two factors: (i) one is the pollination control system. Based on Chinese breeders' experience, the use of the genic male sterile S45AB showed much higher heterosis compared with pol cms and gms 9012. Unfortunately, with this system a 100% male sterile population could not be generated (50:50 segregation in the mother lines). Alternatively, Ogu-INRA cytoplasmic male sterility could be a potential system for China. (ii) Another major factor is the genetic distance between the parents. To enlarge the genetic difference between parents, it might be a useful way to create new sub-genomic materials based on the germplasm from other *Brassicaceae*.

The process of rapeseed production in China today is still basically the traditional manual operation. The labour costs from planting to harvest of rapeseed account for about 60% of production costs. For the decrease of labour costs, the current varieties should be transformed into "mechanization" varieties. While Canada and other countries have realized full mechanization, the labour costs have been drastically reduced and only amount to a few percent of total costs.

1.8.2 Improving the Efficiency of OSR Production Enabling Higher Yield Stability

Besides the mechanization of cultivation, the improvement of disease, pest and stress resistance of OSR varieties is a major general target, whereas abiotic stress and insect pest resistances still present great challenges and not much progress has been made there. On the contrary, substantial progress has been achieved regarding resistance to diseases. For example, Indian colleagues have recently evaluated different genotypes of *B. juncea*, *B. carinata*, *B. napus* and *B. rapa* under natural and artificial inoculation field conditions for their reaction against white rust, Alternaria blight and Sclerotinia rot diseases in several seasons. Various lines exhibited a white rust intensity consistently below 5%. Other brassica germplasm showed resistance against Alternaria blight in 2 years of field testing, and 10 lines exhibited consistently low Sclerotinia rot incidence (<5%) in different years (Gaur et al. 2016).

Based on the current situation and in the light of future development, the rapeseed breeding goals in the near future may be summarized as "highly efficient quality type" of OSR. The need for double low quality in China as everywhere else will lead to a change of the current fatty acid composition favouring high oleic acid, low saturate fatty acids and low linolenic acid. The oil yield based on seed yield and oil content has to be increased. To improve yield stability, resistance to lodging, diseases, herbicides, pod shattering, etc. will have to be improved. All of these improvements are expected to make the whole production process for the whole growing season of rapeseed more efficient and economic.

In general, the efficiency of rapeseed production and use needs to be improved all over the world: (i) by new varieties with efficient root systems using water and nutrients more efficiently; (ii) efficient tolerance against all kinds of stress limiting OSR yield (cold or heat, drought or water-logging); (iii) efficient resistance against OSR pathogens such as clubroot, phoma, Sclerotinia; (iv) more efficient resistance against pests such as cabbage flies, flower beetles, stem or pod weevils; (v) more efficient seed yield due to a better partitioning of phytosynthates between the vegetative plant and the seed; (vi) more efficient formation of valuable compounds such as oil and protein versus dietary fibre. In order to broaden the options for the use of REM as an animal feed or human food, the nutritional value of the meal deserves further improvement, i.e. by reducing the antinutritional fibre (e.g. for poultry feed, Khajali and Slominski 2012) and increasing the content of valuable protein.

1.9 Oilseed Rape Breeding and Major Variety Types

1.9.1 Genetic Diversity as a Basis for Breeding of Oilseed Rape

It is well known and generally accepted that genetic diversity within a species is principally valuable and especially necessary for breeding and crop improvement. It is also a prerequisite for achieving high crop yield of hybrid cultivars by exploiting heterosis. Genetic variance can be created by crossing, sexual hybridization and recombination of registered cultivars, adapted genotypes, primitive landraces or even alien exotic germplasm. In addition, entirely novel variation can be created by interspecific or intergeneric hybridization within the *Brassicaceae*. Since *B. napus* is a natural hybrid between *B. oleracea* and *B. rapa*, completely new rapeseed types can be created by crosses between selected genotypes of the two parental species. It has been demonstrated that the progeny of such interspecific hybrids ("resynthetic [RS] rapeseed") gives rise to novel breeding material, more or less distant to commercial OSR cultivars. At the same time, different interspecific families show obvious genetic distances and form separate sub-clusters within the interspecific progenies. Inbred lines derived from such wide crosses can be used for test crosses and new hybrids (F1) with a high yield potential (Friedt et al. 2004; Gehringer et al. 2007; Seyis et al. 2003, 2006). Therefore, the use of RS material for the development of new parents and F1 combinations can lead to new hybrid varieties which out-yield even modern OP variety types regarding both seed and oil production.

The development of efficient and cost-effective new methods for plant molecular analyses has opened new options for the characterization and more efficient use of genetic resources. In addition, the complete sequencing of an oilseed rape genome (Chalhoub et al. 2014) enables approaches of comparative genomics greatly enhancing the exploitation and use of exotic or alien germplasm. For example, Fu et al. (2015) have aligned a genetic locus (QTL) for silique length with *Brassica* reference genomes and found homologous QTL on chromosomes A09 and C08. The narrowed QTL region provides clues for breeding by marker-assisted selection or possibly gene cloning.

1.9.2 Recent History of Canola and 00-OSR Breeding

The traditional use of rapeseed oil for lamp fuel was largely superseded by petroleum since the end of the 19th century. Only the high quality of rapeseed fat as a lubricant for industrial machinery guaranteed continued production of the crop throughout the 20th century. Oil from early rapeseed varieties until the 1970s was characterized by a high content of bitter-tasting erucic acid (*cis* 13-docosenoic acid, 22:1n-9, up to 50% of fatty acids), which can lead to cardiac damage and related health problems. Therefore, OSR production in Europe only peaked significantly during the wars in the twentieth century when rapeseed oil was used especially for the production of margarine. The poor reputation of rapeseed oil as a foodstuff was overcome only by the development of "0" and "00" rapeseed varieties in the 1970s. The first major breakthrough came with the initial "0-quality" cultivars with erucic acid levels <2% in the seed lipids. Due to major improvements in seed analysis techniques, fatty acid mutants had been identified. A spontaneous mutant of the German spring forage rape cv. Liho led to the release of a first erucic acid-free variety in Canada in the early 1970s. However, the value of the crop was still limited by the presence of high quantities of glucosinolates in the seed and extraction meal (REM), which made REM unsuitable as a feed for monogastric animals, since the digestion of glucosinolates results in the release of toxic by-products (isothiocyanates). In 1969, the Polish spring rape variety Bronowski was discovered to contain low glucosinolates and this provided the basis for international breeding activities to introduce this complex trait (at least three recessive genes involved) into the high-yielding 0-elite material. The first 00-quality spring rapeseed variety, Tower, was released in 1974, and the advance of 00-oilseed rape (canola) began. As mentioned above, modern rapeseed oil is used for various purposes. Besides its use as a food oil, it also provides a raw material for many other products ranging from rapeseed methyl ester (RME, biodiesel) to industrial lubricants and hydraulic oils, tensides for detergent and soap production and biodegradable plastics.

1.9.3 History of Rapeseed Breeding in China

China has a long history of planting rapeseed which started as early as in the Neolithic period. Combining archaeology with ^{14}C isotope analysis of carbonized rapeseeds, they were thought to be 6000–7000 years old and might derive from early cultivars of *B. rapa* or *B. juncea*. These two species were predominantly grown in China until the 1950s. An improved type of *B. napus* was introduced in China in the 1940s and was named "Liberation rapeseed" for the commemoration of the liberation in World War II. After the introduction of Liberation rapeseed, it was improved by breeders with regard to maturation, and some cultivars were quickly accepted by agronomists because of, e.g. high yield, better resistance to plant diseases and better tolerance to winter cold. The rapeseed improvement process can be roughly divided into four stages: (1) in the first stage during the 1950s and 1960s, breeding of *B. napus* varieties began in many institutes set up for rapeseed improvement. Major target was to improve the local adaptability of *B. napus*, including improvements of disease resistance and maturity. (2) During the second stage in the 1970s, the major target was to breed *B. napus* varieties with higher yield and earlier maturation than *B. rapa*. (3) The 1980s and 1990s stand for canola quality breeding in *B. napus*. In the first time, the OP cultivar Zhongyou 821 and the hybrid Qinyou No. 2 showed a good performance in a large planting area in the Yangtze River Basin where the mean yield was almost doubled. In the 1990s, breeding rapeseed for quality and heterosis was intensified. (4) The fourth and current stage started in the 2000s, when canola quality was combined with the systematic utilization of heterosis. Because of canola quality, the cultivation area of rapeseed was greatly increased. The hybrid cultivar Huaza No. 4 bred by Huazhong Agricultural University became very popular in the Yangtze River. In retrospect, rapeseed yield in China showed two historical leap-forward steps: the first one was due to the large-scale extension of high-yielding and multi-resistant varieties like Zhongyou 821 in the 1970s and 1980s; the second occurred in the 1990s due to the promotion of 00-hybrids, such as Huaza No. 4, No. 6 and Zhongyou No. 2, resulting in the rapid expansion of China's rapeseed production. Compared with the 1950–60s, the planting area was extended in the 1990s more than three times and the total production was increased more than 10-fold. However, rapeseed yield has more or less remained on the same level since 2004 (Table 1.2).

1.9.4 Advantage of Hybrids Versus Open-Pollinated Varieties

As a facultative outcrossing species, oilseed rape crops can either represent open-pollinated (OP) varieties or hybrids. In all major growing areas including Europe, an increasing proportion of the registered cultivars represents single-cross or F1 hybrids versus OP line varieties. New varieties registered recently and today are all hybrids. In central Europe, WOSR generates the highest seed and oil yields in comparison with other oil crops; therefore, it is the most important oilseed in this region. In Germany, average farm yields currently vary around 4 t ha^{-1} (Statistisches Bundesamt, www.destatis.de). A total of 88 tested oilseed rape cultivars were recently listed by the Plant Variety Office of Germany (Anonymus 2016); 83 of those are winter types, 5 spring types. One of the WOSR cultivars is a low glucosinolate but high erucic (0+) type, while all others represent 00 types (zero erucic acid, low seed glucosinolates [equivalent to canola quality]). Since the introduction of the first restored WOSR hybrid variety in 1995 (Paulmann and Frauen 1997), the proportion of hybrids has grown steadily; currently, 65 (74%) of the registered OSR varieties in Germany are hybrids (cf. Fig. 1.2, https://www.bundessortenamt.de/internet30/fileadmin/Files/PDF/bsl_getreide_2016.pdf).

In terms of cultivation, there is a trend towards an increased use of hybrids in all major canola and oilseed rape growing areas worldwide. For example, according to breeders' information, about 80% of the rapeseed acreage in Germany today represents hybrid crops (Norddeutsche Pflanzenzucht Georg Lembke KG, pers. comm.). Under real farm conditions, hybrid varieties tend to generally out-yield OP varieties; the yield advantage of hybrids has been estimated at 8–16%, depending on the yield level (Christen and Friedt 2012). The newest generation of WOSR hybrids displays improved yield performance and stability and also achieves high and stable oil contents (cf. Anonymus, 2016). New hybrid varieties such as "Atora" and "Bender" achieving the highest rank in seed yield are the basis for enhanced farm yields and profitable rapeseed oil production. Other major advantages of hybrids versus lines are (i) their faster and more vigorous development in fall leading to a better crop establishment before winter, (ii) their better nutrient (N) uptake ability resulting in more economic application of fertilizers (e.g. N) and (iii) their generally better winter survival and stronger growth in spring.

1.10 Potentials of Variety and Crop Improvement

1.10.1 Improving the Competitiveness of Oilseed Rape

An evaluation of rapeseed yield and stability from 1970 to 2009 in 12 countries representing a wide range of environments and farming systems was carried out by Rondanini et al. (2012). It was shown that the global average yield was doubled during this time (8 t–1.9 t/ha). In the same period, the rapeseed acreage has steadily increased from about 12 mio ha to 31 mio ha in 2009. The authors concluded that rapeseed yields have increased steadily but yield stability has not. Therefore, the further improvement of crop yield stability is a goal of overriding importance. One way to achieve this would be the improvement of environmental adaptation by breeding for better stress tolerance and nutrient efficiency. A more efficient use of resources such as nutrients will be essential for successful and environmentally friendly OSR cultivation. With a focus on the Yangtze River Basin in China, Li et al. (2015) have shown how to optimize the current N rate with a yield response model, which could increase rapeseed yield with the efficient input of N fertilizers under different indigenous soil nitrogen levels. This study provides data to support regional N fertilizer management systems for WOSR.

1.10.2 The Role of Cultivars for Future Crop Enhancement

Estimations on the future crop potential need to be based on the current knowledge and available germplasm, i.e. modern rapeseed varieties. There is no doubt that the yield potential of rapeseed or canola is clearly higher than the average farm yield usually obtained today. Many field trials and practical observations and experience of breeders and farmers show that the seed yield potential of modern WOSR cultivars is certainly higher than 6 t/ha. While maximum yield usually requires high (nitrogen) fertilization and plant protection treatments, reasonable yields can still be achieved at reduced N fertilization levels and sub-optimal field conditions (i.e. soil quality). Whereas the variety type, i.e. OP or hybrid, is less important at marginal sites or low input conditions (poor soil, sub-optimal climate), maximal seed and oil yield under high yield conditions (fertile soil, optimal temperature and precipitation) tend to be rather achieved with hybrids than with OP varieties. While emissions due to N leaching or aerial loss under low input conditions will be naturally low, such emissions can be avoided under high input conditions by highly productive varieties which combine a strong nutrient (N) uptake ability due to an efficient root system and a pronounced N translocation efficiency, i.e. a high N use efficiency (NUE) as a whole. It has been shown that genetic variation for the relevant components of NUE exists. For example, the work of Stahl et al. (2016) has revealed considerable variation for NUE parameters, including positive effects of early flowering and high leaf N concentration on enhanced N utilization under low N input; for an overview cf. Bouchet et al. (2016).

Based on this and other work, there is no doubt that it is possible to develop new rapeseed varieties with improved NUE. According to the latter authors, the following major points have to be considered: (i) the formation of rapeseed seed yield is complex, with overlapping phases of N uptake and remobilization during the crop cycle; (ii) traits related to N uptake, such as root length and the amount of N absorbed after flowering, and traits related to N remobilization, such as "stay green", have been identified as possible levers to improve NUE in rapeseed; (iii) a substantial body of studies on the genetic control of NUE traits has already been published and potential candidate genes identified; and (iv) genetic diversity in rapeseed may be broadened by exploiting interpopulation genetic variation and the available gene pools of closely related *B. rapa* and *B. oleracea* ecotypes and advanced breeding strategies.

Regarding breeding method and variety type, F1 hybrids have repeatedly been shown to be higher yielding in replicated multi-environment field experiments in comparison with OP varieties under all N regimes studied, including zero mineral N fertilization (e.g. Gehringer et al. 2007). Therefore, it is likely that not only the nutrient (e.g. N) uptake but also the translocation and metabolization efficiency of rapeseed can be improved by breeding, making use of the existing broad germplasm. In a study by Ulas et al. (2012), genotypic differences of N efficiency have been attributed to root growth characteristics. By comparing N-efficient cv. Apex with cv. Capitol, these authors found a higher root length density and more living fine roots in the former variety. The authors concluded that genotypes which particularly invest in root growth in the vegetative stage may be more N efficient than others. By using such genotypes for improving N efficiency on the farm level, the sustainability of rapeseed cultivation and use would be greatly enhanced. Furthermore, OSR is known to have beneficial effects as a breaking crop in cereal-dominated rotations (see above). In an extensive evaluation of more than 700 Austrian cases, Vollmann (2001) demonstrated grain yield advantages in winter wheat and winter barley grown in rotation with OSR. Wheat and barley planted after rapeseed yielded 615 kg/ha (wheat) and 430 kg/ha (barley) more than after cereal pre-crops (http://ipp.boku.ac.at/pz/oilseeds/raps2001/). Results of various German studies show even higher effects of rapeseed cultivation on the grain yield of subsequent wheat crops ranging from 0.7 to 2.0 t/ha (for details: www.ufop.de/agrar). These findings basically agree with other observations and farmers' experience and indicate that OSR is a necessary component in farming systems largely dominated by cereals as is often the case today.

1.11 Conclusions

These days, oilseed rape (*B. napus*) is a highly interesting polyploid plant object for basic and applied research, making advantage from its close relationship to model plants such as *A. thaliana* and *B. rapa*. After the genome of these two species had been fully sequenced earlier, the sequence of the *B. napus* genome is now available, too. This is boosting gene discovery, the elucidation of genetic networks and biochemical pathways involved in major characteristics of rapeseed including complex agronomic and quality traits such as seed and oil yield and oil composition (quality). In addition to its outstanding importance as a vegetable oil source, OSR has been an important feedstock for biofuel (rapeseed methyl ester) production (particularly in Germany) and is expected to be so in the future because of the following reasons: (1) rapeseed is an important dicotyledonous crop plant with nutrient requirements widely differing from monocots (grasses); therefore, it is grown as a major alternative to cereals in many parts of the world, e.g. Canada, China, Europe and Australia. (2) Oilseed rape is characterized by a high oil content and valuable oil quality (fatty acid pattern), which allows its alternative use as a food, feed or fuel; this gives the farmer the freedom of action to produce either for the food markets or for industrial production chains. (3) Besides seed oil as valuable major compound, the rapeseed extraction meal or cake are also rich in protein which determines their high nutritional value as a feed for farm animals (cf. Wittkop et al. 2009). Selection can be facilitated by NIRS technology (Wittkop et al. 2012), but indirect marker-assisted selection is also applicable due to specific genetic loci or genomic regions controlling tannin and fibre contents (e.g. Lipsa et al. 2012; Liu et al. 2012; Stein et al. 2013). (4) *B. napus* comprises a broad morpho-physiological diversity including spring and winter growth types, oil as well as forage types, open-pollinated and hybrid cultivars. (5) Therefore, the rapeseed plant is expected to have a high potential for further improvements; novel germplasm can be developed by interspecific and intergeneric hybridization as a basis for the creation of better germplasm, elite breeding lines and hybrid cultivars. This potential of enhancing seed and oil yield is occasionally underestimated in respective studies on the energy costs of production and balance. Taking this potential into account will allow exploiting the agronomical and industrial advantages of the rapeseed plant.

References

Anonymus (2016) Beschreibende Sortenliste 2016. Bundessortenamt, Hannover, Germany

Bouchet AS, Laperche A, Bissuel-Belaygue C, Snowdon R, Nesi N, Stahl A (2016) Nitrogen use efficiency in rapeseed. A review. Agron Sustain Dev 36:38

Chalhoub B, Denoeud F, Liu S, Parkin IA, Tang H, Wang X, Chiquet J, Belcram H, Tong C, Samans B, Correa M, Da Silva C, Just J, Falentin C, Koh CS, Le Clainche I, Bernard M, Bento P, Noel B, Labadie K, Alberti A, Charles M, Arnaud D, Guo H, Daviaud C, Alamery S, Jabbari K, Zhao M, Edger PP, Chelaifa H, Tack D, Lassalle G, Mestiri I, Schnel N, Le Paslier MC, Fan G, Renault V, Bayer PE, Golicz AA, Manoli S, Lee TH, Thi VH, Chalabi S, Hu Q, Fan C, Tollenaere R, Lu Y, Battail C, Shen J, Sidebottom CH, Wang X, Canaguier A, Chauveau A, Berard A, Deniot G, Guan M, Liu Z, Sun F, Lim YP, Lyons E, Town CD, Bancroft I, Wang X, Meng J, Ma J, Pires JC, King GJ, Brunel D, Delourme R, Renard M, Aury JM, Adams KL, Batley J, Snowdon RJ, Tost J, Edwards D, Zhou Y, Hua W, Sharpe AG, Paterson AH, Guan C, Wincker P (2014) Early allopolyploid evolution in the post-Neolithic *Brassica napus* oilseed genome. Science 345(6199):950–953

Christen O, Friedt W (eds) (2012) Winterraps—Das Handbuch für Profis (2nd edn.). DLG-Verlags-GmbH

Delourme R, Falentin C, Huteau V, Clouet V, Horvais R, Gandon B, Specel S, Hanneton L, Dheu JE, Deschamps M, Margale E, Vincourt P, Renard M (2006) Genetic control of oil content in oilseed rape (*Brassica napus* L.). Theor Appl Genet 113:1331–1345

Diederichsen E, Frauen M, Linders EGA, Hatakeyama K, Hirai M (2009) Status and perspectives of clubroot resistance breeding in crucifer crops. J Plant Growth Regul 28:265–281

FAOSTAT (2016) Food and Agriculture Organization of the United Nations—Statistics Division. http://faostat3.fao.org/download/Q/QC/E

Fredua-Agyeman R, Coriton O, Huteau V et al (2014) Molecular cytogenetic identification of B genome chromosomes linked to blackleg disease resistance in *Brassica napus* × *B. carinata* interspecific hybrids. Theor Appl Genet 127:1305

Friedt W, Snowdon RJ (2010) Oilseed rape. In: Vollmann J, Istvan R (eds) Handbook of plant breeding, vol 4. Oil crops breeding. Springer, Dordrecht, pp 91–126

Friedt W, Seyis F, Snowdon RJ, Lühs W (2004) Resynthetic rapeseed germplasm as a source of novel genetic variation for breeding of *Brassica napus*. GCIRC Bull 21:38–46

Fu Y, Wei D, Dong H, He Y, Cui Y, Mei J, Wan H, Li JN, Snowdon R, Friedt W, Li X, Qian W (2015) Comparative quantitative trait loci for silique length and seed weight in *Brassica napus*. Sci Rep 5(14407)

Gaur RB, Sharma RN, Meena RP (2016) Multiple disease resistance in different Brassica genotypes. J Oilseed Brassica 7:98–105

Gehringer A, Snowdon RJ, Spiller TH, Basunanda P, Friedt W (2007) New oilseed rape (*Brassica napus*) hybrids with high levels of heterosis for seed yield under oligotrophic conditions. Breed Sci 57:315–320

Harker KN, O'Donovan JT, Turkington TK, Blackshaw RE, Lupwayi NZ, Smith EG, Johnson EN, Gan Y, Kutcher HR, Dosdall LM, Peng G (2015) Canola rotation frequency impacts canola yield and associated pest species. Can J Plant Sci 95:9–20

Herrmann IT, Jørgensen A, Bruun S, Hauschild MZ (2013) Potential for optimized production and use of rapeseed biodiesel. Based on a comprehensive real-time LCA case study in Denmark with multiple pathways. Int J Life Cycle Assess 18:418–430

Hwang S-F, Strelkov SE, Feng J, Gossen BD, Howard RJ (2012) *Plasmodiophora brassicae*: a review of an emerging pathogen of the Canadian canola (*Brassica napus*) crop. Mol Plant Pathol 3:105–113

Khajali F, Slominski BA (2012) Factors that affect the nutritive value of canola meal for poultry. Poult Sci 91:2564–2575

Körber N, Wittkop B, Bus A, Friedt W, Snowdon RJ, Stich B (2012) Seedling development in a *Brassica napus* diversity set and its relationship to agronomic performance. Theor Appl Genet 125:1275–1287

Li H, Cong R, Ren T, Li X, Ma C, Zheng L, Zhang Z, Lu J (2015) Yield response to N fertilizer and optimum N rate of winter oilseed rape under different soil indigenous N supplies. Field Crops Res 181:52–59

Lipsa FD, Snowdon R, Friedt W (2012) Quantitative genetic analysis of condensed tannins in oilseed rape meal. Euphytica 184:195–205

Liu L, Stein A, Wittkop B, Sarvari P, Li J, Yan X, Dreyer F, Frauen M, Friedt W, Snowdon RJ (2012) A knockout mutation in the lignin biosynthesis gene CCR1explains a major QTL for acid detergent lignin content in *Brassica napus* seeds. Theor Appl Genet 124:1573–1586

Liu S, Liu Y, Yang X, Tong C, Edwards D, Parkin IAP et al (2014) The *Brassica oleracea* genome reveals the asymmetrical evolution of polyploid genomes. Nat Commun 5(3930)

Mason A, Batley J (2015) Creating new interspecific hybrid and polyploid crops. Trends Biotechnol 33:436–441

Mei J, Li Q, Qian L, Fu Y, Li J, Frauen M, Qian W (2011) Genetic investigation of the origination of allopolyploid with virtually synthesized lines: application to the C subgenome of *Brassica napus*. Heredity 106:955–961

Nagaharu U (1935) Genome analysis in *Brassica* with special reference to the experimental formation of *B. napus* and peculiar mode of fertilization. Jpn J Bot 7:389–452

Nesi N, Delourme R, Bregeon M, Falentin C, Renard M (2008) Genetic and molecular approaches to improve nutritional value of *Brassica napus* L. seed. C R Biol 331:763–771

Obermeier C, Hossain MA, Snowdon R, Knüfer J, von Tiedemann A, Friedt W (2013) Genetic analysis of phenylpropanoid metabolites associated with resistance against *Verticillium longisporum* in *Brassica napus*. Mol Breed 31:347–361

Paulmann W, Frauen M (1997) The firstly restored F1-hybrids of winter oilseed rape. GCIRC Bull 13: 24–25

Potter T, Burton W, Edwards J, Wratten N, Mailer R, Salisbury P, Pearce A (2016) Assessing progress in breeding to improve grain yield, quality and blackleg (*Leptosphaeria maculans*) resistance in selected Australian canola cultivars (1978–2012). Crop Pasture Sci 67:308–316

Primard-Brisset C, Poupard JP, Horvais R, Eber F, Pelletier FG, Renard M, Delourme R (2005) A new recombined double low restorer line for the *Ogu*-INRA cms in rapeseed (*Brassica napus* L.). Theor Appl Genet 111:736–746

Rahman H, Harwood J, Weselake R (2013) Increasing seed oil content in Brassica species through breeding and biotechnology. Lipid Technol 25:182–185

Rahman H, Peng G, Yu F, Falk KC, Kulkarni M, Selvaraj G (2014) Genetics and breeding for clubroot resistance in Canadian spring canola (*Brassica napus* L.). Can J Plant Path 36(supl):122–134

Raman H, Raman R, Kilian A, Detering F, Long Y, Edwards D, Parkin IAP, Sharpe AG, Nelson MN, Larkan N, Zou J, Meng J, Aslam MN, Batley J, Cowling WA, Lydiate D (2013) A consensus map of rapeseed (*Brassica napus* L.) based on diversity array technology markers: applications in genetic dissection of qualitative and quantitative traits. BMC Genom 14:277

Rondanini DP, Gomez NV, Agosti MB, Miralles DJ (2012) Global trends of rapeseed grain yield stability and rapeseed-to-wheat yield ratio in the last four decades. Eur J Agronomy 37:56–65

Seyis F, Snowdon RJ, Lühs W, Friedt W (2003) Molecular characterisation of novel resynthesised rapeseed (*Brassica napus* L.) lines and analysis of their genetic diversity in comparison to spring rapeseed cultivars. Plant Breed 122:473–478

Seyis F, Friedt W, Lühs W (2006) Yield of *Brassica napus* L. hybrids developed using resynthesised rapeseed material. Field Crops Res 96:176–180

Stahl A, Friedt W, Wittkop B, Snowdon RJ (2016) Complementary diversity for nitrogen uptake and utilisation efficiency reveals broad potential for increased sustainability of oilseed rape production. Plant Soil 400:245–262

Stein A, Wittkop B, Liu L-Z, Obermeier C, Friedt W, Snowdon RJ (2013) Dissection of a major QTL for seed colour and fibre content in *Brassica napus* reveals colocalisation with candidate genes for phenylpropanoid biosynthesis and flavonoid deposition. Plant Breed 132:382

Ulas A, Schulte auf'm E, Kamh M, Wiesler F, Horst WJ (2012) Root-growth characteristics contributing to genotypic variation in nitrogen efficiency of oilseed rape. J Plant Nutr Soil Sci 175:489–498

Vollmann J (2001) Vorfruchtwert und Produktionstechnik von Winterraps in Österreich - eine empirische Untersuchung. http://ipp.boku.ac.at/pz/oilseeds/raps2001/

Wittkop B, Snowdon J, Friedt W (2009) Status and perspectives of breeding for enhanced yield and quality of oilseed crops for Europe. Euphytica 170:131–140

Wittkop B, Snowdon RJ, Friedt W (2012) New NIRS calibrations for fibre fractions reveal broad genetic variation in *Brassica napus* seed quality. J Agric Food Chem 60:2248–2256

Zhu Y, Cao Z, Xu F, Huang Y, Chen M, Guo W, Zhou W, Zhu J, Meng J, Zou J, Jiang L (2012) Analysis of gene expression profiles of two near-isogenic lines differing at a QTL region affecting oil content at high temperatures during seed maturation in oilseed rape (*Brassica napus* L.). Theor Appl Genet 124:515–531

Cytogenetics, a Science Linking Genomics and Breeding: The *Brassica* Model

2

Anne-Marie Chèvre, Annaliese S. Mason, Olivier Coriton, Laurie Grandont, Eric Jenczewski and Martin A. Lysak

Abstract

Earlier, classical cytogenetics played a key role in taxonomic studies through identification of chromosome number and morphology. Similarly, the first identifications of polyploid species, and the analysis of relationships between different species from interspecific hybrids, were based on the observation of chromosome pairing during metaphase I of meiosis. Cytogenetics subsequently got a boost with the development of mapping and of next-generation sequencing technologies, enabling the development of modern molecular cytogenetics. In this chapter, we present the major impacts of molecular cytogenetics: shedding new light on genome organization and evolution as well as regulation of meiosis in the economically important genus *Brassica* and the tribe Brassicaceae. First, we present how comparative chromosome painting (CCP) using pools of *Arabidopsis thaliana* BAC clones is used to establish genome organization in diploid and polyploid species in conjunction with genotyping and sequencing data. This method complements phylogenetic analyses in establishment of the common ancestral genome and in the description of the three differentially fractionated *Brassica* ancestral subgenomes. Secondly, intergenomic relationships can be determined by BAC-fluorescent in situ hybridization (BAC-FISH) and genomic in situ hybridization (GISH); these techniques allow identification of the different genomes and chromosomes to quantify homologous and non-homologous pairing in haploids and hybrids, identifying structural rearrangements within allopolyploid species and between genomes in interspecific hybrids. Thirdly, meiosis and meiotic recombination in *Brassica napus* and its close relatives can be studied using antibodies developed against *Arabidopsis* proteins. From all these data, we show how molecular cytogenetics is essential for our understanding of genetics and genomics in the genus *Brassica* and how cytogenetics will undoubtedly play a significant role in the times to come.

A.-M. Chèvre (✉) · O. Coriton
INRA, UMR1349 IGEPP, BP35327,
35653 Le Rheu, France
e-mail: anne-marie.chevre@inra.fr

A. S. Mason
Department of Plant Breeding,
IFZ Research Centre for Biosystems,
Land Use and Nutrition, Justus-Liebig University,
Heinrich-Buff-Ring 26-32, 35392 Giessen, Germany

L. Grandont · E. Jenczewski
Institut Jean-Pierre Bourgin, INRA, AgroParisTech,
CNRS, Université Paris-Saclay, RD10, 78026
Versailles Cedex, France

M. A. Lysak
Central European Institute of Technology (CEITEC),
Masaryk University, 625 00 Brno, Czech Republic

S. Liu et al. (eds.), *The Brassica napus Genome*, Compendium of Plant Genomes,
https://doi.org/10.1007/978-3-319-43694-4_2

2.1 Introduction

Cytogenetics refers to the study of genetics at the cellular level, and most particularly to chromosome observations at mitosis and/or meiosis. Despite being an old method, it is still commonly used in high-quality scientific studies. Cytogenetics can deduce chromosome number, genome structure, and relationships between genomes in natural or artificial interspecific hybrids. Data generated by cytogenetics approaches have been widely used in taxonomic studies and to explore genetic diversity in genera, species, and populations and in breeding programs. The recent advent of next-generation sequencing (NGS) technologies has given a fresh boost to cytogenetics, allowing the development of molecular cytogenetics and providing new insights into bioinformatically obtained questions and information related to genome organization, evolution, and regulation.

The Brassicaceae tribe is one of the 49 tribes in the Brassicaceae family (Al-Shehbaz 2012). The first cytogenetic analyses in this tribe described 220 species in 46 genera, with 37 species belonging to the Brassica genus (Gomez-Campo 1980; Al-Shehbaz 2012 for review). These studies revealed chromosome numbers ranging from $n = 7$ to 75 in this tribe, thus including species of various ploidy levels. Within the *Brassica* genus, which is the main focus of this chapter, establishment of the karyotypes of the different species revealed that several chromosomes shared the same morphology in different species, probably due to their common origin (Prakash et al. 2009 for review). The genome structure of the diploid *Brassica* species (with chromosome numbers ranging from 7 to 12) was first analyzed from karyotypes and meiotic behavior in metaphase I of pollen mother cells. The comparison of autosyndetic pairing (non-homologous chromosome pairing between different chromosomes of the same genome) in haploids with the rate of chromosome pairing in interspecific hybrids allowed the first establishment of intra- and intergenomic relationships. These cytogenetic analyses revealed that each genome carried duplications. From chromosome morphology and chromosome pairing data, these studies postulated that genomes of Brassicaceae tribe species were derived from a common ancestor with six or seven chromosomes, with subsequent duplication (Mizushima 1980 for review).

The origin of the allopolyploid species was first depicted in the famous U triangle figure (U 1935). Confirming results from Morinaga (1934), U showed that *Brassica napus* (AACC, $2n = 38$), *B. juncea* (AABB, $2n = 36$), and *B. carinata* (BBCC, $2n = 34$) originated from natural interspecific hybridization events between *B. rapa* (AA, $2n = 20$), *Brassica oleracea* (CC, $2n = 18$), and *B. nigra* (BB, $2n = 16$). All three allotetraploid species show relatively strict disomic inheritance, indicating preferential pairing between homologous chromosomes with the formation of bivalents in metaphase I. However, occasional multivalents were also observed, suggesting that homeologous (non-homologous) pairing between chromosomes of related genomes can also generate exchanges between the two genomes in each of the allotetraploids (Prakash and Hinata 1980; Prakash et al. 2009 for review). These data were confirmed by the establishment of genetic maps with molecular markers (Parkin 2011 for review), and allelic segregation distortion revealed that homeologous exchanges did indeed generate reciprocal (Lombard and Delourme 2001; Osborn et al. 2003; Piquemal et al. 2005) and non-reciprocal (Udall et al. 2005) translocations in different *B. napus* varieties. Comparison of the published *B. rapa* (Wang et al. 2011) and *B. oleracea* (Liu et al. 2014; Parkin et al. 2014) genome sequences to the *B. napus* "Darmor" reference sequence revealed numerous small translocations and other rearrangements between the A and C genomes in the established allopolyploid *B. napus* genome (Chalhoub et al. 2014) relative to the diploid genomes, as already suggested by earlier work (Cheung et al. 2009). Structural variations such as translocations, deletions, duplications, and inversions are not purely of academic interest. Increasing evidence suggests that structural variation may play an important role in genome evolution (Chester et al. 2012; Edwards et al. 2013), gene expression regulation (Wang et al.

2012), and even in crop phenotypes (Zou et al. 2011; Schiessl et al. 2014). These homeologous rearrangements fundamentally result from ancestrally shared homeology between chromosomes from different subgenomes and are mediated by genetic control of chromosome pairing.

Variation for genetic control of chromosome pairing was later described in *B. napus*. Two main meiotic behaviors (high or low frequency of chromosome pairing) were detected in AC haploids of different *B. napus* varieties (Renard and Dosba 1980; Attia and Röbbelen 1986; Cifuentes et al. 2010). After the production of F_1 hybrids from varieties with contrasting meiotic behavior, meiotic analyses of large segregating AC haploid populations combined with genetic mapping allowed identification of QTL for genetic control of homeologous recombination. A major QTL, *PrBn* (for pairing regulator in *B. napus*) was identified, plus minor QTL and epistatic interactions (Jenczewski et al. 2003; Liu et al. 2006). This control was subsequently determined to mainly affect the frequency of crossovers between homeologous chromosomes (Nicolas et al. 2009; 2012).

Interspecific hybrids have frequently been produced in order to introduce agronomic traits from one species to another (Prakash et al. 2009 for review). Different strategies have been developed in *Brassica*: trait introgression via interspecific hybrids can be achieved by either direct crossing between diploid species, crosses between tetraploid species and parental diploid species, or by crosses between tetraploid species (Prakash et al. 2009 for review). Crosses between diploids produce interspecific hybrids (allohaploids or digenomic haploids) that are generally sterile (the few viable gametes produced are generally unreduced); colchicine doubling is classically used to produce new synthetic allopolyploids from these lines. Crosses between tetraploids and progenitor diploids generate hybrids with a diploid genome plus a haploid one; for example, *B. napus* crossed with *B. rapa* produces AAC hybrids, with a majority of cells containing 10 AA bivalents and 9 C univalents at metaphase I (Leflon et al. 2006). These plants can be fully fertile and also show boosted homologous recombination in the A genome (Leflon et al. 2010). Crosses between diploids and tetraploids that do not share a genome are also possible: Such trigenomic ABC hybrids can be used either as bridge species, subsequently backcrossed to introduce new variability in crops, or can be induced by colchicine doubling to generate allohexaploids, which may have potential as a new crop species (Chen et al. 2011). Finally, crosses between pairs of allotetraploid species can also be used with subsequent backcrossing for allotetraploid crop improvement and also increase the chance that homeologous recombination will occur between divergent genomes. Three genome hybrids with one genome at the diploid stage (e.g., AABC, BBAC, and CCAB) generated by crosses between the allotetraploids revealed more bivalents between remaining haploid genomes than the corresponding dihaploid (AC, BC, and AB) hybrids (Nagpal et al. 1996; Mason et al. 2010).

All these data highlight the role of classical cytogenetics in increasing our knowledge of genome structure and evolution, as well as how regulation of chromosome pairing between genomes can be manipulated to introduce new genetic variability into crops. However, in the last two decades, the development of new molecular cytogenetic techniques, including genomic in situ hybridization (GISH), fluorescent in situ hybridization (FISH), and immunolocalization of crossover proteins has opened new avenues of research. In this chapter, we will present the major impacts of molecular cytogenetics: shedding new light on the phylogenetic relationships between Brassica species, genome organization and evolution as well as regulation of meiosis in the agronomically important genus *Brassica*.

2.2 New Insights from Molecular Cytogenetics in Genome Organization and Evolution

Recently published genome assemblies of *B. napus* (Chalhoub et al. 2014) and its parental genomes (Wang et al. 2011; Liu et al. 2014) along with the available paleogenomic hypotheses (Lysak et al. 2006; Schranz et al. 2006; Mandáková and Lysak 2008) have permitted reconstruction of the origin and later evolution of *Brassica* genomes.

Even the first generations of *Brassica* researchers, working only with classical cytogenetics techniques, realized that *Brassica* species with diploid-like chromosome numbers (n = 7–12) were probably "balanced secondary polyploids," characterized by intra-genomic chromosomal homeologies (e.g., Catcheside 1934; Röbbelen 1960). However, it took almost 40 years until a genome triplication theory gained more solid ground due to results from comparative genetic mapping (Lagercrantz and Lydiate 1996; Lagercrantz 1998), and the ancestral hexaploid nature of the diploid *Brassica* and Brassicaceae genomes was eventually confirmed by cross-species (*Arabidopsis thaliana, B. napus*) restriction fragment length polymorphism (RFLP) mapping (Parkin et al. 2005) and, of course, comparative cytogenetic analysis (Lysak et al. 2005, 2007; Ziolkowski et al. 2006). Pools of chromosome-specific BACs from *A. thaliana* (Arabidopsis henceforth) were applied to paint large homeologous chromosome regions on pachytene chromosomes in *Brassica* and Brassicaceae species (Lysak et al. 2005, 2007; Ziolkowski et al. 2006). The technique has become known as comparative chromosome painting (CCP, Lysak et al. 2006). Howell et al. (2005) used *Arabidopsis* BACs as individual in situ probes to analyze a region on *B. oleracea* chromosome O6 which, from genetic mapping, was thought to be composed of two homeologous copies of a ~5 Mb region on the bottom arm of *A. thaliana* chromosome At1. Eleven *Arabidopsis* and three *B. oleracea* BAC clones were applied separately to pachytene spreads with a *B. oleracea* chromosome O6 BAC as a marker. The two copies were shown to be adjacent with the proximal one inverted relative to the homeologous region in *Arabidopsis*. However, this approach was designed specifically to investigate the *Brassica* region, and because only the signals on chromosome O6 could be unequivocally identified, signals seen elsewhere were not analyzed and the presence of further copies of the *Arabidopsis* region was not investigated (E. Howell, pers. comm.). Lysak et al. (2005) analyzed a ~8.7 Mb BAC contig from *Arabidopsis* chromosome At4 (genomic block U in Schranz et al. 2006) by CCP in 21 crucifer species traditionally classified as members of the tribe Brassicaceae or being closely related. Despite the contrasting chromosome numbers ($2n$ = 14–38), the analyzed segment was found as three copies (in 13 species) or as six copies in four species of Brassicaceae (including the six *Brassica* species of U's triangle). The homeologous chromosome segments resembled the *Arabidopsis*-like structure or were modified by paracentric inversions and translocations. To confirm the initial findings, CCP with BAC contigs covering the majority of the longer arm of *Arabidopsis* chromosome At3 (block F in Schranz et al. 2006) was carried out in ten species traditionally treated as members of the Brassicaceae. Three homeologous copies of the contig were identified per haploid chromosome complement in Brassicaceae species with $2n$ = 14, 18, 20, 32, and 36. In high polyploid species ($n \geq 30$; n = 30, 34, and 60), six or 12 copies of the analyzed block have been revealed. Congruent data have been published by Ziolkowski et al. (2006). These authors analyzed BAC contigs from *Arabidopsis* chromosomes At1, At2, and At3 (~8.3 Mb in total) in *B. oleracea*. Except for a short contig from At1, all *Arabidopsis* probes were found to be triplicated in the karyotype of *B. oleracea* (n = 9). The largest analyzed segment (~5.4 Mb) from the bottom arm of At3 was found to occur in three homeologous copies on three different *B. oleracea* chromosomes (O4, O6, and O8).

All the studies reviewed here suggested that single-copy *Arabidopsis* chromosome segments have usually three or six homeologous counterparts within Brassicaceae and *Brassica* genomes. The presence of three copies in "diploid" species with $n = 7$–12 and of six or twelve copies in neopolyploid Brassicaceae species was most parsimoniously explained by descent from a mesohexaploid ancestor. The cytogenetic comparative studies and "the triplication theory" gained further support from comparative genetic mapping. Parkin et al. (2005) mapped over 1000 *B. napus* RFLP markers to *Arabidopsis* to estimate the level of genome colinearity shared by the two species. At least, 21 so-called conserved genomic units (analogous to genomic blocks sensu Schranz et al. 2006) were identified in the *Arabidopsis* genome, making up almost 90% of the *B. napus* genetic map. Conserved segments were present between four and seven times within the *B. napus* genome, with 86% of conserved units found in at least six copies. Again, these findings strongly supported the idea that "diploid" *Brassica* species are descendants of a hexaploid ancestor. In agreement with cytogenetic data, Parkin et al. also showed that "diploid" *Brassica* genomes underwent chromosome reshuffling following the *Arabidopsis–Brassica* split. Some rearrangements were shared by the A and C genomes of *B. napus* and, thus, most likely predated the divergence of *B. rapa* and *B. oleracea*. The authors conclude that "genome triplication followed by a small number of insertions/deletions/translocations would provide the simplest explanation for the present structure of the *Brassica* diploid genome" (Parkin et al. 2005). Indeed, the triplicated nature of *Brassica* genomes has been unambiguously confirmed by whole-genome and transcriptome sequencing in *B. oleracea*, *B. rapa*, *B. juncea*, and *B. napus* (Chalhoub et al. 2014; Liu et al. 2014; Paritosh et al. 2014; Parkin et al. 2014; Wang et al. 2011), and by reconstructing the origin and evolution of an ancestral mesohexaploid genome (Cheng et al. 2013, 2014). Moreover, the recently constructed sequence-based genome maps are to a larger extent congruent with pioneering cytogenetic analyses, as shown for the localization of block U in the *B. nigra* genome (Fig. 2.1).

For phylogenetic studies, Mandáková and Lysak (2008) showed that genomes in several tribes, including the Brassicaceae and falling into the so-called extended lineage II (Franzke et al. 2011), have descended from a common ancestral genome with seven linkage groups—the Proto-Calepineae Karyotype (PCK). The PCK genome is a younger derivative of the older Ancestral Crucifer Karyotype (ACK) with eight chromosomes, differentiated from it by a translocation-based chromosome fusion "that reduced the number of chromosomes from $n = 8$ to $n = 7$". In a younger clade of the extended lineage II, the structure of PCK was altered by a whole-arm reciprocal translocation to form the tPCK ancestral genome (t standing for translocation). When Mandáková and Lysak (2008) compared the A subgenome structure within the *B. napus* genome (Parkin et al. 2005) with the PCK (tPCK), two PCK-specific rearrangements were identified. Later on, whole-genome sequencing of *B. rapa* identified three differentially fractioned subgenomes, subsequently named MF1, MF2, and LF (Wang et al. 2011), which offered a new possibility to substantiate the hypothesis that the PCK is the ancestral genome of the genus *Brassica*. Cheng et al. (2013) showed that all three *B. rapa* subgenomes contain associations of genomic blocks (V/K/L/Wa/Q/X and O/P/W/R) diagnostic for both PCK and tPCK; moreover, associations D/V and M/E pointed to the younger tPCK genome. The authors thus concluded that the quasi-diploid genome of *B. rapa* and most likely genomes of all other Brassica species originated through re-diploidization of the hexaploid ancestor merging together three very similar diploid genomes, structurally resembling the seven chromosomes of tPCK. Considering the most parsimonious scenario for the origin of the *Brassica* mesohexaploid genome, three diploid tPCK-like genomes each with seven chromosomes were merged through hybridization/polyploidization to form a hexaploid genome with 21 chromosome pairs ($2n = 42$). Consequently, the 24 ancestral genomic blocks identified by Schranz et al. (2006) and comprising each

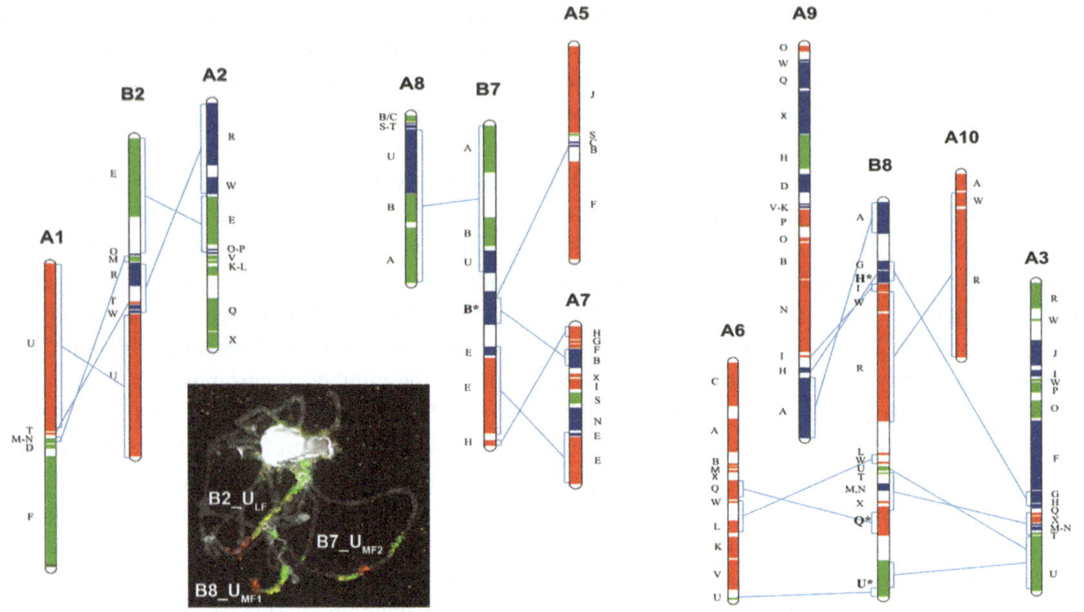

Fig. 2.1 Cross-specific localization of genomic block U on pachytene chromosomes of *Brassica nigra* and its position on chromosomes of *B. nigra* (B2, B7, and B8) and *B. rapa* (A1–A3, A5–A10). Block U was localized based on in situ hybridization of *A. thaliana* BAC contigs (labeled by yellow, red, and green fluorochrome, respectively) homeologous to this block in *B. nigra* (adopted from Lysak et al. 2005). The genomic block arrangement of three B genome chromosomes in *B. juncea* was based on single nucleotide polymorphism/intron polymorphism markers (adopted from Paritosh et al. 2014). Gene blocks on the B genome linkage groups (LGs) that show homeology to corresponding blocks on the A genome LGs are shown by the connecting lines. Genomic blocks assigned to the subgenomes LF (red), MF1 (green), and MF2 (blue) are color coded according to Cheng et al. (2013). Blocks showing variations in their fragmentation pattern in the B genome are shown in bold with an asterisk

of the three hybridizing genomes were multiplied to 72 blocks.

How was this hexaploid *Brassica* ancestor actually formed? Similar to the allohexaploid bread wheat (AABBDD, Marcussen et al. 2014), the origin of the *Brassica* hexaploid was probably via a two-step process of tetraploidization first, later followed by hybridization between this tetraploid and another diploid ancestor (Ziolkowski et al. 2006; Wang et al. 2011; Tang et al. 2012). This scenario is supported by the differential fractionation of the three (MF1, MF2, and LF) subgenomes in *B. rapa* (Wang et al. 2011; Tang et al. 2012; Cheng et al. 2013, 2014) and was also indicated by earlier cytogenetic studies (Lysak et al. 2005, 2007; Ziolkowski et al. 2006). The first to occur was an auto- or allotetraploid (XXXX or XXX^1X^1, $n = 14$) genome formed through a hybridization between two tPCK-like genomes (XX or X^1X^1, $n = 7$), followed by hybridization with a third tPCK-like genome (ZZ, $n = 7$). The second hybridization event was mediated either by the union of unreduced gametes or polyploidization of the primary triploid (XXZ or XX^1Z, $2n = 21$) to form the allohexaploid genome (XXXXZZ or XXX^1X^1ZZ, $n = 21$). This scenario is plausible considering the frequency with which intraspecific autotetraploid "chromosomal races" (sensu Müntzing 1936) and tetraploid species ($n = 14$) apparently originate(d) via autopolyploidy in tribes descending from PCK- to tPCK-like ancestral genome(s) (Mandáková and Lysak 2008; Mandáková et al. 2015). In the autotetraploid *Goldbachia laevigata* ($n = 14$, $2n = 28$), a genome derived from the PCK ancestor, the two duplicated chromosome sets are structurally identical apart from three chromosomes differentiated from their homeologous partners by three pericentric inversions (Mandáková and Lysak 2008). These intra-genomic rearrangements illustrate how re-diploidization of the *Brassica* autotetraploid may have proceeded, to be followed by subsequent hybridization with another tPCK-like diploid genome.

2.3 Molecular Cytogenetics Provides New Insights into Intergenomic Relationships

A range of different molecular cytogenetics techniques exists to distinguish between genomes and individual chromosomes and to identify specific chromosome structural features. These molecular cytogenetic approaches can allow for quantification of homologous and non-homologous chromosome pairing in interspecific hybrids, investigation of genome structural variation such as duplications, deletions, and inversions, and can be used to characterize genomic introgressions and chromosome addition lines for crop improvement.

2.3.1 Using rDNA, GISH, and BAC-FISH Techniques to Distinguish Between Genomes and Individual Chromosomes

Initial molecular cytogenetic experiments on chromosome structure and identification were performed on ribosomal DNA loci, which are an effective cytogenetic marker for *Brassica* chromosomes but are difficult to map and organize after sequencing due to the large number of repeats at each locus. The rDNA units are composed of 500 to more than 40 000 genes copies per genome, arranged in tandem repeats on several loci (Long and Dawid 1980; Rogers and Bendich 1987). Recently, the sequence composition and gene content of the short arm of rye (*Secale cereale*) chromosome 1 (1RS) have been published: The short arm of rye 1R contains genes coding for 45S rRNA. The number of the 45S rDNA genes was estimated to be about 2000, amounting to about 3% of 1RS DNA. The 5S rDNA locus contained about 5000 copies of the 5S rDNA gene, constituting about 0.4% of 1RS DNA (Fluch et al. 2012). Shotgun 454 pyrosequencing of DNA was also obtained from flow-sorted 1RS. This novel approach permitted a detailed description of the gene space and the repetitive portion of this important chromosome arm. However, although NGS technologies offer powerful tools to generate suitable sequence reads, detection of different copies expected in these genomes is still challenging. FISH experiments with 45S and 5S rDNA probes enabled far more reliable identification of individual chromosomes (Maluszynska and Heslop-Harrison 1993; Snowdon et al. 1997; Fukui et al. 1998).

Combined FISH with 5S and 45S rDNA probes enables the discrimination of a number of chromosomes of diploid and tetraploid *Brassica* species of the "U triangle," allowing otherwise indistinguishable chromosomes to be identified. Twelve out of 20 chromosomes can be identified in diploid *B. rapa* (A genome) using a 45S rDNA probe. The strong FISH signal located on the A3 chromosomes likely reflects a large tandem repeat array of 45S rDNA sequences. The A3 locus has been shown to be the only one that carries transcriptionally active genes of the A genome nucleolar-organizing regions (NORs) (Hasterok and Maluszynska 2000). The second gene-rich locus is located on chromosome A1, proximal to the centromere. The remaining sites are located on cytogenetically undistinguishable A5, A6, and A9 chromosomes, collectively grouped as *Brassica* chromosomal type VIII (Hasterok et al. 2006). The 5S probe hybridizes to two major sites on the A10 and A1 chromosomes and to a minor site on chromosome A3 adjacent to the NOR. Based on the signal intensity, the small A10 metacentric chromosome contains the largest number of 5S genes. *B. oleracea* (C genome) has two pairs of chromosomes (C7 and C8) containing 45S loci. The active site is located on chromosome C8 according to Howell et al. (2002); it shows more decondensation, while loci on C7 are fully condensed. The 5S probe hybridizes to a unique locus on chromosome C4. In natural *B. napus* allotetraploids, twelve 45S rDNA signals are observed. Four rDNA sites occur on C genome chromosomes (stained by GISH-like BAC-FISH specific for the C genome) implying that the C and A genomes carry four and eight sites, respectively. The decondensed signals are found on morphologically distinct A3 chromosomes (NORs), which are actively expressed (Chen and Pikaard 1997; Książczyk et al. 2011), while both C8 NORs are always condensed. Eight 5S rDNA sites are visible, with three loci in the A genome and one in the C genome (Fig. 2.2). The origin of the latter is evidenced by staining of chromosome C4 by GISH-like, with dispersed signals along the chromosome, and by the 5S probe which provides a highly condensed signal close to the centromere.

The first step to tie together cytogenetics and molecular genetics was to reconcile chromosome number and linkage group nomenclature. RFLP probes used for genetic map establishment were used either directly or to identify the corresponding BAC to design the different chromosomes; for example, the nomenclature of the different *B. oleracea* chromosomes was redefined by Howell et al. (2002), allowing the nine linkage groups of the *B. oleracea* genetic map to be assigned to the nine chromosomes of the karyotype derived from mitotic metaphase spreads of *B. oleracea* using BAC-FISH.

Use of GISH labeling of DNA from one parent can identify both genomes in interspecific hybrid or allopolyploid Brassica plants. This method is efficient for distinguishing chromosomes from the B genome (*B. nigra*) from A or C chromosomes, but the A and C genomes are too similar to be differentiated using GISH. Two strategies were developed to overcome this difficulty: the use of either rDNA as a blocking agent at the same time than *B. oleracea* DNA labeled (Howell et al. 2008) or a GISH-like method using *B. oleracea* BAC, BoB014O06 which selectively hybridizes to all C genome chromosomes in *B. napus* (Leflon et al. 2006; Nicolas et al. 2007) (Fig. 2.3).

Recently, a new technique for BAC-FISH analysis has been developed to allow identification of each chromosome in *B. napus* (Xiong and Pires 2011). This technique relies on a double

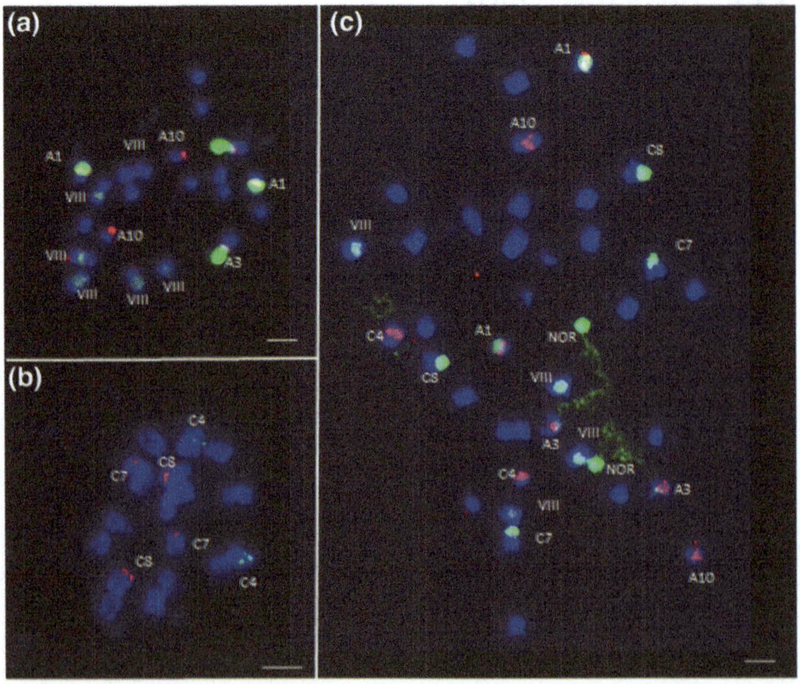

Fig. 2.2 FISH analyses of somatic metaphase chromosomes using 45S rDNA (green) and 5S rDNA (red) on *B. rapa* (**a**), on *B. oleracea* (**b**), and on *B. napus* (**c**). Bar = 5 μm

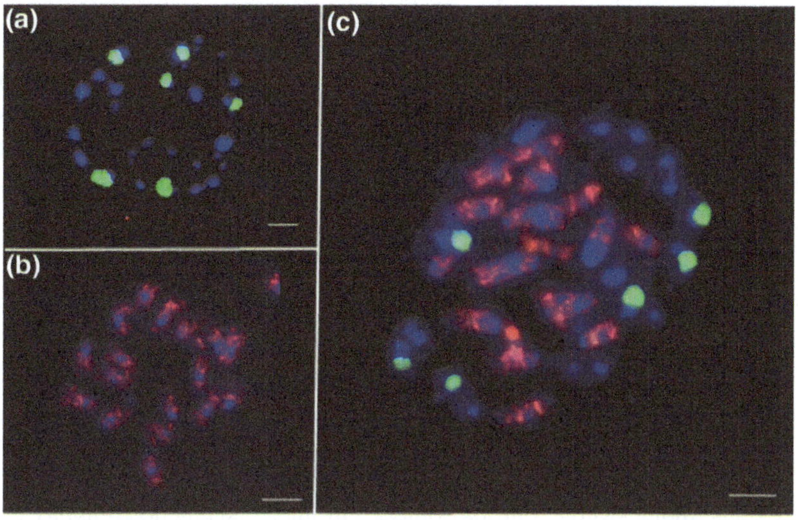

Fig. 2.3 FISH was carried out using BACs BoB014O06 probe (red) and 54B2 (green). FISH analyses of somatic metaphase chromosomes of *B. rapa* (**a**), *B. oleracea* (**b**), and *B. napus* (**c**). Chromosomes were counterstained with DAPI (blue). Bar = 5 μm

hybridization of the same slide, using a different set of three to four probes each time (45S, 5S, centBr1, centBr2, two BACs with multiple hybridization signals and a BAC that contained a C genome repeat to label the C genome chromosomes), and so is technically difficult. However, it was used successfully to identify chromosome rearrangements, duplications, and deletions in advanced resynthesized *B. napus* lines, showing selection for total number of homeologs (e.g., four copies of either A1 or C1, such as A1/C1/C1/C1 or A1/A1/A1/A1) in a "dosage balance" effect (Fig. 2.4) (Xiong et al. 2011; Grandont et al. 2014).

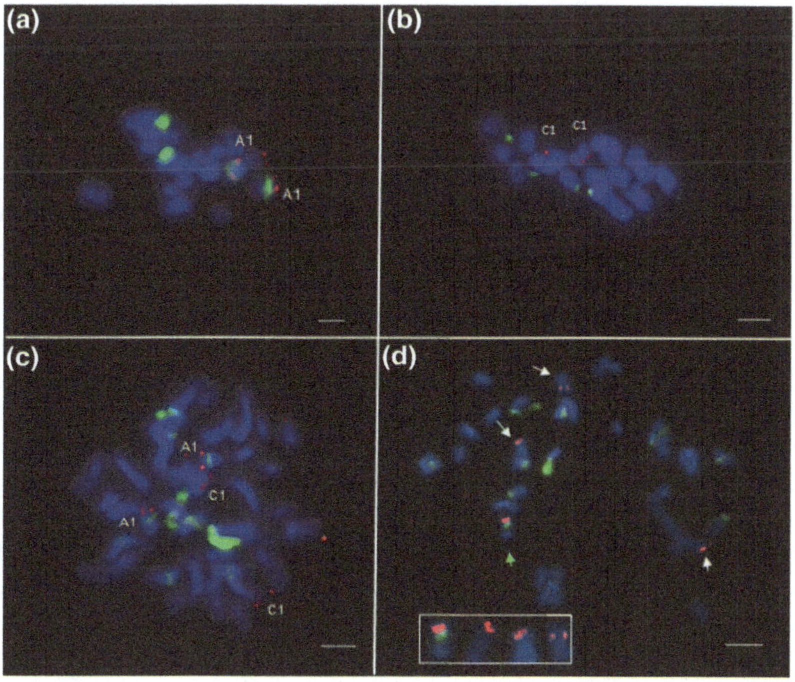

Fig. 2.4 FISH was carried out using BAC KBrB055A02 probe (A1/C1) (red) and 45S rDNA (green). FISH analyses of somatic metaphase chromosomes of *B. rapa* (**a**), *B. oleracea* (**b**), *B. napus* (DN) (**c**), and resynthesized *B. napus* line (**d**). Chromosomes were counterstained with DAPI (blue). Bar = 5 μm

2.3.2 Quantifying Homologous and Non-homologous Pairing Between Genomes in Haploids and Hybrids

Genomic sequence similarity is perhaps the primary factor in determining whether chromosomes will pair and recombine during meiosis I (Bozza and Pawlowski 2008), although genetic factors certainly play a role (Jenczewski and Alix 2004). In allopolyploid *B. napus*, mapping and sequencing data have revealed that the constitutive A and C genomes show highly conserved homeologous genomic regions along whole chromosomes, chromosome arms, and smaller genomic regions (Chalhoub et al. 2014; Parkin et al. 1995, 2003). Similar homeologous regions have been inferred between the less closely related B genome and the A and C genomes (Lagercrantz and Lydiate 1996; Navabi et al. 2013). This sequence homeology can lead to pairing and recombination between chromosomes belonging to related species during meiosis. Using FISH and GISH techniques as described above, the relative frequency of autosyndesis (pairing between chromosomes from the same genome) and of allosyndesis (pairing between the different genomes) can be determined in metaphase I of meiosis.

In haploid AC plants with one copy of each genome, it is possible to determine frequencies of both allosyndesis (AC) and autosyndesis (AA and CC). In these plants, autosyndetic bivalents are always observed to represent close to 20% of bivalents in metaphase I of meiosis; this result suggests that rearrangements between paralogous regions can occur (Nicolas et al. 2007, 2009). Similar analyses using genome labeling FISH and GISH in hybrids between *B. napus*, *B. carinata,* and *B. juncea* (genome compositions AABC, BBAC, and CCAB) also confirmed this result, as well as showing autosyndesis in the haploid B genome and AB, BC, and AC allosyndesis at metaphase I (Mason et al. 2010) (Fig. 2.5). Complementary analyses using BACs specific to particular homeologous chromosome pairs in *B. napus* haploids revealed that the rate of homeologous pairing depends on the pair of homeologs concerned (A1/C1, A3/C3, A10/C9, or A7/C6) and on the *B. napus* variety (Grandont et al. 2014). In AC hybrids produced from crosses between *B. rapa* and *B. oleracea*, the frequency of homeologous pairing is even higher than in *B. napus* haploids (Cifuentes et al. 2010; Szadkowski et al. 2011). This high rate of homeologous exchange was also observed in synthetic *B. napus* obtained by crosses between *B. rapa* and *B. oleracea*. GISH-like methods showed that A and C chromosomes frequently paired during the first meiosis in resynthesized AACC S0 plants, suggesting that the first meiosis acts as a genome blender (Szadkowski et al. 2010). These results were also confirmed by analyzing the dynamics of rDNA loci rearrangements in advanced generations of synthetic *B. napus* (Książczyk et al. 2011). However, when one genome is diploid and the other is haploid, such as in AAC hybrids, GISH-like methods show that the nine C chromosomes generally remain as univalents, with only low frequencies of homeologous pairing observed between the A and C genomes (Leflon et al. 2006). On the contrary, homeologous pairing is promoted between two haploid genomes when a third genome is at the diploid stage in the same plant: These data were confirmed by GISH and GISH-like analysis in AABC, BBAC, and CCBA hybrids by Mason et al. (2010). All of these molecular cytogenetic studies not only contribute to our understanding of meiosis and genomic relationships, but provide useful information for breeders targeting genomic introgressions or promoting non-homologous chromosome exchange for crop improvement.

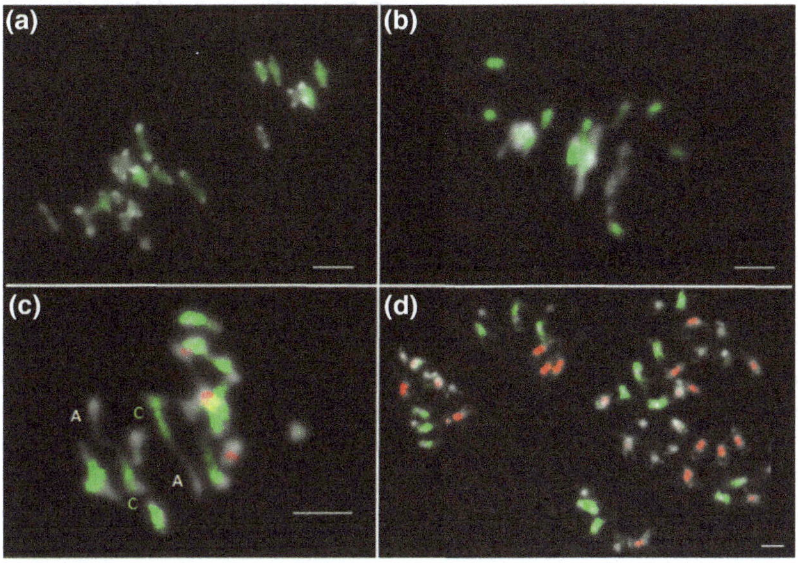

Fig. 2.5 **a–c** FISH analysis of metaphase I PMCs: FISH was carried out using BAC BoB014O06 which identifies all the C chromosomes (green). Nine C bivalent chromosomes and ten unmarked A bivalents (in blue DAPI) on *B. napus* (**a**) and nine C univalent chromosomes on AAC hybrid (**b**). Detection of autosyndesis (AA pairing)/allosyndesis (AC pairing) at metaphase I using BAC BoB014O06 (green) and 54B2 specific of 3 A chromosomes (red) on pollen mother cells at MI from haploid Darmor-bzh (**c**); **d** FISH using genome labels for chromosomes in second-generation progeny derived from a near-allohexaploid *Brassica* ($2n$ = AABBCC) plant [(*B. napus* × *B. carinata*) × *B. juncea*]: A genome chromosomes are blue (DAPI, background stain), B genome chromosomes are labeled red, and C genome chromosomes are labeled green. Bar = 5 μm

2.3.3 Identifying Structural Variation and Homeologous Exchanges Between Genomes

It is now clearly established that homeologous pairing can generate different structural rearrangements in natural allotetraploid species (Schubert and Lysak 2011) and that these rearrangements are difficult to detect only by mapping and sequencing strategies. Specifically, identification of reciprocal translocations and inversions between genotypes is extremely challenging using short-read sequencing technologies (Talkowski et al. 2011). This hinges on the fact that mapping of sequence reads to a reference genome assembly is the preferred method of comparing between genotypes or cultivars within a species. However, with this approach sequence inversions and reciprocal translocations will be almost invisible: Sequences from the target genotype will still map to the reference genome, but in incorrect locations determined by the reference genome rather than by the genome of the target genotype. Genetic mapping can assist with this by analysis of marker distortion, but requires the availability of linkage mapping populations for every genotype of interest. Despite the amount of work and inference involved, this method has been efficient in detecting translocations in several *B. napus* varieties (Lombard and Delourme 2001; Osborn et al. 2003; Piquemal et al. 2005). Similarly in hybrids carrying A, B, and C genomes, structural rearrangements between genomes are large enough to be detected by GISH using *B. nigra* labeled DNA to identify the B genome and GISH using the *B. oleracea* BAC BoB014O06 hybridized specifically to all C genome chromosomes (Howell et al. 2002). In near-allohexaploid lines containing the A, B, and C genomes, this combined labeling technique effectively identified AC, AB, and BC recombinant chromosomes (Mason et al. 2014). It is important to also mention that this molecular cytogenetics technique is the best and perhaps only method available to detect rearrangements at the heterozygous stage (Fig. 2.4).

Loss of genomic regions resulting from non-reciprocal translocations (duplication/deletion events) can be detected in next-generation sequencing by lack of sequences from the target genotype mapping to the reference genome over the deletion region. However, this approach requires high sequence coverage to accurately identify copy number variants: Enough sequencing depth must be generated that absence of mapped reads (or presence of twice as many mapped reads in the case of duplications) in a chromosomal region can be definitively identified as a deletion (or duplication) event, rather than just natural variation in sequence coverage (Alkan et al. 2011). Both SNP array data and fluorescently labeled microsatellite marker data have also been used to assess duplication/deletion events in *Brassica* hybrids, although detection of duplications using relative allele fluorescence is challenging (Mason et al. 2011, 2015). Translocations can be directly physically detected at the homozygous and heterozygous stages by combining GISH and BAC-FISH, if the genomic regions are large and if a BAC is available specific to the translocated regions. Inversions, which are the most challenging of all to identify using sequencing data, can be readily confirmed by the use of several chromosome-specific BAC probes to hybridize to the chromosome of interest, as was done to integrate genetic and physical maps in *B. oleracea* (Howell et al. 2002). In future, technologies currently under development (such as optical mapping) may allow easier identification of structural rearrangements such as non-reciprocal translocations and inversions. Currently, the most direct and unambiguous method to assess these variants is still molecular cytogenetics.

2.3.4 Characterizing Chromosome Addition Lines and Genomic Introgressions for Crop Improvement

So-called chromosome addition lines provide a valuable resource for breeding and genetics analyses. These lines consist of a core genome, usually diploid, with the addition of a single chromosome from another genome, present in either one copy (monosomic addition line) or two copies (disomic addition line). The phenotypic characterization of these chromosome addition lines can allow localization of the trait of interest to particular chromosomes, as has been previously demonstrated for the yellow-seededness trait in *Brassica* AA + C addition lines by Heneen et al. (2012). However, the generation of these lines can be technically challenging. To produce lines with, for example, a complete diploid A genome with single additional C genome chromosomes in *Brassica*, usually a hybrid with $2n$ = AAC, will first be produced from crosses between *B. napus* (or resynthesized *B. napus* from the cross *B. rapa* × *B. oleracea*). This hybrid will then be backcrossed to *B. rapa* in order to eliminate all but one C genome chromosome. However, this approach offers plentiful opportunities for non-homologous recombination to occur, as well as being technically challenging. The use of FISH and GISH techniques to identify which chromosomes are present and in how many copies and to check for large-scale rearrangements such as translocations between the diploid genome and the addition chromosome/s is invaluable. Several studies have reported such analyses (e.g., Snowdon et al. 1997; Chèvre et al. 2007; Heneen et al. 2012; Mason et al. 2014), and BACs specific to different chromosomes allowed their identification in metaphase I (e.g., Suay et al. 2014).

The movement of traits from wild relatives or related species into crops first requires the production of a hybrid between the wild relative and the crop species. This hybrid must then allow recombination between chromosomes from the crop genome and the donor species to occur to produce recombinant chromosomes. These recombinant chromosomes must then be successfully transferred in a backcross to crop to become an integral, stable part of the crop genome to establish the genomic transfer of the trait loci. Depending on the degree of genomic divergence between genomes (as well as genetic factors), a greater or fewer number of recombination events may occur in the hybrid. Currently, only cytogenetic analyses allow assessment of the chromosome number and genomic structure of subsequent introgressed plants. If only very few recombination events occur, then subsequent backcrossing may fail to recover any introgression events in the *B. napus* background. In this case, whole chromosomes may be retained if selection for the phenotype of interest is carried out, but these are generally less desirable as additional chromosomes or chromosome pairs from alien genomes can be frequently lost or selected against (Heneen et al. 2012). If a large number of recombination events take place, then many recombinant chromosomes may be transmitted in backcross generations. This can also cause problems, as many of these recombinants may carry undesirable allelic variants or have a negative effect on crop phenotype.

The larger the population size, the greater the chance of finding an individual with a single recombinant chromosome carrying desirable genomic introgressions for the phenotypic locus or loci of interest. However, the screening method is also critical: If individual lines of interest can be identified early and accurately, chances of success and cost-effectiveness and efficiency are greatly increased. For most species crossed to Brassica crop genomes, GISH is an efficient technique to determine the number and location of independent introgression events, as well as the size of the introgression in some cases (Snowdon et al. 1997; Fredua-Agyeman et al. 2014). In future, the development of BAC-FISH probes specific to introgression regions may offer a solution to detect smaller introgressions.

2.4 New Insights into Meiosis

Cytology and cytogenetics are central to all meiotic studies. Meiosis consists of two rounds of cell division during which chromosome number is halved (from diploid to haploid) and gametes are generated. In most organisms, accurate separation of homologous chromosomes during the first division requires that they first be connected to one another by crossovers (COs), which are one of the products of meiotic recombination. Meiotic COs are the reciprocal exchange of genetic material between chromosomes; they are formed during prophase I between all pairs of homologous chromosomes and start to become visible as chiasmata from diakinesis (they are more clearly resolved at metaphase I). Non-reciprocal recombination events, the so-called non-crossovers (NCOs), can also be recovered genetically but are not amenable to cytological analysis.

Despite its small chromosomes, *A. thaliana* has become a powerful model system for the analysis of meiosis and characterization of meiotic mutants (Mercier et al. 2015 for review). This has paved the way to meiotic studies in *Brassica*. Most notably, the high degree of primary sequence similarity between *A. thaliana* and *B. napus* has recently made possible the use of the antibodies developed against *Arabidopsis* proteins to study meiosis and meiotic recombination in *B. napus* and its close relatives. The first of these studies aimed to decipher the meiotic behavior of *Brassica* allotetraploid (AACC) and allotriploid (AAC) hybrids (Leflon et al. 2006). Polyclonal *Arabidopsis* antibodies that recognize the meiotic proteins ASY1, which associates with chromosome axis, and ZYP1, which is involved in the synaptonemal complex (SC) formation, were successfully used to demonstrate that some chromosomes were completely synapsed at pachytene (intimately associated by a fully formed SC along their length) while other remained unsynapsed even at a latter meiotic stage (i.e., diplotene). These unsynapsed chromosomes likely correspond to the C genome chromosomes that remained as univalents (chromosomes that failed to form COs) at metaphase I, while the synapsed bivalents most probably correspond to pairs of homologous A chromosomes between which COs are formed. Actually, COs get a boost in the AAC hybrids compared to AA diploids (Leflon et al. 2010). Immunolocalization of *Arabidopsis* MLH 1 antibodies, which specifically mark the main fraction of COs (also called class I CO or CO I; Chelysheva et al. 2010), indicated that these crossovers undergo a 1.7-fold increase during male meiosis in the AAC hybrids compared to diploid controls. An even higher increase was detected in female meiosis by comparing genetic map distances (Leflon et al. 2010), suggesting either a sex-specific effect or a greater boost of the otherwise *minority* type of COs (which are not marked by MLH1). Although this has yet to be resolved, this study confirms the importance of combining both genetic and cytological approaches to study such an intricate biological process as meiotic recombination.

Finally, a wider range of cytological and cytogenetical tools was recently used to investigate the formation, progression, and completion of several key hallmarks of meiosis in *B. napus* allotetraploids (AACC) and allohaploids (AC, 19 chromosomes) (Grandont et al. 2014). This study has provided a thorough comparative description and analysis of sister chromatid cohesion, chromosome axes, the synaptonemal complex, and meiotic recombination in two representative *B. napus* accessions (Cifuentes et al. 2010). Analyses of surface-spread prophase I nuclei with electron microscopy have demonstrated a precocious and efficient sorting of homologous versus homeologous chromosomes during early prophase I in the two *B. napus* varieties that otherwise show a genotypic difference in the progression of homologous recombination. Most notably, the spatial-temporal localization of HEI10, an essential protein that can be used to follow the progressive channeling of recombination intermediates into the class I CO pathway (see Chelysheva et al. 2012 for details), was shown to vary from one genotype to another and to correlate with the two main meiotic behavior described at the haploid stage (see introduction). Moreover, the detailed comparison of meiosis in allohaploid and allotetraploid plants showed that the mechanism(s) promoting efficient chromosome sorting in allotetraploids is adjusted to promote crossover formation between homeologs in allohaploids. This suggests that, in contrast to other polyploid species, the threshold for committing a pair of chromosomes to form a CO is not fixed once and for all in *B. napus*, but depends on the operating chromosomes (Grandont et al. 2014). This probably remains an error-prone process, and a few COs can be expected to occasionally form between homeologs and generate homeologous exchanges in the allotetraploid plants (see Sect. 2.3.3). FISH analyses were also carried out to characterize CO formation between individual chromosomes during meiosis in allohaploid *B. napus*; these analyses indicated that both chromosome- and genotype-specific effects change the odds of forming a CO between a given pair of homeologous regions in these plants.

Interestingly, the cytological survey of meiosis in *B. napus* has tentatively pointed toward some of the genomic features of the *B. napus* genome. First, observation of genotype-specific bivalents in the allohaploids has led to the assumption that several chromosome(s) may carry the products of homeologous exchanges (HEs) (Grandont et al. 2014); the presence of numerous HEs was confirmed when assembling the *B. napus* genome sequence (see Sect. 2.3.3; Chalhoub et al. 2014). Likewise, the limited extent to which synaptic multivalents persisted to pachytene in *B. napus* allotetraploids has led to suppose that most interhomeolog recombination intermediates abort early and are redirected into intersister or non-crossover pathways (Grandont et al. 2014). Evidence for very short non-reciprocal exchanges between homeologous sequences, which possibly originated from meiotic non-crossovers, was recently obtained by Chalhoub et al. (2014). More work is needed to confirm that all these HEs, whether they are large or small, have originated during or affect the meiotic behavior of *B. napus*.

2.5 Conclusions

Although cytogenetics as a technique significantly predates molecular genetics, let alone genome sequencing, this methodology and subsequent advances to molecular cytogenetic techniques provide extremely valuable information on physical organization of genomes that complements both mapping and sequencing data. Cytogenetics can provide new information on the genomic structure of ancient and recent polyploid species and shed light on evolutionary processes of genomic rearrangements, genome regulation, and genome structure variation. Cytogenetics can also be used to detect homeologous exchanges between constitutive genomes in interspecific hybrids or addition lines and to track chromosome introgressions from one genome to another. Comparative cytogenetics can be used for phylogenetic and taxonomic analysis, complementary to sequencing-based approaches, and can show evolutionary changes in functional features such as rDNA loci. Cytogenetics is of course essential for our basic understanding of meiosis, and progressively more sensitive molecular biology techniques are allowing visualization of this important biological process in *Brassica*. Use of interspecific hybridization coupled with cytogenetic analysis allows a deep insight into the occurrence and control of homologous and non-homologous chromosome pairing. Cytogenetics techniques are also currently the primary means of validating chromosome rearrangements such as duplications, deletions, and inversions, acting in a complementary fashion to confirm the sequence-based analysis. From the past to the future, cytogenetics has contributed a great deal to our understanding of genetics and genomics in the Brassica genus, and undoubtedly, cytogenetics will play a significant role in the times to come.

Acknowledgements M. A. L. was supported by a research grant from the Czech Science Foundation (P501/12/G090). ASM is supported by an Emmy Noether grant from the Deutsche Forschungsgemeinschaft (MA 6473/1-1).

References

Alkan C, Coe BP, Eichler EE (2011) Applications of next-generation sequencing genome structural variation discovery and genotyping. Nat Rev Genet 12:363–375

Al-Shehbaz IA (2012) A generic and tribal synopsis of the Brassicaceae (Cruciferae). Taxon 61(5):931–954

Attia T, Röbbelen G (1986) Meiotic pairing in haploids and amphihaploids of spontaneous versus synthetic origin in rape, *Brassica napus* l. Can J Genet Cytol 28:330–334

Bozza CG, Pawlowski WP (2008) The cytogenetics of homologous chromosome pairing in meiosis in plants. Cytogenet Genome Res 120:313–319

Catcheside DG (1934) The chromosomal relationships in the Swede and turnip groups of *Brassica*. Ann Bot 48:601–633

Chalhoub B, Denoeud F, Liu SY, Parkin IAP, Tang HB, Wang XY, Chiquet J, Belcram H, Tong CB, Samans B, Correa M, Da Silva C, Just J, Falentin C, Koh CS, Le Clainche I, Bernard M, Bento P, Noel B, Labadie K, Alberti A, Charles M, Arnaud D, Guo H, Daviaud C, Alamery S, Jabbari K, Zhao MX, Edger PP, Chelaifa H, Tack D, Lassalle G, Mestiri I, Schnel N, Le Paslier MC, Fan GY, Renault V, Bayer PE, Golicz AA, Manoli S, Lee TH, Thi VHD, Chalabi S, Hu Q, Fan CC, Tollenaere R, Lu YH, Battail C, Shen JX, Sidebottom CHD, Wang XF, Canaguier A, Chauveau A, Berard A, Deniot G, Guan M, Liu ZS, Sun FM, Lim YP, Lyons E, Town CD, Bancroft I, Wang XW, Meng JL, Ma JX, Pires JC, King GJ, Brunel D, Delourme R, Renard M, Aury JM, Adams KL, Batley J, Snowdon RJ, Tost J, Edwards D, Zhou YM, Hua W, Sharpe AG, Paterson AH, Guan CY, Wincker P (2014) Early allopolyploid evolution in the post-Neolithic *Brassica napus* oilseed genome. Science 345:950–953

Chelysheva L, Grandont L, Vrielynck N, le Guin S, Mercier R, Grelon M (2010) An easy protocol for studying chromatin and recombination protein dynamics during *Arabidopsis thaliana* meiosis: immunodetection of cohesins, histones and MLH1. Cytogenet Genome Res 129(1–3):143–153

Chelysheva L, Vezon D, Chambon A, Gendrot G, Pereira L, Lemhemdi A, Vrielynck N, Le Guin S, Novatchkova M, Grelon M (2012) The *Arabidopsis* HEI10 is a new ZMM protein related to Zip3. PLoS Genet 8(7):e1002799

Chen ZJ, Pikaard CS (1997) Transcriptional analysis of nucleolar dominance in polyploid plants: biased expression/silencing of progenitor rRNA genes is developmentally regulated in *Brassica*. Proc Natl Acad Sci U S A 94:3442–3447

Chen S, Nelson MN, Chèvre AM, Jenczewski E, Li Z, Mason AS, Meng J, Plummer JA, Pradhan A, Siddique KH (2011) Trigenomic bridges for *Brassica* improvement. Crit Rev Plant Sci 30(6):524–547

Cheng F, Mandáková T, Wu J, Xie Q, Lysak MA, Wang X (2013) Deciphering the diploid ancestral genome of the mesohexaploid *Brassica rapa*. Plant Cell 25(5):1541–1554

Cheng F, Wu J, Wang X (2014) Genome triplication drove the diversification of *Brassica* plants. Hortic Res 1:14024. https://doi.org/10.1038/hortres.2014.24

Chester M, Gallagher JP, Symonds VV, da Silva AVC, Mavrodiev EV, Leitch AR, Soltis PS, Soltis DE (2012) Extensive chromosomal variation in a recently formed natural allopolyploid species, *Tragopogon miscellus* (Asteraceae). Proc Natl Acad Sci USA 109:1176–1181

Cheung F, Trick M, Drou N, Lim YP, Park J-Y, Kwon SJ, Kim JA, Scott R, Pires JC, Paterson AH, Town C, Bancroft I (2009) Comparative analysis between homoeologous genome segments of *Brassica napus* and its progenitor species reveals extensive sequence-level divergence. Plant Cell 21:1912–1928

Chèvre AM, Adamczyk K, Eber F, Huteau V, Coriton O, Letanneur JC, Laredo C, Jenczewski E, Monod H (2007) Modelling gene flow between oilseed rape and wild radish. 1. Evolution of chromosome structure. Theor Appl Genet 114:209–221

Cifuentes M, Eber F, Lucas MO, Lode M, Chèvre AM, Jenczewski E (2010) Repeated polyploidy drove different levels of crossover suppression between homoeologous chromosomes in *Brassica napus* allohaploids. Plant Cell 22(7):2265–2276

Edwards D, Batley J, Snowdon RJ (2013) Accessing complex crop genomes with next-generation sequencing. Theor Appl Genet 126:1–11

Fluch S, Kopecky D, Burg K, Šimková H, Taudien S, Petzold A, Kubaláková M, Platzer M, Berenyi M, Krainer S, Doležel J, Lelley T (2012) Sequence composition and gene content of the short arm of Rye (*Secale cereale*) chromosome 1. PLoS ONE 7(2):e30784

Franzke A, Lysak MA, Al-Shehbaz IA, Koch MA, Mummenhoff K (2011) Cabbage family affairs: the evolutionary history of Brassicaceae. Trends Plant Sci 16:108–116

Fredua-Agyeman R, Coriton O, Huteau V, Parkin IAP, Chevre AM, Rahman H (2014) Molecular cytogenetic identification of B genome chromosomes linked to blackleg disease resistance in *Brassica napus* × *B. carinata* interspecific hybrids. Theor Appl Genet 127(6):1305–1318

Fukui K, Nakayama S, Ohmido N, Yoshiaki H, Yamabe M (1998) Quantitative karyotyping of three diploid *Brassica* species by imaging methods and localization of 45s rDNA loci on the identified chromosomes. Theor Appl Genet 96(3–4):325–330

Gomez-Campo C (1980) Morphology and morphotaxonomy of the tribe Brassiceae. In: Tsunoda S, Hinata K, Gomez-Campo C (eds) *Brassica* crops and wild allies: biology and breeding. Japan Scientific Society Press, Tokyo, pp 3–31

Grandont L, Cuñado N, Coriton O, Huteau V, Eber F, Chèvre AM, Grelon M, Chelysheva L, Jenczewski E (2014) Homoeologous chromosome sorting and progression of meiotic recombination in *Brassica napus*: ploidy does matter! Plant Cell 26(4):1448–1463

Hasterok R, Maluszynska J (2000) Nucleolar dominance does not occur in root tip cells of allotetraploid *Brassica* species. Genome 43:574–579

Hasterok R, Wolny E, Hosiawa M, Kowalczyk M, Kulak-Ksiazczyk S, Ksiazczyk T, Heneen WK, Maluszynska J (2006) Comparative analysis of rDNA distribution in chromosomes of various species of Brassicaceae. Ann Bot 97:205–216

Heneen WK, Geleta M, Brismar K, Xiong ZY, Pires JC, Hasterok R, Stoute AI, Scott RJ, King GJ, Kurup S (2012) Seed colour loci, homoeology and linkage groups of the C genome chromosomes revealed in *Brassica rapa-B. oleracea* monosomic alien addition lines. Ann Bot 109:1227–1242

Howell EC, Barker G, Jones G, Kearsey MJ, King G, Kop EP, Ryder CD, Teakle GR, Vicente JG, Armstrong SJ (2002) Integration of the cytogenetic and genetic linkage maps of *Brassica oleracea*. Genetics 161:1225–1234

Howell EC, Armstrong SJ, Barker GC, Jones GH, King GJ, Ryder CD, Kearsey MJ (2005) Physical organization of the major duplication on *Brassica oleracea* chromosome O6 revealed through fluorescence in situ hybridization with *Arabidopsis* and *Brassica* BAC probes. Genome 48(6):1093–1103

Howell EC, Kearsey MJ, Jones GH, King GJ, Armstrong SJ (2008) A and C genome distinction and chromosome identification in *Brassica napus* by sequential fluorescence in situ hybridization and genomic in situ hybridization. Genetics 180(4): 1849–1857

Jenczewski E, Alix K (2004) From diploids to allopolyploids: the emergence of efficient pairing control genes in plants. Crit Rev Plant Sci 23:21–45

Jenczewski E, Eber F, Grimaud A, Huet S, Lucas MO, Monod H, Chèvre AM (2003) *Prbn*, a major gene controlling homoeologous pairing in oilseed rape (*Brassica napus*) haploids. Genetics 164:645–653

Książczyk T, Kovarik A, Eber F, Huteau V, Khaitova L, Tesarikova Z, Coriton O, Chèvre AM (2011) Epigenetic silencing precedes rDNA loci rearrangements during the stabilization of a polyploid species *Brassica napus*. Chromosoma 120:557–571

Lagercrantz U (1998) Comparative mapping between *Arabidopsis thaliana* and *Brassica nigra* indicates that *Brassica* genomes have evolved through extensive genome replication accompanied by chromosome fusions and frequent rearrangements. Genetics 150:1217–1228

Lagercrantz U, Lydiate D (1996) Comparative genome mapping in *Brassica*. Genetics 144:1903–1910

Leflon M, Eber F, Letanneur JC, Chelysheva L, Coriton O, Huteau V, Ryder CD, Barker G, Jenczewski E, Chèvre AM (2006) Pairing and recombination at meiosis of *Brassica rapa* (AA) × *Brassica napus* (AACC) hybrids. Theor Appl Genet 113(8):1467–1480

Leflon M, Grandont L, Eber F, Huteau V, Coriton O, Chelysheva L, Jenczewski E, Chèvre AM (2010) Crossovers get a boost in *Brassica* allotriploid and allotetraploid hybrids. Plant Cell 22(7):2253–2264

Liu Z, Adamczyk K, Manzanares-Dauleux M, Eber F, Lucas MO, Delourme R, Chevre AM, Jenczewski E (2006) Mapping PrBn and other quantitative trait loci responsible for the control of homeologous chromosome pairing in oilseed rape (*Brassica napus* L.) haploids. Genetics 174:1583–1596

Liu S, Liu Y, Yang X, Tong C, Edwards D, Parkin IA, Zhao M, Ma J, Yu J, Huang S, Wang X, Wang J, Lu K, Fang Z, Bancroft I, Yang TJ, Hu Q, Wang X, Yue Z, Li H, Yang L, Wu J, Zhou Q, Wang W, King GJ, Pires JC, Lu C, Wu Z, Perumal S, Wang Z, Guo H, Pan S, Yang L, Min J, Zhang D, Jin D, Li W, Belcram H, Tu J, Guan M, Qi C, Du D, Li J, Jiang L, Batley J, Sharpe AG, Park BS, Ruperao P, Cheng F, Waminal NE, Huang Y, Dong C, Wang L, Li J, Hu Z, Zhuang M, Huang Y, Huang J, Shi J, Mei D, Liu J, Lee TH, Wang J, Jin H, Li Z, Li X, Zhang J, Xiao L, Zhou Y, Liu Z, Liu X, Qin R, Tang X, Liu W, Wang Y, Zhang Y, Lee J, Kim HH, Denoeud F, Xu X, Liang X, Hua W, Wang X, Wang J, Chalhoub B, Paterson AH (2014) The *Brassica oleracea* genome reveals the asymmetrical evolution of polyploid genomes. Nat Commun 5:3930

Lombard V, Delourme R (2001) A consensus linkage map for rapeseed (*Brassica napus* L.): construction and integration of three individual maps from DH populations. Theor Appl Genet 103:491–507

Long EO, Dawid IB (1980) Repeated genes in eukaryotes. Annu Rev Biochem 49:727–764

Lysak MA, Koch MA, Pecinka A, Schubert I (2005) Chromosome triplication found across the tribe Brassiceae. Genome Res 15:516–525

Lysak MA, Berr A, Pecinka A, Schmidt R, McBreen K, Schubert I (2006) Mechanisms of chromosome number reduction in *Arabidopsis thaliana* and related Brassicaceae species. Proc Natl Acad Sci U S A 103:5224–5229

Lysak MA, Cheung K, Kitschke M, Bures P (2007) Ancestral chromosomal blocks are triplicated in Brassiceae species with varying chromosome number and genome size. Plant Physiol 145:402–410

Maluszynska J, Heslop-Harrison JS (1993) Physical mapping of rDNA loci in *Brassica* species. Genome 36:774–781

Mandáková T, Lysak MA (2008) Chromosomal phylogeny and karyotype evolution in x = 7 crucifer species (Brassicaceae). Plant Cell 20:2559–2570

Mandáková T, Singh V, Krämer U, Lysak MA (2015) Genome structure of the heavy metal hyperaccumulator *Noccaea caerulescens* and its stability on metalliferous and nonmetalliferous soils. Plant Physiol 169:674–689

Marcussen T, Sandve SR, Heier L, Spannagl M, Pfeifer M; International Wheat Genome Sequencing Consortium, Jakobsen KS, Wulff BB, Steuernagel B, Mayer KF, Olsen OA (2014) Ancient hybridizations among the ancestral genomes of bread wheat. Science 18;345(6194):1250092

Mason AS, Huteau V, Eber F, Coriton O, Yan G, Nelson MN, Cowling WA, Chèvre AM (2010) Genome structure affects the rate of autosyndesis and allosyndesis in AABC, BBAC and CCAB *Brassica* interspecific hybrids. Chromosome Res 18:655–666

Mason AS, Nelson MN, Castello MC, Yan G, Cowling WA (2011) Genotypic effects on the frequency of homoeologous and homologous recombination in *Brassica napus* × *B. carinata* hybrids. Theor Appl Genet 122:543–553

Mason AS, Nelson MNN, Takahira J, Cowling WA, Moreira Alves G, Chaudhuri A, Chen N, Ragu ME, Dalton-Morgan J, Coriton O, Huteau V, Eber F, Chèvre AM, Batley J (2014) The fate of chromosomes and alleles in an allohexaploid *Brassica* population. Genetics 197:273–283

Mason AS, Takahira J, Atri C, Samans B, Hayward A, Cowling WA, Batley J, Nelson MN (2015) Microspore culture reveals complex meiotic behaviour in a trigenomic *Brassica* hybrid. BMC Plant Biol 15:173

Mercier R, Mézard C, Jenczewski E, Macaisne N, Grelon M (2015) The molecular biology of meiosis in plants. Annu Rev Plant Biol 66:297–327

Mizushima U (1980) Genome analysis in *Brassica* and allied Genera. In: Tsunoda S, Hinata K, Gómez-Campo C (eds) *Brassica* crops and wild allies: biology and breeding. Japan Scientific Society Press, Tokyo, pp 89–106

Morinaga T (1934) Interspecific hybridization in Brassica VI. The cytology of F1 hybrdis of *B. juncea* and *B. nigra*. Cytologia 6:62–67

Müntzing A (1936) The evolutionary significance of autopolyploidy. Hereditas 21:263–378

Nagaharu U (1935) Genome analysis in *Brassica* with special reference to the experimental formation of *B. napus* and peculiar mode of fertilization. Jap J Bot 7:389–452

Nagpal R, Raina SN, Sodhi YS, Mukhopadhyay A, Arumugam N, Pradhan AK, Pental D (1996) Transfer of *Brassica tournefartii* (TT) genes to allotetraploid oilseed *Brassica* species (*B. juncea* AABB, *B. napus* AACC, *B. carinata* BBCC): homoeologous pairing is more pronounced in the three-genome hybrids (TACC, TBAA, TCAA, TCBB) as compared to allodiploids (TA, TB, TC). Theor Appl Genet 92:566–571

Navabi ZK, Huebert T, Sharpe AG, O'Neill CM, Bancroft I, Parkin IAP (2013) Conserved microstructure of the *Brassica* B genome of *Brassica nigra* in relation to homologous regions of *Arabidopsis*

thaliana, *B. rapa* and *B. oleracea*. BMC Genom 14:250
Nicolas S, Le Mignon G, Eber F, Coriton O, Monod H, Clouet V, Huteau V, Lostanlen A, Delourme R, Chalhoub B, Ryder C, Chevre AM, Jenczewski E (2007) Homoeologous recombination plays a major role in chromosome rearrangements that occur during meiosis of *Brassica napus* haploids. Genetics 175:487–503
Nicolas S, Leflon M, Monot H, Eber F, Coriton O, Huteau V, Chèvre AM, Jenczewski E (2009) Genetic regulation of meiotic crossovers between related genomes in *Brassica napus* haploids and hybrids. Plant Cell 21:373–385
Nicolas S, Monot H, Eber F, Chèvre AM, Jenczewski E (2012) Non random distribution of extensive chromosome rearrangements in *Brassica napus* depends on genome organization. Plant J 70:691–703
Osborn TC, Butrulle DV, Sharpe AG, Pickering KJ, Parkin IA, Parker JS, Lydiate DJ (2003) Detection and effects of a homeologous reciprocal transposition in *Brassica napus*. Genetics 165:1569–1577
Paritosh K, Gupta V, Yadava SK, Singh P, Pradhan AK, Pental D (2014) RNA-seq based SNPs for mapping in *Brassica juncea* (AABB): synteny analysis between the two constituent genomes A (from *B. rapa*) and B (from *B. nigra*) shows highly divergent gene block arrangement and unique block fragmentation patterns. BMC Genom 15:396
Parkin I (2011) Chasing ghosts: comparative mapping in the Brassicaceae. In: RA Jorgensen (ed) Plant genetics and genomics: crops and models, vol 9, pp 153–170
Parkin IAP, Sharp AG, Keith DJ, Lydiate DJ (1995) Identification of the A and C genomes of amphidiploid *Brassica napus* (oilseed rape). Genome 38:1122–1131
Parkin IAP, Sharpe AG, Lydiate DJ (2003) Patterns of genome duplication within the *Brassica napus* genome. Genome 46:291–303
Parkin IA, Gulden SM, Sharpe AG, Lukens L, Trick M, Osborn TC, Lydiate DJ (2005) Segmental structure of the *Brassica napus* genome based on comparative analysis with *Arabidopsis thaliana*. Genetics 171: 765–781
Parkin IA, Koh C, Tang H, Robinson SJ, Kagale S, Clarke WE, Town CD, Nixon J, Krishnakumar V, Bidwell SL, Denoeud F, Belcram H, Links MG, Just J, Clarke C, Bender T, Huebert T, Mason AS, Pires JC, Barker G, Moore J, Walley PG, Manoli S, Batley J, Edwards D, Nelson MN, Wang X, Paterson AH, King G, Bancroft I, Chalhoub B, Sharpe AG (2014) Transcriptome and methylome profiling reveals relics of genome dominance in the mesopolyploid *Brassica oleracea*. Genome Biol 15(6):R77
Piquemal J, Cinquin E, Couton F, Rondeau C, Seignoret E, Doucet E, Perret D, Villeger M-J, Vincourt P, Blanchard P (2005) Construction of an oilseed rape (*Brassica napus* L.) genetic map with SSR markers. Theor Appl Genet 111:1514–1523
Prakash S, Hinata K (1980) Taxonomy, cytogenetics and origin of crop *Brassica*, a review. Opera Botanica 55:1–57
Prakash SR, Bhat SR, Quiros CF, Kirti PB, Chopra VL (2009) *Brassica* and its close allies: cytogenetics and evolution. Plant Breed Rev 31:21–187
Renard M, Dosba F (1980) Etude de l'haploidie chez le colza (*Brassica napus* L. var *oleifera* metzger). Annales de l'amélioration des Plantes 30:191–209
Röbbelen G (1960) Beiträge zur Analyse des *Brassica*-Genoms. Chromosoma 11:205–228
Rogers SO, Bendich AJ (1987) Ribosomal RNA genes in plants: variability in copy number and in the intergenic spacers. Plant Mol Biol 9:509–520
Schiessl S, Samans B, Huttel B, Reinhard R, Snowdon RJ (2014) Capturing sequence variation among flowering-time regulatory gene homologs in the allopolyploid crop species *Brassica napus*. Front Plant Sci 5
Schranz ME, Lysak MA, Mitchell-Olds T (2006) The ABC's of comparative genomics in the Brassicaceae: building blocks of crucifer genomes. Trends Plant Sci 11:535–542
Schubert I, Lysak MA (2011) Interpretation of karyotype evolution should consider chromosome structural constraints. Trends Genet 27:207–216
Snowdon RJ, Kohler W, Friedt W, Kohler A (1997) Genomic in situ hybridization in *Brassica* amphidiploids and interspecific hybrids. Theor Appl Genet 95:1320–1324
Suay L, Zhang D, Eber F, Jouy H, Lodé M, Huteau V, Coriton O, Szadkowski E, Leflon M, Martin OC, Falque M, Jenczewski E, Paillard S, Chèvre AM (2014) Crossover rate between homologous chromosomes and interference are regulated by the addition of specific unpaired chromosomes in *Brassica*. New Phytol 201(2):645–656
Szadkowski E, Eber F, Huteau V, Lodé M, Huneau C, Belcram H, Coriton O, Manzanares-Dauleux MJ, Delourme R, King GJ, Chalhoub B, Jenczewski E, Chèvre AM (2010) The first meiosis of resynthesized *Brassica napus*, a genome blender. New Phytol 186:102–112
Szadkowski E, Eber F, Huteau V, Lodé M, Coriton O, Jenczewski E, Chèvre AM (2011) Polyploid formation pathways have an impact on genetic rearrangements in resynthesized *Brassica napus*. New Phytol 191:554–894
Talkowski ME, Ernst C, Heilbut A, Chiang C, Hanscom C, Lindgren A, Kirby A, Liu ST, Muddukrishna B, Ohsumi TK, Shen YP, Borowsky M, Daly MJ, Morton CC, Gusella JF (2011) Next-generation sequencing strategies enable routine detection of balanced chromosome rearrangements for clinical diagnostics and genetic research. Am J Hum Genet 88:469–481
Tang H, Woodhouse MR, Cheng F, Schnable JC, Pedersen BS, Conant G, Wang X, Freeling M, Pires JC

(2012) Altered patterns of fractionation and exon deletions in *Brassica rapa* support a two-step model of paleohexaploidy. Genetics 190:1563–1574

Udall JA, Quijada PA, Osborn TC (2005) Detection of chromosomal rearrangements derived from homologous recombination in four mapping populations of *Brassica napus* L. Genetics 169:967–979

Wang X, Wang H, Wang J, Sun R, Wu J, Liu S, Bai Y, Mun JH, Bancroft I, Cheng F, Huang S, Li X, Hua W, Wang J, Wang X, Freeling M, Pires JC, Paterson AH, Chalhoub B, Wang B, Hayward A, Sharpe AG, Park BS, Weisshaar B, Liu B, Li B, Liu B, Tong C, Song C, Duran C, Peng C, Geng C, Koh C, Lin C, Edwards D, Mu D, Shen D, Soumpourou E, Li F, Fraser F, Conant G, Lassalle G, King GJ, Bonnema G, Tang H, Wang H, Belcram H, Zhou H, Hirakawa H, Abe H, Guo H, Wang H, Jin H, Parkin IA, Batley J, Kim JS, Just J, Li J, Xu J, Deng J, Kim JA, Li J, Yu J, Meng J, Wang J, Min J, Poulain J, Wang J, Hatakeyama K, Wu K, Wang L, Fang L, Trick M, Links MG, Zhao M, Jin M, Ramchiary N, Drou N, Berkman PJ, Cai Q, Huang Q, Li R, Tabata S, Cheng S, Zhang S, Zhang S, Huang S, Sato S, Sun S, Kwon SJ, Choi SR, Lee TH, Fan W, Zhao X, Tan X, Xu X, Wang Y, Qiu Y, Yin Y, Li Y, Du Y, Liao Y, Lim Y, Narusaka Y, Wang Y, Wang Z, Li Z, Wang Z, Xiong Z, Zhang Z; *Brassica rapa* Genome Sequencing Project Consortium (2011) The genome of the mesopolyploid crop species *Brassica rapa*. Nat Genet 43(10):1035–1039

Wang YP, Wang XY, Paterson AH (2012) Genome and gene duplications and gene expression divergence: a view from plants. Ann NY Acad Sci 1256:1–14

Xiong Z, Pires JC (2011) Karyotype and identification of all homoeologous chromosomes of allopolyploid *Brassica napus* and its diploid progenitors. Genetics 187:37–49

Xiong ZY, Gaeta RT, Pires JC (2011) Homoeologous shuffling and chromosome compensation maintain genome balance in resynthesized allopolyploid *Brassica napus*. Proc Natl Acad Sci USA 108:7908–7913

Ziolkowski PA, Kaczmarek M, Babula D, Sadowski J (2006) Genome evolution in *Arabidopsis/Brassica*: conservation and divergence of ancient rearranged segments and their breakpoints. Plant J 47:63–74

Zou J, Fu DH, Gong HH, Qian W, Xia W, Pires JC, Li RY, Long Y, Mason AS, Yang TJ, Lim YP, Park BS, Meng JL (2011) *De novo* genetic variation associated with retrotransposon activation, genomic rearrangements and trait variation in a recombinant inbred line population of *Brassica napus* derived from interspecific hybridization with *Brassica rapa*. Plant J 68:212–224

Genes and Quantitative Trait Loci Mapping for Major Agronomic Traits in *Brassica napus* L.

3

Régine Delourme, Anne Laperche, Anne-Sophie Bouchet, Mélanie Jubault, Sophie Paillard, Maria-J. Manzanares-Dauleux and Nathalie Nesi

Abstract

The advent of high-throughput genomic technologies and the availability of a reference genome sequence of *Brassica napus* and its diploid parental species, *B. rapa* and *B. oleracea*, open new insights into the genomic localization of agronomic trait-associated quantitative trait loci (QTL), the identification of underlying genes and their sequence variation. Over the last 20 years, many genetic maps of *B. napus* have been built, progressively integrating various types of markers. Large single-nucleotide polymorphism (SNP) arrays allowed the construction and integration of high density maps and their anchorage to the *B. napus* sequence. Increasingly, precise genetic analyses of agronomic traits could thus be carried out, either through linkage analysis or through genome-wide association mapping. Comparative genomics allowed the genomic localization of the genes and QTL controlling agronomic traits, as well as an assessment of the impact that the high level of duplications present in this polyploid species has on the genetic architecture of the traits and on the structural and functional diversity of the genes involved. This chapter reviews the evolution of *B. napus* genetic and genomic resources and their use in gene and QTL mapping for several major traits and then shows how the availability of the *B. napus* genome sequence allows more accurate investigation of the genomic regions and underlying genes involved.

3.1 Introduction

The demand for vegetable oils and proteins is projected to increase by more than 40% by 2030, as a result of an increasing world population and rising living standards. Therefore, high yield and high seed quality are major goals for crop production. At the same time, there is a need to ensure seed yield and quality under fluctuating environments with various biotic and abiotic stresses, and a need to reduce the environmental impacts of agriculture by lowering chemical inputs (fertilizers, pesticides, herbicides). Many genetic studies of various agronomic traits, including developmental traits, seed quality, oil yield, yield components, nutrient use efficiency, and disease resistance, have been performed. Linkage analyses (LAs) with denser and denser genetic maps or genome-wide association studies (GWASs) were carried out to decipher the genetic determinism of these traits and optimize

R. Delourme (✉) · A. Laperche · A.-S. Bouchet · M. Jubault · S. Paillard · M.-J. Manzanares-Dauleux · N. Nesi
IGEPP, AGROCAMPUS OUEST, INRA, Univ Rennes, 35650 Le Rheu, France
e-mail: regine.delourme@inra.fr

S. Liu et al. (eds.), *The Brassica napus Genome*, Compendium of Plant Genomes,
https://doi.org/10.1007/978-3-319-43694-4_3

the mining and use of new genetic diversity. Most of the traits of interest are complex, quantitative, and often influenced by environment (E) and genotype (G) x environment (GxE) interactions, requiring the use of various genetic backgrounds and multiple environments to strengthen the results of the genetic analyses. The comparison and integration of the results of these studies now benefit from the advent of high-throughput genomic technologies and the availability of reference genome sequences, which allow comparative mapping. In this review, we outline the evolution of *Brassica napus* genetic and genomic resources, along with their use for gene and quantitative trait loci (QTL) mapping for several major traits. We also show how the availability of the *B. napus* genome sequence allows more accurate investigation of the genomic regions and underlying genes involved.

3.2 Linkage Maps

3.2.1 Development of Genetic Maps

Genetic linkage maps are highly valuable tools for the identification of genomic regions carrying major genes and QTL controlling agronomic traits. Dense genetic maps allow finer comparative genome analyses with sequenced related genomes and faster map-based cloning of major genes and QTL. In recent years, the establishment of genetic maps has benefited from the development of new types of molecular markers which take advantage of automated sequencing and genotyping technologies. Over the last 20 years, many *B. napus* genetic maps have been built, progressively integrating various types of markers (Gali and Sharpe 2011). The first marker-based genetic maps were built with restriction fragment length polymorphisms (RFLPs) (Landry et al. 1991; Ferreira et al. 1994; Parkin et al. 1995, 2005; Toroser et al. 1995; Sharpe et al. 1995; Uzunova et al. 1995), random amplified polymorphic DNAs (RAPDs) (Foisset et al. 1996), amplified fragment length polymorphisms (AFLPs) (Lombard and Delourme 2001; Badani et al. 2006; Liu et al. 2006a; Basunanda et al. 2007), simple sequence repeats (SSRs) (Lowe et al. 2004; Piquemal et al. 2005; Qiu et al. 2006; Long et al. 2007; Radoev et al. 2008; Suwabe et al. 2008; Wang et al. 2011a). More recently, dense genetic maps included sequence-related amplified polymorphism (SRAP) (Sun et al. 2007), diversity arrays technology (DArT) (Raman et al. 2013), and single-nucleotide polymorphisms (SNPs). As reviewed by Kaur et al. (2012), SNP discovery is challenging in allopolyploid species such as *B. napus*. SNPs may arise both between allelic (homologous) sequences within each diploid genome and between homoeologous sequences between diploid genomes, but also from polymorphisms between paralogous duplicated sequences. SNP discovery has been based on sequence analysis of *B. napus* expressed sequence tags (ESTs) (Durstewitz et al. 2010; Hu et al. 2012b) or on second-generation high-throughput sequencing (Trick et al. 2009; Bancroft et al. 2011; Bus et al 2012; Clarke et al. 2013; Huang et al. 2013). Different infinium arrays were developed and further used to build genetic maps. An 8K array including 7322 SNP markers was used to build an integrated genetic map comprising 5764 SNP and 1603 PCR markers from four doubled haploid segregating populations (Delourme et al. 2013). A 6K array (Dalton-Morgan et al. 2014) allowed 631 and 1667 SNPs to be incorporated into two genetic maps built by Raman et al. (2014a) and Cai et al. (2014), respectively. Another 6K *B. napus* Illumina infinium array was developed from various next-generation sequencing (NGS) data (http://aafc-aac.usask.ca/ASSYST/. A genetic map with 14,675 SNPs corresponding to 895 genetic bins (marker pairs with zero recombination were assigned to the same genetic bin) was constructed using a modified double-digested restriction-site associated DNA (ddRAD) sequencing technique (Chen et al. 2013). In 2012, an international Brassica SNP consortium produced a 60K SNP infinium genotyping array containing 52,157 SNPs for *B. napus*, in cooperation with Illumina Inc., San Diego, CA, USA (Snowdon and Iniguez-Luy 2012; Edwards et al.

2013), and maps comprising ca. 9K or 11K SNPs were built (Liu et al. 2013; Zhang et al. 2014). Bancroft et al. (2011, 2015) developed genetic maps comprising 21,323 SNPs (arranged in 887 bins) derived from transcriptome sequence analyses, which were then used to infer the order of unigenes and to position the probe flanking sequences from the Brassica research community's 60K SNP infinium array, thus integrating these two genotyping resources.

3.2.2 Map Integration

Few studies reported the integration of genetic maps developed from different *B. napus* segregating populations. Lombard and Delourme (2001) derived a consensus map from three crosses, and a translocation between A7 and C6 linkage groups was identified in a cross derived from two spring oilseed rape lines. The integration of genetic maps derived from crosses between oilseed rape lines and from crosses between resynthesized and natural *B. napus* parents showed that marker orders were positively correlated between the individual maps, allowing confidence for the subsequent use of bridge markers for the map integration (Wang et al. 2011a). Another *B. napus* consensus map consisting of a 1,359 DArT and non-DArT markers was built from six populations originating from Australia, Canada, China, and Europe (Raman et al. 2013). However, for some linkage groups (LGs), the number of shared markers was very low between populations. In these cases, it was difficult to judge the overall consistency of marker order among maps. A major advantage of genetic maps based on genome-wide SNP screening arrays is the frequent occurrence of consensus markers for integration and alignment of maps and then of QTL, both with each other and also with reference genome sequences. The integration of four individual maps—DYDH ('Darmor-*bzh*' × 'Yudal' DH), TNDH ('Tapidor' × 'Ningyou7' DH), AADH ('Aviso' × 'Aburamasari' DH), and AMDH ('Aviso' × 'Montego' DH)—showed a very good collinearity overall between all the maps, with the exception of three inversions that were observed at the bottom of A2 on the TNDH map, at the top of C8 on the AADH map, and at the bottom of C8 on the AMDH map (Delourme et al. 2013). More recently, the combination of the 8 K and 60K arrays with a 20K array and RADseq-SNPs obtained in the course of the *B. napus* genome sequencing project allowed the construction of an integrated genetic map comprising 41,001 markers mapped into 7287 genetic bins, which was used for the *B. napus* genome anchoring (Chalhoub et al. 2014).

3.2.3 Comparative Mapping

The construction of genetic maps of diploid and amphidiploid *Brassica* species, and their comparison and alignment to the *Arabidopsis* genome sequence, has provided insights into *Brassica* genome organization and evolution following the different rounds of polyploidization and diploidization that have occurred in the history of these species. These comparative mapping studies revealed the complex organization of the *Brassica* genomes resulting from extensive duplications and rearrangements. The collinearity observed between *A. thaliana* and *B. napus* led to the description of a genomic block system determined by Parkin et al. (1995), who demonstrated that the structure of the *Brassica* A and C genomes could be described with approximately 21 conserved blocks. A framework consisting of 24 genomic blocks (A-X) was then built within the ancestral karyotype (Schranz et al. 2006; Parkin 2011). The comparisons were refined using dense genetic maps with SSRs (Wang et al. 2011a) and SNPs (Bancroft et al. 2011; Cai et al. 2014). However, divergence of gene content and relative positions as well as discontinuous collinearity between *Arabidopsis* and Brassicas have been reported (Parkin et al. 2005; Town et al. 2006; Trick et al. 2009), which can make the transfer of information from *Arabidopsis* to *Brassica* species difficult (e.g., for the identification of candidate genes). This difficulty can now be circumvented through the availability of the sequences of

B. napus and its diploid parental species genomes (Wang et al. 2011c; Chalhoub et al. 2014; Liu et al. 2014; Parkin et al. 2014).

3.3 Linkage Disequilibrium Assessment

By integrating genetic map data with genotyping data generated from collections of accessions/varieties, the linkage disequilibrium (LD) pattern can be investigated along the genome of a given species. The success of GWAS in locating genes responsible for complex traits depends on the extent of LD, the number and the distribution of markers, as well as on the collection structure. Since the extent of LD may vary depending on the history of the collections, it should be investigated prior to GWAS design. The first published studies relied on either a low number of lines or a low number of markers (Ecke et al. 2010; Bus et al. 2011; Xiao et al. 2012), preventing precise estimation of LD. The availability of SNPs thereafter enabled an increasing number of markers and thus the reliability of LD measures. Harper et al. (2012) carried out associative transcriptomics on 53 *B. napus* lines using >60K SNPs. Delourme et al. (2013), Fopa Fomeju et al. (2014), and Li et al. (2014a) assessed the LD on panels consisting of 313, 116, and 472 lines corresponding to different germplasms genotyped with 4329, 3228, and 26,841 SNPs. In all these studies, the average LD extent ranged from 0.5 to 1.2 or 2 cM, with some variation between collections and between and within linkage groups. For example, the LD extent in winter oilseed rape is higher on LGs A2, A6, A8, C5, and C6 (Delourme et al. 2013; Fopa Fomeju et al. 2014). With the availability of the *B. napus* genome sequence, the extent of LD can be related to physical distances. Hatzig et al. (2015) showed, in a collection of 218 lines from diverse origins genotyped with 22,169 SNPs, that LD (as measured by r^2 values between each pair of marker) declined to 0.1 for markers distant from 480 kbp (A1) to 1283 kpb (A9). Stronger LD patterns were observed on C3, C8 and particularly on C1 and C4, with r^2 values above 0.1 for up to 6651 and 4048 kbp, respectively. Overall, the number of available SNPs is large enough to perform genome-wide association studies, but depending on the panels and the genomic regions the accuracy might not be sufficient, at least at the regional genomic level. Therefore, regional association mapping after development of new SNPs in a given genomic region might allow further refinement of the genetic mapping of trait-associated markers and identification of candidate genes (Snowdon et al. 2010; Li et al. 2014c; Shi et al. 2015).

3.4 Gene and QTL Mapping for Different Agronomic Traits

Most gene and QTL mapping efforts in *B. napus* are focused on agronomic traits, with the exception of some miscellaneous studies on other traits, such as homoeologous chromosome pairing (Liu et al. 2006a) or paternal inheritance of a mitochondrial plasmid (Oshima and Handa 2012). We review here the main results on agronomic traits focusing on seed oil content, seed quality, flowering time, yield and yield components under different abiotic environments, and disease or pest resistance (Table 3.1). Other studies report mapping of genes involved in preharvest sprouting (Feng et al. 2009), pod shattering (Hu et al. 2012a; Wen et al. 2012; Raman et al. 2014b; Liu et al. 2016a), and seed dormancy (Schatzki et al. 2013). In addition, many studies report the mapping or cloning of male sterility and male fertility restoration genes, but these will not be presented here.

3.4.1 Seed Oil Content

QTLs for seed oil content were identified mainly by linkage analyses using doubled haploid (DH) populations (Ecke et al. 1995; Zhao et al. 2006a, 2012a; Delourme et al. 2006b; Qiu et al. 2006; Wu et al. 2006; Cao et al. 2010; Chen et al. 2010; Würschum et al. 2012; Sun et al. 2012; Wang et al. 2013a; Bouchet et al. 2014; Javed et al. 2014; Jiang et al. 2014; Teh and Möllers 2016;

Table 3.1 Gene and QTL mapping for different agronomic traits in *Brassica napus*, rapeseed

Trait category	Detail of the traits studied	Mapping population, size, number of environment(s) studied	Maps Number and type of markers Length of the genetic maps	Method (linkage or association mapping)	References
Seed quality traits	Oil content, erucic acid content	Mansholt's Hamburger Raps × Samourai, 151 DH, 1 environment (field)	205—RFLP (1441 cM)	LA	Ecke et al. (1995)
	Glucosinolate content	Major × Stellar, 105 DH, 1 environment (field)	135—RFLP (1343 cM)	LA	Toroser et al. (1995)
	Glucosinolate content	Mansholt's Hamburger Raps × Samourai, 151 DH, 1 environment (field)	205—RFLP (1441 cM)	LA	Uzunova et al. (1995)
	Oil content	4 BC populations, 2 environments (field)	276—RFLP	LA	Butruille et al. (1999)
	Oil content	Victor × Tapidor, 22 substitution lines, 9 environments (field)	158—RFLP (1204 cM)	LA	Burns et al. (2003)
	Glucosinolate content	Victor × Tapidor, 195 B1, 1 environment (greenhouse) Tapidor × Bienvenu, 60 BC1, 1 environment (greenhouse)	158—(1204 cM)	LA	Howell et al. (2003)
	Glucosinolate content, aliphatic, indolic, and aromatic glucosinolate content	H5200 × NingRS-1, 128 F2:3, 1 environment (field)	107—RFLP, AFLP (1620 cM)	LA	Zhao and Meng (2003)
	Tocopherol content	Mansholt's Hamburger Raps × Samourai, 144DH, 2 environments (field)	205—RFLP (1441 cM)	LA	Marwede et al. (2005)
	Seed color, acid detergent fiber	25629-3 × K26-96, 105 DH Express 617 × 1012/98, 179 F2 Express 617 × 1012/98, 166 DH	193—AFLP, SSR (923 cM) 263—AFLP, SSR (1186 cM) 347—AFLP, SSR (1721 cM)	LA	Badani et al. (2006)
	Oil content, flowering time	Darmor-*bzh* × Yudal, 445 DH, 2 environments (field)	305—SSR (2690 cM)	LA	Delourme et al. (2006b)
	Seed color	Youyan 2 × GH06, 132 F2, 1 environment	164—AFLP, SSR, RAPD, SCAR (2550 cM)	LA	Liu et al. (2006b)
	Oil content, erucic acid content	Tapidor × Ningyou7, 188 DH, 4 environments (field)	277—AFLP, RFLP, SSR, STS (1685 cM)	LA	Qiu et al. (2006)
	Glucosinolate content	MF216 × P1804, 150 DH RV128 × P1804, 150 DH, 4 environments (field)	218—RFLP (1398 cM) 250—RFLP (1453 cM)	LA	Quijada et al. (2006)
	Oil content, protein content	Sollux × Gaoyou, 282 DH, 4 environments (field)	125—SSR (1196 cM)	LA	Zhao et al. (2006a)

(continued)

Table 3.1 (continued)

Trait category	Detail of the traits studied	Mapping population, size, number of environment(s) studied	Maps Number and type of markers Length of the genetic maps	Method (linkage or association mapping)	References
	Erucic acid content, glucosinolate content	Express × V8, 262 DH, 2 environments	476—AFLP, SSR (1673 cM)	LA	Basanunda et al. (2007)
	Seed color	GH 06 × Zhongyou 821, 185 RILs, 4 environments (field) GH 06 × Youyan 2, 183 RILs, 4 environments (field)	420—SSR, RAPD, SRAP (1744 cM) 265—SSR, SRAP (1135 cM)	LA	Fu et al. (2007a)
	Husk proportion, lignin content	GH06 × Zhongyou, 821 RILs (field)	509—SSR, RAPD, SRAP (1923 cM)	LA	Fu et al. (2007b)
	Oil content, hull content	GH 06 × P 147, 188 RILs, 1 environment (field)	300—AFLP, SRAP, SSR, TRAP (1248.5 cM)	LA	Jin et al. (2007)
	Phytosterol content, sinapate ester content	Mansholt's Hamburger Raps × Samourai, 148 DH, 4 environments (field)	185—RFLP, AFLP (1739 cM)	LA	Amar et al. (2008)
	Glucosinolate content	94 AND 46 accessions 1 environment	348—SSR	GWAS	Hasan et al. (2008)
	Oil content, hull content, seed color	GH06 × P174, 3 environments (field)	451—AFLP, TRAP, SRAP, SSR (1589 cM)	LA	Yan et al. (2009)
	Oil content	High × low oil content lines, 150 DH, 3 environments (field)	387—SRAP, SSR (1868 cM)	LA	Chen et al. (2010)
	Start of flowering, duration and end of flowering, maturity, plant height, length of main raceme, pods per main raceme, pod density, oil, protein, glucosinolate, sulfur, oleic acid, and linolenic acid contents	84 WOSR accessions, 7 environments (field)	684—AFLP	GWAS	Honsdorf et al. (2010)
	Seed color, acid detergent fiber, acid detergent lignin, neutral detergent fiber, polymeric proanthocyanidins	49 WOSR accessions, 3 environments (field)	114—SSR	Regional association mapping	Snowdon et al. (2010)
	Oil content	69 cultivars, 103 new-type *B. napus*, 2 environments (field)	116—SSR	GWAS	Zou et al. (2010)

(continued)

Table 3.1 (continued)

Trait category	Detail of the traits studied	Mapping population, size, number of environment(s) studied	Maps Number and type of markers Length of the genetic maps	Method (linkage or association mapping)	References
	Phenolic acids, monomer and oligomers PAs, polymers PAs and other pigmented complex, sinapoyl glucose content, sinapine content	49 WOSR accessions	138—SSR	GWAS	Rezaeizad et al. (2011)
	Flowering time, seed yield, plant height, protein, oil, and glucosinolate contents	Nine crosses from ten parents, 391 DH, 4 environments (field)	253—SNP	LA	Würschum et al. (2012)
	Seed color	Quantum × No. 2127-17, 247 DH, 3 environments (field) 2127-17 × 94-570, 118 F2:3, 3 environments (field)	397—SSR, SRAP (1747 cM) 143—SSR, AFLP (1208 cM)	LA	Zhang et al. (2011a)
	Seed color, total, oligomeric and polymeric proanthocyanidins	Express 617 × 1012/98, 166 DH, 3 environments	191—AFLP, SSR (1171 cM)	LA	Lipsa et al. (2012)
	Tocopherol content	TNDH: Tapidor × Ningyou7, 202 DH, 404 reconstructed F2, 142 accessions, 3 environments (field)	621—SSR, RFLP, SNP, MS-AFLP, STS (2060 cM) 327 loci	LA GWAS	Wang et al. (2012)
	Oil content	Sollux × Gaoyou, 11 environments (field)	481—CAPS, SSR, SCAR, STS, SRAP (1949 cM)	LA	Zhao et al. (2012a)
	Oil content	KenC-8 × N53-2, 348 DH, 8 environments (field)	403—SSR, SRAP, STS, IFLP (1784 cM)	LA	Wang et al. (2013a)
	Acid detergent lignin, seed color, cellulose, and hemicellulose contents	GH06 × P174, 172 RIL, 4 environments (field)	9164—SNP (1833 cM)	LA	Liu et al. (2013)
	Oil content	TNDH: Tapidor × Ningyou7, 202 DH, 404 reconstructed F2, 12 environments (field)	786—RFLP, SSR, STS, IGF, AFLP (2117 cM)	LA	Jiang et al. (2014)
	Cruciferin and napin contents cruciferin/napin ratio, protein, oil, and glucosinolate contents	Express × R53, 229 DH, 2 environments	229—SSR, AFLP (2283 cM)	LA	Schatzki et al. (2014)
	Glucosinolate content	101 accessions	144,131—SNP 100,534—GEM	Associative transcriptomics	Lu et al. (2014)

(continued)

Table 3.1 (continued)

Trait category	Detail of the traits studied	Mapping population, size, number of environment(s) studied	Maps Number and type of markers Length of the genetic maps	Method (linkage or association mapping)	References
	Thousand seed weight, oil, seed glucosinolate, and seed erucic acid contents	472 accessions (spring, winter, semi-winter), 2 environments (field)	26,841—SNP	GWAS	Li et al. (2014a)
	Seed color, oil content	217 accessions	389—SSR	GWAS	Qu et al. (2015)
	Seed glucosinolates content	Express × SWU07, 261 DH, 2 environments, 5 and 2 years, respectively	316—SSR (1198 cM)	LA	Fu et al. (2015)
	Oil, protein, seed linolenic acid, seed glucosinolates, seed hemicellulose, and seed cellulose contents	WOSR diversity set, 89 individuals, 4 environments (field)	5606—SNP	GWAS and QTL analysis (regional association mapping)	Gajardo et al. (2015)
	Glucosinolate content Erucic acid content	TNDH: Tapidor × Ningyou7 DHs × Tapidor, BC1 DHs × Ningyou7, BC2, 2 environments (field)	786—AFLP, RFLP, SNP, SSR, STS, SSCP, CAPS (2117 cM)	LA	Xu et al. (2015b)
	Amino acid: lysine, threonine, methionine	TNDH: Tapidor × Ningyou7 DHs × Tapidor, BC1 DHs × Ningyou7, BC2, 2 environments (field)	786—AFLP, RFLP, SNP, SSR, STS, SSCP, CAPS (2117 cM)	LA	Xu et al. (2015a)
	Fatty acid composition	Tapidor × Ningyou7, 202 DH, 6 environments (field)	932—AFLP, RFLP, SNP, SSR, STS, SSCP, CAPS, genes	LA	Wang et al. (2015c)
	Oil, protein, glucosinolate, sulfur, oleic acid, linolenic acid, erucic acid, neutral detergent fiber, acid detergent fiber, acid detergent lignin, and cellulose contents	405 OSR accessions	6 K SNP array	GWAS	Körber et al. (2016)
	Phytosterol content, fatty acid composition, seed oil content, protein content of the defatted meal	Sansibar × Oase, 226 DH, 6 environments (field)	1638—AFLP, SSR, SNP, DArT, KASP, candidate gene-based markers (2350 cM)	LA	Teh and Möllers (2016)

(continued)

Table 3.1 (continued)

Trait category	Detail of the traits studied	Mapping population, size, number of environment(s) studied	Maps Number and type of markers Length of the genetic maps	Method (linkage or association mapping)	References
	Oil content	6F313 × 51070 (G1) and 61616 × 51070 (G2) F_2 populations, 105 lines per population, 3 environments (field); 227 OSR accessions, 2 environments (field)	G1: 9711 SNP (1987 cM) G2: 9854 SNP (1836 cM)	LA	Sun et al. (2016)
			34292 SNP	GWAS	
	Nonessential amino acids (arginine, histidine, glutamic acid, glycine, proline, alanine, aspartic acid)	Two backcross populations BC1F1: TN DH × Tapidor; TN DH × Ningyou7	786—RFLP, AFLP, MS-AFLP, SSR, SNP, STS, SSCP, CAPS (2117 cM)	LA	Wen et al. (2016)
	Seed oil and seed protein content	KenC-8 × N53-2, 300 DH, 12 environments (field)	3207—SSR, STS, SRAP, SNP	LA	Chao et al. (2017)
	Seed oil content, silique length, seeds per silique, see weight	M201 × 352, 192 RIL, 3 environments (field)	1526—ddRAD, INDEL, SSR (1610 cM)	LA	Chen et al. (2017)
	Erucic acid content, glucosinolate content, seed oil content	189 OSR accessions, 4 growing seasons (field)	3.82 million markers—SNP, INDEL	GWAS	Wang et al. (2017c)
	Erucic acid and tocopherol content	383 accessions	355,536 markers—SNP, transcript abundance	Associative transcriptomics	Havlickova et al. (2018)
Flowering time	Flowering time, response to vernalization	Major × Stellar, 104 DH, 3 environments	132—RFLP (1116 cM)	LA	Ferreira et al. (1995)
	Flowering time	Major × Stellar, 89 DH, 3 environments	480—RFLP	LA	Osborn et al. (1997)
	Flowering time, bolting time, budding time	Tapidor × Ningyou7, 202 DH, 404 F2, 11 environments (field)	621—SSR, RFLP, SNP, MS-AFLP, STS (2060 cM)	LA	Long et al. (2007)
	Flowering time, plant height	2091 × 99CDAM (dwarf), 145 F2:3, 2 environments (field)	241—AFLP, SSR (2095 cM)	LA	Mei et al. (2009)
	Flowering time, response to vernalization	Skipton × Ag-spectrum, 186 DH, 4 environments (field, greenhouse)	674—DArT, SSRs, SRAP, SCARs (4290 cM)	LA	Raman et al. (2013)
	Flowering time	Spring late flowering × early flowering, 207 DH, 3 environments (field)	256—SSR, AFLP, SRAP (2374 cM)	LA	Luo et al. (2014)

(continued)

Table 3.1 (continued)

Trait category	Detail of the traits studied	Mapping population, size, number of environment(s) studied	Maps Number and type of markers Length of the genetic maps	Method (linkage or association mapping)	References
	Flowering time, thermal time to flowering, leaf nodes to flowering	Lynx-037DH × Monty-028, 131 DH, 2 environments	329—SSR, DArT (2288 cM)	LA	Nelson et al. (2014)
	Flowering time	523 OSR accessions and lines, 8 environments	26,024—SNPs	GWAS	Xu et al. (2016)
Seed yield and aerial architecture traits	Seed yield, flowering time, plant height, thousand seed weight	4 BC populations, 2 environments (field)	276—RFLP	LA	Butruille et al. (1999)
	Seed yield, flowering time, plant height, lodging, thousand seed weight	MF216 × P1804, 150 DH RV128 × P1804,150 DH, 4 environments (field)	218—RFLP (1398 cM) 250—RFLP (1453 cM)	LA	Quijada et al. (2006)
	Seed yield, flowering time, plant height, thousand seed weight	MF216 × P1804, 150 DH RV128 × P1804, 150 DH, 3 environments (field)	218—RFLP (1398 cM 205—RFLP (1353 cM)	LA	Udall et al. (2006)
	Plant height, height of lowest primary branch, number of primary silique, length of main inflorescence, number of primary branches, silique density	Quantum × No. 2127-17, 258 DH, immortalized F2, 258 crosses, 3 environments (field)	397—SSR, SRAP (1747 cM)	LA	Chen et al. (2007)
	Seed yield, thousand seed weight, number of seeds per pod, number of pods per unit area	Express 617 × R53, 250 DH and 250 testcross, 4 environments (field)	377—AFLP, SSR (2045 cM)	LA	Radoev et al. (2008)
	Flowering time, maturity time, plant height, branch number, pod number, number of seeds per pod, thousand seed weight, biomass yield per plant, seed yield per plant	Tapidor × Ningyou7, 202 DH, reconstructed F2 (RC-F2) between 3 and 10 environments (field)	786—AFLP, RFLP, SNP, SSR, STS, SSCP, CAPS (2117 cM)	LA	Shi et al. (2009)

(continued)

Table 3.1 (continued)

Trait category	Detail of the traits studied	Mapping population, size, number of environment(s) studied	Maps Number and type of markers Length of the genetic maps	Method (linkage or association mapping)	References
	Plant height, shoot fresh weight, number of pods per unit area, thousand seed weight, seed yield	Express 617 × V8 Express 617 × R53 250 DH and 250 testcross, 4 environments (greenhouse)	476—AFLP, SSR (1673 cM) 377—AFLP, SSR (2045 cM)	LA	Basunanda et al. (2010)
	Seed yield, flowering time	High × low oil content lines, 150 DH, 3 environments (field)	387—SRAP, SSR (1868 cM)	LA	Chen et al. (2010)
	Thousand seed weight	SW Hickory × JA177, 238 DH, 2 environments (field)	327—SSR (2011 cM)	LA	Fan et al. (2010)
	Pod length, number of seeds per pod, thousand seed weight	HZ396 × Y106, 140 DH, 6 environments (field)	345—SSR, AFLP (1759 cM)	LA	Zhang et al. (2011b)
	Pod length, thousand seed weight	S1 × S2, 186 RILs, 2 environments (field) No2127 × S1, 192 RILs, 1 environment (field)	289—SSR (1381 cM)	LA	Yang et al. (2012)
	Plant height, first branch height, inflorescence length, pod length, number of seeds per pod, thousand seed weight	192 accessions, 3 environments (field)	1191—SSR, AFLP	GWAS	Cai et al. (2014)
	Pod length, thousand seed weight	Zhongshuang11 × No. 73290, 184 F2, F2:3 and F2:4 individuals/lines, 5 environments, 576 accessions, 1 environment	529—SSR, STS, SNP (1934 cM)	LA Regional association mapping	Li et al. (2014c)
	Silique length, silique number, thousand seed weight	8008 9 × 4942C-5, 181 DH, 4 environments (field)	385—IPs,SSRs, AFLP (1979 cM)	LA	Qi et al. (2014)

(continued)

Table 3.1 (continued)

Trait category	Detail of the traits studied	Mapping population, size, number of environment(s) studied	Maps Number and type of markers Length of the genetic maps	Method (linkage or association mapping)	References
	Net photosynthesis rate, transpiration rate, stomatal conductance, internal CO_2 concentration	GH06 × P174, 172 RILs, 1 environment (field)	9164—SNP (1833 cM)	LA	Yan et al. (2015)
	Seed yield, plant height, branch number, biomass yield per plant, harvest index	155 accessions, 4 environments (field)	35,791—SNP	GWAS	Luo et al. (2015)
	Petalous degree (mean number of petals per flower)	APL01 × Holly, 189 F9 RIL, 5 environments (field)	17,414—SNP, 81 SSR (2027.53 cM)	LA	Wang et al. (2015a)
	Plant height	KN: KenC-8xN53-2, 348 DH, 6 environments (field)	275 SSRs, 117 SRAPs, 10 STSs, 1 IFLP (1783.9 cM)	LA	Wang et al. (2015b)
	Plant height, number of primary branches	472 accessions, 2 environments (field)	26,841—SNP	GWAS	Li et al. (2016a)
	Flowering time and seed yield	DH, 2 environments	7716 DArTseq	LA	Raman et al. (2016a)
	Seed yield	KN: KenC-8xN53-2, 348 DH, 8 environments (field)	275 SSRs, 117 SRAPs, 10 STSs, 1 IFLP (1783.9 cM)	LA	Zhao et al. (2016b)
	Branch angle	143 OSR accessions, 3 environments (field)	Illumina Brassica 60K infinium® SNP array	GWAS	Liu et al. (2016b)
	Main inflorescence length proportion, branch height proportion, branch segment proportion, branch average length, total number of plant siliques, plant yield, main inflorescence yield, plant seed number, thousand seed weight, silique seed number, main inflorescence seed number	Huashuang × J7005, 254 DH, 4 environments (field)	SSR and 6K SNP infinium HD array	LA	Cai et al. (2016)

(continued)

Table 3.1 (continued)

Trait category	Detail of the traits studied	Mapping population, size, number of environment(s) studied	Maps Number and type of markers Length of the genetic maps	Method (linkage or association mapping)	References
	Emergence, development after emergence, stem elongation after winter, winter hardiness, phoma leaves, flowering, seed yield, seed quality estimated by NIRS	Set1: 217 inbred lines, 1 environment Set2: 188 inbred lines, 1 environment	6K SNP array	GWAS	Körber et al. (2016)
	Seed number per pod	Zhongshuang11 × 73290, 184 RILs, 4 environments	Illumina Brassica 60K infinium® SNP array	LA	Yang et al. (2016)
	Seed yield and seed yield general combining ability	175 inbred lines and 525 hybrids (NC design II)	Illumina Brassica 60K infinium® SNP array	Breeding signature	Zhao et al. (2016a)
	Pod shatter resistance index	143 accessions (3 environments) R1xR2: 96 DH lines (2 environments) 124 immortalized F2 (1 environment)	Illumina Brassica 60K infinium® SNP array	LA and GWAS	Liu et al. (2016a)
	Plant height, branching height, main raceme length, and average branching interval	Bndwf/dcl1x 'ZS11' BC1, 1 environment	Illumina Brassica 60K infinium® SNP array	LA	Wang et al. (2016a)
	Branch angle	Purler x Huyou19, 277 F2 2 pools of 30 F2	QTLseq	Bulked segregant analysis	Wang et al. (2016b)
	Silique number, seed number per silique, silique length, silique breadth, silique thickness, seed density, and silique volume	KN: KenC-8xN53-2, 348 DH, 4 environments (field)	403—SSRs, SRAPs, STSs, IFLP (1783.9 cM)	LA	Wang et al. (2016c)
	Days to flower and seed yield	Hi-Q and RIL-144, 110 DH, 9 environments (field)	79—SSR, 93 AFLP (1262.1 cM)	LA	Rahman et al. (2017)
	Lamina length, lamina width, petiole length, leaf total length, lobe number, and the lamina size ratio	GH06 × P174, 172 F11 RIL, 1 environment (field) 2 pools of 5 extremes RIL	2795 SNP (1832.9 cM) RNAseq	LA	Jian et al. (2017)

(continued)

Table 3.1 (continued)

Trait category	Detail of the traits studied	Mapping population, size, number of environment(s) studied	Maps Number and type of markers Length of the genetic maps	Method (linkage or association mapping)	References
	Pod number	Zhongshuang11 × 73290, 184 RILs, 2 parents	Illumina Brassica 60K infinium® SNP array; RNAseq	LA	Ye et al. (2017)
	Seed yield	TNRC-F: immortalized F2 population from TNDH, 318 hybrids and 180 parental lines, 3 environments	Illumina Brassica 60K infinium® SNP array	GWAS, genomic prediction	Liu et al. (2017)
	Seed yield and yield component traits, flowering time, glucosinolate presence in different tissues, seed erucic acid and oil content, resistance to *Leptosphaeria maculans*, resistance to *Sclerotinia sclerotiorum*, and vitamin E content	TNDH: 182 DH, 19 environments	Illumina Brassica 60K infinium® SNP array	LA and meta-analysis	Luo et al. (2017)
	Seeds per silique, seed weight, silique length	1167 × HZ396: 167 DH	Illumina Brassica 60K infinium® SNP array (2209.1 cM)	LA	Yang et al. (2017)
	Branch number	327 OSR accessions SWU07 × express, 261DH RC-F2: 233 immortalized F2, 2 environments (field)	Illumina Brassica 60K infinium® SNP array	GWAS LA	He et al. (2017)
	Branch angle	Y689 × Zhongyou 821, 208 DH, 6 environments	Illumina Brassica 60K infinium® SNP array (2242.1 cM)	LA	Shen et al. (2018)

(continued)

Table 3.1 (continued)

Trait category	Detail of the traits studied	Mapping population, size, number of environment(s) studied	Maps Number and type of markers Length of the genetic maps	Method (linkage or association mapping)	References
Abiotic stress/nutrient constraint and root architecture traits	Maturity date, bolting date, boron efficiency coefficient	Qingyou 10 × Bakow, 128 F2, 2 environments (2 boron treatments, field)	120—RFLP, AFLP (1833 cM)	LA	Xu et al. (2001)
	Winter survival, flowering time	Major × Stellar, 90 DH, 7 environments (field) +in vitro	480 (2007 cM)	LA	Kole et al. (2002)
	Pod length, number of seeds per pod, thousand seed weight	Mansholt's Hamburger Raps × Samouraï, 142 DH, 8 environments (2 N treatments, field)	185—RFLP (1739 cM)	LA	Gül et al. (2003)
	Number of leaves per plant at rosette stage, crown wet/dry weight, crown water content	Winter cold resistant × spring susceptible to low temperature, 200 F2:3, 1 environment (field)	42—RAPD (870 cM)	LA	Asghari et al. (2007)
	Freezing tolerance	SLMO46 × quantum, 199 F3, 1 environment (field)	70—RAPD, SSR (1199 cM)	LA	Asghari et al. 2008
	Seed mineral concentration	B104-2 × Eyou Changjia, 124 RILs, 4 environments (2 phosphorus treatments, field)	553—SSR, AFLP, SRAP (1593 cM)	LA	Ding et al. (2010)
	Seed yield, plant height, branch number, pod number, seed number, thousand seed weight	B104-2 × Eyou Changjia, 124 RILs, 4 environments (2 phosphorus treatments, field)	840—SSR, AFLP, SRAP, gene-based (1914 cM)	LA	Ding et al. (2012)
	Flowering time, seed yield, harvest index, plant height, branch number, number of seeds per pod, seed nitrogen content, NUpE, NUtE, NUE, nitrogen harvest index, total above ground biomass, seed number/m^2, stem nitrogen content, chlorophyll in bracts, chlorophyll in leaves	Tapidor × Ningyou7, 174/94 DH, 2 environments (2 nitrogen treatments, field)	786—AFLP, RFLP, SNP, SSR, STS, SSCP, CAPS (2117 cM)	LA	Miro (2010)
	Dry weight, root length, root surface area, root volume, plant phosphorus uptake	Eyou Changjia × B104-2, 124 RILs, 6 environments (2 phosphorous treatments, field)	553—SSR, AFLP, SRAP, FM (1593 cM)	LA	Yang et al. (2010)

(continued)

Table 3.1 (continued)

Trait category	Detail of the traits studied	Mapping population, size, number of environment(s) studied	Maps Number and type of markers Length of the genetic maps	Method (linkage or association mapping)	References
	Mineral concentrations	Tapidor × Ningyou7, 162 DH, 2 environments (2 boron treatments, controlled conditions)	621—RFLP, SSR, SNP, AFLP (2060 cM)	LA	Liu et al. (2009)
	Primary root length, shoot/root dry weight and ratio, shoot/root boron accumulation and ratio	Tapidor × Ningyou7, 181 DH, 2 environments (2 boron treatments, controlled conditions)	621—RFLP, SSR, SNP, AFLP (2060 cM)	LA	Shi et al. (2012)
	Seed yield, plant height, branch number, pod number, seed number, seed weight	Tapidor × Ningyou7, 181 DH, 2 environments (2 boron treatments, field) QY10 × Bakow, 200 DH, 3 environments (2 boron treatments, field)	621—RFLP, SSR, SNP, AFLP (2060 cM) 486—SSR (1874 cM)	LA	Zhao et al. (2012b)
	Seed mineral concentration	B104-2 × Eyou Changjia, 124 RILs, 4 environments (2 phosphorus treatments, field)	840—SSR, AFLP, SRAP, gene-based (1914 cM)	LA	Ding et al. (2013)
	Seed yield, plant height, number of primary branches, height to the first primary branch, relative first primary branch height, pod number per plant, seed number per pod, thousand seed weight	Tapidor × Ningyou7, 188 DH, 6 environments (2 phosphorus treatments, field)	798—AFLP, RFLP, SNP, SSR, STS, SSCP, CAPS (2050 cM)	LA	Shi et al. (2013a)
	Flowering time, seed yield, seed number/m^2, thousand seed weight, oil content, protein content, oil yield, protein yield	Aviso × Montego, 115 DH, 6 environments (2 nitrogen treatments, field)	2301—SNP (1947 cM)	LA	Bouchet et al. (2014)
	Shoot mineral concentration	509 accessions, 1 environment	3910—SNP	GWAS	Bus et al. (2014)

(continued)

Table 3.1 (continued)

Trait category	Detail of the traits studied	Mapping population, size, number of environment(s) studied	Maps Number and type of markers Length of the genetic maps	Method (linkage or association mapping)	References
	Seed yield, plant height, branch number, pod number, seed number, boron efficiency coefficient, phosphorous efficiency coefficient, thousand seed weight	QY10 × Bakow, 200 DH, 6 environments (2 boron treatments, field) B104-2 × Eyou Changjia,124 RILs, 4 environments (2 phosphorous treatments, field)	486—SSR, AFLP, SRAP, gene-based (1874 cM) 840—SSR, AFLP, SRAP, gene-based (1914 cM)	LA	Ding et al. (2014)
	Boron efficiency coefficient, increment of primary root length, shoot dry weight, root dry weight, shoot boron accumulation, root boron accumulation	Qingyou 10 × Westar 10, 70 DH, 3 environments (2 boron treatments, controlled conditions)	11,080—SNP (2139 cM)	LA	Zhang et al. (2014)
	Plant height, root length, root/shoot dry weight, total dry weight	6-1035 × 6-1169, 150 DH, 3 environments (controlled conditions)	340—AFLP, SSR (1489 cM)	LA	Li et al. (2014b)
	Flowering time, seed yield, root pulling force	Wichita × IMC106RR, 225 DH, 2 environments (field)	1179—SNP (2041 cM)	LA	Fletcher et al. (2015)
	Shoot dry weight, ion accumulation	85 accessions, 2 environments (controlled conditions)	51,109—SNP	GWAS	Yong et al. (2015)
	Days to flower, root pulling force	MC106RR × Wichita, 225 DH, 2 environments (field: dry vs wet)	SNP obtained by resequencing	LA	Fletcher et al. (2016)
	Plant height, straw yield, harvest index, seed yield under high N and no N fertilization	Alesi-*bzh* × H30 242 DH × male sterile tester, 8 environments (2 N levels) field	471—SNPs, SSRs (1301 cM)	LA	Miersch et al. (2016)
	Primary root length, total lateral root length, lateral root number, total root length, mean lateral root length, lateral root density	Tapidor × Ningyou, 202 DH Low and normal phosphate conditions/hydroponic system	Illumina Brassica 60K infinium® SNP array	LA	Zhang et al. (2016)

(continued)

(continued)

Table 3.1 (continued)

Trait category	Detail of the traits studied	Mapping population, size, number of environment(s) studied	Maps Number and type of markers Length of the genetic maps	Method (linkage or association mapping)	References
	Days to flowering, seed yield, thousand seed weight, seed number/m², seed oil content, seed protein content, seed oil content/seed protein content ratio	69 WOSR lines 92 WOSR lines Aviso × Montego,112 DH Tenor × express, 75 DH, 14 environments (field) with low and high N conditions	Illumina Brassica 60K infinium® SNP array	LA and GWAS	Bouchet et al. (2016)
	Boron efficiency coefficient, root fresh weight, root elongation length	Q10 × W10, 190 DH Hydroponic culture system High and low boron conditions	Whole genome sequencing and QTLseq	Bulked segregant analysis	Hua et al. (2016)
	Root vigor, days to flowering	F2 population	3090 SNP from genotyping by sequencing	LA	ArifUzZaman et al. (2017)
	Salt tolerance index, shoot length, taproot length, and shoot fresh weight	368 OSR accessions Hydroponic conditions (with NaCl and without NaCl)	Illumina Brassica 60K infinium® SNP array	GWAS	Wan et al. (2017)
	Root length, leaf fresh weight, leaf dry weight, root dry weight, electrical conductivity, superoxide dismutase, soluble protein, chlorophyll content, salt tolerance rating, and seedling height	2205 × 1423, 196 F2:3 population, 3 environments: hydroponic conditions (with salt)	AFLP + SSR	LA	Lang et al. (2017)
	Shoot dry weight, root dry weight, primary root length, lateral root number, mean lateral root, lateral root length, (total root length, lateral root density)	405 OSR accessions Low and high P hydroponic conditions	Illumina Brassica 60K infinium® SNP array	GWAS	Wang et al. (2017a)

(continued)

Table 3.1 (continued)

Trait category	Detail of the traits studied	Mapping population, size, number of environment(s) studied	Maps Number and type of markers Length of the genetic maps	Method (linkage or association mapping)	References
	Primary root length, total root length, total root number, root dry weight, total dry weight, root nitrogen content, shoot nitrogen content, N uptake of root, N uptake of shoot, total N uptake, N utilization efficiency	Zhongshuang11 × 73290, 184 lines Hydroponic conditions (high N and low N)	Illumina Brassica 60K infinium® SNP array (2107 cM)	LA	Wang et al. (2017b)
	Manganese tolerance Fresh shoot biomass, leaf surface, cotyledon leaf chlorosis	Darmor-*bzh* × Yudal, 191 DH Hydroponic conditions 9 M versus 134 of $MnCl_2$, $4H_2O$	7805—DArTseq	LA	Raman et al. (2017)
Disease/pest resistance traits	Stem canker	Cresor × Westar, 98 DH, 4 environments (field)	175—RFLP	LA	Dion et al. (1995)
	Stem canker	Darmor-*bzh* × Yudal, 152 DH, 2 environments (field)	288—RFLP, RAPD (1954 cM)	LA	Pilet et al. (1998a)
	Stem canker	Darmor × Samourai, 134 DH, 185 F2:3, 2 environments (field)	DH: 257—RFLP, RAPD (1975 cM) F2/3: 85—RFLP, RAPD (609 cM)	LA	Pilet et al. (2001)
	Stem canker	Caiman3 × Westar10, 91 DH Canberrra4 × Westar10, 76 DH Sapphire5 × Westar10, 133 DH Rainbow4 × Sapphire5, 91 DH, 2 environments	436—SSR, AFLP (1611 cM) 468—SSR, AFLP, SNP (1801 cM) 404—SSR, AFLP, SNP (1718 cM) 219—SSR, AFLP, SNP (1173 cM)	LA	Kaur et al. (2009)
	Stem canker	128 WOSR accessions, 1 environment (field)	71—SSR, SCAR	GWAS	Jestin et al. (2011)
	Stem canker	Darmor-*bzh* x Yudal, 279 DH, 1 environment (field)	549—SSR (3293 cM)	LA	Jestin et al. (2012)
	Stem canker	Skipton × AG-spectrum, 177 DH, 3 environments (greenhouse, field)	256—SSR (2672 cM)	LA	Raman et al. (2012)
	Stem canker	Skipton × AG-spectrum, 177 DH, 3 environments (greenhouse)	841—SNP, DArT, SSR, SRAP, gene-based (2515 cM)	LA	Raman et al. (2014a)

(continued)

Table 3.1 (continued)

Trait category	Detail of the traits studied	Mapping population, size, number of environment(s) studied	Maps Number and type of markers Length of the genetic maps	Method (linkage or association mapping)	References
	Stem canker	116 WOSR accessions, 2 environments (field)	3228—SNP	GWAS	Fopa-Fomeju et al. (2014)
	Stem canker	Aviso x Bristol, 115 F2:3 Canberra × Bristol, 118 F2:3 Darmor × Bristol, 91 F2:3 Grizzly × Bristol, 117 F2:3, 3 environments (field)	154—SSR (1119 cM) 143—SSR (1115 cM) 190—SSR (1421 cM) 199—SSR (1627 cM)	LA	Jestin et al. (2015)
	Stem canker	Topas × AG-Castle, 242 DH Topas × AV-Sapphire, 109 DH, 2 environments (field)	503—SSR, DArT (2182.3 cM) 212—SSR (1714.97 cM)	LA	Larkan et al. (2016)
	Stem canker	179 OSR accessions	18,804—SNP	GWAS	Raman et al. (2016b)
	Light leaf spot	Darmor-*bzh* × Yudal, 152 DH, 2 environments (field)	288—RFLP, RAPD (1954 cM)	LA	Pilet et al. (1998b)
	Clubroot	Darmor-*bzh* × Yudal, 110 DH, 2 isolates (greenhouse)	388—RFLP, RAPD (2000 cM)	LA	Manzanares-Dauleux et al. (2000)
	Clubroot	Darmor-*bzh* × Yudal, 152 DH, 3 isolates (greenhouse) Stellar × Drakkar, 94 DH, 3 isolates (greenhouse)	388—RFLP, RAPD (2000 cM) 340—RAPD (1900 cM)	LA	Manzanares-Dauleux et al. (2003)
	Clubroot	263/11 × express, 151 DH, 7 isolates (greenhouse)	394—AFLP, SSR (1570 cM)	LA	Werner et al. (2008)
	Clubroot	472 accessions, 2 environments (field, greenhouse)	26,841—SNP	GWAS	Li et al. (2016b)
	Clubroot	Darmor-*bzh* × Yudal, 356 DH, 92 OSR accessions, 2 isolates × 2 N levels (greenhouse)	3592 SNP—(2128.2 cM) Illumina Brassica 60K infinium® SNP array	LA GWAS	Laperche et al. (2017)
	Sclerotinia stem rot	H5200 × NingRS-1, 128 F2:3, 1 environment (field)	107—RFLP, AFLP (1620 cM)	LA	Zhao and Meng (2003)
	Sclerotinia stem rot		243—RFLP (1460 cM)	LA	Zhao et al. (2006b)

(continued)

Table 3.1 (continued)

Trait category	Detail of the traits studied	Mapping population, size, number of environment(s) studied	Maps Number and type of markers Length of the genetic maps	Method (linkage or association mapping)	References
		RV289 × P1804, 152DH, 3 environments (greenhouse) Stellar × Major, 104 DH, 1 environment (greenhouse)	132—RFLP (1016 cM)		
	Sclerotinia stem rot	DH821 × DHBao604, 72 DH, 4 years (field), 3 inoculation methods	251—SSR, RAPD, SRAP (1746 cM)	LA	Yin et al. (2010)
	Sclerotinia stem rot	Huashuang5 × J7005, 190 DH, 3 years	272—SSR (1579 cM)	LA	Wu et al. (2013)
	Sclerotinia stem, flowering time	Express × SWU07, 161 DH, 2 environments (field, greenhouse)	322—SSR (1175 cM)	LA	Wei et al. (2014)
	Sclerotinia stem rot	152 SOSR accessions, 1 environment (plastic greenhouse)	690—SSR	GWAS	Gyawali et al. (2016)
	Sclerotinia stem rot	347 OSR accessions, 2 environments (field)	30,932—SNP	GWAS	Wei et al. (2016)
	Sclerotinia stem rot	448 OSR accessions, 2 environments (field)	25,573—SNP	GWAS	Wu et al. (2016)
	Verticillium	307-406-1 × 307-230-2, 163 DH, 4 environments (greenhouse)	304—RFLP, AFLP, SSR (1793 cM)	LA	Rygulla et al. (2008)
	Verticillium	Express617 × R53, 214 DH, environments (greenhouse)	201—SSR	LA	Obermeier et al. (2013)
	Diamondback moth: number of eggs, larvae per leaf, intensity of damage	SLMO46 × quantum, 180, F2:4, 1 environment (field)	70—SSR, RAPD (1199 cM)	LA	Asghari et al. (2009)

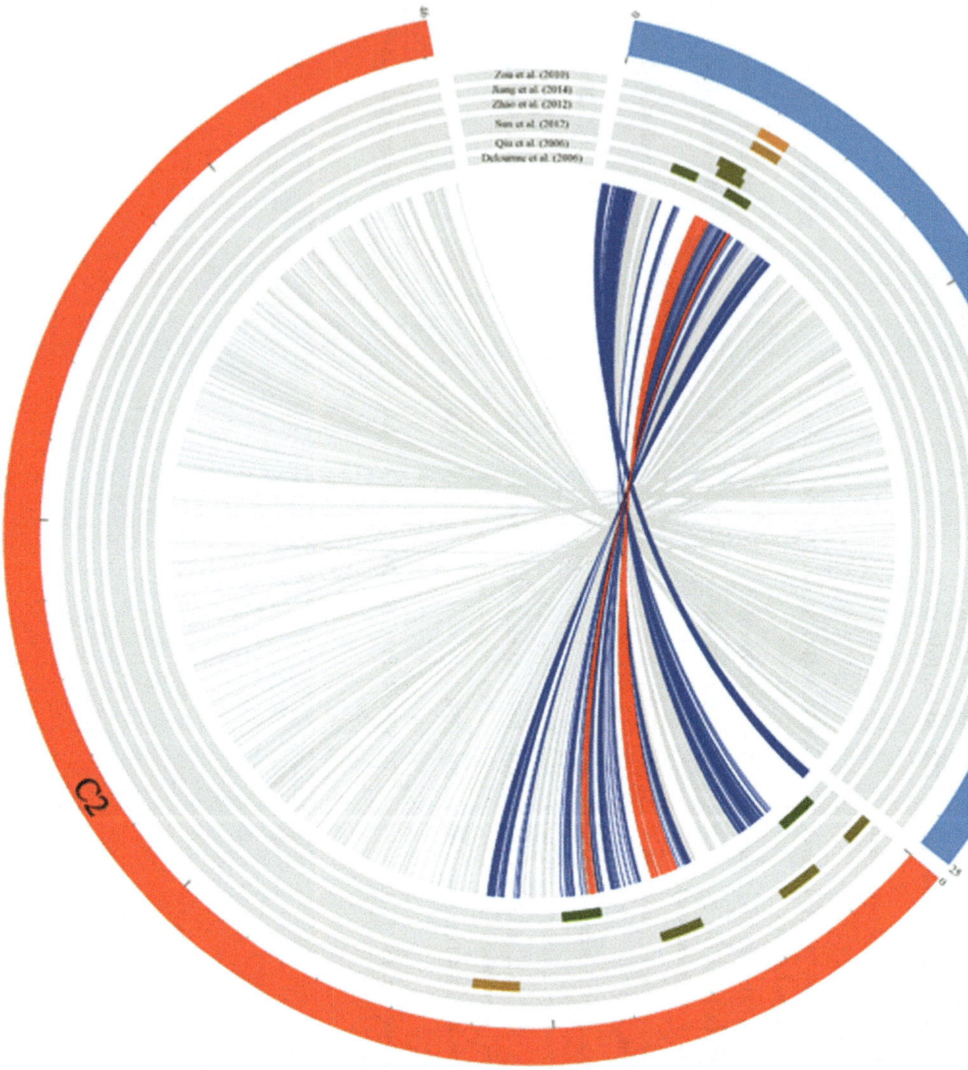

Fig. 3.1 Homoeologous relationships between genes underlying oil-related QTL on the A2 and/or C2 linkage groups of the *B. napus* genome. The QTL for seed oil content arising from six independent studies published from 2006 and 2014 that mapped on the A2 and/or C2 chromosomes is reported on each chromosome (in Mbp). Each pair of homoeologous genes is represented by a line with the following code: in red, the two homoeologous genes are located in oil QTL both on A2 and C2; in blue, only one homoeologous gene is located in a oil QTL either on A2 or C2; in gray, the homoeologous genes do not match with any oil QTL on A2 or C2

Chao et al. 2017). Three of the DH populations were used as references for subsequent QTL analyses, i.e., the Darmor-*bzh* × Yudal (DY, Delourme et al. 2006b), Tapidor × Ningyou7 (TN, Qiu et al. 2006), and Sollux × Gaoyou (SG, Zhao et al. 2005) biparental crosses. Other types of mapping populations were also useful to reveal LA-QTLs for oil content including inbred backcross lines (Butruille et al. 1999), substitution lines (Burns et al. 2003), F2 populations (Sun et al. 2016), or recombinant inbred lines (Jin et al. 2007, Yan et al. 2009; Chen et al. 2017). Independent studies each reported from tens of genomic regions involved in the genetic control of oil content spanning all the rapeseed linkage groups. However, these huge numbers of loci were shown to be considerably lowered when the stability of the QTLs over at least two environments in the same study was taken into account (Zhao et al. 2012a; Wang et al. 2013a). Variations in oil content were controlled mainly by additive effects, with the estimated phenotypic variation ranging from 1 to 20%, and to a lesser extent by epistatic interaction (Delourme et al. 2006b; Qiu et al. 2006; Zhao et al. 2006a; Jiang et al. 2014). More recently, several studies reported the identification of associations between markers and oil content by linkage disequilibrium analyses (LDAs) in large populations (Honsdorf et al. 2010; Zou et al. 2010; Li et al. 2014a; Gajardo et al. 2015; Qu et al. 2015; Körber et al. 2016; Sun et al. 2016; Wang et al. 2017d). The number of LDA-QTLs varied from 1 to 54 depending on the populations and the models used to eventually correct the effects of population structure and kinship. Substantial overlaps between LA and LDA-QTLs were demonstrated (Zou et al. 2010; Sun et al. 2016).

The large body of QTL information arising from individual studies prompted researchers to unravel the effects of the QTL × G and QTL E interactions on the QTL stability. However, due to the variety of populations and markers used in the different studies, only limited comparisons were possible up to now. By comparing LA-QTLs for seed oil content obtained in the DY, RSNL, TN, SG, and TV populations, Delourme et al. (2006b) suggested that the loci on A1, A3, A8, and C3 were common to at least two mapping populations, with co-localizations with the *BnFAE1* genes on A8 and C3. On the other hand, some other loci appeared to be population specific. More recently, QTL meta-analyses were conducted using several data sets and the resulting consensus loci (meta-QTLs) were aligned onto a consensus map (Wang et al. 2013b) or projected onto a selected map (Jiang et al. 2014). Wang et al. (2013b) identified 47 oil-related meta-QTLs with hotspots on the A1, A3, A6, A7, and C2 linkage groups, and Jiang et al. (2014) reported 46 distinct loci in the control of oil content, with critical genomic regions on the A1, A3, A8, C2, and C3 linkage groups. Based on these data, consensus genomic regions are likely to be targeted zones for fine mapping. For instance, the oil A1 locus is of particular interest since the favorable allele is lacking in most of the Chinese varieties. By designing near-isogenic lines and using the collinearity between the *Brassica* and *Arabidopsis* genomes in this particular key region, the confidence interval of the QTL was significantly reduced to 1.4 Mb (Chen et al. 2013), which theoretically included around 300 genes.

3.4.2 Seed Quality

For many years, oilseed rape breeding activities have been largely focused on improvement of its nutritional value mainly, driven by human and animal nutrition needs as well as by non-food uses. This included modification of fatty acid (FA) balance in the oil and improvement of meal value (Nesi et al. 2008). Low erucic acid content in the oil was obtained through mutation at two additive loci (E_A and E_C) that were mapped to A8 and C3 (Ecke et al. 1995; Jourdren et al. 1996b; Basunanda et al. 2007). They correspond to *BnFAE1.1* and *BnFAE1.2*, two homoeologous copies of the Arabidopsis fatty acid elongase gene (Barret et al. 1998; Fourmann et al. 1998; Wu et al. 2008). These two loci were recently identified through genome-wide association mapping (Li et al. 2014b) and through associative transcriptomics (Harper et al. 2012; Havlickova et al. 2018). Reduced level of polyunsaturated FA (especially linolenic acid, C18:3) and increased content of monounsaturated FA (oleic acid, C18:1) provide higher oil stability, and the resulting product can be used for salad dressings as well as for food and non-food (biofuel) uses that require high temperatures. Genetic analyses revealed that, depending on the mutants, one locus on A5 (Schierholt et al. 2001) or two major loci on A5 and C5 (Falentin et al. 2007), corresponding to *FAD2* (fatty acid desaturase) genes, controlled C18:1 content, and two loci on A4 and C4 corresponding to *FAD3* genes controlled C18:3 content (Jourdren et al. 1996a; Barret et al. 1999;

Hu et al. 2006). Using a population derived from a cross with a synthetic *B. napus* line, Smooker et al (2011) identified 34 QTL for fatty acid content of seed oil, of which 13 loci showed novel alleles inherited from the progenitors of the resynthesized *B. napus*, which may prove useful for modulating the content or extent of desaturation of polyunsaturated fatty acids. Only one QTL coincided with the position of an ortholog of *FAD2*. In silico mapping assays of candidate genes encoding enzymes or regulatory factors involved in fatty acid synthesis in Arabidopsis with QTL for fatty acid content in *B. napus* have been reported (Wang et al. 2015c; Raboanatahiry et al. 2017). New potential genes emerged which need experimental approach for functional validation.

Breeding efforts aiming at increasing the quality of oilseed rape meal included increasing essential amino acid content and reducing the amount of antinutritional factors such as glucosinolates, sinapate esters, phytic acid, tannins, and crude fiber (Nesi et al. 2008; Wittkop et al. 2009). Four QTLs located on chromosomes A9, C2, C7, and C9 have been recurrently mapped for total seed glucosinolates (GSL) in different studies of *B. napus* (Toroser et al. 1995, Uzunova et al. 1995; Howell et al. 2003; Zhao and Meng 2003; Quijada et al. 2006; Basunanda et al. 2007; Würschum et al. 2012). QTL involved in the control of individual glucosinolates content (Zhao and Meng 2003) has also been identified. A total of 11 peak SNPs significantly associated with GS content were detected on *B. napus* chromosomes A08, A09, C03, and C09, respectively (Qu et al. 2015). Feng et al. (2012) constructed an advanced metabolic network and associated epistatic interactions responsible for the GSL composition in both leaves and seeds of *B. napus*, through the measurement of total and individual glucosinolates in seeds and leaves within a DH mapping population. Nonetheless, all of the above studies utilized crosses involving at least one parent with high seed GSL content. Reliable detection of minor QTL that segregates between different low-GSL materials was, therefore, masked by the strong effects of a few major QTLs. Other minor effect QTL for total seed glucosinolates was identified through association mapping (Gajardo et al. 2015; Xu et al. 2015b) or through linkage analyses performed in crosses between two low-GSL oilseed rape lines (Fu et al. 2015).

For sinapate ester content, four QTLs were detected and the QTL with the strongest additive effects was mapped on linkage groups A8 and C3, within the confidence intervals of the two erucic acid genes as well as two main QTLs controlling total phytosterol content. This might result from a pleiotropic effect of the two erucic acid genes on phytosterol and sinapate ester content, with alleles for low erucic acid content increasing phytosterol and sinapate ester content as suggested by Amar et al. (2008) or from closely linked genes. Yellow-seeded (YS) oilseed rape lines are considered advantageous for the meal quality due to a thinner seed coat and higher protein content. YS materials that are deprived of condensed tannins have been developed through interspecific introgression of yellow seed coat color genes from related species (*B. rapa*, *B. carinata*, *B. juncea*) or through EMS-mutagenized oilseed rape populations (Nesi et al. 2008). Numerous genetic mapping studies of seed color loci or seed color-related trait loci (proanthocyanidin, fiber, and lignin content) have been reported in *B. napus*. These used different biparental populations, or more recently, association mapping, and detected major loci on A8, A9, C1, C4, C5, or C8 (Badani et al. 2006; Fu et al. 2007b; Yan et al. 2009; Snowdon et al. 2010; Zhang et al. 2011a, Lipsa et al. 2012; Liu et al. 2013; Gajardo et al. 2015). Liu et al. (2013) also mapped QTL for cellulose and hemicellulose content, which did not co-localize with seed color QTL in the studied population.

3.4.3 Flowering Time

Oilseed rape includes both spring types with little or no requirement for vernalization in order to flower and winter types that do require a vernalization period to flower. Therefore, understanding of the genetic control of flowering time and vernalization response in *B. napus* is important for plant breeders, in order to optimize crop performance in specific environments. QTL studies were performed in crosses within winter or spring varieties, but also often in crosses between winter and spring varieties. The first studies, conducted in a limited number of environments, detected from four to nine regions involved in flowering time (Ferreira et al. 1995; Osborn et al. 1997; Butruille et al. 1999). In these studies, two main regions on A2 and A10 and a third one on A3 were detected for the vernalization response. Subsequent studies performed on diverse sets of materials and using more complete genetic maps or GWAS revealed many regions involved in flowering time (Delourme et al. 2006a; Long et al. 2007; Raman et al. 2013; Luo et al. 2014; Nelson et al. 2014; Bouchet et al. 2014; Xu et al. 2016). Long et al. (2007) detected 36 major QTL in a study performed on a cross between a winter type and a semi-spring type evaluated in 11 environments; 31 of these QTLs were found in at least two environments, of which a major locus on A10 was only detected in spring environments and was associated with vernalization requirement. Raman et al. (2013) identified at least 20 QTLs for flowering time in a cross between two semi-spring oilseed rape varieties, of which the majority overlapped with QTL previously identified on chromosomes A2, A3, C2, C6, and C8 (Long et al. 2007) and with a circadian period QTL on chromosomes A2, A3, and A7 (Lou et al. 2011). Except for two regions on A2 and A10 that clearly carry vernalization-associated QTL, as well as another likely region on A3, the other regions are involved in earliness of flowering and were more or less consistently detected depending on the genetic backgrounds and the environmental conditions.

3.4.4 Yield and Yield-Related Traits Under Abiotic Stress and/or Nutritional Constraints

Grain yield is a very complex trait in oilseed rape compared with other crops. This complexity is related mainly to the potential of oilseed rape for growth and branching after flowering, which enables the crop to use one yield component to compensate for limitations in another one. As a consequence, a given final yield may result from different combinations of yield components (number of plants per m^2, branch number, number of pods per plant, number of seeds per pod, seed weight, etc.) and many studies reported QTL for seed yield and yield components (Table 3.1). For instance, Shi et al. (2009) reported 870 QTL (85 for seed yield and 785 for eight yield-associated traits) on a single genetic map of *B. napus* in two related populations grown in 10 natural environments. Thanks to the use of common genomic resources such as the Illumina Brassica 60K SNP array, comparison and meta-analyses of QTL were developed (Luo et al. 2017; Zhao et al. 2016b; Shi et al. 2009). For instance, a trait-by-trait meta-analysis revealed 401 consensus QTLs, of which 82.5% were clustered and integrated into 111 pleiotropic unique QTLs and 47 of which were relevant for seed yield. A high percentage (85%) of QTL for seed yield co-localized with QTL for other yield-associated traits. This indicates that, in addition to pleiotropy, the effect of the QTL for seed yield could be a synthetic effect of several underlying tightly linked QTLs of different yield-associated traits. Some QTL were consistently identified in different studies (Shi et al. 2009; Ding et al. 2012; Cai et al. 2014; Bouchet et al. 2016), of which a region on A5 was shown to carry QTL for many yield-related traits. Recently, aerial architecture was densely characterized, as a main determinant of yield compensation. Major QTLs were detected for branch angle (Shen et al. 2018; Wang et al. 2016b; Liu et al. 2016a), branch number (He et al. 2017; Li et al. 2016a; Luo et al. 2015), leaf morphology (Jian et al. 2017), plant architecture factors (Cai et al. 2016), dwarf architecture (Wang et al. 2016c), or apetalous phenotype (Wang et al. 2015a). These QTLs co-localized with seed yield and seed yield components QTL (Cai et al. 2016) or involved in genotype by environment interactions (Shen et al. 2018), thus indicating a putative role in the compensation phenomena. A large number of studies revealed QTL for silique-related traits, (seeds per siliques, seed weight, silique length, silique number; Yang et al. 2017; Ye et al. 2017; Yang et al. 2016; Wang et al. 2016a; Shi et al. 2009; Yang et al. 2012; Cai et al. 2014; Li et al. 2014c; Qi et al. 2014), thus highlighting the processes that occur at the end of crop cycle. Seed number, plant density, earliness, biomass dynamics are less documented.

Developmental traits (flowering time, seed filling duration) and environmental conditions (climatic conditions, water, and fertilizer availability) impacted the yield components. Abiotic constraints or stresses can be summarized in two categories: (1) nutrition limitations that are more likely to occur under sustainable production systems including low-input practices (low nitrogen (N), phosphorus (P), boron (B), or sulfur (S) input) and (ii) abiotic stresses that are direct or indirect consequences of global warming (drought, temperature, and salt stress, for instance). Most studies have focused on yield stability under limited conditions or on the genetic control of plant mechanisms involved in the response to the abiotic constraints such as nutrient use efficiency or root morphology.

Up to now, genetic studies were mainly conducted for nitrogen, phosphorus, and boron and were carried out by comparing QTL detected under either low or high nutrient conditions, or by detecting QTL for efficiency traits defined as yield or the biomass ratio between low and high conditions. Most studies reported for yield under abiotic constraints were performed through linkage analyses (Table 3.1). GWAS were carried out for oilseed rape response to N and phosphorus constraints (Bus et al. 2014; Bouchet et al. 2016; Wang et al. 2017b). Few QTL studies under different N nutrition levels have been performed so far and have revealed either few (Gül et al. 2003; Bouchet et al. 2014; Bouchet et al. 2016; Wang et al. 2017b; Miersch et al. 2016) or many (Miro 2010) genotype × N or few QTL × N nutrition level interactions. QTL × P or B regime interactions as well as epistatic QTL were identified in many studies (Ding et al. 2012, 2013, 2014; Shi et al. 2013a; Zhao et al. 2010, 2012b). Common yield or yield-related trait QTL in different genetic backgrounds was identified either under low P or under low B conditions, but only a few main effect QTLs can be detected both under low P or low B environments for the same traits. Thus, different genetic mechanisms for B and P efficiency may be involved in these processes (Shi et al. 2013a; Ding et al. 2014; Wang et al. 2017b). Whole genome sequencing approaches and use of the high density 60K SNP array permitted the identification of candidate genes, for instance, on chromosome C2 for boron efficiency (Hua et al. 2016), on A3 for phosphorus efficiency (Wang et al. 2017b) as well as the identification of genomic regions suitable for marker-assisted purposes for nitrogen use efficiency (Wang et al. 2017b), such as the dwarfing gene *bzh* on chromosome A06 (Miersch et al. 2016). Genetic study of the tolerance of oilseed rape to abiotic stresses has been limited to a few abiotic stresses such as drought (Li et al. 2014b; Fletcher et al. 2016), freezing (Asghari et al. 2007, 2008; Kole et al. 2002) and recently salt (Lang et al. 2017; Wan et al. 2017) and manganese (Raman et al. 2017) tolerance. To our knowledge, no QTL was identified in response to high temperature, CO_2 concentration. Most studies relied on field-based phenotyping. For most stresses, the efficiency coefficient (Ding et al. 2014) or index (Yong et al. 2015; Li and Sillanpää 2015) is defined as the ratio between a global trait (usually yield or total biomass) estimated under stressed conditions and the same trait estimated under non-limiting conditions. This definition has been used in studies on salt tolerance (Yong et al. 2015) and drought tolerance (Li et al. 2014b).

To further elucidate plant responses to abiotic stresses, some authors have identified more 'functional' traits involved in the plant response and compared the QTL detected for these functional traits to the QTL detected for yield-related traits and indices. The capacity of the plant to absorb nutrients from the soil, and therefore the nutrient uptake efficiency, is a key process for improving nutrient use efficiency. Therefore, particular attention has been paid to the roots and the identification of QTL for root architecture traits (Shi et al. 2012, 2013b; Yang et al. 2010;

Wang et al. 2017a; ArifUzZaman et al. 2017). For instance, Yang et al. (2010) and Shi et al. (2012) detected clusters of QTLs for root length, lateral root number, root surface area, root biomass, and root volume specific to low phosphorous conditions. The co-localization of QTL for flowering time and root pulling force on A10 and C2 was assessed by Fletcher et al. (2015), in a study that suggested pleiotropic effects on the genetic control of root traits in *B. napus*. Fletcher et al. (2015) considered flowering time and the root morphology as traits involved in drought tolerance. Yong et al. (2015) completed their study of salt tolerance using an ion homeostasis approach in order to obtain clues as to the physiological processes involved in the plant response to salt. Finally, Asghari et al. (2007, 2008) completed their analysis of winter survival using a characterization of LT50, the temperature at which 50% of the plants died.

Recently, hybrid varieties have been greatly increasing as a proportion of the total oilseed rape cropping area. Hybrids are mainly grown for their yield stability and adaptability across environments relative to inbred lines. This indicates a major role of heterosis for yield stability across pedo-climatic conditions and cultural practices, resulting generally in various nutrient constraints and abiotic stresses. Different studies have been carried out to decipher the genetic architecture of oilseed rape heterosis (Shen et al. 2006; Radoev et al. 2008; Basunanda et al. 2010; Liu et al. 2017), identifying a strong contribution of epistatic interactions to heterosis. The use of immortalized F2 populations greatly helped developing QTL analyses on dominance effects (Liu et al. 2017; Shi et al. 2009; He et al. 2017; Liu et al. 2016a). These studies, however, focused mainly on non-limiting environmental conditions or studied heterosis using adjusted means across different environments. In contrast, Shi et al. (2011) compared the heterosis genetic components in different environments and identified a great impact of the environment on the genes and QTL underlying the heterosis measured by yield-related traits. A specific study was dedicated to general combining ability (Zhao et al. 2016a) and identified that the subgenome C was more involved in GCA control than the subgenome A and gave perspective to breed for GCA by marker-assisted selection. Moreover, these researchers showed that the mode of inheritance of heterotic QTL could change according to the environment. These results have opened new ways to consider QTL × E interactions and yield stability analyses across various abiotic conditions.

3.4.5 Disease and Pest Resistance

During its life cycle, oilseed rape may be challenged by aerial and root pathogens causing numerous diseases, as well as by insects and other pests. International efforts have been undertaken to describe the genetic architecture of disease resistance, but these have been focused only on a small number of plant aggressors (Table 3.1).

Both qualitative and quantitative resistances have been identified for clubroot, caused by the obligate protist *Plasmodiophora brassicae* Woron. The qualitative resistance of the five *B. napus* hosts included in the European Clubroot Differential set (Buczacki et al. 1975), and of several swede genotypes (Crute et al. 1980; Gustafsson and Fält 1986), has been studied. However, the genetic factors involved in these host genotypes remain unknown. Mapping of resistance genes revealed one major locus and two recessive genes in the clubroot-resistant oilseed rape variety 'Mendel' (Diederichsen et al. 2006; Fredua-Agyeman and Rahman 2016) and one major gene in swedes (Hasan and Rahman 2016). The genetic determinism of partial quantitative resistance to clubroot in *B. napus* was studied in DH populations (Manzanares-Dauleux et al. 2000, 2003; Werner et al. 2008). At least, 22 QTLs (broad-spectrum and isolate-specific) distributed on 9 chromosomes have been identified as involved in resistance against 11 isolates of the clubroot pathogen.

Manzanares-Dauleux et al. (2000, 2003) found that either major QTL (*Pb-Bn1*, *Pb-Bn2,* and *Pb-Bn3*) or QTL with moderate effects was detected in the same genomic regions on A3, C9, and A5, respectively, depending on the isolates used. While the same QTL was detected under high and low nitrogen levels, their effects were altered and this modulation was isolate-specific (Laperche et al. 2017). GWAS performed on a panel of 472 accessions revealed nine genomics regions involved in clubroot resistance in the field or in controlled conditions, of which some were common to previously identified regions (Li et al. 2016b).

For phoma stem canker (blackleg) caused by *Leptosphaeria maculans*, both qualitative and quantitative resistances have been identified in *B. napus* or in related species (Delourme et al. 2006b; Rimmer 2006; Hayward et al. 2012; Raman et al. 2012). More than ten specific resistance genes were identified in *B. napus* or introgressed in *B. napus* from the related *Brassica* species *B. rapa*, *B. juncea,* and *B. nigra* (*Rlm1–11*; *LepR1-4*) (Delourme et al. 2006a; Rimmer 2006; Balesdent et al. 2013). They were mapped on A2, A6, A7, and A10, and some of these genes are organized in clusters on A7 (Delourme et al. 2004; 2006b; Long et al. 2011a; Yu et al. 2005, 2008, 2013). Two recently cloned genes, *LepR3* and *Rlm2*, were shown to be allelic and correspond to a receptor-like protein (Larkan et al. 2013, 2015). The first study concerning quantitative resistance to blackleg was realized in the Crésor × Westar DH population and allowed the identification of two regions on a linkage map of 175 RFLP markers (Dion et al. 1995). Several subsequent studies were published in DH populations or $F_{2:3}$ populations (Pilet et al. 1998a, 2001; Kaur et al. 2009; Jestin et al. 2012; Raman et al. 2012; Raman et al. 2014a; Larkan et al. 2016), or in a multi-parental connected design (Jestin et al. 2015). Combining GWAS (Jestin et al. 2011; Fopa Fomeju et al. 2014, 2015; Raman et al. 2016b) and LA, QTL for blackleg resistance was identified on most *B. napus* chromosomes. QTL was identified on A1, A2, A5, A6, C2, C1, C3, C4, C5, C6, and C8 from some different resistance sources (Jestin et al. 2015). The linkage analyses conducted with field phenotypic data revealed strong QTL × E interactions. Huang et al. (2016) identified QTL that was less sensitive to environmental factors in a multi-year study involving five environments.

QTL for sclerotinia stem rot resistance was investigated in DH populations (Zhao et al. 2006b; Yin et al. 2010; Wu et al. 2013; Wei et al. 2014) or $F_{2:3}$ populations (Zhao and Meng 2003), either in field or in greenhouse experiments. QTL was distributed on all the chromosomes except C3 and C5, and different QTLs were involved in leaf and stem resistance. Homoeologous QTL was found on A2 and C2 (Wei et al. 2014). In another cross, Zhao et al. (2006b) detected QTL in these regions, which corresponded to a homoeologous non-reciprocal translocation from A2 to C2, leading these researchers to conclude that the resistance allele on C2 may be identical to the *B. rapa* allele from A2. GWAS performed various panels in greenhouse, and field conditions confirmed that sclerotinia stem rot is controlled by multiple minor QTL (Gyawali et al. 2016; Wei et al. 2016; Wu et al. 2016).

Efforts to determine the genetic factors involved in quantitative resistance to light leaf spot, caused by *Pyrenopeziza brassicae*, and Verticillium wilt, caused by *Verticillium longisporum*, have been limited, with only one or two studies published to date, respectively (Pilet et al. 1998b; Rygulla et al. 2008; Obermeier et al. 2013). For light leaf spot, three genomic regions, located on A2, A7, and A9, are common to the resistance assessed on leaves and stems. In addition, two major resistance genes to *P. brassicae*, *PBR1* and *PBR2*, were, respectively, introgressed from *B. rapa* and *B. oleracea* on A1 and C6 in *B. napus* (Bradburne et al. 1999). With respect to Verticillium wilt, the experiments on DH populations, carried out under greenhouse conditions, revealed six regions on chromosomes A6, C1, C4, C5, and C8, of which one region on C5 was common to the two studies. In addition, the regions on C1 and C5 were shown to be major genomic regions for phenylpropanoid synthesis or modification in oilseed rape, indicating a potential role of these compounds in defense against *V. longisporum* (Obermeier et al. 2013).

To our knowledge, only two linkage analysis studies have been published for insect pest resistance. Three regions were shown to be involved in resistance to diamondback moth (*Plutella xylostella*) (Asghari et al. 2009). One region on A7 revealed a QTL for kaempferol 3-O-sinapoyl-sophoroside 7-O-glucoside (KSSG) metabolite accumulation, which was previously shown to be correlated with a reduction of cabbage seedpod weevil (*Ceutorhynchus obstrictus*) larval infestation in a *Sinapis alba* × *B. napus* derived material (Lee et al. 2014).

3.5 The Genome Sequence Allows Further Insights into the Genes Underlying the QTLs

The recent release of the genome sequences of several *Brassica* species, together with use of the latest DNA sequencing technologies (Edwards et al. 2013), has enabled new insights into the genomic localization of trait-associated QTL, identification of the underlying genes, and evaluation of their sequence variation.

3.5.1 QTL Genomic Localization

Genetic analyses of the aforementioned agronomic traits were conducted in various mapping populations, but the QTL positions were often difficult to compare due to the low number of common markers. The availability of a *B. napus* reference genome sequence provides now a tool to directly compare the physical positions of the QTL. For instance, in silico integration of seed yield and yield-related trait QTL in *B. napus* has been performed using *B. rapa* and *B. oleracea* sequences, and 736 QTLs coming from a number of independent studies were mapped to 283 loci in the A and C genomes of *B. napus* (Zhou et al. 2014). This showed an uneven distribution of QTL on the A and C genomes (with more QTL on the A genome) or on the chromosomes (A3 had the highest number of QTL, and C6 had the fewest QTL). Overall, 142 loci were detected with conserved QTL across genetic backgrounds or environments, including 25 multi-functional loci mostly for traits such as flowering time, plant height, seed weight, maturity time, and seed yield. Li et al. (2015) integrated 35 *S. sclerotiorum* resistance QTL from five different studies using 353 markers mapped on the *B. napus* sequence. Two conserved QTLs identified in multiple studies were identified on the chromosomes A9 (from 22.5 to 27.5 Mb) and C6 (from 29.5 to 36.1 Mb), spanning two clusters of candidate NBS-LRR genes. C6 region was further confirmed after integration of last GWAS studies (Wu et al. 2016).

The more precise genomic localization of *B. napus* regions involved in polygenic traits provides an opportunity to study the impact of genome duplications on the structural and functional organization of such regions in a highly duplicated genome. Homoeologous or paralogous regions were shown to be involved in the control of important traits such as seed glucosinolate content on A9, C2, and C9 (Howell et al. 2003), flowering time on A2 and C6 (Wang et al. 2009), yield-related components (Chen et al. 2007), and resistance to sclerotinia stem rot on A2 and C2 (Zhao et al. 2006; Wei et al. 2014). Recently, in an attempt to compare the results from several studies related to oil content, we demonstrated that an oil QTL identified on A2 from six QTL-independent populations displayed homoeologous relationships with an oil QTL located on C2 (Fig. 3.1; Bouchet, unpublished data). Fopa Fomeju et al. (2014, 2015) investigated quantitative resistance to phoma stem canker in *B. napus* through genome-wide association analyses and found that more than 44% of the resistance-associated regions are duplicated homoeologous regions. These regions are mainly located in duplications of five of the 24 ancestral blocks that constitute the *B. napus* genome. Furthermore, Fopa Fomeju et al. (2015) showed that around 60% of the genes identified in these duplicated regions are single-copy genes, while less than 5% are retained in all the duplicated copies of a given ancestral block. Genes retained in several copies are involved mainly in response to stress, signaling, or transcriptional regulation. Genes with resistance-associated markers were mainly retained in more than two copies. These results suggested that some genes underlying quantitative resistance to stem canker might be duplicated genes, but further analysis is required to determine the extent to which duplicated genes contribute to the expression of the resistance phenotype. Segmental exchanges between homoeologous chromosomes are frequent throughout the *B. napus* genome. Numerous small-scale homoeologous exchanges (HEs) are observed throughout the genomes of natural *B. napus* accessions, whereas large-scale HEs are common in synthetic accessions (Chalhoub et al. 2014; Rousseau-Gueutin et al. 2017). Such HE might influence agronomic traits, as shown for seed quality traits (Stein et al. 2017) or sclerotinia stem rot (Zhao et al. 2006b).

3.5.2 Candidate Gene Identification

The cloning of the trait causal genes is the most advanced and effective step for improving both knowledge of the mechanisms involved and the efficiency of breeding programs. Map-based cloning can be used to identify the causal gene. Using the results of fine mapping and targeted regional association, Liu et al. (2015) identified seven putative ORFs for a QTL explaining ca. 30% of the phenotypic variation for seed weight. Based on the genome sequence of *B. napus*, the seven genes including the upstream regulatory and coding regions were cloned and their sequence compared between the two parental lines. Finally, a 165-bp deletion in the auxin response factor 18 (ARF18) gene was associated with increased seed weight and silique length, which was further validated through gene expression and overexpression analyses (Liu et al. 2015).

Nevertheless, QTL map-based cloning and fine mapping are often laborious with a low success rate. As a consequence, the high conservation rate in coding sequences between *B. napus* and *A. thaliana* is often exploited to find orthologous genes related to traits of interests. For instance, 14 orthologous genes involved in a lipid synthesis pathway in Arabidopsis co-localized with six oil loci in the SG DH population (Zhao et al. 2012a) and thus represented good candidate genes for seed oil content. In a similar way, Sun et al. (2012) mapped oil-related markers on a DH population. These markers were selected from differentially expressed genes between individuals exhibiting extreme variation in seed oil content and showed four co-localizations with oil QTL. More recently, interactions between key genes of the oil pathway, as well as with master regulators of seed filling such as the transcription factor WRINKLED 1 (WRI1), were identified in a large population of elite rapeseed inbred lines (Würschum et al. 2013). The identification of candidate genes for boron and phosphorus use efficiency was carried out by identifying genes that were involved in mineral homeostasis in Arabidopsis and which were located in syntenic regions where QTL was identified for yield and efficiency traits in oilseed rape (Ding et al. 2010, 2012, 2013; Liu et al. 2009). Wu et al. (2013) identified the Arabidopsis homolog of an indole glucosinolate methyltransferase gene, *IGMT5* (At1g76790), as a candidate gene for a QTL on C6 that explained ca. 30% of sclerotinia stem rot resistance over three environments in a DH population.

As a result of the physical anchorage of SNP markers on the *B. napus* sequence, GWAS is now increasingly used to identify candidate genes. Yong et al. (2015) carried out a GWAS to elucidate the genetic control of salt tolerance. These researchers identified 62 QTL involved in plant response to salt stress, and identified a short list of 10 candidate genes located in the QTL regions. Of these ten candidates, one was investigated further (*BnaaTSN1)* and validated for its impact on salt tolerance. *BnaaTSN1* is an ortholog of *TSN1* (RNA-binding protein Tudor-SN) in *A. thaliana.* Fine mapping or regional association mapping, together with the use of *B. rapa* and *B. napus* sequences and synteny with Arabidopsis, also led to the identification of candidate genes for seed color-related traits. Transparent testa (TT) genes as well as key phenylpropanoid biosynthesis genes (CCR1; CAD2/CAD3) were hypothesized for the major loci on A9, indicating either a pleiotropic effect of some genes on different traits or linkage between different genes acting on these traits (Fu et al. 2007a, b; Snowdon et al. 2010; Lipsa et al. 2012; Liu et al. 2013; Stein et al. 2013). The results indicate that independent mutations in different phenylpropanoid genes may cause similar seed coat phenotypes in different genetic backgrounds (Liu et al. 2013).

Another association mapping methodology, which is termed 'associative transcriptomics,' was applied to panels of *B. napus* lines. Information on gene sequence (as SNP markers) and gene expression (as gene expression markers; GEMs) obtained from mRNA-Seq data was used to define genetic regions and also identify candidate genes controlling total seed GSL content (Harper et al. 2012; Lu et al. 2014). Within the associated GSL content peaks, 26 genes were inferred to be involved in GSL biosynthesis. In the low-GSL accessions, genomic deletions were identified in two GSL content QTLs on A9 and C2, and the deleted segments contained orthologs of the MYB transcription factor, *HAG1* (At5g61420; also known as MYB28) (Harper et al. 2012). This transcription factor controls aliphatic glucosinolates biosynthesis in *A. thaliana* (Hirai et al. 2007). Two other genes, *BnaA.GTR2* (a proton-dependent glucosinolate-specific transporter) and *BnaC.HAG3b*, located within the respective association peaks on A2 and C9, were of particular interest as they jointly explained 25.8% of the trait variation (Lu et al. 2014). Very recently, the associative transcriptomics platform was validated for other traits such as the variation in the relative proportions of tocopherol forms in seeds (Havlickova et al. 2018). Combined association mapping and gene expression analyses for 14 seedling traits revealed several candidate genes (germin-like protein (*GER1*), aluminum-induced protein (*AILP1*), ethanolamine-phosphate cytidylyltransferase (*PECT*),and fructose-1,6-bisphosphatase precursor (*FBP*)) that were strongly correlated to seedling development traits (Körber et al. 2015). The results also suggested that the studied genes ribulose 1,5-bisphosphate carboxylase/oxygenase small subunit (*RBC*) on the chromosomes A4 and C4, and fructose-1,6-bisphosphatase precursor (*FBP*) on the chromosomes A9 and C8, are cis-regulated.

Due to the large confidence intervals of QTL, genes underlying trait-associated loci often remain numerous and are still difficult to analyze. One possible strategy to get a first overview of the biological pathways underlying those regions is to use the gene ontology (GO) network to identify the enriched GO terms. Recently, Bargsten et al. (2014) proposed a method of candidate gene prioritization to unravel a set of 1591 QTLs associated with 231 different traits in rice. For each trait, they compared the occurrence of the biological functions linked to the genes found in the QTLs with the rest of the genome and retained the genes with a significant overrepresentation of a particular function. This method was used to successfully reduce the number of putative candidate genes to tenfold. Thus, an approach that combines a wide and precise gene annotation with a large QTL data set could help decipher the mechanisms underlying the QTLs.

3.5.3 Gene Sequence Variation

Knowledge of the flowering process in *Brassica* species is largely based on studies with *A. thaliana*. Orthologs of the main genes involved in flowering time in *A. thaliana* have been found in *Brassica* crops, and some of these were identified as candidate genes for flowering time QTL, including *FLC* (FLOWERING LOCUS C), *FT* (FLOWERING LOCUS T), and *FRI* (FRIGIDA) (Wang et al. 2009, 2011b; Nelson et al. 2014). *BnA2.FT* and two paralogs on chromosome C6, *BnC6.FT.a* and *BnC6.FT.b*, were associated with two major QTL clusters for flowering time, indicating that the 'florigen' of *B. napus* may be functionally differentiated between winter- and spring-type cultivars (Wang et al. 2009). *BnaA.FRI.a* mapped to a region on chromosome A3 that co-localized with a major flowering time QTL, which was identified in multiple environments in a DH mapping population. Association analysis of *BnaA.FRI.a* revealed that six SNPs, including at least one at a putative functional site, and one haplotype block were associated with flowering time variation in 248 accessions, with flowering times differing by 13–19 days between extreme haplotypes (Wang et al. 2011b). It was shown that a tourist-like MITE insertion in the promoter region of a *FLC* homolog in the *B. napus* A genome was associated with vernalization requirement in European winter rapeseed morphotypes (Hou et al. 2012).

Given the different polyploidization steps that have occurred in *Brassica* species, a huge expansion of active flowering time regulatory genes is expected in the allopolyploid *B. napus* genome relative to *A. thaliana*, where most of these genes are represented by only a single active copy. Throughout the evolution of *Brassica* species, homoeologous exchanges between the A and C genomes, genome fractionation, and sequence evolution have led to copy number variations (CNVs), presence/absence variations (PAVs), or sequence variations (INDELs, SNPs), which create enormous potential for functional differentiation and regulatory plasticity across all pathways that influence the flowering time gene expression network. Schiessl et al. (2014) investigated the genetic variation in all homologous and paralogous copies of 29 selected flowering time genes in four *B. napus* types, based on sequences derived from the *B. napus* genome and from the diploid progenitors *B. rapa* and *B. oleracea*. In total, 160 individual homologs/paralogs were identified for the 29 genes of the target panel. Most of the variations corresponded to SNPs, with many non-synonymous mutations in vernalization-related genes and in genes from the temperature and the photoperiod pathways. In addition, Schiessl et al. (2014) identified CNVs for the genes *Bna.CDF1.unk*, *Bna.CO.C09*, *Bna.CO-like2.A10*, *Bna.FLC.A10*, *Bna.FLC.C09*, and *Bna.TEM1.C02*, and PAVs for *Bna.TEM1.C02*. In the swede type, which is strongly vernalization-dependent and flowers later than winter-type oilseed types, a copy number reduction affecting two *Bna.FLC* paralogs was observed on chromosome C9, which was mirrored by a corresponding copy number increase on A10 potentially deriving from homoeologous exchange. De novo variation in copies of some important flowering time genes in *B. napus* that arose during allopolyploidization, enabling sub-functionalization, was shown to lead to different morphotypes in *B. napus* (Schiessl et al. 2017). This example illustrates the high potential for genomic plasticity in *B. napus*, as a highly duplicated polyploid species. These variations in gene content and gene sequence, as well as in regulatory networks, result in a high potential for functional divergence and regulation differentiation between the numerous genes involved in the important traits for oilseed rape yield components and biotic and abiotic environmental adaptation.

3.6 Conclusion and Perspectives

Relatively large numbers of genes and QTL involved in the expression of important agronomic traits have been identified and localized on the *B. napus* genome. The availability today of genome-wide SNPs and of the genomic sequences of *B. napus* and its two parental species, along with the exploitation of the synteny between *B. napus* and Arabidopsis, facilitates the genetic analyses and discovery of causal candidate genes explaining agronomic traits. Beyond SNP polymorphisms, more attention should be paid to the different types of structural variants, ranging from small variations like INDELs, CNVs, and PAVs, to larger rearrangements of genomes (Rousseau-Gueutin et al. 2017; Stein et al. 2017). Recently, by assembling the pangenome of *B. napus*, Hurgobin et al. (2018) showed that 38% of the genes display PAV behavior, with some of these variable genes predicted to be involved in important agronomic traits including flowering time, disease resistance, acyl lipid metabolism, and glucosinolate metabolism.

Most genetic analyses (both LA and LDA) have been performed by trait category and/or under only a single or a few environmental conditions. In the future, it will be necessary to establish more precisely the relationships between traits and to analyze the plant response to combined abiotic and biotic variations in fluctuating environments. Multi-trait and multi-environment approaches have been successfully implemented in other crops to study the architecture of complex traits (Na et al. 2013) and to gain a greater understanding of genotype or QTL × E interactions or of pleiotropic effects across traits (Verbyla et al. 2014). The current development of high-throughput and/or automated phenotype platforms, involving both automated recording and screening of phenotypes by various imaging techniques, allows the production of time-course phenotypic data and thereby facilitates the study of the genetic control of plant development and/or growth-related traits. A new approach called 'functional mapping' is emerging to detect QTL associated with the whole developmental process of the traits, instead of being associated with any single observation. Functional mapping is developed based on the assumption of function-valued traits: Phenotypic values at discrete time points are 'snapshots' of a continuous function/curve over time. Thus, by integrating information over multiple time points, it is possible to study the genetic architecture of dynamical complex traits (Li and Sillanpää 2015).

Nonetheless, there is still a need to develop new or more precise methods for phenotyping certain traits. For example, phenotyping different components of disease resistance would be worthwhile in order to identify the mode of action of QTL on the different steps of the pathogen life cycle and to better understand the mechanisms underlying resistance. For some traits, such as plant/insect interaction traits that are difficult or even impossible to phenotype via high-throughput approaches, indirect phenotyping with metabolites that are related to insect attraction, feeding, egg production and deposition or larval development has been suggested (Hervé et al. 2014).

These structural genetic studies have to be completed and correlated with the functional organization of genomes. The advent of newer and more effective gene expression (Bancroft et al. 2011), proteomics (Chen et al. 2015), and high-throughput metabolomics (Feng et al. 2012) technologies in *B. napus* is making it feasible to study a large number of genes and metabolites and their regulatory interactions (reviewed by Snodown et al. 2012). Small RNAs including microRNAs (miRNAs) play an indispensable role in cell signaling mechanisms and have been shown to be key regulators of various stress responses in plants, including responses to low nutrient availability and pathogen infection (Rajwanshi et al. 2014; Paul et al. 2015). Therefore, the role of miRNAs is worthy of investigation in *B. napus* through high-throughput miRNome sequencing. It has been shown already that miRNAs may play a role in response to *B. napus* to *V. longisporum* (Shen et al. 2014) and *P. brassicae* (Verma et al. 2014) infection, as well as in *B. napus* sulfur or cadmium homeostasis (Huang et al. 2010; Zhou et al. 2012). More recently, a full catalog of the miRNA and their targets involved in fatty acids and lipids metabolism was described in *B. napus* seeds (Wang et al. 2017d). There is also increasing evidence that epigenetic marks such as DNA methylation contribute to phenotypic variation by regulating gene transcription, developmental plasticity, and interactions with the environment. Long et al. (2011b) detected between two and 17 epiQTLs associated with seven agronomic traits in regions having a greater density of DNA methylation markers. The authors concluded that studies relying solely on polymorphism of conventional gene markers may consistently underestimate the quantity and distribution of QTL effects and that epimarkers can increase the ability to resolve previously 'cryptic' QTL.

The availability of whole genome sequence data provides an exhaustive catalog of polymorphic sites segregating within and across oilseed rape populations, which may be used in combination with QTL data in the near future to establish genomic selection or to better predict hybrid performance. The first results obtained in oilseed rape have shown that genomic selection is a valuable approach for complex agronomic traits, facilitating the process of knowledge-based breeding (Würschum et al. 2014; Zou et al. 2016).

Acknowledgements The authors gratefully acknowledge Stephen Strelkov for English editing of the manuscript.

References

Amar S, Ecke W, Becker HC, Moellers C (2008) QTL for phytosterol and sinapate ester content in *Brassica napus* L. collocate with the two erucic acid genes. Theor Appl Genet 116:1051–1061

ArifUzZaman M, Mamidi S, McClean P, Rahman M (2017) QTL mapping for root vigor and days to flowering in *Brassica napus* L. Can J Plant Sci 97:99–109

Asghari A, Mohammadi SA, Moghaddam M, Mohammaddust H (2007) Identification of QTLs controlling winter survival in *Brassica napus* using RAPD markers. Biotech Biotech Equip 21:413–416

Asghari A, Mohammadi SA, Moghaddam M, Toorchi M, Mohammadinasa AD (2008) Analysis of quantitative trait loci associated with freezing tolerance in rapeseed (*Brassica napus* L.). Biotech Biotech Equip 22:548–552

Asghari A, Asghar Fathi A, Mohammadi SA, Mohammaddust H (2009) QTL analysis for diamondback moth resistance in canola (*Brassica napus* L.). Int J Plant Prod 3:29–34

Badani AG, Snowdon RJ, Wittkop B, Lipsa FD, Baetzel R, Horn R, De Haro A, Font R, Luehs W, Friedt W (2006) Colocalization of a partially dominant gene for yellow seed colour with a major QTL influencing acid detergent fibre (ADF) content in different crosses of oilseed rape (*Brassica napus*). Genome 49:1499–1509

Balesdent MH, Fudal I, Ollivier B, Bally P, Grandaubert J, Eber F, Chèvre AM, Leflon M, Rouxel T (2013) The dispensable chromosome of *Leptosphaeria maculans* shelters an effector gene conferring avirulence towards *Brassica rapa*. New Phytol 198:887–898

Bancroft I, Morgan C, Fraser F, Higgins J, Wells R, Clissold L, Baker D, Long Y, Meng J, Wang X, Liu S, Trick M (2011) Dissecting the genome of the polyploid crop oilseed rape by transcriptome sequencing. Nat Biotech 29:762–768

Bancroft I, Fraser F, Morgan C, Trick M (2015) Collinearity analysis of Brassica A and C genomes based on an updated inferred unigene order. Data in Brief 3:51–55

Bargsten JW, Nap JP, Sanchez-Perez GF, van Dijk ADJ (2014) Prioritization of candidate genes in QTL regions based on associations between traits and biological processes. BMC Plant Biol 14:330

Barret P, Delourme R, Renard M, Domergue F, Lessire R, Delseny M, Roscoe TJ (1998) A rapeseed FAEI gene is linked to the E1 locus associated with variation in the content of erucic acid. Theor Appl Genet 96:177–186

Barret P, Delourme R, Brunel D, Jourdren C, Horvais R, Renard M (1999) Low linolenic acid level in rapeseed can be easily assessed through the detection of two single base substitution in fad3 genes. In: Proceedings of 10th international rapeseed congress, Canberra, Australia, pp 26–29

Basunanda P, Spiller TH, Hasan M, Gehringer A, Schondelmaier J, Luehs W, Friedt W, Snowdon RJ (2007) Marker-assisted increase of genetic diversity in a double-low seed quality winter oilseed rape genetic background. Plant Breed 126:581–587

Basunanda P, Radoev M, Ecke W, Friedt W, Becker HC, Snowdon RJ (2010) Comparative mapping of quantitative trait loci involved in heterosis for seedling and yield traits in oilseed rape (*Brassica napus* L.). Theor Appl Genet 120:271–281

Bouchet AS, Nesi N, Bissuel C, Bregeon M, Lariepe A, Navier H, Ribiere N, Orsel M, Grezes-Besset B, Renard M, Laperche A (2014) Genetic control of yield and yield components in winter oilseed rape (*Brassica napus* L.) grown under nitrogen limitation. Euphytica 199:183–205

Bouchet AS, Laperche A, Bissuel-Belaygue C, Baron C, Morice J, Rousseau-Gueutin M, Dheu JE, George P, Pinochet X, Foubert T, Maes O, Dugue D, Guinot F, Nesi N (2016) Genetic basis of nitrogen use efficiency and yield stability across environments in winter rapeseed. BMC Genet 17:131

Bradburne R, Majer D, Magrath R, Werner CP, Lewis B, Mithen R (1999) Winter oilseed rape with high levels of resistance to *Pyrenopeziza brassicae* derived from wild Brassica Species. Plant Pathol 48:550–558

Buczacki ST, Toxopeus H, Mattusch P, Johnston TD, Dixon GR, Hobolth LA (1975) Study of physiologic specialization in *Plasmodiophora brassicae*: proposals for attempted rationalization through an international approach. Trans British Mycol Soc 65:295–303

Burns MJ, Barnes SR, Bowman JG, Clarke MHE, Werner CP, Kearsey MJ (2003) QTL analysis of an intervarietal set of substitution lines in *Brassica napus*: (i) seed oil content and fatty acid composition. Heredity 90:39–48

Bus A, Koerber N, Snowdon RJ, Stich B (2011) Patterns of molecular variation in a species-wide germplasm set of *Brassica napus*. Theor Appl Genet 123:1413–1423

Bus A, Hecht J, Huettel B, Reinhardt R, Stich B (2012) High-throughput polymorphism detection and genotyping in *Brassica napus* using next-generation RAD sequencing. BMC Genom 13:281

Bus A, Koerber N, Parkin IAP, Samans B, Snowdon RJ, Li J, Stich B (2014) Species- and genome-wide dissection of the shoot ionome in *Brassica napus* and its relationship to seedling development. Front Plant Sci 5:485

Butruille DV, Guries RP, Osborn TC (1999) Linkage analysis of molecular markers and quantitative trait loci in populations of inbred backcross lines of *Brassica napus* L. Genetics 153:949–964

Cai D, Xiao Y, Yang W, Ye W, Wang B, Younas M, Wu J, Liu K (2014) Association mapping of six yield-related traits in rapeseed (*Brassica napus* L.). Theor Appl Genet 127:85–96

Cai GQ, Yang QY, Chen H, Yang Q, Zhang CY, Fan CC, Zhou YM (2016) Genetic dissection of plant architecture and yield-related traits in *Brassica napus*. Sci Rep 6:21625

Cao Z, Tian F, Wang N, Jiang C, Lin B, Xia W, Shi J, Long Y, Zhang C, Meng J (2010) Analysis of QTLs for erucic acid and oil content in seeds on A8 chromosome and the linkage drag between the alleles for the two traits in *Brassica napus*. J Genet Genom 37:231–240

Chalhoub B, Denoeud F, Liu S, Parkin IAP, Tang HB, Wang XX, Chiquet J, Belcram H et al (2014) Early allopolyploid evolution in the post-Neolithic *Brassica napus* oilseed genome. Science 354:950–953

Chao H, Wang H, Wang X, Guo L, Gu J, Zhao W, Li B, Chen D, Raboanatahiry N, Li M (2017) Genetic dissection of seed oil and protein content and identification of networks associated with oil content in *Brassica napus*. Sci Rep 7:46295

Chen W, Zhang Y, Liu X, Chen B, Tu J, Fu T (2007) Detection of QTL for six yield-related traits in oilseed rape (*Brassica napus*) using DH and immortalized F-2 populations. Theor Appl Genet 115:849–858

Chen G, Geng J, Rahman M, Liu X, Tu J, Fu T, Li G, McVetty PBE, Tahir M (2010) Identification of QTL for oil content, seed yield, and flowering time in oilseed rape (*Brassica napus*). Euphytica 175:161–174

Chen X, Li X, Zhang B, Xu J, Wu Z, Wang B, Li H, Younas M, Huang L, Luo Y, Wu J, Hu S, Liu K (2013) Detection and genotyping of restriction fragment associated polymorphisms in polyploid crops with a pseudo-reference sequence: a case study in allotetraploid *Brassica napus*. BMC Genom 14:346

Chen S, Ding G, Wang Z, Cai H, Xu F (2015) Proteomic and comparative genomic analysis reveals adaptability of *Brassica napus* to phosphorus-deficient stress. J Proteomics 117:106–119

Chen J, Wang B, Zhang Y, Yue X, Li Z, Liu K (2017) High-density ddRAD linkage and yield-related QTL mapping delimits a chromosomal region responsible for oil content in rapeseed (*Brassica napus* L.). Breed Sci 67:296–306

Clarke WE, Parkin IAP, Gajardo HA, Gerhardt DJ, Higgins E, Sidebottom C, Sharpe AG, Snowdon RJ, Federico ML, Iniguez-Luy FL (2013) Genomic DNA enrichment using sequence capture microarrays: a novel approach to discover sequence nucleotide polymorphisms (SNP) in *Brassica napus* L. PLoS ONE 8:e81992

Crute IR, Gray AR, Crisp P, Buczacki ST (1980) Variation in *Plasmodiophora brassicae* and resistance to clubroot disease in Brassicas and allied crops—a critical review. Plant Breed Abstr 50:91–104

Dalton-Morgan J, Hayward A, Alamery S, Tollenaere R, Mason AS, Campbell E, Patel D, Lorenc MT, Yi B, Long Y, Meng J, Raman R, Raman H, Lawley C, Edwards D, Batley J (2014) A high-throughput SNP array in the amphidiploid species *Brassica napus* shows diversity in resistance genes. Funct Integ Genomics 14:643–655

Delourme R, Pilet-Nayel ML, Archipiano M, Horvais R, Tanguy X, Rouxel T, Brun H, Renard M, Balesdent MH (2004) A cluster of major specific resistance genes to *Leptosphaeria maculans* in *Brassica napus*. Phytopathology 94:578–583

Delourme R, Chèvre AM, Brun H, Rouxel T, Balesdent MH, Dias JS, Salisbury P, Renard M, Rimmer SR (2006a) Major gene and polygenic resistance to *Leptosphaeria maculans* in oilseed rape (*Brassica napus*). Eur J Plant Pathol 114:41–52

Delourme R, Falentin C, Huteau V, Clouet V, Horvais R, Gandon B, Specel S, Hanneton L, Dheu JE, Deschamps M, Margale E, Vincourt P, Renard M (2006b) Genetic control of oil content in oilseed rape (*Brassica napus* L.). Theor Appl Genet 113:1331–1345

Delourme R, Falentin C, Fomeju BF, Boillot M, Lassalle G, Andre I, Duarte J, Gauthier V, Lucante N, Marty A, Pauchon M, Pichon JP, Ribiere N, Trotoux G, Blanchard P, Riviere N, Martinant JP, Pauquet J (2013) High-density SNP-based genetic map development and linkage disequilibrium assessment in *Brassica napus* L. BMC Genom 14:120

Diederichsen E, Beckmann J, Schondelmeier J, Dreyer F (2006) Genetics of clubroot resistance in *Brassica napus* 'Mendel'. Acta Hortic 706:307–311

Ding G, Yang M, Hu Y, Liao Y, Shi L, Xu F, Meng J (2010) Quantitative trait loci affecting seed mineral concentrations in *Brassica napus* grown with contrasting phosphorus supplies. Ann Bot 105:1221–1234

Ding G, Zhao Z, Liao Y, Hu Y, Shi L, Long Y, Xu F (2012) Quantitative trait loci for seed yield and yield-related traits, and their responses to reduced phosphorus supply in *Brassica napus*. Ann Bot 109:747–759

Ding G, Shi L, Zhao H, Cai H, Liu K, Xu F (2013) Genetic analysis of seed mineral accumulation affected by phosphorus deprivation in *Brassica napus*. Euphytica 193:251–264

Ding G, Zhao Z, Wang L, Zhang D, Shi L, Xu F (2014) Identification and multiple comparisons of QTL and epistatic interaction conferring high yield under boron and phosphorus deprivation in *Brassica napus*. Euphytica 198:337–351

Dion Y, Gugel RK, Rakow GFW, Seguin-Swartz G, Landry BS (1995) RFLP mapping of resistance to the blackleg disease causal agent, *Leptosphaeria maculans* (Desm) Ces, et de Not in canola (*Brassica napus* L.). Theor Appl Genet 91:190–1194

Durstewitz G, Polley A, Plieske J, Luerssen H, Graner EM, Wieseke R, Ganal MW (2010) SNP discovery by amplicon sequencing and multiplex SNP genotyping in the allopolyploid species *Brassica napus*. Genome 53:948–956

Ecke W, Uzunova M, Weissleder K (1995) Mapping the genome of rapeseed (*Brassica napus* L.) 2: localization of genes controlling erucic acid synthesis and seed oil content. Theor Appl Genet 91:972–977

Ecke W, Clemens R, Honsdorf N, Becker HC (2010) Extent and structure of linkage disequilibrium in canola quality winter rapeseed (*Brassica napus* L.). Theor Appl Genet 120:921–931

Edwards D, Batley J, Snowdon RJ (2013) Accessing complex crop genomes with next-generation sequencing. Theor Appl Genet 126:1–11

Falentin C, Brégeon M, Lucas MO, Deschamps M, Leprince F, Fournier MT, Delourme R, Renard M (2007) Identification of fad2 mutations and development of allele-specific markers for high oleic acid content in rapeseed (*Brassica napus* L.). In: Proceedings of 12th international rapeseed congress, Wuhan, China p 126

Fan C, Cai G, Qin J, Li Q, Yang M, Wu J, Fu T, Liu K, Zhou Y (2010) Mapping of quantitative trait loci and development of allele-specific markers for seed weight in *Brassica napus*. Theor Appl Genet 121:1289–1301

Feng F, Liu P, Hong D, Yang G, Feng FQ, Liu PW, Hong DF, Yang GS (2009) A major QTL associated with preharvest sprouting in rapeseed (*Brassica napus* L.). Euphytica 169:57–68

Feng J, Long Y, Shi L, Shi J, Barker G, Meng J (2012) Characterization of metabolite quantitative trait loci and metabolic networks that control glucosinolate concentration in the seeds and leaves of *Brassica napus*. New Phytol 193:96–108

Ferreira ME, Williams PH, Osborn TC (1994) RFLP mapping of *Brassica napus* using doubled haploid lines. Theor Appl Genet 89:615–621

Ferreira ME, Satagopan J, Yandell BS, Williams PH, Osborn TC (1995) Mapping loci controlling vernalization requirement and flowering time in *Brassica napus*. Theor Appl Genet 90:727–732

Fletcher RS, Mullen JL, Heiliger A, McKay JK (2015) QTL analysis of root morphology, flowering time, and yield reveals trade-offs in response to drought in *Brassica napus*. J Exp Bot 66:245–256

Fletcher RS, Herrmann D, Mullen JL, Li QF, Schrider DR, Price N, Lin JJ, Grogan K, Kern A, Mckay JK (2016) Identification of polymorphisms associated with drought adaptation QTL in *Brassica napus* by resequencing G3. Genes Genomes Genet 6:793–803

Foisset N, Delourme R, Barret P, Hubert N, Landry BS, Renard M (1996) Molecular mapping analysis in *Brassica napus* using isozyme, RAPD and RFLP markers on a doubled-haploid progeny. Theor Appl Genet 93:1017–1025

Fopa Fomeju B, Falentin C, Lassalle G, Manzanares-Dauleux MJ, Delourme R (2014) Homoeologous duplicated regions are involved in quantitative resistance of *Brassica napus* to stem canker. BMC Genom 15:498

Fopa Fomeju B, Falentin C, Lassalle G, Manzanares-Dauleux MJ, Delourme R (2015) Comparative genomic analysis of duplicated homoeologous regions involved in the resistance of *Brassica napus* to stem canker. Front Plant Sci 6:772

Fourmann M, Barret P, Renard M, Pelletier G, Delourme R, Brunel D (1998) The two genes homologous to Arabidopsis FAE1 cosegregate with the two loci governing erucic acid content in *Brassica napus*. Theor Appl Genet 96:852–858

Fredua-Agyeman R, Rahman H (2016) Mapping of the clubroot disease resistance in spring Brassica napus canola introgressed from European winter canola cv. 'Mendel'. Euphytica 211:201–213

Fu FY, Liu LZ, Chai YR, Chen L, Yang T, Jin MY, Ma AF, Yan XY, Zhang ZS, Li JN (2007a) Localization of QTLs for seed color using recombinant inbred lines of *Brassica napus* in different environments. Genome 50:840–854

Fu FY, Liu LZ, Chai YR, Chen L, Yang T, Ma AF, Qu CN, Jiang L, Zhang ZS, Li JN (2007b) Localization of QTLs for husk proportion and lignin content using a high-density genetic linkage map of *Brassica napus*. Korean J Genet 29:343–353

Fu Y, Wei D, Dong H, He Y, Cui Y, Mei J, Wan H, Li J, Snowdon R, Friedt W, Li X, Qian W (2015) Comparative quantitative trait loci for silique length and seed weight in *Brassica napus*. Sci Rep 5:14407

Gajardo HA, Wittkop B, Soto-Cerda B, Higgins EE, Parkin IAP, Snowdon RJ, Federico ML, Iniguez-Luy FL (2015) Association mapping of seed quality traits in *Brassica napus* L. using GWAS and candidate QTL approaches. Mol Breed 35:143

Gali KK, Sharpe AG (2011) Molecular linkage maps: strategies, resources and achievements In: Edwards D, Batley J, Parkin IAP, Kole C (eds) Genetics, genomics and breeding of oilseed Brassicas. Science Publishers, CRC Press, USA, pp 85–129

Gül MK, Becker H, Ecke W (2003) QTL mapping and analysis of QTL × nitrogen interactions for protein and oil contents in *Brassica napus* L. In: Proceedings of 11th international rapeseed congress, Copenhagen, Denmark, pp 91–93

Gustafsson M, Fält AS (1986) Genetic studies on resistance to clubroot in *Brassica napus*. Ann Appl Biol 108:409–415

Gyawali S, Harrington M, Durkin J, Horner K, Parkin IAP, Hegedus DD, Bekkaoui D, Buchwaldt L (2016) Microsatellite markers used for genome-wide association mapping of partial resistance to *Sclerotinia sclerotiorum* in a world collection of *Brassica napus*. Mol Breed 36:72

Harper AL, Trick M, Higgins J, Fraser F, Clissold L, Wells R, Hattori C, Werner P, Bancroft I (2012) Associative transcriptomics of traits in the polyploid crop species *Brassica napus*. Nat Biotech 30:798–802

Hasan MJ, RahmanH (2016) Genetics and molecular mapping of resistance to *Plasmodiophora brassicae* pathotypes 2, 3, 5, 6, and 8 in rutabaga (*Brassica napus* var. *napobrassica*). Genome 59:805–815

Hasan M, Friedt W, Pons-Kuehnemann J, Freitag NM, Link K, Snowdon RJ (2008) Association of gene-linked SSR markers to seed glucosinolate content in oilseed rape (*Brassica napus* ssp. *napus*). Theor Appl Genet 116:1035–1049

Hatzig SV, Frisch M, Breuer F, Nesi N, Ducoumau S, Wagner MH, Leckband G, AbbadiA Snowdon RJ (2015) Genome-wide association mapping unravels the genetic control of seed germination and vigor in *Brassica napus*. Front Plant Sci 6:221

Havlickova L, He Z, Wang L, Langer S, Harper AL, Kaur H, Broadley MR, Gegas V, Bancroft I (2018) Validation of an updated associative transcriptomics platform for the polyploid crop species *Brassica napus* by dissection of the genetic architecture of erucic acid and tocopherol isoform variation in seeds. Plant J 93:181–192

Hayward A, Mason AS, Dalton-Morgan J, Zander M, Edwards D, Batley J (2012) SNP discovery and applications in *Brassica napus*. J Plant Biotech 39:49–61

He YJ, Wu DM, Wei DY, Fu Y, Cui YX, Dong HL, Tan CD, Qian W (2017) GWAS QTL mapping and gene expression analyses in *Brassica napus* reveal genetic control of branching morphogenesis. Sci Rep 7:15971

Hervé MR, Delourme R, Gravot A, Marnet N, Berardocco S, Cortesero AM (2014) Manipulating feeding stimulation to protect crops against insect pests? J Chem Ecol 40:1220–1231

Hirai MY, Sugiyama K, Sawada Y et al (2007) Omics based identification of Arabidopsis Myb transcription factors regulating aliphatic glucosinolate biosynthesis. Proc Natl Acad Sci USA 104:6478–6483

Honsdorf N, Becker HC, Ecke W (2010) Association mapping for phenological, morphological, and quality traits in canola quality winter rapeseed (*Brassica napus* L). Genome 53:899–907

Hou J, Long Y, Raman H, Zou X, Wang J, Dai S, Xiao Q, Li C, Fan L, Liu B, Meng J (2012) A tourist-like MITE insertion in the upstream region of the *BnFLCA10* gene is associated with vernalization requirement in rapeseed (*Brassica napus* L.). BMC Plant Biol 12:238

Howell PM, Sharpe AG, Lydiate DJ (2003) Homoeologous loci control the accumulation of seed glucosinolates in oilseed rape (*Brassica napus*). Genome 46:454–460

Hu X, Sullivan-Gilbert M, Gupta M, Thompson SA (2006) Mapping of the loci controlling oleic and linolenic acid contents and development of *fad2* and *fad3* allele-specific markers in canola (*Brassica napus* L.). Theor Appl Genet 113:497–507

Hu Z, Hua W, Huang S, Yang H, Zhan G, Wang X, Liu G, Wang H (2012a) Discovery of pod shatter-resistant associated SNPs by deep sequencing of a representative library followed by bulk segregant analysis in rapeseed. PLoS ONE 7:e34253

Hu Z, Huang S, Sun M, Wang H, Hua W (2012b) Development and application of single nucleotide polymorphism markers in the polyploid *Brassica napus* by 454 sequencing of expressed sequence tags. Plant Breed 131:293–299

Hua YP, Zhou T, Ding GD, Yang QY, Shi L, Xu FS (2016) Physiological genomic and transcriptional diversity in responses to boron deficiency in rapeseed genotypes. J Exp Bot 67:5769–5784

Huang SQ, Xiang AL, Che LL, Chen S, Li H, Song JB, Yang ZM (2010) A set of miRNAs *Brassica napus* in response to sulphate deficiency and cadmium stress. Plant Biotechnol J 8:887–899

Huang S, Deng L, Guan M, Li J, Lu K, Wang H, Fu D, Mason AS, Liu S, Hua W (2013) Identification of genome-wide single nucleotide polymorphisms in allopolyploid crop *Brassica napus*. BMC Genom 14:717

Huang YJ, Jestin C, Welham SJ, King GJ, Manzanares-Dauleux MJ, Fitt BDL, Delourme R (2016) Identification of environmentally stable QTL for resistance against *Leptosphaeria maculans* in oilseed rape (*Brassica napus*). Theor Appl Genet 129:169–180

Hurgobin B, Golicz AA, Bayer PE, Chan CK, Tirnaz S, Dolatabadian A, Schiessl SV, Samans B et al (2018) Homoeologous exchange is a major cause of gene presence/absence variation in the amphidiploid *Brassica napus*. Plant Biotech J. https://doi.org/10.1111/pbi.12867

Javed N, Tahir M, Geng J, Li G, McVetty PBE (2014) Identification of Brassica genotypes and molecular markers for increased seed oil content. Can J Plant Sci 94:1103–1108

Jestin C, Lode M, Vallee P, Domin C, Falentin C, Horvais R, Coedel S, Manzanares-Dauleux MJ, Delourme R (2011) Association mapping of quantitative resistance for *Leptosphaeria maculans* in oilseed rape (*Brassica napus* L.). Mol Breed 27:271–287

Jestin C, Vallée P, Domin C, Manzanares-Dauleux MJ, Delourme R (2012) Assessment of a new strategy for selective phenotyping applied to complex traits in *Brassica napus*. Open J Genet 2:190–201

Jestin C, Bardol N, Lodé M, Duffé P, Domin C, Vallée P, Mangin B, Manzanares-Dauleux MJ, Delourme R (2015) Connected populations for detecting quantitative resistance factors to Phoma stem canker in oilseed rape (*Brassica napus* L.). Mol Breed 35:167

Jian HJ, Yang B, Zhang AX, Zhang L, Xu XF, Li JN, Liu LZ (2017) Screening of candidate leaf morphology genes by integration of QTL mapping and RNA sequencing technologies in oilseed rape (*Brassica napus* L.). PLoS ONE 12:e0169641

Jiang C, Shi J, Li R, Long Y, Wang H, Li D, Zhao J, Meng J (2014) Quantitative trait loci that control the oil content variation of rapeseed (*Brassica napus* L.). Theor Appl Genet 127:957–968

Jin M, Li J, Fu F, Zhang Z, Zhang X, Liu L, Jin MY, Li JN, Fu FY, Zhang ZS, Zhang XK, Liu LZ (2007) QTL analysis of oil and hull content in *Brassica napus* L. Sci Agric Sin 40:677–684

Jourdren C, Barret P, Horvais R, Delourme R, Renard M (1996a) Identification of RAPD markers linked to linolenic acid genes in rapeseed. Euphytica 90:351–357

Jourdren C, Barret P, Horvais R, Foisset N, Delourme R, Renard M (1996b) Identification of RAPD markers linked to the loci controlling erucic acid level in rapeseed. Mol Breed 2:61–71

Kaur S, Cogan NOI, Ye G, Baillie RC, Hand ML, Ling AE, McGearey AK, Kaur J, Hopkins CJ, Todorovic M, Mountford H, Edwards D, Batley J, Burton W, Salisbury P, Gororo N, Marcroft S, Kearney G, Smith KF, Forster JW, Spangenberg GC (2009) Genetic map construction and QTL mapping of resistance to blackleg (*Leptosphaeria maculans*) disease in Australian canola (*Brassica napus* L.) cultivars. Theor Appl Genet 120:71–83

Kaur S, Francki MG, Forster JW (2012) Identification, characterization and interpretation of single-nucleotide sequence variation in allopolyploid crop species. Plant Biotech J 10:125–138

Kole C, Thormann CE, Karlsson BH, Palta JP, Gaffney P, Yandell B, Osborn TC (2002) Comparative mapping of loci controlling winter survival and related traits in oilseed *Brassica rapa* and *B. napus*. Mol Breed 9:201–210

Körber N, Bus A, Li J, Higgins J, Bancroft I, Higgins EE, Parkin IAP, Salazar-Colqui B, Snowdon RJ, Stich B (2015) Seedling development traits in *Brassica napus* examined by gene expression analysis and association mapping. BMC Plant Biol 15:136

Körber N, Bus A, Li J, Parkin IA, Wittkop B, Snowdon RJ, Stich B (2016) Agronomic and seed quality traits dissected by genome-wide association mapping in *Brassica napus*. Front Plant Sci 7:386

Landry BS, Hubert N, Etoh T, Harada JJ, Lincoln SE (1991) A genetic map of *Brassica napus* based on restriction fragment length polymorphism detected with expressed DNA sequences. Genome 34:543–552

Lang LN, Xu AX, Ding J, Zhang Y, Zhao N, Tian ZS, Liu YP, Wang Y, Liu X, Liang FH, Zhang BB, Qin MF, Dalelhan J, Huang Z (2017) Quantitative trait locus mapping of salt tolerance and identification of salt-tolerant genes in *Brassica napus* L. Front Plant Sci 8:1000

Laperche A, Aigu Y, Jubault M, Ollier M, Guichard S, Glory P, Strelkov SE, Gravot A, Manzanares-Dauleux MJ (2017) Clubroot resistance QTL are modulated by nitrogen input in *Brassica napus*. Theor Appl Genet 130:669–684

Larkan NJ, Lydiate DJ, Parkin IAP, Nelson MN, Epp DJ, Cowling WA, Rimmer SR, Borhan MH (2013) The *Brassica napus* blackleg resistance gene *LepR3* encodes a receptor-like protein triggered by the *Leptosphaeria maculans* effector AVRLM1. New Phytol 197:595–605

Larkan NJ, Ma L, Borhan MH (2015) The *Brassica napus* receptor-like protein RLM2 is encoded by a second allele of the *LepR3/Rlm2* blackleg resistance locus. Plant Biotech J 13:983–992

Larkan NJ, Raman H, Lydiate DJ, Robinson SJ, Yu F, Barbulescu DM, Raman R, Luckett DJ, Burton W, Wratten N, Salisbury PA, Rimmer SR, Borhan MH (2016) Multi-environment QTL studies suggest a role for cysteine-rich protein kinase genes in quantitative resistance to blackleg disease in *Brassica napus*. BMC Plant Biol 16:183

Lee RWH, Malchev IT, Rajcan I, Kott LS (2014) Identification of putative quantitative trait loci associated with a flavonoid related to resistance to cabbage seedpod weevil (*Ceutorhynchus obstrictus*) in canola derived from an intergeneric cross, *Sinapis alba* × *Brassica napus*. Theor Appl Genet 127:419–428

Li Z, Sillanpää MJ (2015) Dynamic quantitative trait locus analysis of plant phenomic data. Trends Plant Sci 20:12

Li F, Chen B, Xu K, Wu J, Song W, Bancroft I, Harper AL, Trick M, Liu S, Gao G, Wang N, Yan G, Qiao J, Li J, Li H, Xiao X, Zhang T, Wu X (2014a) Genome-wide association study dissects the genetic architecture of seed weight and seed quality in rapeseed (*Brassica napus* L.). DNA Res 21:355–367

Li Z, Mei S, Mei Z, Liu X, Fu T, Zhou G, Tu J (2014b) Mapping of QTL associated with waterlogging tolerance and drought resistance during the seedling stage in oilseed rape (*Brassica napus*). Euphytica 197:341–353

Li N, Shi J, Wang X, Liu G, Wang H, Li N, Shi JQ, Wang XF, Liu GH, Wang HZ (2014c) A combined linkage and regional association mapping validation and fine mapping of two major pleiotropic QTLs for seed weight and silique length in rapeseed (*Brassica napus* L.). BMC Plant Biol 14:114

Li J, Zhao Z, Hayward A, Cheng H, Fu D (2015) Integration analysis of quantitative trait loci for resistance to *Sclerotinia sclerotiorum* in *Brassica napus*. Euphytica 205:483–489

Li F, Chen B, Xu K, Gao G, Yan G, Qiao J, Li J, Hao L, Li L, Xiao X, Zhang T, Nishio T, Wu X (2016a) A genome-wide association study of plant height and primary branch number in rapeseed (*Brassica napus*). Plant Sci 242:169–177

Li L, Luo Y, Chen B, Xu K, Zhang F, Li H, Huang Q, Xiao X, Zhang T, Hu J, Li F, Wu X (2016b) A genome-wide association study reveals new loci for resistance to clubroot disease in *Brassica napus*. Front Plant Sci 7:1483

Lipsa FD, Snowdon R, Friedt W (2012) Quantitative genetic analysis of condensed tannins in oilseed rape meal. Euphytica 184:195–205

Liu Z, Adamczyk K, Manzanares-Dauleux M, Eber F, Lucas MO, Delourme R, Chevre AM, Jenczewski E (2006a) Mapping *PrBn* and other quantitative trait loci responsible for the control of homeologous chromosome pairing in oilseed rape (*Brassica napus* L.) haploids. Genetics 174:1583–1596

Liu L, Meng J, Lin N, Chen L, Tang Z, Zhang X, Li J, Liu LZ, Meng JL, Lin N, Chen L, Tang ZL, Zhang XK, Li JN (2006b) QTL mapping of seed coat color for yellow seeded *Brassica napus*. Acta Genet Sin 33:181–187

Liu J, Yang J, Li R, Shi L, Zhang C, Long Y, Xu F, Meng J (2009) Analysis of genetic factors that control shoot mineral concentrations in rapeseed (*Brassica napus*) in different boron environments. Plant Soil 320:255–266

Liu L, Qu C, Wittkop B, Yi B, Xiao Y, He Y, Snowdon RJ, Li J (2013) A high-density SNP map for accurate mapping of seed fibre QTL in *Brassica napus* L. PLoS ONE 8:e83052

Liu SY, Liu YM, Yang XH, Tong CB, Edwards D, Parkin IAP et al (2014) The *Brassica oleracea* genome reveals the asymmetrical evolution of polyploid genomes. Nat Comm 5:3930

Liu J, Hua W, Hu Z, Yang H, Zhang L, Li R, Deng L, Sun X, Wang X, Wang H (2015) Natural variation in *ARF18* gene simultaneously affects seed weight and silique length in polyploid rapeseed. PNAS E5123–E5132

Liu J, Wang J, Wang H, Wang WX, Zhou RJ, Mei DS, Cheng HT, Yang J, Raman H, Hu Q (2016a) Multigenic control of pod shattering resistance in Chinese rapeseed germplasm revealed by genome-wide association and linkage analyses. Front Plant Sci 7:1058

Liu J, Wang WX, Mei DS, Wang H, Fu L, Liu DM, Li YC, Hui Q (2016b) Characterizing variation of branch angle and genome-wide association mapping in rapeseed (*Brassica napus* L.). Front Plant Sci 7:21

Liu PF, Zhao YS, Liu GZ, Wang M, Hu DD, Hu J, Meng JL, Reif JC, Zou J (2017) Hybrid performance of an immortalized F-2 rapeseed population is driven by additive dominance and epistatic effects. Front Plan Sci 8:815

Lombard V, Delourme R (2001) A consensus linkage map for rapeseed (*Brassica napus* L.): construction and integration of three individual maps from DH populations. Theor Appl Genet 103:491–507

Long Y, Shi J, Qiu D, Li R, Zhang C, Wang J, Hou J, Zhao J, Shi L, Park BS, Choi SR, Lim YP, Meng J (2007) Flowering time quantitative trait loci analysis of oilseed Brassica in multiple environments and genome wide alignment with Arabidopsis. Genetics 177:2433–2444

Long Y, Wang Z, Sun Z, Fernando DWG, McVetty PBE, Li G (2011a) Identification of two blackleg resistance genes and fine mapping of one of these two genes in a *Brassica napus* canola cultivar 'Surpass 400'. Theor Appl Genet 122:1223–1231

Long Y, Xia W, Li R, Wang J, Shao M, Feng J, King GJ, Meng J (2011b) Epigenetic QTL Mapping in *Brassica napus*. Genetics 189:1093–U585

Lou P, Xie Q, Xu X, Edwards C, Brock M, Weinig C, McClung C (2011) Genetic architecture of the circadian clock and flowering time in *Brassica rapa*. Theor Appl Genet 123:397–409

Lowe AJ, Moule C, Trick M, Edwards KJ (2004) Efficient large-scale development of microsatellites for marker and mapping applications in Brassica crop species. Theor Appl Genet 108:1103–1112

Lu G, Harper AL, Trick M, Morgan C, Fraser F, O'Neill C, Bancroft I (2014) Associative transcriptomics study dissects the genetic architecture of seed glucosinolate content in *Brassica napus*. DNA Res 21:613–625

Luo YX, Luo CY, Du DZ, Fu Z, Yao YM, Xu CC, Zhang HS (2014) Quantitative trait analysis of flowering time in spring rapeseed (*B. napus* L.). Euphytica 200:321–335

Luo X, Ma C, Yue Y, Hu K, Li Y, Duan Z, Wu M, Tu J, Shen J, Yi B, Fu T (2015) Unravelling the complex trait of harvest index in rapeseed (*Brassica napus* L.) with association mapping. BMC Genom 16:379

Luo ZL, Wang M, Long Y, Huang YJ, Shi L, Zhang CY, Liu X, Fitt BDL, Xiang JX, Mason AS, Snowdon RJ, Liu PF, Meng JL, Zou J (2017) Incorporating pleiotropic quantitative trait loci in dissection of complex traits: seed yield in rapeseed as an example. Theor Appl Genet 130:1569–1585

Manzanares-Dauleux MJ, Delourme R, Baron F, Thomas G (2000) Mapping of one major gene and of QTLs involved in resistance to clubroot in *Brassica napus*. Theor Appl Genet 101:885–891

Manzanares-Dauleux MJ, Delourme R, Glory P, Giboulot A, Thomas G (2003) Mapping QTLs and major resistance genes to clubroot (*Plasmodiophora brassicae*) in *Brassica napus*. In: Proceedings of 11th international rapesesed congress, Copenhagen, Denmark, pp 105–107

Marwede V, Gul MK, Becker HC, Ecke W (2005) Mapping of QTL controlling tocopherol content in winter oilseed rape. Plant Breed 124:20–26

Mei DS, Wang HZ, Hu Q, Li YD, Xu YS, Li YC (2009) QTL analysis on plant height and flowering time in *Brassica napus*. Plant Breed 128:458–465

Miersch S, Gertz A, Breuer F, Schierholt A, Becker HC (2016) Influence of the semi-dwarf growth type on seed yield and agronomic parameters at low and high nitrogen fertilization in winter oilseed rape. Crop Sci 56:1573–1585

Miro B (2010) Identification of traits for nitrogen use efficiency in oilseed rape (*Brassica napus* L.). Ph.D. thesis, University of Newcastle, 280 pp

Na A, Bink MC, Dieleman JA, Magán JJ, Wubs AM, Palloix A, van Eeuwijk FA (2013) Multi-trait and multi-environment QTL analyses of yield and a set of physiological traits in pepper. Theor Appl Genet 126:2597–2625

Nelson MN, Rajasekaran R, Smith A, Chen S, Beeck CP, Siddique KHM, Cowling WA (2014) Quantitative trait loci for thermal time to flowering and photoperiod responsiveness discovered in summer annual-type *Brassica napus* L. PLoS ONE 9:e102611

Nesi N, Delourme R, Bregeon M, Falentin C, Renard M (2008) Genetic and molecular approaches to improve nutritional value of *Brassica napus* L. seed. C R Biol 331:763–771

Obermeier C, Hossain MA, Snowdon R, Knuefer J, von Tiedemann A, Friedt W (2013) Genetic analysis of phenylpropanoid metabolites associated with resistance against *Verticillium longisporum* in *Brassica napus*. Mol Breed 31:347–361

Osborn TC, Kole C, Parkin IAP, Sharpe AG, Kuiper M, Lydiate DJ, Trick M (1997) Comparison of flowering time genes in *Brassica rapa*, *B. napus* and *Arabidopsis thaliana*. Genetics 146:1129

Oshima M, Handa H (2012) The identification of quantitative trait loci that control the paternal inheritance of a mitochondrial plasmid in rapeseed (*Brassica napus* L.). Gen Genet Syst 87:19–27

Parkin IAP (2011) Chasing ghosts: Comparative mapping in the Brassicaceae. In: Schmidt R, Bancroft I (eds), Genetics and genomics of the Brassicaceae, plant genetics and genomics: crop and models 9 (Chapter 5). Springer Science+Business Media, LLC, pp 153–170

Parkin IAP, Sharpe AG, Keith DJ, Lydiate DJ (1995) Identification of the A and C genomes of amphidiploids *Brassica napus* (oilseed rape). Genome 38:1122–1131

Parkin IAP, Gulden SM, Sharpe AG, Lukens L, Trick M, Osborn TC, Lydiate DJ (2005) Segmental structure of the *Brassica napus* genome based on comparative analysis with *Arabidopsis thaliana*. Genetics 171:765–781

Parkin IAP, Koh C, Tang H, Robinson SJ, Kagale S, Clarke WE et al (2014) Transcriptome and methylome profiling reveals relics of genome dominance in the mesopolyploid *Brassica oleracea*. Genome Biol 15: R77

Paul S, Datta SK, Datta K (2015) miRNA regulation of nutrient homeostasis in plants. Front Plant Sci 6:232

Pilet ML, Delourme R, Foisset N, Renard M (1998a) Identification of loci contributing to quantitative field resistance to blackleg disease, causal agent *Leptosphaeria maculans* (Desm) Ces et de Not, in winter rapeseed (*Brassica napus* L.). Theor Appl Genet 96:23–30

Pilet ML, Delourme R, Foisset N, Renard M (1998b) Identification of QTL involved in field resistance to light leaf spot (*Pyrenopeziza brassicae*) and blackleg resistance (*Leptosphaeria maculans*) in winter rapeseed (*Brassica napus* L.). Theor Appl Genet 97:398–406

Pilet ML, Duplan G, Archipiano M, Barret P, Baron C, Horvais R, Tanguy X, Lucas MO, Renard M, Delourme R (2001) Stability of QTL for field resistance to blackleg across two genetic backgrounds in rapeseed. Crop Sci 41:197–205

Piquemal J, Cinquin E, Couton F, Rondeau C, Seignoret E, Doucet I, Perret D, Villeger MJ, Vincourt P, Blanchard P (2005) Construction of an oilseed rape (*Brassica napus* L.) genetic map with SSR markers. Theor Appl Genet 111:1514–1523

Qi L, Mao L, Sun C, Pu Y, Fu T, Ma C, Shen J, Tu J, Yi B, Wen J (2014) Interpreting the genetic basis of silique traits in *Brassica napus* using a joint QTL network. Plant Breed 133:52–60

Qiu D, Morgan C, Shi J, Long Y, Liu J, Li R, Zhuang X, Wang Y, Tan X, Dietrich E, Weihmann T, Everett C, Vanstraelen S, Beckett P, Fraser F, Trick M, Barnes S, Wilmer J, Schmidt R, Li J, Li D, Meng J, Bancroft I (2006) A comparative linkage map of oilseed rape and its use for QTL analysis of seed oil and erucic acid content. Theor Appl Genet 114:67–80

Qu C, Hasan M, Lu K, Liu LZ, Zhang K, Fu FY, Wang M, Liu SY, Bu HD, Wang R, Xu XF, Chen L, Li JN (2015) Identification of QTL for seed coat colour and oil content in *Brassica napus* by association mapping using SSR markers. Can J Plant Sci 95:387–395

Quijada PA, Udall JA, Lambert B, Osborn TC (2006) Quantitative trait analysis of seed yield and other complex traits in hybrid spring rapeseed (*Brassica napus* L.): 1 Identification of genomic regions from winter germplasm. Theor Appl Genet 113:549–561

Raboanatahiry N, Chao H, Guo L, Gan J, Xiang J, Yan M, Zhang L, Yu L, Li M (2017) Synteny analysis of genes and distribution of loci controlling oil content and fatty acid profile based on QTL alignment map in *Brassica napus*. BMC Genom 18:776

Radoev M, Becker HC, Ecke W (2008) Genetic analysis of heterosis for yield and yield components in rapeseed (*Brassica napus* L.) by quantitative trait locus mapping. Genetics 179:1547–1558

Rahman H, Bennett RA, Kebede B (2017) Mapping of days to flower and seed yield in spring oilseed *Brassica napus* carrying genome content introgressed from *Brassica oleracea*. Mol Breed 37:5

Rajwanshi R, Chakraborty S, Jayanandi K, Deb B, Lightfoot DA (2014) Orthologous plant microRNAs: microregulators with great potential for improving stress tolerance in plants. Theor Appl Genet 127:2525–2543

Raman R, Taylor B, Marcroft S, Stiller J, Eckermann P, Coombes N, Rehman A, Lindbeck K, Luckett D, Wratten N, Batley J, Edwards D, Wang X, Raman H (2012) Molecular mapping of qualitative and quantitative loci for resistance to *Leptosphaeria maculans* causing blackleg disease in canola (*Brassica napus* L.). Theor Appl Genet 125:405–418

Raman H, Raman R, Eckermann P, Coombes N, Manoli S, Zou X, Edwards D, Meng J, Prangnell R, Stiller J, Batley J, Luckett D, Wratten N, Dennis E (2013) Genetic and physical mapping of flowering time loci in canola (*Brassica napus* L.). Theor Appl Genet 126:119–132

Raman H, Dalton-Morgan J, Diffey S, Raman R, Alamery S, Edwards D, Batley J (2014a) SNP markers-based map construction and genome-wide linkage analysis in *Brassica napus*. Plant Biotech J 12:851–860

Raman H, Raman R, Kilian A, Detering F, Carling J, Coombes N, Diffey S, Kadkol G, Edwards D, McCully M, Ruperao P, Parkin IAP, Batley J, Luckett DJ, Wratten N (2014b) Genome-wide delineation of natural variation for pod shatter resistance in *Brassica napus*. PLoS ONE 9:e101673

Raman R, Diffey S, Carling J, Cowley RB, Kilian A, Luckett DJ, Raman H (2016a) Quantitative genetic analysis of grain yield in an Australian *Brassica napus* doubled-haploid population. Crop Pasture Sci 67:298–307

Raman H, Raman R, Coombes N, Song J, Diffey S, Kilian A, Lindbeck K, Barbulescu DM, Batley J, Edwards D, Salisbury PA, Marcroft S (2016b) Genome-wide association study identifies new loci for resistance to *Leptosphaeria maculans* in Canola. Front Plant Sci 7:1513

Raman H, Raman R, McVittie B, Orchard B, Qiu Y, Delourme R (2017) A major locus for manganese tolerance maps on chromosome A09 in a doubled haploid population of *Brassica napus* L. Front Plant Sci 8:1952

Rezaeizad A, Wittkop B, Snowdon R, Hasan M, Mohammadi V, Zali A, Friedt W (2011) Identification of QTLs for phenolic compounds in oilseed rape (*Brassica napus* L.) by association mapping using SSR markers. Euphytica 177:335–342

Rimmer SR (2006) Resistance genes to *Leptosphaeria maculans* in *Brassica napus*. Can J Plant Pathol 28: S288–S297

Rousseau-Gueutin M, Morice J, Coriton O, Huteau V, Trotoux G, Negre S, Falentin C et al (2017) The impact of open pollination on the structural evolutionary dynamics Allotetraploid *Brassica napus* L. Genes Genomes Genet 7:705–717

Rygulla W, Snowdon RJ, Friedt W, Happstadius I, Cheung WY, Chen D (2008) Identification of quantitative trait loci for resistance against *Verticillium longisporum* in oilseed rape (*Brassica napus*). Phytopathol 98:215–221

Schatzki J, Schoo B, Ecke W, Herrfurth C, Feussner I, Becker HC, Moellers C (2013) Mapping of QTL for seed dormancy in a winter oilseed rape doubled haploid population. Theor Appl Genet 126:2405–2415

Schatzki J, Ecke W, Becker HC, Moellers C (2014) Mapping of QTL for the seed storage proteins cruciferin and napin in a winter oilseed rape doubled haploid population and their inheritance in relation to other seed traits. Theor Appl Genet 127:1213–1222

Schierholt A, Rücker B, Becker HC (2001) Inheritance of high oleic acid mutations in winter oilseed rape (*Brassica napus* L.). Crop Sci 41:1444–1449

Schiessl S, Samans B, Hüttel B, Reinhard R, Snowdon RJ (2014) Capturing sequence variation among flowering-time regulatory gene homologs in the allopolyploid crop species *Brassica napus*. Front Plant Sci 5:404

Schiessl S, Huettel B, Kuehn D, Reinhardt R, Snowdon R (2017) Post-polyploidisation morphotype diversification associates with gene copy number variation. Sci Rep 7:41845

Schranz ME, Lysak MA, Mitchell-Olds T (2006) The ABC's of comparative genomics in the Brassicaceae: building blocks of crucifer genomes. Trends Plant Sci 11:535–542

Sharpe AG, Parkin IAP, Keith DJ, Lydiate DJ (1995) Frequent non-reciprocal translocations in the amphiploid genome of oilseed rape (*Brassica napus*). Genome 38:1112–1121

Shen JX, Fu TD, Yang GS, Tu JX, Ma CZ (2006) Prediction of heterosis using QTLs for yield traits in rapeseed (*Brassica napus* L.). Euphytica 151:165–171

Shen D, Suhrkamp I, Wang Y, Liu S, Menkhaus J, Verreet JA, Fan L, Cai D (2014) Identification and characterization of microRNAs in oilseed rape (*Brassica napus*) responsive to infection with the pathogenic fungus *Verticillium longisporum* using Brassica AA (*Brassica rapa*) and CC (*Brassica oleracea*) as reference genomes. New Phytol 204:577–594

Shen YS, Yang Y, Xu ES, Ge XH, Xiang Y, Li ZY (2018) Novel and major QTL for branch angle detected by using DH population from an exotic introgression in rapeseed (*Brassica napus* L.). Theor Appl Genet 131:67–78

Shi J, Li R, Qiu D, Jiang C, Long Y, Morgan C, Bancroft I, Zhao J, Meng J, Shi JQ, Li RY, Qiu D, Jiang CC, Long Y, Zhao JY, Meng JL (2009) Unraveling the complex trait of crop yield with quantitative trait loci mapping in *Brassica napus*. Genetics 182:851–861

Shi J, Li R, Zou J, Long Y, Meng J, Shi JQ, Li RY, Zou J, Long Y, Meng JL (2011) A dynamic and complex network regulates the heterosis of yield-correlated traits in rapeseed (*Brassica napus* L.). PLoS ONE 6: e21645–e21645

Shi L, Yang J, Liu J, Li R, Long Y, Xu F, Meng J (2012) Identification of quantitative trait loci associated with low boron stress that regulate root and shoot growth in *Brassica napus* seedlings. Mol Breed 30:393–406

Shi T, Li R, Zhao Z, Ding G, Long Y, Meng J, Xu F, Shi L (2013a) QTL for yield traits and their association with functional genes in response to phosphorus deficiency in *Brassica napus*. PLoS ONE 8:e54559

Shi L, Shi T, Broadley MR, White PJ, Long Y, Meng J, Xu F, Hammond JP (2013b) High-throughput root phenotyping screens identify genetic loci associated with root architectural traits in *Brassica napus* under contrasting phosphate availabilities. Ann Bot 112:381–389

Shi J, Zhan J, Yang Y, Ye J, Huang S, Li R, Wang X, Liu G, Wang H (2015) Linkage and regional association analysis reveal two new tightly-linked major-QTLs for pod number and seed number per pod in rapeseed (*Brassica napus* L.). Sci Rep 5:14481

Smooker AM, Wells R, Morgan C, Beaudoin F, Cho K, Fraser F, Bancroft I (2011) The identification and mapping of candidate genes and QTL involved in the fatty acid desaturation pathway in *Brassica napus*. Theor Appl Genet 122:1075–1090

Snowdon RJ, Iniguez-Luy FL (2012) Potential to improve oilseed rape and canola breeding in the genomics era. Plant Breed 131:351–360

Snowdon RJ, Wittkop B, Rezaidad A, Hasan M, Lipsa F, Stein A, Friedt W (2010) Regional association analysis delineates a sequenced chromosome region influencing antinutritive seed meal compounds in oilseed rape. Genome 53:917–928

Stein A, Wittkop B, Liu L, Obermeier C, Friedt W, Snowdon RJ (2013) Dissection of a major QTL for seed colour and fibre content in *Brassica napus* reveals colocalization with candidate genes for phenylpropanoid biosynthesis and flavonoid deposition. Plant Breed 132:382–389

Stein A, Coriton O, Rousseau-Gueutin M, Samans B, Schiessl SV, Obermeier C, Parkin IAP, Chèvre AM, Snowdon RJ (2017) Mapping of homoeologous chromosome exchanges influencing quantitative trait variation in *Brassica napus*. Plant Biotech J 15:1478–1489

Sun Z, Wang Z, Tu J, Zhang J, Yu F, McVetty PBE, Li G (2007) An ultradense genetic recombination map for *Brassica napus*, consisting of 13551 SRAP markers. Theor Appl Genet 114:1305–1317

Sun M, Hua W, Liu J, Huang S, Wang X, Liu G, Wang H (2012) Design of new genome- and gene-sourced primers and identification of QTL for seed oil content in a specially high-oil *Brassica napus* cultivar. PLoS ONE 7:e47037

Sun F, Liu J, Hua W, Sun X, Wang X, Wang H (2016) Identification of stable QTLs for seed oil content by combined linkage and association mapping in *Brassica napus*. Plant Sci 252:388–399

Suwabe K, Morgan C, Bancroft I (2008) Integration of Brassica A genome genetic linkage map between *Brassica napus* and *B. rapa*. Genome 51:169–176

Teh L, Möllers C (2016) Genetic variation and inheritance of phytosterol and oil content in a doubled haploid population derived from the winter oilseed rape Sansibar × Oase cross. Theor Appl Genet 129:181–199

Toroser D, Thormann CE, Osborn TC, Mithen R (1995) RFLP mapping of quantitative trait loci controlling seed aliphatic glucosinolate content in oilseed rape (*Brassica napus* L.). Theor Appl Genet 91:802–808

Town CD, Cheung F, Maiti R, Crabtree J, Haas BJ, Wortman JR, Hine EE, Althoff R, Arbogast TS, Tallon LJ, Vigouroux M, Trick M, Bancroft I (2006) Comparative genomics of *Brassica oleracea* and *Arabidopsis thaliana* reveal gene Loss, fragmentation, and dispersal after polyploidy. Plant Cell 18:1348–1359

Trick M, Long Y, Meng J, Bancroft I (2009) Single nucleotide polymorphism (SNP) discovery in the polyploid *Brassica napus* using Solexa transcriptome sequencing. Plant Biotech J 7:334–346

Udall JA, Quijada PA, Lambert B, Osborn TC (2006) Quantitative trait analysis of seed yield and other complex traits in hybrid spring rapeseed (*Brassica napus* L.): 2 identification of alleles from unadapted germplasm. Theor Appl Genet 113:597–609

Uzunova M, Ecke W, Weissleder K, Robbelen G (1995) Mapping the genome of rapeseed (*Brassica napus* L.) 1: construction of an RFLP linkage map and localization of QTLs for seed glucosinolate content. Theor Appl Genet 90:194–204

Verbyla AP, Cavanagh CR, Verbyla KL (2014) Whole-genome analysis of multienvironment or multitrait QTL in MAGIC. G3 (Bethesda) 18:1569–1584

Verma SS, Rahman MH, Deyholos MK, Basu U, Kav NNV (2014) Differential expression of miRNAs in *Brassica napus* root following infection with *Plasmodiophora brassicae*. PLoS ONE 9:e86648

Wan HP, Chen LL, Guo JB, Li Q, Wen J, Yi B, Ma CZ, Tu JX, Fu TD, Shen JX (2017) Genome-wide association study reveals the genetic architecture underlying salt tolerance-related traits in rapeseed (*Brassica napus* L.). Front Plant Sci 8:593

Wang J, Long Y, Wu B, Liu J, Jiang C, Shi L, Zhao J, King GJ, Meng J (2009) The evolution of *Brassica napus FLOWERING LOCUST* paralogues in the context of inverted chromosomal duplication blocks. BMC Evol Biol 9:271

Wang J, Lydiate DJ, Parkin IAP, Falentin C, Delourme R, Carion PWC, King GJ (2011a) Integration of linkage maps for the Amphidiploid *Brassica napus* and comparative mapping with Arabidopsis and *Brassica rapa*. BMC Genom 12:101

Wang N, Qian W, Suppanz I, Wei L, Mao B, Long Y, Meng J, Mueller AE, Jung C (2011b) Flowering time variation in oilseed rape (*Brassica napus* L.) is associated with allelic variation in the *FRIGIDA* homologue *BnaAFRIa*. J Exp Bot 62:5641–5658

Wang X, Wang H, Wang J, Sun R, Wu J, Liu S et al (2011c) The genome of the mesopolyploid crop species *Brassica rapa*. Nature Genet 43:1035–U157

Wang X, Zhang C, Li L, Fritsche S, Endrigkeit J, Zhang W, Long Y, Jung C, Meng J (2012) Unraveling the genetic basis of seed tocopherol content and composition in rapeseed (*Brassica napus* L.). PLoS ONE 7:e50038

Wang J, Chen H, Chen Y, Zhang X, Xu H, Zhao J, Wang JB, Chen HC, Chen YB, Zhang XY, Xu HM, Zhao JY (2013a) Molecular dissection of oil content with respect to fatty acid compositions by conditional QTL analysis in oilseed rape. J Zhejiang Univ 39:504–512

Wang X, Wang H, Long Y, Li D, Yin Y, Tian J, Chen L, Liu L, Zhao W, Zhao Y, Yu L, Li M (2013b) Identification of QTLs associated with oil content in a high-oil *Brassica napus* cultivar and construction of a high-density consensus map for QTLs comparison in *B. napus*. PLoS ONE 8:e80569

Wang XD, Yu KJ, Li HG, Peng Q, Chen F, Zhang W, Chen S, Maolong HL, Zhang JF (2015a) High-density SNP map construction and QTL identification for the apetalous character in *Brassica napus* L. Front Plant Sci 6:1164

Wang XD, Wang H, Long Y, Liu LZ, Zhao YJ, Tian JH, Zhao WG, Li BJ, Chen L, Chao HB, Li MT (2015b) Dynamic and comparative QTL analysis for plant height in different developmental stages of *Brassica napus* L. Theor Appl Genet 128:1175–1192

Wang X, Long Y, Yin Y, Zhang C, Gan L, Liu L, Yu L, Meng J, Li M (2015c) New insights into the genetic networks affecting seed fatty acid concentrations in *Brassica napus*. BMC Plant Biol 15:91

Wang XD, Chen L, Wang AN, Wang H, Tian JH, Zhao XP, Chao HB, Zhao YJ, Zhao WG, Xiang J, Gan JP, Li MT (2016a) Quantitative trait loci analysis and genome-wide comparison for silique related traits in *Brassica napus*. BMC Plant Biol 16:71

Wang H, Cheng HT, Wang WX, Liu J, Hao MY, Mei DS, Zhou RJ, Fu L, Hu Q (2016b) Identification of BnaYUCCA6 as a candidate gene for branch angle in *Brassica napus* by QTL-seq. Sci Rep 6:38493

Wang YK, Chen WJ, Chu P, Wan SB, Yang M, Wang MM, Guan RZ (2016c) Mapping a major QTL responsible for dwarf architecture in *Brassica napus* using a single-nucleotide polymorphism marker approach. BMC Plant Biol 16:178

Wang XH, Chen YL, Thomas CL, Ding GD, Xu P, Shi DX, Grandke F, Jin K, Cai HM, Xu FS, Yi B, Broadley MR, Shi L (2017a) Genetic variants associated with the root system architecture of oilseed rape (*Brassica napus* L.) under contrasting phosphate supply. DNA Res 24:407–417

Wang J, Dun XL, Shi JQ, Wang XF, Liu GH, Wang HZ (2017b) Genetic dissection of root morphological traits related to nitrogen use efficiency in *Brassica napus* L. under two contrasting nitrogen conditions. Front Plant Sci 8:1709

Wang Z, Qiao Y, Zhang J, Shi W, Zhang J (2017c) Genome wide identification of microRNAs involved in fatty acid and lipid metabolism of *Brassica napus* by small RNA and degradome sequencing. Gene 619:61–70

Wang B, Wu Z, Li Z, Zhang Q, Hu J, Xiao Y, Cai D, Wu J, King GJ, Li H, Liu K (2017d) Dissection of the genetic architecture of three seed-quality traits and consequences for breeding in *Brassica napus*. Plant Biotechnol J. https://doi.org/10.1111/pbi.12873

Wei D, Mei J, Fu Y, Disi JO, Li J, Qian W (2014) Quantitative trait loci analyses for resistance to *Sclerotinia sclerotiorum* and flowering time in *Brassica napus*. Mol Breed 34:1797–1804

Wei L, Jian H, Lu K, Filardo F, Yin N, Liu L, Qu C, Li W, Du H, Li J (2016) Genome-wide association analysis and differential expression analysis of resistance to Sclerotinia stem rot in *Brassica napus*. Plant Biotech J 14:1368–1380

Wen YC, Zhang SF, Yi B, Wen J, Wang JP, Zhu JC, He JP, Cao JH (2012) Identification of QTLs involved in pod-shatter resistance in *Brassica napus* L. Crop Past Sci 63:1082–1089

Wen J, Xu JF, Long Y, Wu JG, Xu HM, Meng JL, Shi CH (2016) QTL mapping based on the embryo and maternal genetic systems for non-essential amino acids in rapeseed (*Brassica napus* L.) meal. J Sci Food Agric 96:465–473

Werner S, Diederichsen E, Frauen M, Schondelmaier J, Jung C (2008) Genetic mapping of clubroot resistance genes in oilseed rape. Theor Appl Genet 116:363–372

Wittkop B, Snowdon RJ, Friedt W (2009) Status and perspectives of breeding for enhanced yield and quality of oilseed crops for Europe. Euphytica 170:131–140

Wu JG, Shi CH, Zhang HZ (2006) Partitioning genetic effects due to embryo, cytoplasm and maternal parent for oil content in oilseed rape (*Brassica napus* L.). Genet Mol Biol 29:533–538

Wu G, Wu Y, Xiao L, Li X, Lu C (2008) Zero erucic acid trait of rapeseed (*Brassica napus* L.) results from a deletion of four base pairs in the *fatty acid elongase 1* gene. Theor Appl Genet 116:491–499

Wu J, Cai G, Tu J, Li L, Liu S, Luo X, Zhou L, Fan C, Zhou Y (2013) Identification of QTLs for resistance to sclerotinia stem rot and *BnaCIGMT5a* as a candidate gene of the major resistant QTL SRC6 in *Brassica napus*. PLoS ONE 8:e67740

Wu J, Zhao Q, Liu S, Shahid M, Lan L, Cai G, Zhang C, Fan C, Wang Y, Zhou Y (2016) Genome-wide association study identifies new loci for resistance to Sclerotinia stem rot in *Brassica napus*. Front Plant Sci 7:1418

Würschum T, Liu W, Maurer HP, Abel S, Reif JC (2012) Dissecting the genetic architecture of agronomic traits in multiple segregating populations in rapeseed (*Brassica napus* L.). Theor Appl Genet 124:153–161

Würschum T, Maurer HP, Dreyer F, Reif JC (2013) Effect of inter- and intragenic epistasis on the heritability of oil content in rapeseed (*Brassica napus* L.). Theor Appl Genet 126:435–441

Würschum T, Abel S, Zhao Y (2014) Potential of genomic selection in rapeseed (*Brassica napus* L.) breeding. Plant Breed 133:45–51

Xiao Y, Cai D, Yang W, Ye W, Younas M, Wu J, Liu K (2012) Genetic structure and linkage disequilibrium pattern of a rapeseed (*Brassica napus* L.) association mapping panel revealed by microsatellites. Theor Appl Genet 125:437–447

Xu FS, Wang YH, Meng J (2001) Mapping boron efficiency gene(s) in *Brassica napus* using RFLP and AFLP markers. Plant Breed 120:319–324

Xu JF, Long Y, Wu JG, Xu HM, Wen J, Meng JL, Shi CH (2015a) QTL mapping and analysis of the embryo and maternal plant for three limiting amino acids in rapeseed meal. Eur Food Res Technol 240:147–158

Xu JF, Long Y, Wu JG, Xu HM, Zhao ZG, Wen J, Meng JL, Shi CH (2015b) QTL identification on two genetic systems for rapeseed glucosinolate and erucic acid contents over two seasons. Euphytica 205:647–657

Xu L, Hu K, Zhang Z, Guan C, Chen S, Hua W, Li J, Wen J, Yi B, Shen J, Ma C, Tu J, Fu T (2016) Genome-wide association study reveals the genetic architecture of flowering time in rapeseed (*Brassica napus* L.). DNA Res 23:43–52

Yan XY, Li JN, Fu FY, Jin MY, Chen L, Liu LZ (2009) Co-location of seed oil content, seed hull content and seed coat color QTL in three different environments in *Brassica napus* L. Euphytica 170:355–364

Yang M, Ding G, Shi L, Feng J, Xu F, Meng J (2010) Quantitative trait loci for root morphology in response to low phosphorus stress in *Brassica napus*. Theor Appl Genet 121:181–193

Yang P, Shu C, Chen L, Xu J, Wu J, Liu K (2012) Identification of a major QTL for silique length and seed weight in oilseed rape (*Brassica napus* L.). Theor Appl Genet 125:285–296

Yan XY, Qu CM, Li JN, Chen L, Liu LZ (2015) QTL analysis of leaf photosynthesis rate and related physiological traits in *Brassica napus*. J Integ Agric 14:1261–1268

Yang YH, Shi JQ, Wang XF, Liu GH, Wang HZ (2016) Genetic architecture and mechanism of seed number per pod in rapeseed: elucidated through linkage and near-isogenic line analysis. Sci Rep 6:24124

Yang Y, Shen YS, Li SD, Ge XH, Li ZY (2017) High density linkage map construction and QTL detection for three silique-related traits in *Orychophragmus violaceus* derived *Brassica napus* population. Front Plant Sci 8:1512

Ye J, Yang YH, Chen B, Shi JQ, Luo MZ, Zhan JP, Wang XF, Liu GH, Wang HZ (2017) An integrated analysis of QTL mapping and RNA sequencing provides further insights and promising candidates for pod number variation in rapeseed (*Brassica napus* L.). BMC Genom 18:71

Yin X, Yi B, Chen W, Zhang W, Tu J, Fernando WGD, Fu T (2010) Mapping of QTLs detected in a *Brassica napus* DH population for resistance to *Sclerotinia sclerotiorum* in multiple environments. Euphytica 173:25–35

Yong HY, Wang C, Bancroft I, Li F, Wu X, Kitashiba H, Nishio T (2015) Identification of a gene controlling variation in the salt tolerance of rapeseed (*Brassica napus* L.). Planta 242:313–326

Yu F, Lydiate DJ, Rimmer SR (2005) Identification of two novel genes for blackleg resistance in *Brassica napus*. Theor Appl Genet 110:969–979

Yu F, Lydiate DJ, Rimmer SR (2008) Identification and mapping of a third blackleg resistance locus in *Brassica napus* derived from *B. rapa* subsp *sylvestris*. Genome 51:64–72

Yu F, Gugel RK, Kutcher HR, Peng G, Rimmer SR (2013) Identification and mapping of a novel blackleg resistance locus *LepR4* in the progenies from *Brassica napus* × *B. rapa* subsp *sylvestris*. Theor Appl Genet 126:307–315

Zhang Y, Li X, Chen W, Yi B, Wen J, Shen J, Ma C, Chen B, Tu J, Fu T (2011a) Identification of two major QTL for yellow seed color in two crosses of resynthesized *Brassica napus* line No 2127-17. Mol Breed 28:335–342

Zhang L, Yang G, Liu P, Hong D, Li S, He Q (2011b) Genetic and correlation analysis of silique-traits in *Brassica napus* L. by quantitative trait locus mapping. Theor Appl Genet 122:21–31

Zhang D, Hua Y, Wang X, Zhao H, Shi L, Xu F (2014) A high-density genetic map identifies a novel major QTL for boron efficiency in oilseed rape (*Brassica napus* L.). PLoS ONE 9:e112089

Zhang Y, Thomas CL, Xiang JX, Long Y, Wang XH, Zou J, Luo ZL, Ding GD, Cai HM, Graham NS, Hammond JP, King GJ, White PJ, Xu FS, Broadley MR, Shi L, Meng JL (2016) QTL meta-analysis of root traits in *Brassica napus* under contrasting phosphorus supply in two growth systems. Sci Rep 6:33113

Zhao J, Meng J (2003) Detection of loci controlling seed glucosinolate content and their association with Sclerotinia resistance in *Brassica napus*. Plant Breed 122:19–23

Zhao JY, Becker HC, Zhang DQ, Zhang YF, Ecke W (2005) Oil content in a European × Chinese rapeseed population: QTL with additive and epistatic effects and their genotype-environment interactions. Crop Sci 45:51–59

Zhao JY, Becker HC, Zhang DQ, Zhang YF, Ecke W (2006a) Conditional QTL mapping of oil content in rapeseed with respect to protein content and traits related to plant development and grain yield. Theor Appl Genet 113:33–38

Zhao JW, Udall JA, Quijada PA, Grau CR, Meng JL, Osborn TC (2006b) Quantitative trait loci for resistance to *Sclerotinia sclerotiorum* and its association with a homeologous non-reciprocal transposition in *Brassica napus* L. Theor Appl Genet 112:509–516

Zhao H, Liu J, Shi L, Xu F, Wang Y (2010) Development of boron-efficient near isogenic lines of *Brassica napus* and their response to low boron stress at seedling stage. Genetika 46:66–72

Zhao J, Huang J, Chen F, Xu F, Ni X, Xu H, Wang Y, Jiang C, Wang H, Xu A, Huang R, Li D, Meng J (2012a) Molecular mapping of Arabidopsis thaliana lipid-related orthologous genes in *Brassica napus*. Theor Appl Genet 124:407–421

Zhao Z, Wu L, Nian F, Ding G, Shi T, Zhang D, Shi L, Xu F, Meng J (2012b) Dissecting quantitative trait loci for boron efficiency across multiple environments in *Brassica napus*. PLoS ONE 7:e45215

Zhao XW, Li B, Zhang K, Hu KN, Yi B, Wen J, Ma CZ, Shen JX, Fu TD, Tu JX (2016a) Breeding signature of combining ability improvement revealed by a genomic variation map from recurrent selection population in *Brassica napus*. Sci Rep 6:29553

Zhao WG, Wang XD, Wang H, Tian JH, Li BJ, Chen L, Chao HB, Long Y, Xiang J, Gan JP, Liang WS, Li MT (2016b) Genome-wide identification of QTL for seed yield and yield-related traits and construction of a high-density consensus map for QTL comparison in *Brassica napus*. Front Plant Sci 7:17

Zhou ZS, Song JB, Yang ZM (2012) Genome-wide identification of *Brassica napus* microRNA and their targets in response to cadmium. J ExpBot 63:4597–4613

Zhou QH, Fu DH, Mason AS, Zeng YJ, Zhao CX, Huang YJ (2014) In silico integration of quantitative trait loci for seed yield and yield-related traits in *Brassica napus*. Mol Breed 33:881–894

Zou J, Jiang C, Cao Z, Li R, Long Y, Chen S, Meng J (2010) Association mapping of seed oil content in *Brassica napus* and comparison with quantitative trait loci identified from linkage mapping. Genome 53:908–916

Zou J, Zhao Y, Liu P, Shi L, Wang X, Wang M, Meng J, Reif JC (2016) Seed quality traits can be predicted with high accuracy in *Brassica napus* using genomic data. PLoS ONE 11:e0166624

Deciphering Genome Organization of the Polyploid *Brassica napus*

4

Fengming Sun, Boulos Chalhoub, Shengyi Liu and Wei Hua

Abstract

Allopolyploidy species oilseed rapa ($A_nA_nC_nC_n$; *Brassica napus* L.) were formed in a relatively short time (less than 10,000 years) by hybridization between *Brassica rapa* (A_r) and *Brassica oleracea* (C_o). The *Brassica* species provide an opportunity to study the evolution of genome organization over a short timescales. It has been proved that progenitor A and C genomes are highly intact in *B. napus*. Based on the previous study, small-scale chromosomal changes, like homeologous exchanges (HEs), were found to be happened between two subgenomes using BAC sequencing or physical mapping methods. With the development of sequencing technology, the comparative analysis of the genome-wide level becomes feasible. Recently, the genome assemblies of two parental species A_r and C_o were both completed. Therefore, in this chapter, we mainly discuss the assembly and annotation of the genome of a winter phonotype of 'Darmor-*bz*' and a semi-winter phonotype 'ZS11', and also investigate the subtle changes of genomic structures, including segmental and microstructure (gene order and content) changes, between the *B. napus* and its parental species.

4.1 Introduction

Polyploidy plants are widespread in angiosperms and reveal some phenotypic advantages during genome evolution (Leitch and Bennett 1997; Wendel 2000). Understanding the mechanisms involved in the structural and functional evolution of polyploidy genomes is greatly important to plant biology. Oilseed rapa (*Brassica napus* L.; $A_nA_nC_nC_n$, $2n = 38$) was a recent allopolyploid species derived from hybridization between two close *Brassiceae* species, *Brassica rapa* (A_rA_r, $2n = 20$) and *Brassica oleracea* (C_oC_o, $2n = 18$) in less than 10,000 years ago (Rana et al. 2004). Both two parental species underwent several events of whole-genome duplication, including two triplications and two diploidizations ($\times$ 3 $\times$ 2 $\times$ 2 $\times$ 3). After a combination

F. Sun · S. Liu · W. Hua (✉)
Key Laboratory of Biology and Genetic Improvement of Oil Crops, the Ministry of Agriculture, Oil Crops Research Institute of the Chinese Academy of Agricultural Sciences, Wuhan 430062, China
e-mail: huawei@oilcrops.cn

B. Chalhoub
Institut National de Recherche Agronomique (INRA)/Université d'Evry Val d'Essone, Unité de Recherche en Génomique Végétale, UMR1165, Organization and Evolution of Plant Genomes, 2 rue Gaston Crémieux, 91057 Evry, France

S. Liu et al. (eds.), *The Brassica napus Genome*, Compendium of Plant Genomes,
https://doi.org/10.1007/978-3-319-43694-4_4

of *B. rapa* and *B. oleracea* (× 3×2 × 2×3 2), *B. napus* has 72 × genome and more gene content than the origin of angiosperms. Within the *Brassicaceae* family, the *brassica* species are the most closely related to the *Arabidopsis*, and the divergence time was estimated in ~20 million years ago (Mya) (Yang et al. 1999). The progenitor species *B. rapa* and *B. oleracea* have been estimated to have diverged time of ~3.7 Mya (Inaba and Nishio 2002). The *Brassica* species have a relative short evolutionary time, which is meaningful for studying the genomic evolution and the organization after formation polyploidy during a relatively short time. It has been proved that the progenitor A and C genomes are essentially intact in *B. napus* using genetic mapping methods (Parkin et al. 1995). The organization of *B. napus* genome has undergone a series of segmental or genome-wide duplicates events, chromosomal rearrangement, gene fractionation. It has been proved that homeologous non-reciprocal transpositions within *B. napus* are great affect genetic changes and are correlated with qualitative changes in the expression of specific genes and with phenotypic variation (Gaeta et al. 2007). In microstructure level, physical mapping approaches show that natural *B. napus* appears to show relatively little change in gene content and order (Rana et al. 2004). Based on the previous study, genome organization (segmental variation, gene fractionation, or single-nucleotide variation) during polyploidization in genome-wide level has not been reported. Due to the limitation of genome assembly, many studies have been performed based on BAC clones sequencing or other technologies that have no ability to cover whole-genome level such as genetic or physical mapping (Rana et al. 2004; Parkin et al. 1995; Cheung et al. 2009; Park et al. 2005). Therefore, a relatively complete genome of *B. napus* is necessary to trace the chromosomal changes. In this chapter, we mainly discuss the assembly of whole genome of a winter type of 'Darmor-*bzh*' (Chalhoub 2014) and a semi-winter type of 'ZS11' (Sun et al. 2017), annotation of the genome, and the evolutionary processes shaping the structure of polyploid genomes over relatively short timescale.

4.2 Genome Assembly of *B. napus*

4.2.1 Background of Assembly for Allopolyploid

As the allopolyploid plant genomes are always large and contain highly homeologous subgenomes, high-quality assembly of genomes is still a challenging task. Many genomes of diploid animals and plants have been successfully assembled using whole-genome shotgun (WGS) approach based on the next-generation sequencing technology (NGS). Although NGS is a cost-effective approach to large-scale sequencing, which also can provide an opportunity to get access to perform whole-genome comparative studies, relatively short read length and insert size of WGS libraries makes it is difficult to assemble the complicated genome regions, like highly duplicated and highly heterozygous segments. Therefore, in order to conquer such deficiencies, combination methods of integrating different sequencing platforms have been applied to improve the quality of assembly, such as combination of WGS and the Sanger sequencing technology, combing of WGS and bacterial artificial chromosome (BAC)-end sequences technology. Previous study revealed that the quality of assembly is greatly improved by using combination method than only WGS. Based on WGS approach, a draft assembly representing 9.1-gigabase of total 16-gigabase genome of monocotyledonous allohexaploid *Triticum aestivum* (AABBDD) had been reported before assembly of *B. napus* (Chapman 2015). Due to the huge genome size and limitation of sequencing technology, the assembly of *T. aestivum* is dissatisfactory. Recently, the allotetraploid *Gossypium hirsutum* L. TM-1 (AADD) was successfully assembled by two different groups through integrating whole-genome shotgun reads and bacterial artificial chromosome (BAC)-end sequences, which obtained genome size of about 2.2 and 2.4 Gb, respectively (total estimated is ~2.25–2.43 Gb) (Li et al. 2015; Zhang et al. 2015). Although the combination is effective to assembly the highly duplicated subgenomes of polyploidy, it is still a time-consuming and money-consuming method.

4.2.2 Strategy for Sequencing and Assembly of 'Darmor-*bzh*'

The methods were designed by the 'Darmor-*bzh*' genome article. The sequencing of *B. napus* genome was expected to face great difficulties in differentiating the A_n and C_n homeologous subgenomes. Fortunately, *B. rapa* and *B. oleracea* have been successfully assembled through WGS methods, which provided an opportunity to distinguish two subgenomes, and also distinguish the duplicated segments resulted from genome triplication or mesoploidy of *B. napus* (Wang et al. 2011; Liu 2014; Parkin 2014). The homozygous *B. napus* genome of European winter oilseed cultivar 'Darmor-*bzh*' was selected to be sequenced (Chalhoub 2014). Different sequencing platforms were used in the process of assembly, including Sanger BAC-end sequencing (~7.8×), GS FLX Titanium 454 sequencing that included long reads of 700 bases (~21.2×), as well as the Illumina SBS technology (~53.9×). In addition, about 5 × 454 sequencing data of *B. rapa* ('Chiifu') or *B. oleracea* ('TO1000') were used to distinguish the A and C subgenomes. Finally, about 849.7 Mb with scaffold N50 size of 763.7 Kb, accounting 75% of estimated genome size of 1130 Mb, was obtained, and total 18,288 of 20,702 scaffolds were successfully assigned to either the A_n (8294) or the C_n (9984) subgenomes (Table 4.1).

In order to construct a combined genetic linkage map with a high quality, three populations of 'Darmor-bzh' × 'Yudal' (DY), 'Darmor' × 'Bristol' (DB), and 'Avisol' × 'Aburamasari' (AA) were used to develop single-nucleotide polymorphism (SNP) markers by using the Infinium 20K BeadChip (Illumina). For DY populations, a total of 5738 genetic bins were developed and covered 2807 cM. Correspondingly, the genetic maps of DB and AA populations contained 2350 and 2692 genetic bins, covered 1959 and 4048 cM, respectively (Chalhoub 2014). Taking the genetic map of DY population as reference, we next integrated the DB and AA map step by step using BioMercator V4.2 program (Arcade et al. 2004). Final consensus map contained 7287 bins and covered 2881 cM. Through allele sequence matching, 384 anchored scaffolds were aligned with 19 pseudochromosomes including A01–A10 A-subgenome and C01–C09 C-sungenome. As a result, 712.3 Mb (84%) of the genome was successfully anchored and the other unanchored scaffolds were also grouped based on marker alignment with genetic maps and orthologous alignment with parental genomes. The anchored C_n subgenome (525.8 Mb) is larger than the A_n subgenome (314.2 Mb) consistent with the size

Table 4.1 Comparison of the assembly results between 'ZS11' and 'Darmor-*bzh*'

	'ZS11'				'Darmor-bzh'			
	Contig		Scaffold		Contig		Scaffold	
Size (bp)	Size (bp)	Number	Size (bp)	Number	Size (bp)	Number	Size (bp)	Number
N90	10,208	23,955	163,802	1654	6338	22,223	29,731	2586
N80	17,270	17,188	275,878	1202	13,999	14,562	109,022	1043
N50	39,572	6846	602,215	495	38,893	5319	763,688	299
Longest	343,759	–	2,872,714	–	349,037	–	5,197,798	–
Total size	910,856,700	–	976,047,003	–	738,357,862	–	848,760,698	–

of assembled C_o genome of *B. oleracea* (540 Mb of total ~630 Mb) and the A_r genome of *B. rapa* (312 Mb of total ~530 Mb).

4.2.3 Strategy for Sequencing and Assembly of 'ZS11'

The methods were designed by the 'ZS11' genome article. For the semi-winter type *B. napus* cultivar 'ZS11' (Sun et al. 2017), genomic DNA was isolated using standard techniques and nine libraries with insert sized of 180 bp, 250 bp, 500 bp, 800 bp, 2 kb, 5 kb, 10 kb, 20 kb, or 40 kb were constructed. All of the libraries were sequenced using the Illumina HiSeq 2000 sequencing platform (Illumina). In total, 137 Gb of high-quality reads were retained for assembly. In order to improve the quality of the assembly, a BAC-to-BAC strategy was also applied in the process of assembly. In total, 59,904 BAC clones with an average insert size of 120 Kb were created for Illumina sequencing. For each BAC clone, two paired-end libraries with 250-bp or 500-bp inserts were constructed for the HiSeq 2000 sequencing platform. Each BAC clone was sequenced to >118-fold coverage, representing >851 Gb of data on average.

In general, a BAC-to-BAC sequencing strategy was mainly used to assemble the 'ZS11' genome, and the NGS data were used to improve the assembly. Each BAC was assembled using SOAPdenovo (Version2) (Li et al. 2010). All assembled scaffolds of WGS BAC sequences were then pooled for overlap-layout-consensus assembly. Identical sequences were merged, and redundant bases were filtered out. All super-contigs obtained from overlap-layout-consensus assembly were linked to final scaffolds using the WGS-based long-insert library reads (2–40 kb) step by step using SSPACE (Boetzer et al. 2011). To fill gaps (regions comprised of 'N' bases) within the scaffold, all WGS-based short-insert library reads were mapped to the scaffold. Short reads were locally assembled to fill the gaps. All scaffolds were then anchored to pseudochromosomes. Five genetic maps of *B. napus* were collected firstly, including 'ZS11' × '73290' (Z7) (Li et al. 2014), 'Darmor-*bzh*' × 'Yudal' (DY) (Chalhoub et al. 2014), 'GH06' × 'P174' (GP) (Liu et al. 2013), 'Tapidor' × 'Ningyou7' (TN) (Zhang et al. 2016), and 'M083' × '888-5' (M8), and then merged using ALLMAPS (Tang et al. 2015). Our strategy resulted in ~976 Mb of genomic sequence with a scaffold N50 of 602.22 kb and a contig N50 of 39.57 kb (Table 4.1). The anchored C_n subgenome (520.24 Mb) was larger than the A_n subgenome (334.74 Mb), which also coincided with the sizes of the assembled genomes of 'Darmor-*bzh*' (anchored with ordered and non-ordered, A_n: 314.2 Mb, C_n: 525.8 Mb) (Chalhoub et al. 2014), and 'Tapidor' (A_n: 246 Mb, C_n: 381 Mb) (Bayer et al. 2017). The BUSCO assessment of two assembles showed that 'ZS11' assembly has fewer missing and fragmented genes than that in 'Darmor-*bzh*' (Sun et al. 2017) (Table 4.1).

4.2.4 Genome Syntenic Analysis and Homeologous Exchanges in *B. napus*

A total of three types of *B. napus* have been sequenced by now, including two winter types of 'Darmor-bzh' and 'Tapidor' (Bayer et al. 2017), one semi-winter type of 'ZS11'. Three genomes revealed a high collinearity (Figs. 4.1, 4.2 and 4.3). For details, the 'Darmor-bzh' and 'ZS11' had more orthologous with 75,788 gene pairs, 'Darmor-*bzh*' and 'Tapidor' had 60,936 gene pairs, and 'Tapidor' and 'ZS11' had 58,534 gene pairs.

Homeologous exchanges (HEs) are frequently happened in two *B. napus* subgenomes and range in size from big segments to SNPs (Udall et al. 2005; Wang and Paterson 2011). At the chromosomal segment level, HEs were characterized by the loss of a segment of chromosome that was replaced by a corresponding homeologous region

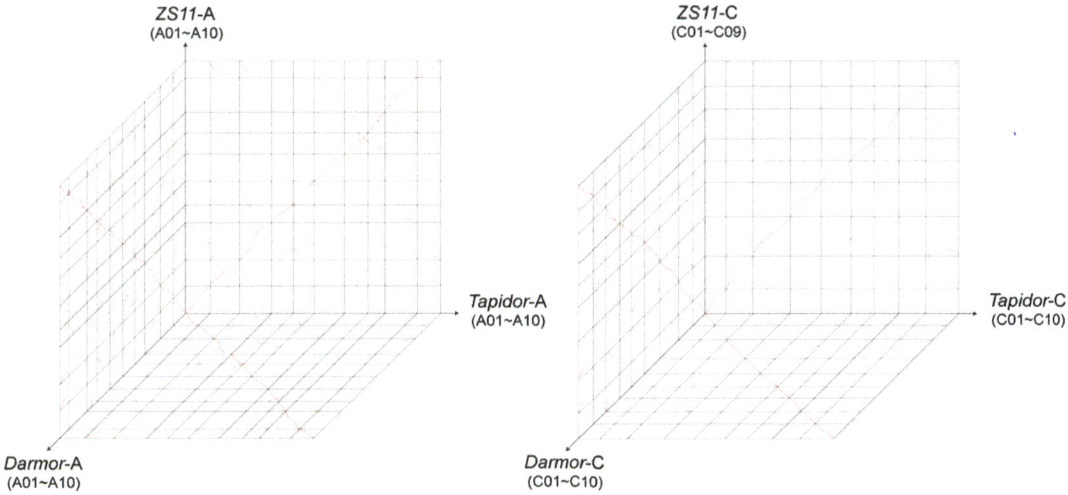

Fig. 4.1 Syntenic blocks between 'ZS11', 'Darmor-*bzh*', and 'Tapidor'. Genome-wide syntenic alignments could identify large chromosomal rearrangements, segmental duplicates, and so on. Many segmental of subgenomes of *B. napus* have three copies in *A. thaliana* could be obviously identified, which consist with the fact that the *brassica* species underwent genome-wide triploidization event after divergence from *Arabidopsis* (Fig. 4.2). The genome of *B. napus* maintains a high colinearity with its progenitor species. Only few segments happen to be obviously reversed in chromosome A02, A09, A10, C06, C07, and C09 (Fig. 4.3)

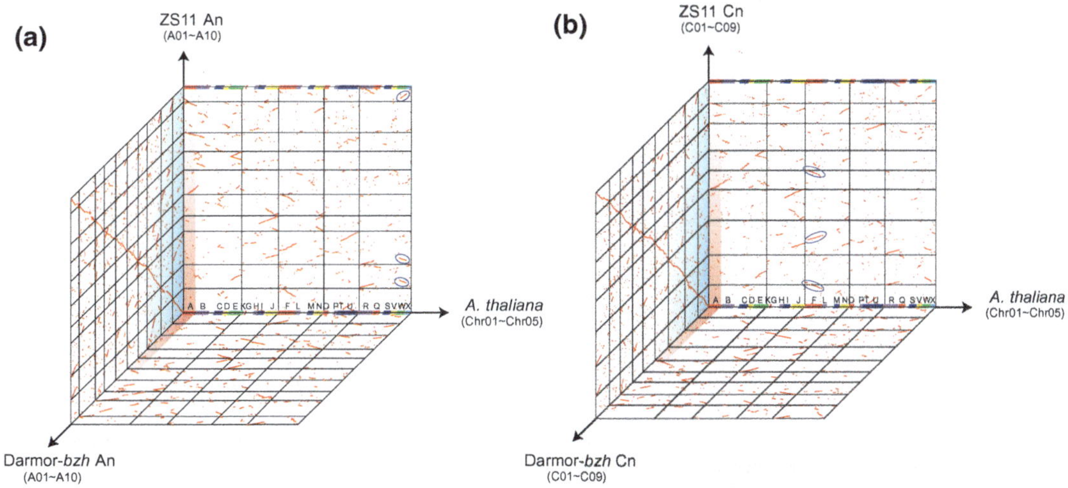

Fig. 4.2 Syntenic analysis of the *B. napus* genome and the *A. thaliana*. **a**, **b** Collinear relationships between *A. thaliana* with the *B. napus* of 'ZS11' and 'Darmor-*bzh*'. Ancestral crucifer blocks (A–X) in *A. thaliana* are clearly observed as triplicated copies in genomes of *B. napus* [This figure was adopted from the 'ZS11' genome article (Sun et al. 2017)]

(homeologous non-reciprocal translocations; NHRT). For *B. napus*, more than 15 HEs were identified using mapping sequencing reads to progenitor genomes and some of them were shared with other seven different diverse *B. napus* genotypes (Chalhoub 2014; Sun et al. 2017). Functional enrichment suggests that glucosinolate biosynthesis (p-value = 0.006) and plant hormone signal transduction (p-value = 0.02) are signally enriched in C_n to A_n HEs, while steroid

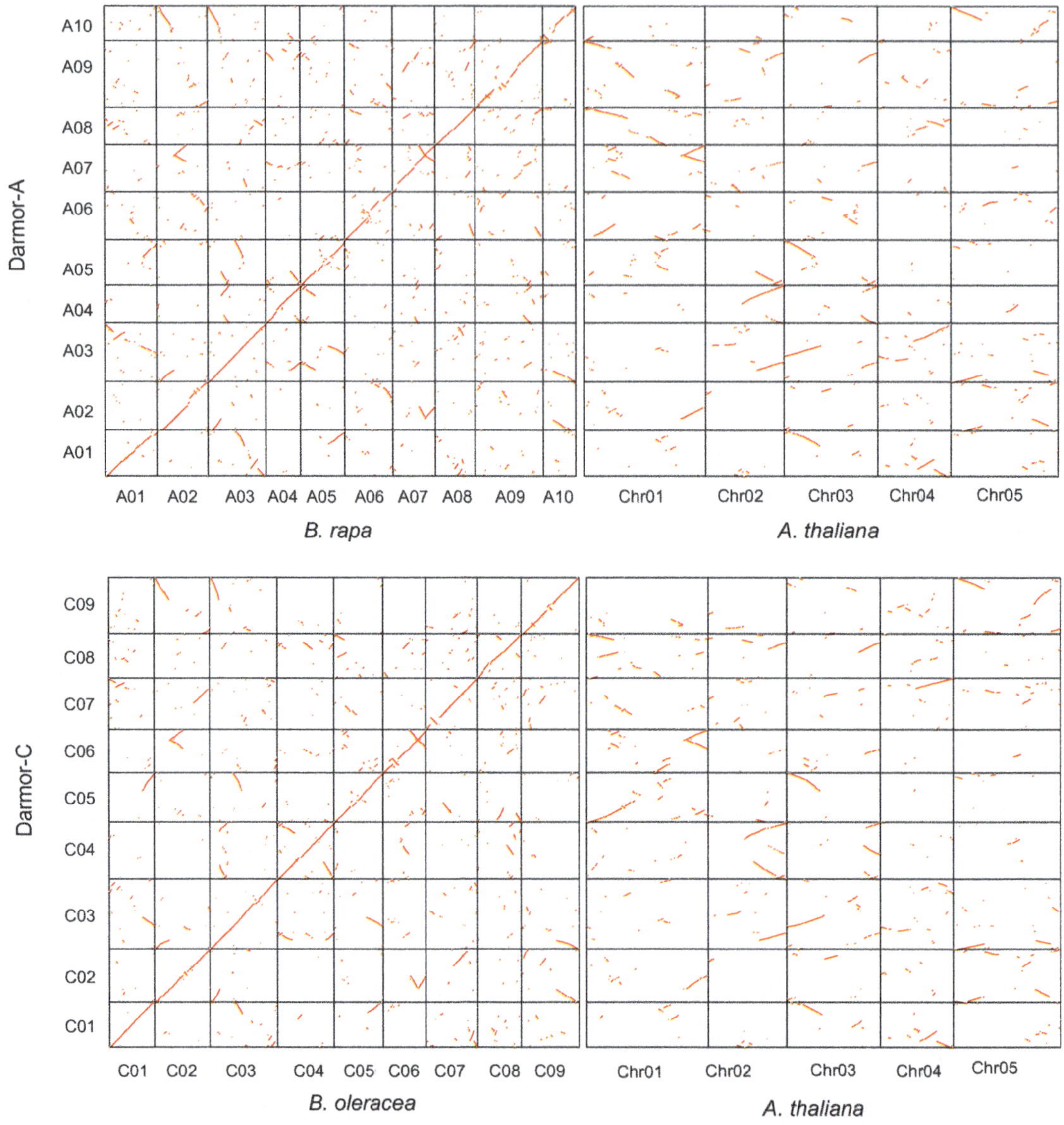

Fig. 4.3 Syntenic blocks among various *Brassicaceae* species. Left and right figure is the A and C subgenomes, respectively [This figure was adopted from the 'ZS11' genome article (Sun et al. 2017)]

biosynthesis (p-value = 0.006) and zeatin biosynthesis (p-value = 0.06) are signally enriched in A_n to C_n HEs. Meanwhile, some important genes that related to seed oil content and lipid composition, nutritionally undesirable erucic acid and glucosinolates (GSLs), flowering behavior, and pathogen resistance were also found in some HE regions that were maintained to optimize favorable trait of the rapeseed during intensive breeding (Chalhoub 2014; Sun et al. 2017).

4.3 Genomic Annotations of *B. napus* Genome

4.3.1 General Genomic Annotation Methods

Although annotation pipelines differ in their details on various projects, they share a common core set of properties. In general, genome-wide annotation contains two distinct core ways. In a first way, the homeologous search, a series of known sequences, such as expressed sequence tags (ESTs), proteins, transposable elements (TEs), are aligned to the target genome, and then the exact structure and locations of genes or other elements are predicted based on those known information. The other important way is ab initio prediction, where so many different tools or programs are developed to perform prediction based on effective algorithms and statistical models instead of existing sequences. For gene structure annotation, current pipelines are only focus on the protein-coding genes, and the results are always shown some differences due to using different methods or parameters by different researchers.

4.3.2 Repeat Elements Annotation

The methods were originally designed by the 'Darmor-*bzh*' and 'ZS11' genome articles. Repetitive sequences, including interspersed repetitive sequences and tandem repetitive sequences, make up a major part of eukaryotic genomes especially in most plant genome. Transposable elements (TEs) of *B. napus* were annotated by integrating both of de novo and homology-based approaches. The local querying database of TEs was built by three different programs LTR_FINDER (Xu and Wang 2007), PILER (Edgar and Myers 2005), and Repeat Scout (Price et al. 2005), and then the raw sequences of this database were classified by RepeatModeler (http://www.repeatmasker.org/RepeatModeler.html). The software RepeatMasker was used to search TEs against whole genome based on the existing repeat database Repbase (Bao et al. 2015) and the local database we constructed. As a complementary of homology-based methods, we performed PepeatProteinMask (http://www.repeatmasker.org/) to search the TE-related proteins against the genome based on existing TE proteins of Repbase. The tandem repeat elements of *B. napus* were annotated by using TRF-finder software. For 'Darmor-*bzh*', about 37.48% of the genome sequences (318 Mb) are repetitive sequences (Table 4.2), while only about 4.01% of them are tandem repeats. For 'ZS11', about 49.78% (485 Mb) of the genome sequences (318.65 Mb) are repetitive sequences, while only about 5.81% of them are tandem repeats, which revealed a more repetitive sequences in 'ZS11' that was proved to be the main reason of longer genomic length of 'ZS11'.

Table 4.2 Transposable elements (TEs) of *B. napus* and its parental species

Class	*B. napus* 'ZS11'		*B. napus* 'Darmor-*bzh*'		*B. rapa* (A_r)		*B. oleracea* (C_o)	
Type	Length (Bp)	% in genome	Length (Bp)	% in genome	Length (Bp)	% in genome	Length (Bp)	% in genome
DNA	123,665,905	12.67	72,775,042	8.56	26,923,608	9.48	46,493,747	9.51
LINE	66,009,487	6.76	48,441,399	5.70	17,344,782	6.11	31,356,353	6.41
SINE	2,782,466	0.29	892,949	0.11	418,772	0.14	976,450	0.19
LTR	326,032,933	33.40	201,508,020	23.70	43,023,735	15.15	133,866,847	27.39
Other	972,487	0.10	578,371	0.07	201,977	0.07	345,393	0.07
Unknown	10,402,870	1.07	19,574,410	2.30	23,104,332	8.13	14,964,238	3.06
Total	485,999,732	49.78	318,651,235	37.48	101,422,228	35.73	210,551,261	43.09

The results reveal that the ratio of TE contents between *B. napus* and its parental species is similar (A_n and A_r, C_n and C_o). Total size of LTR content is significantly different between A_n and C_n, or A_r and C_o genomes

4.3.3 Gene Annotations for 'Darmor-*bzh*'

The methods were originally designed by the 'Darmor-*bzh*' genome article. Functional genes of 'Darmor-*bzh*' were identified iteratively using a combination of homology-based and de novo prediction algorithms (Chalhoub 2014). (1) Protein mapping. Proteomes of five species were collected firstly to perform homology searches, including *Arabidopsis thaliana* (TAIR 10), *B. rapa* (Wang et al. 2011), *B. oleracea* (Liu 2014; Parkin 2014) and *O. sativa* (plantGBD, release 186). In order to shorten the calculation time of Genewise (Birney et al. 2004), the proteins were aligned to genome and extracted the candidate hits using BLAT (Kent 2002). Each candidate match was refined using Genewise to confirm the exact gene structure and the open reading frame (ORF). (2) De novo prediction. The refined curate gene models were used to train the parameters of Hidden Markov Model (HMM) for ab initio software Geneid (Parra et al. 2000) and SNAP (Korf 2004) using *A. thaliana* gene models, then two programs were used to predict gene models in Darmor-*bzh*. (3) Transcriptome sequence mapping. We firstly collected 643,937 cDNAs and 41,165 unigenes from EMBL and public URL (http://www.brassica.info/resource/transcriptomics/BrasEX1s.unigene.public.fasta). All cDNAs and unigenes were aligned to reference genome by BLAT to identify the best matches (identity >90%) and the initial gene structure. Then, Est2genome software was used to realign each match to cDNA sequences. In addition, RNA-Seq reads of major tissue and developmental stages for Darmor-*bzh* were obtained by Illumina technology. We next mapped all filtered reads and identified transcript models by using SOAP2 (Li et al. 2009) and Gmorse software (Denoeud 2008). Finally, we obtained 162,177 loci that were clustered from 930,181 models. (4) Integration of all predicted gene models. Consensus gene set of Darmor-*bzh* was obtained by integrating all predicted gene models using GAZE (Howe et al. 2002). Final non-redundant gene set contains the total number of 101,040 genes, which were consisted with the combination of *B. rapa* and *B. oleracea*. According to three public assemblies, a total number of 41,174 non-redundancy gene models predicted in *B. rapa* accession Chiifu-401-42, while 45,758 and 59,225 gene models in *B. oleracea* accession var. capitata line 02-12 and accession TO1000, respectively.

4.3.4 Gene Annotations for 'ZS11'

The methods were originally designed by the 'ZS11' genome article. Firstly, the de novo gene predictors including Augustus (Stanke and Waack 2003) and Glimmer-HMM (Majoros et al. 2004) were used to identify candidate gene models on repeat-masking genomic sequences. The Hidden Markov Model matrix trained from *A. thaliana* was chosen in the prediction process. Secondly, for homology-based predictions, protein sequences from five sequenced genomes (*A. thaliana*, *B. oleracea* 'TO1000', *B. oleracea* '*capitata*', *B. rapa* 'Chiifu-401-42', and *B. napus* 'Darmor-*bzh*') were aligned to the *B. napus* genome using tblastn (E-value $\leq$ 1e−5). Transcript sequences assembled from the RNA-Seq reads were also aligned to the genome sequence using blast (identity $\geq$0.95, and coverage $\geq$0.90). All candidate gene regions were predicted accurately spliced alignments and accurate gene structures using Genewise (Birney et al. 2004). Thirdly, we used GLEAN (Elsik et al. 2007) to integrate data from the two methods described above and generate a consensus set of genes. Finally, the genome of the semi-winter type *B. napus* cultivar 'ZS11' includes 101,942 non-redundant genes, which 42,416 (41.61%) and 51,042 (50.07%) genes could be assigned to the A_n and C_n subgenomes, respectively. The gene number of 'ZS11' is similar to that of the 'Darmor-*bzh*' assembly (101,040).

4.3.5 Gene Syntenic and Families Clustering Between 'Darmor-*bzh*' and 'ZS11'

About 97.35% of the genes in 'ZS11' were homeologous to genes in *B. rapa* (A_r), *B. oleracea* (C_o), *B. napus* 'Darmor-*bzh*', and *A. thaliana* (Sun et al. 2017) (Figs. 4.2 and 4.3). Based on homology information, 92,668 of the genes (91%) of 'ZS11', 86,197 the genes (85%) of 'Darmor-*bzh*', 38,678 of the genes (94%) of A_r, 39,786 of the genes (87%) of C_o ('*capitata*'), and 50,104 of the genes (85%) of C_o ('TO1000') could be further clustered into OrthoMCL gene families (Li et al. 2003). A total of 43,521 families shared between 'ZS11' and 'Darmor-*bzh*' represent potential core gene families in *B. napus* (all these families were also supported by the raw reads of 'Tapidor'). Chi-squared tests showed that only 10 families differed in gene number between the two cultivars ($p < 0.01$), which was indicative of a high degree of gene number conservation between 'ZS11' and 'Darmor-*bzh*'.

4.3.6 Gene Families Clustering Between *B. napus* and Its Parental Species

Due to a highly conserved relationship between 'ZS11' and 'Darmor-*bzh*', we here selected 'Darmor-*bzh*' as the *B. napus* for the next analysis. A total of 91,167 genes showed highly homeologous with the parental species *B. rapa* and *B. oleracea*. Meanwhile, two subgenomes of *B. napus* are largely gene colinear to the corresponding diploid A_r and C_o, 34,255 of total 42,320 genes and 38,661 of total 48,874 genes, respectively. More than 80% of gene families (38,622 of total 47,659 families), amounting to 75,947 genes, are shared by two parental species using OrthoMCL (Li et al. 2003) clustering methods (Fig. 4.4). Gene functional enrichment analysis for 9037 specific families show that five pathways, ribosome (p-value = 4.17e−17), oxidative phosphorylation (p-value = 2.07e−12), spliceosome (p-value = 7.68e−07), and photosynthesis (p-value = 4.37e−03), are significantly enriched (p-value $\leq$ 0.01). Meanwhile, ribosome biogenesis in eukaryotes (p-value = 8.14e−17), RNA degradation (p-value = 4.33e−08), and glucosinolate biosynthesis (p-value = 1.19e−02) are significantly enriched in parental species. The results show that *B. napus* underwent different gene retention during intensive breeding, such as to decrease nutritionally undesirable erucic acid and glucosinolates (GSLs). In *B. napus*, GSLs are unpopular in the seed because of the toxicological effects of their products, which severely hinder the use of the seedcake (Wittkop et al. 2009). Due to the defects of breakdown products, the GSL content of *B. napus* has been decreased from 60–100 to 10–15 μmol/g during decades of breeding (Nesi et al. 2008).

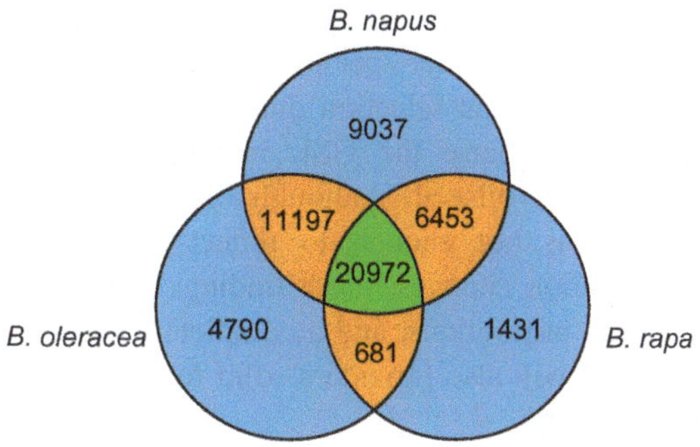

Fig. 4.4 Gene families clustering among three-related species

4.4 Conclusion

A whole genome of *B. napus* was successfully assembled using a combination strategy. The progenitor A_r and C_o genomes are relatively intact in *B. napus*, and there are small scales chromosomal organization changes after formation of *B. napus*, including some obviously segmental inversion and homeologous exchanges (HEs). Some genes that related to dominant agronomic trait were found to be participant in HEs, like genes that related flowering behavior, and pathogen resistance. Most gene families are shared by *B. napus* and its progenitor, and only small ratio families that are specific existed in those species, such as glucosinolate biosynthesis-related genes, are significantly enriched in parental species, which may be the results of human selection during long-breeding activities because the GSL is harmful to the nutrition of oilseed. Although this preliminary analysis is not enough to reflect all genome organization changes, these findings will surely bring us inspirations, and a complete genome of rapeseed will also provide a solid foundation for future study.

References

Arcade A et al (2004) BioMercator: integrating genetic maps and QTL towards discovery of candidate genes. Bioinformatics 20:2324–2326. https://doi.org/10.1093/bioinformatics/bth230

Bao WD, Kojima KK, Kohany O (2015) Repbase update, a database of repetitive elements in eukaryotic genomes. Mobile DNA-Uk 6. https://doi.org/10.1186/s13100-015-0041-9 (UNSP 11)

Bayer PE et al (2017) Assembly and comparison of two closely related *Brassica napus* genomes. Plant Biotechnol J 15:1602–1610. https://doi.org/10.1111/pbi.12742

Birney E, Clamp M, Durbin R (2004) Genewise and genomewise. Genome Res 14:988–995. https://doi.org/10.1101/gr.1865504

Boetzer M, Henkel CV, Jansen HJ, Butler D, Pirovano W (2011) Scaffolding pre-assembled contigs using SSPACE. Bioinformatics 27:578–579

Chalhoub B (2014) Early allopolyploid evolution in the post-neolithic *Brassica napus* oilseed genome (vol 348, 1260782, 2014). Science 345:1255

Chapman JA et al (2015) A whole-genome shotgun approach for assembling and anchoring the hexaploid bread wheat genome. Genome Biol 16. https://doi.org/10.1186/s13059-015-0582-8 (UNSP 26)

Cheung F et al (2009) Comparative analysis between homoeologous genome segments of *Brassica napus* and its progenitor species reveals extensive sequence-level divergence. Plant Cell 21:1912–1928. https://doi.org/10.1105/tpc.108.060376

Denoeud F et al (2008) Annotating genomes with massive-scale RNA sequencing. Genome Biol 9. https://doi.org/10.1186/Gb-2008-9-12-R175 (Artn R175)

Edgar RC, Myers EW (2005) PILER: identification and classification of genomic repeats. Bioinformatics 21: I152–I158. https://doi.org/10.1093/bioinformatics/bti1003

Elsik CG et al (2007) Creating a honey bee consensus gene set. Genome Biol 8:R13. https://doi.org/10.1186/gb-2007-8-1-r13

Gaeta RT, Pires JC, Iniguez-Luy F, Leon E, Osborn TC (2007) Genomic changes in resynthesized *Brassica napus* and their effect on gene expression and phenotype. Plant Cell 19:3403–3417. https://doi.org/10.1105/tpc.107.054346

Howe KL, Chothia T, Durbin R (2002) GAZE: a generic framework for the integration of gene-prediction data by dynamic programming. Genome Res 12:1418–1427. https://doi.org/10.1101/gr.149502

Inaba R, Nishio T (2002) Phylogenetic analysis of Brassiceae based on the nucleotide sequences of the S-locus related gene, SLR1. Theor Appl Genet 105:1159–1165. https://doi.org/10.1007/s00122-002-0968-3

Kent WJ (2002) BLAT—the BLAST-like alignment tool. Genome Res 12:656–664. https://doi.org/10.1101/gr.229202

Korf I (2004) Gene finding in novel genomes. Bmc Bioinformatics 5. https://doi.org/10.1186/1471-2105-5-59 (Artn 59)

Leitch IJ, Bennett MD (1997) Polyploidy in angiosperms. Trends Plant Sci 2:470–476. https://doi.org/10.1016/S1360-1385(97)01154-0

Li L, Stoeckert CJ Jr, Roos DS (2003) OrthoMCL: identification of ortholog groups for eukaryotic genomes. Genome Res 13:2178–2189. https://doi.org/10.1101/gr.1224503

Li RQ et al (2009) SOAP2: an improved ultrafast tool for short read alignment. Bioinformatics 25:1966–1967. https://doi.org/10.1093/bioinformatics/btp336

Li Y, Hu Y, Bolund L, Wang J (2010) State of the art de novo assembly of human genomes from massively parallel sequencing data. Hum Genomics 4:271–277

Liu L, Qu C, Wittkop B, Yi B, Xiao Y, He Y, Snowdon RJ, Li J (2013) A high-density SNP map for accurate mapping of seed fibre QTL in *Brassica napus* L. PloS one 8:e83052

Li N, Shi J, Wang X, Liu G, Wang H (2014) A combined linkage and regional association mapping validation and fine mapping of two major pleiotropic QTLs for seed weight and silique length in rapeseed (*Brassica napus* L.). BMC Plant Biol 14:114

Li FG et al (2015) Genome sequence of cultivated upland cotton (*Gossypium hirsutum* TM-1) provides insights into genome evolution. Nat Biotechnol 33:U524–U242. https://doi.org/10.1038/nbt.3208

Liu SY et al (2014) The *Brassica oleracea* genome reveals the asymmetrical evolution of polyploid genomes. Nat Commun 5. https://doi.org/10.1038/Ncomms4930 (Artn 3930)

Majoros WH, Pertea M, Salzberg SL (2004) TigrScan and GlimmerHMM: two open source ab initio eukaryotic gene-finders. Bioinformatics 20:2878–2879. https://doi.org/10.1093/bioinformatics/bth315

Nesi N, Delourme R, Bregeon M, Falentin C, Renard M (2008) Genetic and molecular approaches to improve nutritional value of *Brassica napus* L. seed. CR Biol 331:763–771. https://doi.org/10.1016/j.crvi.2008.07.018

Park JY et al (2005) Physical mapping and microsynteny of *Brassica rapa* ssp. pekinensis genome corresponding to a 222 kbp gene-rich region of Arabidopsis chromosome 4 and partially duplicated on chromosome 5. Mol Genet Genomics 274:579–588. https://doi.org/10.1007/s00438-005-0041-4

Parkin IAP, Sharpe AG, Keith DJ, Lydiate DJ (1995) Identification of the A and C genomes of amphidiploid *Brassica napus* (oilseed rape). Genome/Nat Res Counc Can (Genome/Conseil national de recherches Canada) 38:1122–1131

Parkin IAP et al (2014) Transcriptome and methylome profiling reveals relics of genome dominance in the mesopolyploid *Brassica oleracea*. Genome Biol 15. https://doi.org/10.1186/Gb-2014-15-6-R77 (Artn R77)

Parra G, Blanco E, Guigo R (2000) GeneID in Drosophila. Genome Res 10:511–515. https://doi.org/10.1101/Gr.10.4.511

Price AL, Jones NC, De Pevzner PA (2005) Novo identification of repeat families in large genomes. Bioinformatics 21:I351–I358. https://doi.org/10.1093/bioinformatics/bti1018

Rana D et al (2004) Conservation of the microstructure of genome segments in *Brassica napus* and its diploid relatives. Plant J 40:725–733. https://doi.org/10.1111/j.1365-313X.2004.02244.x

Stanke M, Waack S (2003) Gene prediction with a hidden Markov model and a new intron submodel. Bioinformatics 19(Suppl 2):ii215–ii225

Sun F et al (2017) The high-quality genome of *Brassica napus* cultivar 'ZS11' reveals the introgression history in semi-winter morphotype. Plant J 92:452–468. https://doi.org/10.1111/tpj.13669

Tang H, Zhang X, Miao C, Zhang J, Ming R, Schnable JC, Schnable PS, Lyons E, Lu J (2015) ALLMAPS: robust scaffold ordering based on multiple maps. Genome Bio 16:3

Udall JA, Quijada PA, Osborn TC (2005) Detection of chromosomal rearrangements derived from homeologous recombination in four mapping populations of *Brassica napus* L. Genetics 169:967–979. https://doi.org/10.1534/genetics.104.033209

Wang XY, Paterson AH (2011) Gene conversion in angiosperm genomes with an emphasis on genes duplicated by polyploidization. Genes-Basel 2:1–20. https://doi.org/10.3390/genes2010001

Wang XW et al (2011) The genome of the mesopolyploid crop species *Brassica rapa*. Nat Genet 43:U1035–U1157. https://doi.org/10.1038/ng.919

Wendel JF (2000) Genome evolution in polyploids. Plant Mol Biol 42:225–249. https://doi.org/10.1023/A:1006392424384

Wittkop B, Snowdon RJ, Friedt W (2009) Status and perspectives of breeding for enhanced yield and quality of oilseed crops for Europe. Euphytica 170:131–140. https://doi.org/10.1007/s10681-009-9940-5

Xu Z, Wang H (2007) LTR_FINDER: an efficient tool for the prediction of full-length LTR retrotransposons. Nucleic Acids Res 35:W265–W268. https://doi.org/10.1093/nar/gkm286

Yang YW, Lai KN, Tai PY, Li WH (1999) Rates of nucleotide substitution in angiosperm mitochondrial DNA sequences and dates of divergence between Brassica and other angiosperm lineages. J Mol Evol 48:597–604. https://doi.org/10.1007/Pl00006502

Zhang TZ et al (2015) Sequencing of allotetraploid cotton (*Gossypium hirsutum* L. acc. TM-1) provides a resource for fiber improvement. Nat Biotechnol 33:U531–U252. https://doi.org/10.1038/nbt.3207

Zhang Y, Thomas CL, Xiang J et al. (2016) QTL meta-analysis of root traits in *Brassica napus* under contrasting phosphorus supply in two growth systems. Sci Rep 6:33113

From Alpha-Duplication to Triplication and Sextuplication

5

Haibao Tang, Xingtan Zhang, Chaobo Tong, Boulos Chalhoub, Shengyi Liu and Eric Lyons

Abstract

The *Brassica napus* is a truly great genome for the study of genome and gene family evolution, with a rich history of past whole-genome duplication (WGD) events. The genome has undergone a total of 5 rounds of duplications since the common ancestor with the basal angiosperm Amborella. This level of genetic redundancy is unparalleled by any other flowering plant genome that was sequenced prior to the release of the *B. napus* genome assembly. Three recent WGDs that occurred within the lineage of Brassicaceae are of significant value to polyploid research, namely the 'alpha' duplication event, the *Brassica* triplication and *B. napus* allotetraploidization. These events occurred at different evolutionary times and are representatives of paleo-polyploidy, meso-polyploidy, and neo-polyploidy, respectively. Studies of evolutionary changes and transcriptional regulation of duplicate genes derived from these WGD events have led to groundbreaking discoveries in the dynamics of polyploid genomes, including genome reorganization, gene fractionation (loss of duplicated genes), and genome dominance. These breakthroughs were largely facilitated by a number of innovations in computational methods, databases, and interconnected cyberinfrastructure that are devoted to plant comparative genomics research. With its genome now fully deciphered, *B. napus* continues to be one of the most important model organisms in post- polyploidy genome evolution research.

H. Tang · X. Zhang
Center for Genomics and Biotechnology, Fujian Agriculture and Forestry University, Fuzhou 350002, Fujian Province, China

H. Tang · E. Lyons (✉)
School of Plant Sciences, BIO5 Institute, iPlant Collaborative, University of Arizona, Tucson, AZ, USA
e-mail: elyons.uoa@gmail.com

C. Tong · S. Liu
Key Laboratory of Biology and Genetic Improvement of Oil Crops, the Ministry of Agriculture, Oil Crops Research Institute of the Chinese Academy of Agricultural Sciences, Wuhan 430062, China

B. Chalhoub
Institute of Crop Science, Zhejiang University, Hangzhou 310058, China

B. Chalhoub
Institut National de Recherche Agronomique (INRA)/Université d'Evry Val d'Essone, Unité de Recherche en Génomique Végétale, UMR1165, Organization and Evolution of Plant Genomes, 2 rue Gaston Crémieux, 91057 Evry, France

5.1 Genome Duplication and Polyploidy

Whole-genome duplications (WGDs) are prominent and recurring features during the evolution of flowering plant genomes (Bowers et al. 2003; Tang et al. 2008, 2010). WGDs can induce genome

S. Liu et al. (eds.), *The Brassica napus Genome*, Compendium of Plant Genomes,
https://doi.org/10.1007/978-3-319-43694-4_5

instability, as characterized by massive gene rearrangements, losses, and genome reorganization (Wolfe 2001). Initially following polyploidy, many of the gene redundancies are retained, providing a wealth of new genetic materials for developing novel and modified functions. Three evolutionary outcomes are possible for duplicated genes: the initial functions divided in the daughter copies (subfunctionalization), novel functions acquired (neofunctionalization), or function lost in one of the redundant gene copies (nonfunctionalization) (Force et al. 1999).

Conserved synteny is the conservation of gene contents and order between chromosomes with a shared evolutionary history and is often used to describe relationships between different genomes, but can also be used to describe such patterns within the same genome (Tang et al. 2008). Conserved synteny is evident when large sets of genes or genomic features are preserved in close proximity and often in the same order and orientations (also known as 'collinearity') (Tang et al. 2008). Recursive genome duplications generate multiple collinear gene patterns within a genome, which are derived from a single ancestral region, and are evident in syntenic comparisons both within and between plant genomes. Such collinear patterns of gene arrangements underlie most of the computational methods to identify the duplicated genes that originated from WGD events (Lyons and Freeling 2008; Tang et al. 2008, 2010, 2011).

Brassica napus has been one of the most important taxa for the study of polyploids and is considered a hallmark for polyploidy research (Chalhoub et al. 2014). Autopolyploids refer to the taxa having two or more haploid sets of chromosomes ('subgenomes') derived from the same species, where subgenomes from allopolyploids, such as *B. napus*, are derived from different, diverged species. The 'triangle of U,' initially proposed in 1935 by Korean–Japanese botanist Woo Jang-choon, stated that pairwise mergers between the diploid Brassicas (*Brassica rapa* AA, *B. nigra* BB, and *Brassica oleracea* CC) created three allotetraploid vegetables and oil crops, including *B. juncea* (AABB), *B. napus* (AACC), and *B. carinata* (BBCC) (Chalhoub et al. 2014).

The *B. napus* has experienced a number of WGDs in its lineage, giving rise to a genome with $\sim$72× more redundancy than the basal flowering plant *Amborella*. The recurring WGDs include two genome triplications and three genome doublings, giving rise to a 72× ($72 = 3 \times 2 \times 2 \times 3 \times 2$) genome (Chalhoub et al. 2014) (Fig. 5.1). Herein, we review three relatively recent rounds of genome duplications that are of particular evolutionary significance in the lineage of *B. napus*. The three genome duplications include the alpha event occurred 50–65 million years ago (Bowers et al. 2003), the Brassica genome triplication event occurred 24–29 million years ago (Moghe et al. 2014) and the genome merger event that gave rise to *B. napus* occurred 7500 thousand years ago (Chalhoub et al. 2014). These WGD events occurred at different times and are representatives of paleo-polyploidy, meso-polyploidy, and neo-polyploidy, respectively. The chronological gradient in WGDs along with distinct gene duplicates of different ages represents unique materials for understanding the patterns and mechanism of polyploid evolution.

5.2 Brassicaceae Alpha Event

The Bowers et al. (2003) study of WGDs within the Arabidopsis genome was groundbreaking in polyploidy research (Bowers et al. 2003). The study uncovered unambiguous evidence that Arabidopsis, a modern-day diploid genome, was not only an ancient polyploid ('paleo-polyploid') but had been impacted by three WGDs in its lineage (Bowers et al. 2003). While dating of the three WGDs was later revised and refined, the occurrence and nature of these WGDs were proven to be accurate by the comparisons of many flowering plant genomes (Bowers et al. 2003; Jaillon et al. 2007; Tang et al. 2008; The Tomato Genome Consortium 2012). The most recent event, coined 'alpha' (α) event, occurred in the shared lineage of Brassicaceae; the β

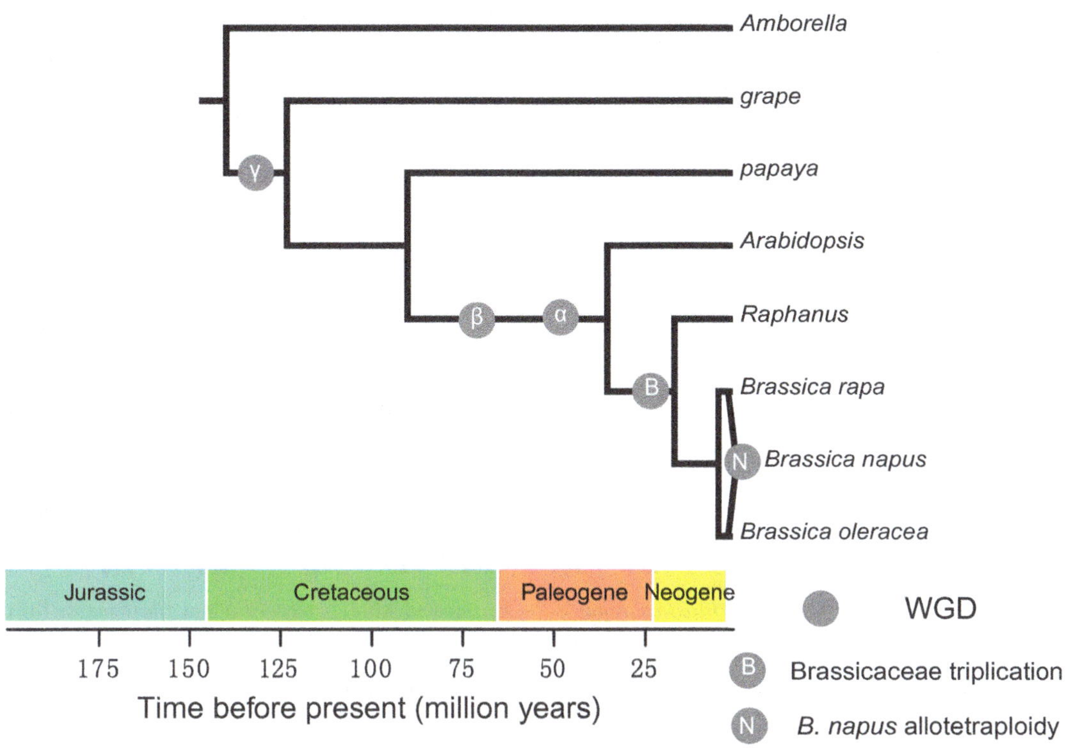

Fig. 5.1 Dating of whole-genome duplication (WGD) events in the lineage of *Brassica napus* in relationship to selected flowering plant taxaCircles represent known WGDs identified previously. The phylogeny and divergence time between taxa were derived from (Magallon et al. 2015; Moghe et al. 2014)

event, occurred within Brassicales, but not affecting the papaya linage; the deepest γ event occurred in the shared lineage of core eudicots (Fig. 5.1).

The alpha (α) WGD event, being the most recent event among the three WGDs in the Arabidopsis lineage, immediately spurred progress in theoretical works on paleo-polyploid research. Approximately 90% of the Arabidopsis genes are found to be contained in α duplicated blocks and 28.6% of the genes have a retained pair (Thomas et al. 2006). The retained gene duplicates are enriched in regulatory functions including members of proteasome, signal transduction machinery, and transcription factors. This has lent support to the 'gene balance hypothesis,' which predicts that genes that are highly connected are more likely to be retained following polyploidy (Freeling 2009; Thomas et al. 2006). The surviving α duplicates were also found to be clustered in the metabolic network, suggesting an association between duplicate retention, reactions with high metabolic flux, and dosage constraints of gene products (Bekaert et al. 2011).

5.3 Brassica Genome Triplication

Following the α event shared with Arabidopsis, the diploid *Brassica* species experienced a hexaploidy event in the Brassicaceae lineage after its divergence from Arabidopsis. This genome triplication was first suggested based on early comparative mapping studies (Parkin et al. 2003, 2005). The triplicated, ancestral Brassica blocks in relation to the corresponding Arabidopsis blocks (numbered A through X) are commonly referred to as 'Parkin blocks' (Parkin et al. 2003, 2005). The same genome triplication was

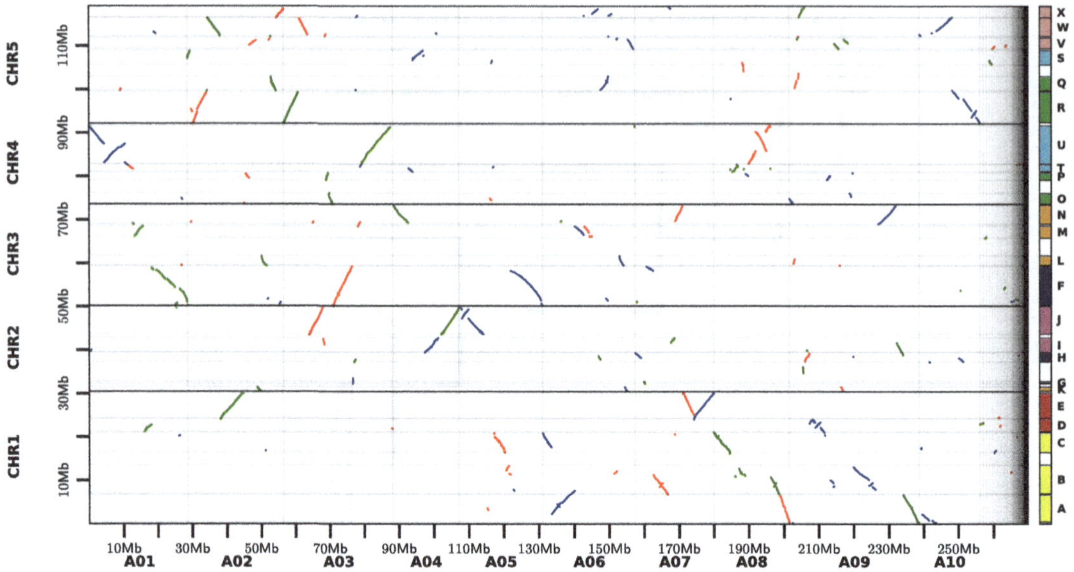

Fig. 5.2 Syntenic dot plot between *Brassica rapa* (*x*-axis) and *Arabidopsis thaliana* (*y*-axis), illustrating triplicated blocks in BrassicaSyntenic blocks are colored based on the differences in fractionation—red (most fractionated MF2), green (medium fractionated: MF1), and blue (least fractionated: LF) (Wang et al. 2011). The "Parkin blocks" (Parkin et al. 2005), numbered A–X, are shown to the right of the plot

inferred to have occurred in the shared lineage of Brassicinae, Moricandiinae, Raphaninae, and other subtribes of Brassicaeceae (Lysak et al. 2007).

The genome sequence of *B. rapa* confirmed this genome triplication event (chronologically speaking, a 'meso-polyploidy' event), that occurred in the common ancestor of all *Brassica* species. The *B. rapa* genome can be partitioned into triplicated blocks when compared to Arabidopsis, with each Parkin block showing up to three matching regions in *B. rapa* (Fig. 5.2). The initially near-identical subgenomes generated by the whole-genome triplication events did not fractionate equally, and each subgenome has a relatively consistent bias in the proportion of genes lost following polyploidy. For *B. rapa*, subgenome LF (least fractionated) has retained almost two thirds of *A. thaliana* orthologous genes, while subgenomes MF1 (medium fractionated) and MF2 (most fractionated) have retained significantly fewer genes (Wang et al. 2011) (Fig. 5.2). While detailed patterns of the Brassicaceae genome triplication was first comprehensively studied in *B. rapa*, the genomes of *B. oleracea* (Liu et al. 2014; Parkin et al. 2014) and wild radish (Moghe et al. 2014) independently validated the observed triplicated genomic patterns—one subgenome consistently has more genes retained on it than the other two subgenomes. Retained duplicates differed significantly from the singleton genes in rates of evolution, expression patterns, and network connectivity, providing a foundation for a statistical model for the predicting the likelihood of retention after the shared Brassica triplication event in both *Brassica* and *Raphanus* (Moghe et al. 2014) (Fig. 5.1).

5.4 *Brassica napus* Allotetraploidy

Brassica napus genome provides a unique opportunity to study the dynamics after WGD following recent genome merger events ('neo-polyploidy'). The availability of the tetraploid genome as well as its diploid progenitors allow genomic regions to be unambiguously aligned and compared—using both synteny- and sequence-based approaches. Structural differences between the progenitor genomes and tetraploid *B. napus* revealed patterns of gene loss as well as homeologous chromosomal exchange, after extensive care in identification and validation of such genomic events (Chalhoub et al. 2014).

The *B. napus* A_n and C_n subgenomes are largely collinear with the two progenitor diploid genomes of *B. rapa* (A_r) and *B. oleracea* (C_o), respectively. With the exception of a few small regions that failed to be incorporated into the pseudomolecules and several regions that appear translocated, gene contents and order on *B. napus* chromosomes are generally very similar to those on the corresponding chromosomes of the diploid progenitors. Much of the diploid progenitor gene space were contained within orthologous synteny blocks to the tetraploid *B. napus* (Chalhoub et al. 2014). A list of 47,080 'quartet' orthologous families was compiled that collectively contain genes from progenitor genomes *B. rapa* and *B. oleracea* (A_r and C_o) and two subgenomes in *B. napus* (A_n and C_n), on the basis of synteny as well as reciprocal best-hit evidence for genes located on the unplaced scaffolds due to fragmented genome assembly (Chalhoub et al. 2014). Each 'quartet' family contains at most one gene from each of A_r, C_o, A_n, and C_n. The curated quartet list provides all possible 9 cases of orthology and homeology between A_r, C_o, A_n, and C_n (Fig. 5.3). A total of 27,360 fully retained quartets A_r-C_o-A_n-C_n were identified (Case 9), while all other instances represented potential gene loss (Fig. 5.3).

As the progenitor genomes and *B. napus* genome were sequenced, assembled and annotated using different methods, an exhaustive search and classification method beyond the simple 'quartet' analysis was applied for discovering gene loss at the DNA sequence level with high confidence. True gene loss was identified and sequentially validated using three approaches. First, all missing syntenic genes for which BLASTN matches were found at orthologous positions with no annotation, or where a gene could be predicted if the same annotation method was used, or those that are at orthologous positions but not retained by the synteny search criteria, were all invalidated with visual aid from CoGe tools (Fig. 5.4). Following this stringent analysis, 663 candidate lost genes were identified in the *B. napus* assembly as compared to the corresponding parental genomes. This putative *B. napus* missing genes were carefully checked for confirmation based on sequence coverage from raw genomic reads. The read coverage confirmed the deletion of 176 genes, 71 of which (~35%) constitute segments of two to four adjacent deleted genes that represent segmental rather than individual gene losses. Finally, PCR validation by non-amplification of a subset of those inferred deleted genes confirmed 22 out of 23 genes in *B. napus*, whereas they could be amplified in the corresponding parents *B. rapa* and *B. oleracea* (Chalhoub et al. 2014). The three-tiered approach represented the most diligent efforts for validation of gene losses reported in any genome study (Chalhoub et al. 2014).

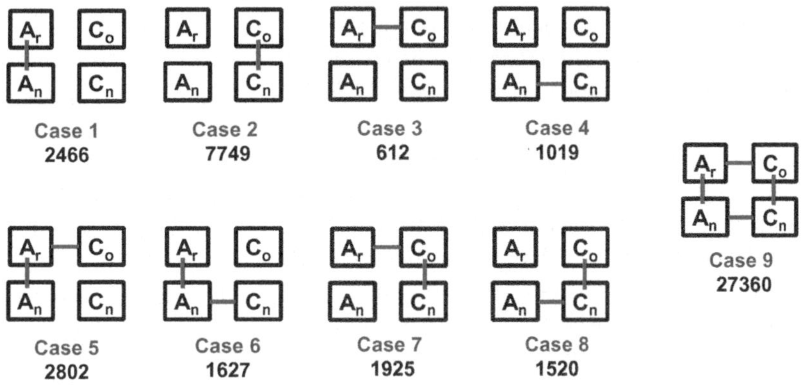

Fig. 5.3 All nine possible gene retention cases (red lines) of the quartet genes between *B. rapa* (A_r genome), *B. oleracea* (C_o genome), and *B. napus* (A_n and C_n subgenomes)

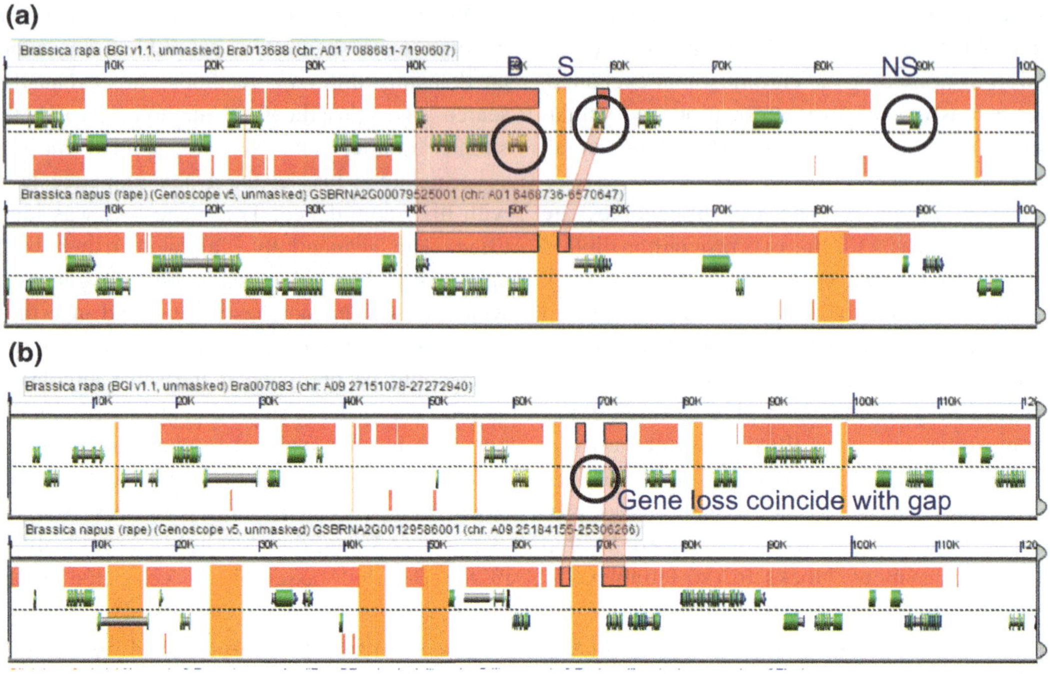

Fig. 5.4 CoGe validation of various categories in comparing gene structures between diploid (*B. rapa* and *B. oleracea*) and tetraploid (*B. napus*) orthologs**a** Examples showing each of the category B, S, and NS (B: has syntenic ortholog; S: has syntenic sequence match without a gene model in *B. napus*; NS: has nonsyntenic sequence match), CoGe link: http://genomevolution.org/r/93qq; **b** Example showing a potential gene loss which coincides with an assembly gap that was invalidated in our stringent pipeline. CoGe link: http://genomevolution.org/r/93s2

5.5 Biased Fractionation

Despite the initial expansion of gene numbers immediately following WGDs, most lineages have since experienced drastic gene loss (gene fractionation), genome down-sizing, and ultimately genomic 'diploidization' (Wolfe 2001). Diploidization was often facilitated by gene fractionation, where each homeologous region retains a 'fraction' of the ancestral gene contents (Langham et al. 2004; Thomas et al. 2006). During the fractionation process, many gene copies with redundant functions, and whose product levels were not under stringent control, tend to be lost (Freeling 2009; Langham et al. 2004; Thomas et al. 2006). Gene fractionations often result in a reduction of gene family members that largely offsets the initial expansion from WGDs. Several theoretical models were proposed to suggest the molecular mechanism of gene fractionation. One mechanism, with support from empirical data, is through sequence deletion in relatively small sequence chunks, or 'bites' (Tang et al. 2012; Woodhouse et al. 2010). In *B. rapa*, the fractionation mechanism was shown to be predominantly short deletions, probably via intra-chromosomal recombination at direct repeats, and not nonfunctionalization by randomization of sequence through nucleotide substitutions (Tang et al. 2012). This is in stark contrast to the mode of gene loss in mammalian genomes, where pseudogenes appear to have a larger contribution to the process of gene removals (Woodhouse et al. 2010).

Gene deletions following the α event, Brassica triplication, and allotetraploidy in the *B. napus* lineage were well studied in past investigations (Chalhoub et al. 2014; Moghe et al. 2014; Parkin et al. 2014; Thomas et al. 2006;

Wang et al. 2011). Indeed, the phenomenon of fractionation bias was first discovered in the analysis of alpha duplicates in the Arabidopsis genome (Thomas et al. 2006). The subgenome with more retained homoelogs shows consistent 'dominance' over the other subgenomes. A two-step fractionation hypothesis was suggested to explain the biased fractionation following the Brassica triplication event. The least fractionated genome, subgenome LF, was postulated as the last genome among the three subgenomes that entered the hypothesized 'two-staged' genome triplication event (Tang et al. 2012).

While many genomes undoubtedly show varied fractionation bias from one subgenome over another, exceptions to this bias still exist in some polyploidy lineages. For example, a close relative of Arabidopsis and Brassicas is Camelina. The Camelina genome has a genome triplication in its lineage, which is unrelated to the Brassica triplication. However, the genome structure of Camelina is highly undifferentiated, showing no evidence of biased fractionation among the triplicated blocks (Kagale et al. 2014). Similarly, no apparent fractionation bias is observed in the *B. napus* genome during the allotetraploid event (Chalhoub et al. 2014), possibly due to the relatively young age and an early stage in gene fractionation.

5.6 Transcriptional Bias and Genome Dominance

Expression correlations between gene duplicates provide useful metrics to indicate functional divergence, complementing the sequence divergence. Cases of functional divergence as measured by tissue expression between duplicate genes in Arabidopsis were identified, with each homeolog participating in parallel networks (Blanc and Wolfe 2004). For duplicates genes that are retained in the same network, the level of gene expression differs between the duplicates in a nonrandom manner. When gene expression across duplicated genomic regions is compared with their respective gene deletion bias, the under-fractionated subgenome expresses its genes to a higher mRNA level than does the other subgenomes, showing substantial deviation from parental additivity (Parkin et al. 2014; Schnable et al. 2011; Wang et al. 2011; Woodhouse et al. 2014). Non-additivity in gene regulation can be shown in three ways: (1) polyploid expression similar to parents (dominance); (2) polyploid expression lower or higher than parents (transgression); (3) biased expression between the homeologs (Yoo et al. 2014). The non-additive homeolog-specific expression is genetically similar to allele-specific expression in diploid organisms and may provide a link between hybrid vigor and polyploid vigor (Chen 2010).

In *B. rapa*, subgenome LF expresses its genes to a higher level than does either subgenome MF1 or MF2 (Wang et al. 2011). The difference has persisted through millions of years following the initial genome duplication event (Tang et al. 2012). The same trend is also seen in its close relative *B. oleracea*, where the cytosine methylation levels of the three subgenomes are also substantially differentiated, suggesting an important role of cytosine methylation in the functional diversification of duplicated genes, perhaps serving as epigenetic 'marks' that maintain the differentiation between subgenomes over time (Parkin et al. 2014). More recently, Woodhouse et al. found that 24-nt small RNA (smRNAs) preferentially target MF1 or MF2 subgenomes, indicating that silencing of transposons near genes causes position-effect down-regulation in a similar manner analogous to a 'rheostat' (Woodhouse et al. 2014). This key observation has led to the best model proposed thus far to explain genome dominance in polyploids, as well as proposing testable solutions to heterosis and the C-value paradox (Woodhouse et al. 2014).

In the allotetraploid *B. napus*, RNA-Seq-based analyses showed that for the majority of the gene pairs (58.3% of the homeologs), A_n and C_n homoeologs contribute equally to gene expression in root and leaf. Biased gene expression was observed where the A_n homoeolog contributed more than the C_n homeolog for

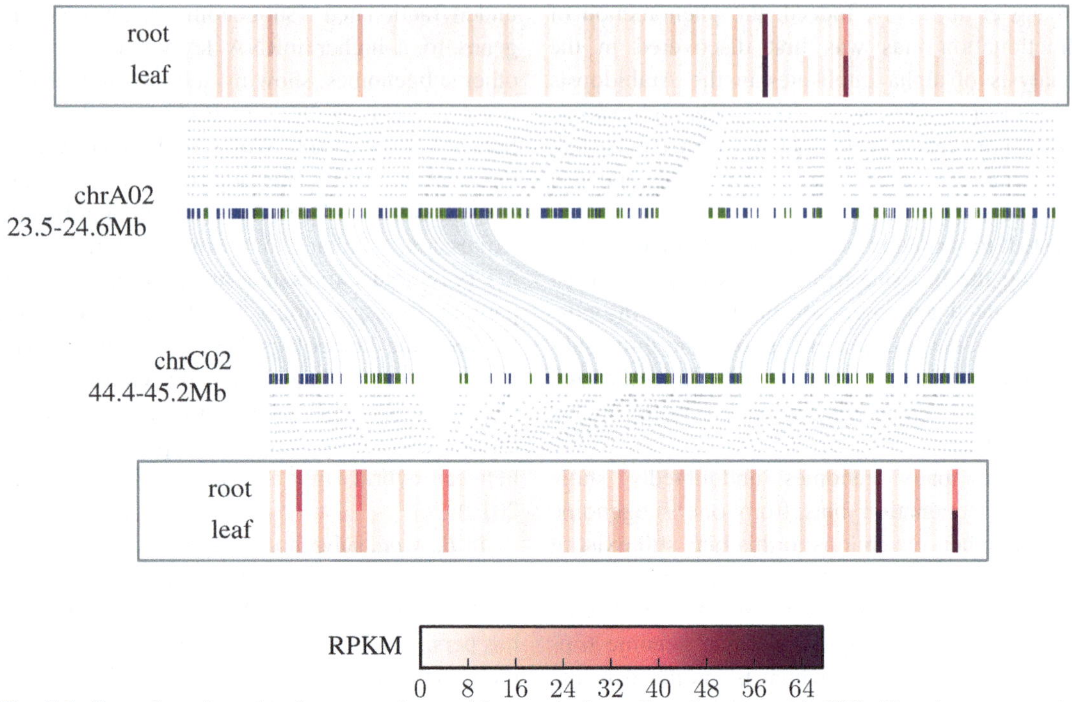

Fig. 5.5 Example regions showing expression partitioning between homeologous gene pairs in *Brassica napus* genomeRNA-Seq transcripts were quantified in leaf and root tissue libraries for each homeologous gene pairs. For each homeologous gene pair shown in this example, one gene is from A_n subgenome (chrA09), and the other gene is from C_n subgenome (chrC08). Two heat maps—A_n genes on top and C_n genes on bottom—show that expression profiles are largely similar between the homeologous regions, with several gene pairs showing expression divergence between A_n and C_n subgenomes

15.7% of the gene pairs; similarly, 17.3% of the gene pairs were found where the C_n homoeolog contributed more than the A_n homoelog in both tissues (Chalhoub et al. 2014). No evidence for pronounced genome dominance was observed within the *B. napus* genome, although there are slightly more gene pairs where the C_n homoeolog contributes more than the A_n homoeolog in both tissues (Chalhoub et al. 2014). A typical exemplar region from *B. napus* is shown in Fig. 5.5, where expression values, as measured in RPKM, are simultaneously compared between homeologs.

5.7 Computational Infrastructure to Study Genome Duplication, Fractionation and Dominance

A key prerequisite for study of WGD events and subsequent fractionation and dominance is an accurate compilation of duplicated genes that are distinctly originated from a specific event. The 'specificity' of the gene pairs is important for making robust evolutionary inference. The computational pipeline for identifying these gene duplicates and associated events is a natural extension from a comparative genomics pipeline that has been heavily engineered in the past few years, following the initial approach by Bowers et al. (2003). The comparative genomics pipeline includes a number of computational steps to extract, enrich, and visualize conserved synteny 'signals' contained in a dot plot (e.g., Arabidopsis vs. *B. rapa* comparison in Fig. 5.2). The pipeline has a modular implementation for added flexibility, with the following components listed in their order of execution: generating anchors, filtering anchors, generating blocks by chaining neighboring anchors, filtering blocks, curating blocks by adding in additional nearby anchors, and finally visualizing the blocks.

Due to recursive WGD events in the Brassicaceae and many other flowering plant lineages, extensive filtering of synteny blocks is required for higher precision to analyze a specific WGD event without contamination of blocks derived from older events. The algorithm, QUOTA-ALIGN, performs depth-based filtering on synteny blocks and can be used to separate blocks that are overlaid on one another as a result of recursive WGD events, even when conventional *Ks*-based filtering is less effective due to overlapping peaks or, for older events, when third-codon substitutions are saturated (Tang et al. 2011). For any pairwise genome comparison combined with QUOTA-ALIGN filtering, researchers can confidently infer the number of *lineage-specific* and *shared* duplication events. The 'quota' parameter, or the maximum allowed depth covered by syntenic block, offers a way to select a subset of blocks derived from a specific evolutionary event by limiting the number of times a genomic region is syntenic to regions in another genome (Tang et al. 2011). Restricting gene duplicates similarly in self-comparison allow the separation of gene duplicates of different ages and origin. For example, Arabidopsis has undergone three paleo-polyploidy events. The synteny blocks, if unscreened, will include mixture of both recent duplicates (α event) and older duplicates (β and γ event) (Bowers et al. 2003). With QUOTA-ALIGN, we can specify a quota of 1:1 when extracting synteny blocks specifically from α event within the Arabidopsis genome (see CoGe links: https://genomevolution.org/r/hkxn and https://genomevolution.org/r/hkxo).

A number of Web sites and database host information related to polyploid evolution, building on similar pipelines as detailed above. CoGe has a data storage system that allows easy access to thousands of genomes as well as to downstream analysis tools such as to GEvo and SynMap for micro- and whole-genome analysis, respectively (Lyons and Freeling 2008; Lyons et al. 2008). GEvo was used extensively for proofing the gene loss events in *B. napus* (Fig. 5.3) (Chalhoub et al. 2014). CoGe's job execution framework (JEX) facilitates parallel processing of queries against multiple genomes. Plant genome duplication database (PGDD) is a public database that identifies and catalogs of plant genes in terms of intra-genomic and inter-genomic relationships (Lee et al. 2013). A key advantage of PGDD over other plant comparative genomics platforms is that the synteny blocks stored in PGDD were carefully curated by experts, providing the basis for many studies of gene family evolution and consequences of polyploidy. PLAZA offers queries and visualization of rich information regarding gene annotation, families, domains, trees, and genome organization (Proost et al. 2015). Phytozome serves both as a hub for storing public releases for many plant genomes and also as a comparative genomics hub, offering views of plant genes at levels of sequence, family, and genome organization (Goodstein et al. 2012).

The development of comparative genomics web portals, including CoGe, PGDD, PLAZA, Phytozome, and many others are successful in bringing about a democratization of bioinformatics research, permiting any researcher to make novel discoveries through access to genomics data, computational tools, and visualization systems. Such ecosystem requires the deployment of modular analysis pipelines that allow new tools to exploit existing computational resources, architectures, and curated datasets. Such portals have leveraged novel algorithmic approaches and access to high-performance computational resources, thereby addressing the challenge of scales in comparative genomics and polyploidy research that consist of mostly multi-dimensional problems.

5.8 Perspective

The study of recurring polyploidies and genomic changes following these events shows not only theoretical value, but also has important applications in the breeding of the *B. napus* oilseed. The genome merger has led to the expansion of lipid biosynthesis genes that greatly exceed other oilseed plants. Gene deletions and homeologous exchanges (HEs) between subgenomes have led to the reduction of seed glucosinolate (GSL) content. Genes responsible for the flowering time (FLC) were greatly expanded from a single copy in Arabidopsis to nine copies in the oilseed genome with some HEs co-localizing with QTLs for vernalization requirements and flowering time. Through detailed analysis of syntenic gene duplicates, we can link much of the genomic changes that have occurred in *B. napus* to its unique biological, adaptive, and agronomic traits. Resequencing of *B. napus* varieties could reveal additional genome structuring events that might have been consciously selected for during human cultivation and improvements.

Acknowledgements We thank the Fujian provincial government for a Fujian '100 Talent Plan' award to HT. EL is supported by USDA 2013-00984; NSF IOS—1339156; IOS—444490.

Competing Interests The authors declare that they have no competing interests.

References

Bekaert M, Edger PP, Pires JC, Conant GC (2011) Two-phase resolution of polyploidy in the Arabidopsis metabolic network gives rise to relative and absolute dosage constraints. Plant Cell 23:1719–1728

Blanc G, Wolfe KH (2004) Functional divergence of duplicated genes formed by polyploidy during Arabidopsis evolution. Plant Cell 16:1679–1691

Bowers JE, Chapman BA, Rong J, Paterson AH (2003) Unravelling angiosperm genome evolution by phylogenetic analysis of chromosomal duplication events. Nature 422:433–438

Chalhoub B et al (2014) Plant genetics. Early allopolyploid evolution in the post-Neolithic *Brassica napus* oilseed genome. Science 345:950–953

Chen ZJ (2010) Molecular mechanisms of polyploidy and hybrid vigor. Trends Plant Sci 15:57–71

Force A et al (1999) Preservation of duplicate genes by complementary, degenerative mutations. Genetics 151:1531–1545

Freeling M (2009) Bias in plant gene content following different sorts of duplication: tandem, whole-genome, segmental, or by transposition. Annu Rev Plant Biol 60:433–453

Goodstein DM et al (2012) Phytozome: a comparative platform for green plant genomics. Nucleic Acids Res 40:D1178–D1186

Jaillon O et al (2007) The grapevine genome sequence suggests ancestral hexaploidization in major angiosperm phyla. Nature 449:463–467

Kagale S et al (2014) The emerging biofuel crop *Camelina sativa* retains a highly undifferentiated hexaploid genome structure. Nat Commun 5:3706

Langham RJ et al (2004) Genomic duplication, fractionation and the origin of regulatory novelty. Genetics 166:935–945

Lee TH, Tang H, Wang X, Paterson AH (2013) PGDD: a database of gene and genome duplication in plants. Nucleic Acids Res 41:D1152–D1158

Liu S et al (2014) The *Brassica oleracea* genome reveals the asymmetrical evolution of polyploid genomes. Nat Commun 5:3930

Lyons E, Freeling M (2008) How to usefully compare homologous plant genes and chromosomes as DNA sequences. Plant J 53:661–673

Lyons E, Pedersen B, Kane J, Freeling M (2008) The value of nonmodel genomes and an example using SynMap within CoGe to dissect the hexaploidy that predates the rosids. Trop Plant Biol 1:181–190

Lysak MA, Cheung K, Kitschke M, Bures P (2007) Ancestral chromosomal blocks are triplicated in Brassiceae species with varying chromosome number and genome size. Plant Physiol 145:402–410

Magallon S, Gomez-Acevedo S, Sanchez-Reyes LL, Hernandez-Hernandez T (2015) A metacalibrated time-tree documents the early rise of flowering plant phylogenetic diversity. New Phytol 207:437–453

Moghe GD et al (2014) Consequences of whole-genome triplication as revealed by comparative genomic analyses of the wild radish *Raphanus raphanistrum* and three other Brassicaceae species. Plant Cell 26:1925–1937

Parkin IA, Sharpe AG, Lydiate DJ (2003) Patterns of genome duplication within the *Brassica napus* genome. Genome 46:291–303

Parkin IA et al (2005) Segmental structure of the *Brassica napus* genome based on comparative analysis with *Arabidopsis thaliana*. Genetics 171:765–781

Parkin IA et al (2014) Transcriptome and methylome profiling reveals relics of genome dominance in the mesopolyploid *Brassica oleracea*. Genome Biol 15: R77

Proost S et al (2015) PLAZA 3.0: an access point for plant comparative genomics. Nucleic Acids Res 43:D974–D981

Schnable JC, Springer NM, Freeling M (2011) Differentiation of the maize subgenomes by genome dominance and both ancient and ongoing gene loss. Proc Natl Acad Sci U S A 108:4069–4074

Tang H et al (2008) Synteny and collinearity in plant genomes. Science 320:486–488

Tang H, Bowers JE, Wang X, Paterson AH (2010) Angiosperm genome comparisons reveal early polyploidy in the monocot lineage. Proc Natl Acad Sci U S A 107:472–477

Tang H et al (2011) Screening synteny blocks in pairwise genome comparisons through integer programming. BMC Bioinf 12:102

Tang H et al (2012) Altered patterns of fractionation and exon deletions in *Brassica rapa* support a two-step model of paleohexaploidy. Genetics 190:1563–1574

The Tomato Genome Consortium (2012) The tomato genome sequence provides insights into fleshy fruit evolution. Nature 485:635–641

Thomas BC, Pedersen B, Freeling M (2006) Following tetraploidy in an Arabidopsis ancestor, genes were removed preferentially from one homeolog leaving clusters enriched in dose-sensitive genes. Genome Res 16:934–946

Wang X et al (2011) The genome of the mesopolyploid crop species *Brassica rapa*. Nat Genet

Wolfe KH (2001) Yesterday's polyploids and the mystery of diploidization. Nat Rev Genet 2:333–341

Woodhouse MR et al (2010) Following tetraploidy in maize, a short deletion mechanism removed genes preferentially from one of the two homologs. PLoS Biol 8:e1000409

Woodhouse MR et al (2014) Origin, inheritance, and gene regulatory consequences of genome dominance in polyploids. Proc Natl Acad Sci USA 111:5283–5288

Yoo MJ, Liu X, Pires JC, Soltis PS, Soltis DE (2014) Nonadditive gene expression in polyploids. Annu Rev Genet 48:485–517

Quantity, Distribution, and Evolution of Major Repeats in *Brassica napus*

6

Nomar Espinosa Waminal, Sampath Perumal, Shengyi Liu, Boulos Chalhoub, Hyun Hee Kim and Tae-Jin Yang

Abstract

Repetitive elements (REs) play major roles in genome organization, size, and evolution, but are often underrepresented in genome assemblies. The recent genome assembly of the allotetraploid *Brassica napus* genome revealed that 48% of the genome comprised REs, including transposons and tandem repeats. In the present work, we show the overall quantity and comparative analyses of major repeat families in both the assembled and unassembled portions of the reference *B. napus* genome. We surveyed the abundance, distribution, diversity, and dynamics of ten major REs in the *B. napus* genome, which represented less than 1% of the total 1130 Mb *B. napus* genome in the current assembly. However, in silico mapping of raw whole-genome sequence reads from nine *B. napus* accessions revealed about 11% of the genome as represented by these ten repeat families. Comparative analyses of these major repeats showed their evolutionary dynamics in the *B. rapa* (A_r), *B. oleracea* (C_o), and *B. napus* (A_nC_n) genomes as well as a considerable inter- and intraspecies repeat diversity among different *B. napus* accessions. Cytogenetic mapping of these major repeats showed their genomic abundance and distribution, with some families having a conserved subgenomic distribution pattern in the *B. napus* genome. Finally, the impact of genetic changes to REs and their corresponding epigenetic readjustments during *B. napus* evolution are also discussed in this chapter.

N. E. Waminal · S. Perumal · T.-J. Yang (✉)
Department of Plant Science, Plant Genomics and Breeding Institute, and Research Institute of Agriculture and Life Sciences, College of Agriculture and Life Sciences, Seoul National University, Seoul 08826, Republic of Korea
e-mail: tjyang@snu.ac.kr

N. E. Waminal · H. H. Kim
Department of Life Science, Chromosome Research Institute, Sahmyook University, Seoul 01795, Republic of Korea

S. Liu
Key Laboratory of Biology and Genetic Improvement of Oil Crops, the Ministry of Agriculture, Oil Crops Research Institute of the Chinese Academy of Agricultural Sciences, Wuhan 430062, China

B. Chalhoub
Institute of Crop Science, Zhejiang University, Hangzhou 310058, China

B. Chalhoub
Institut National de Recherche Agronomique (INRA)/Université d'Evry Val d'Essone, Unité de Recherche en Génomique Végétale, UMR1165, Organization and Evolution of Plant Genomes, 2 rue Gaston Crémieux, 91057 Evry, France

6.1 Introduction

We have come to a pinnacle in the history of genomics when enormous volumes of nucleotide sequence information can be gathered cost

S. Liu et al. (eds.), *The Brassica napus Genome*, Compendium of Plant Genomes,
https://doi.org/10.1007/978-3-319-43694-4_6

effectively in a short time (Soltis et al. 2013; Schatz et al. 2012; Zhou et al. 2010), allowing the genome assembly of many important crop species at an unprecedented pace and accuracy. This permits a better understanding of the overall genome landscape and provides sufficient evidence leading to the negation of the pejorative 'junk' status of repetitive DNA fractions in a genome (Eddy 2012). In contrast, these repeats are fundamental components for the holistic function of a cell (Fedoroff 2012; Plohl et al. 2008; Freeling et al. 2015).

Despite advancements in next-generation sequencing (NGS) technology and assembly algorithms, efficient genome assemblies of repetitive sequences are still hindered especially when using short NGS sequences (Alkan et al. 2011; Claros et al. 2012; Hamilton and Buell 2012). This is even more difficult in plants, which often harbor abundant repetitive DNA and experienced various extents of polyploidization, (for reviews, see Schatz et al. 2012; Claros et al. 2012; Mihai et al. 2002; Renny-Byfield and Wendel 2014). Due to abundant REs such as tandem repeats (TRs) and transposable elements (TE), NGS-based genome assembly causes shrinkage of the actual repeat copies in a genome, even down to a few copies in an assembly (Macas et al. 2007; Alkan et al. 2011; Claros et al. 2012; Mihai et al. 2002; Schatz et al. 2012; Tang et al. 2015). Moreover, polyploidy or whole-genome duplication (WGD) and small-scale duplications further exacerbate this challenge by creating duplicate copies of genes or larger genomic regions. This redundancy drives mis-assemblies that could occur along these large-scale duplications (Paterson and Wendel 2015; Claros et al. 2012).

For example, in the two diploid progenitors of *Brassica napus*, *B. rapa* and *B. oleracea*, about 40 and 38% of respective genomes have not been included in pseudo-molecules, mainly due to REs (Waminal et al. 2015, 2016). Evidently, REs and polyploidy greatly influence the quality of genome assemblies and ultimately the acquisition of high-resolution pseudo-molecules. Although longer reads are offered by single-molecule, or third-generation sequencing technologies, they are still insufficient in resolving mega-base length tandem repeat regions (Schatz et al. 2012; Schadt et al. 2010). Consequently, densely heterochromatic regions such as the centromere and pericentromere have very little representation, or none at all, in some reference genome assemblies, even for model plants such as rice and *Arabidopsis* (Gao et al. 2015).

Although REs remain largely unassembled and unexplored in many sequenced plant genomes (Michael and Jackson 2013; Liu et al. 2014; Wang et al. 2011b), it holds a plethora of information about chromosome and genome dynamics, gene regulation, genome evolution, and epigenetic control (Biemont 2010; Biémont and Vieira 2006; Nowak 1994; Chadwick 2009; Melters et al. 2013; Mehrotra and Goyal 2014). Hence, they deserve a fair genome-wide analysis. Understanding their genomic distribution would provide a more enhanced comprehension of the holistic genome landscape and origin. Moreover, they complement studies in structural and functional genomics (Biemont 2010; Wang et al. 2011a; Choi et al. 2014).

B. napus ($2n = 4x = 38$, $A_nA_nC_nC_n$ genome) is an allopolyploid oilseed crop that formed within the past 7,500 years through hybridization between its progenitor genomes, *B. rapa* ($2n = 2x = 20$, A_rA_r genome) and *B. oleracea* ($2n = 2x = 18$, C_oC_o genome). The recent release of the allopolyploid *B. napus* genome revealed an aggregated 72 × genome multiplication since the origin of angiosperms (Chalhoub et al. 2014). This advance has provided a suitable foundation for deeper understanding of the dynamics of its REs through comparative studies with its progenitor diploid species, *B. rapa* and *B. oleracea*. In the present study, we surveyed the genomic abundance, chromosomal distribution, diversity, and dynamics of the major *Brassica* repeats in nine *B. napus* accessions (Table 6.1). We further discussed the role of epigenetic readjustments and its interplay with genetic changes in response to allopolyploidization.

Table 6.1 Summary of *Brassica napus* accessions used for the survey of major repeats[a]

ID	Accession/cultivar	Origin/type	Amounts (Mbp)	Genome coverage (x)
Bn-1	Zhongsuang11	Winter rapeseed	2,126.6	1.9
Bn-2	M083	Semi-winter rapeseed	1,273.8	1.1
Bn-3	Aburamasari	Asian (Japan) oilseed rape	13,900.7	12.3
Bn-4	Aviso	European oilseed rape	8,679.2	7.7
Bn-5	Darmor-bzh	European winter oilseed rape	6,029.7	5.3
Bn-6	Siberian kale	Kale	13,323.1	11.8
Bn-7	*B. napus* 'H165'	Resynthesized	15,928.4	14.1
Bn-8	Rutabaga	Swede sensation	14,923.5	13.2
Bn-9	Yudal	Asian (Korea) oilseed rape	14,105.3	12.5

[a]All the sequences above were provided by Shengyi Liu and Boulos Chalhoub

6.2 Repeats in the A_r, C_o, and A_nC_n Reference Genomes

Assembled sequences represented 58, 82, and 75% of total genome sizes of 485 Mb, 630 Mb, and 1130 Mb for *B. rapa*, *B. oleracea*, and *B. napus*, respectively (Fig. 6.1) (Wang et al. 2011a; Chalhoub et al. 2014; Liu et al. 2014). Of these sequences, repetitive DNAs represented 23, 41, and 35% for *B. rapa*, *B. oleracea*, and *B. napus*, respectively (Table 6.2). Relative to their respective estimated genome sizes, these values were reduced to 13, 32, and 23% (Table 6.2 and Fig. 6.1). Non-REs representing euchromatic regions covered 45, 50, and 52% of total estimated genome sizes of *B. rapa*, *B. oleracea*, and *B. napus*, respectively (Fig. 6.1).

Of the repeats in the reference assemblies, TEs represented the bulk fraction with 97, 96, and 97% followed by TR with 2, 3, and 2%. The rest were unclassified sequences (Fig. 6.1). While retrotransposons (Class I TE) were more abundant than DNA transposons (Class II TE) in *B. oleracea* and *B. napus*, the reverse was observed in *B. rapa* (Table 6.2). Among Class I TEs, LTR retrotransposons represented the majority with *Ty1/Copia* being more abundant than *Ty3/Gypsy* in both diploids (4.13 vs. 3.42% in *B. rapa* and 10.85 vs. 8.86% in *B. oleracea*), while about the same amount was present in *B. napus* (8.05 vs. 8.18%). Both diploid progenitors had different major Class II TEs. *Helitrons* were more abundant than *CACTA* elements in *B. rapa* (3.74 vs. 1.94%), but the reverse was observed in *B. oleracea* (3.96 vs. 5.55%). Accordingly, both elements had a similar representation in *B. napus* (3.69 vs. 3.83%).

Obviously, a considerable proportion of REs have not yet been included in the assembled sequences. In the following sections, we analyzed the genome proportion of several reported RE families in the current assembly and in the total whole-genome sequences (WGS), and we checked for types of repeats captured in both assembled and unassembled fractions. The RE families that were used in this analysis represented less than 1% of the current assembly; therefore, the values we obtained in this work mostly reflect the portions in the unassembled fraction. While TEs were more abundant in the assembled fractions, TRs were most prevalent in the unassembled, representing 41, 47, and 35% in *B. rapa*, *B. oleracea*, and *B. napus*, respectively (Fig. 6.1).

6.3 The Major Repeats of the *Brassica* Genomes

The difficulty in accounting for RE sequences in genome assembly scaffolds often leaves them in the 'hidden fraction' of a genome. REs can be subcategorized into dispersed or tandem repeats based on genomic distribution (Cizkova et al. 2013; Heslop-Harrison 2000; Plohl et al. 2012).

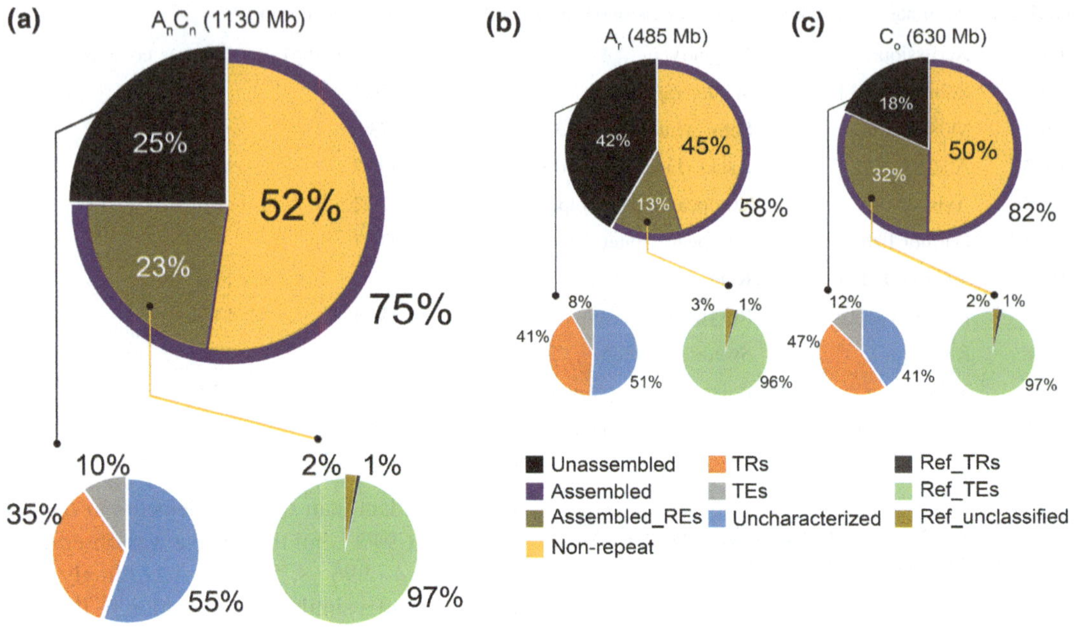

Fig. 6.1 Genomic proportions of both the assembled and unassembled sequences in the *Brassica napus* and its diploid progenitors. **a** Large inner pie chart represents the estimated total non-repeat genic fraction (*yellow slice*) and repeat fraction (*brown slice*) of *B. napus* genome. The outer doughnut chart represents the percentage of assembled (*purple slice*) and unassembled (*black slice*) fractions relative to the estimated genome size of 1130 Mb as calculated by Chalhoub et al. (2014). Smaller pie charts at the bottom left and right summarize the REs in the unassembled and assembled genome fractions, respectively. **b**, **c** Same diagrams for *B. rapa* and *B. oleracea*, respectively

Dispersed repeats include TEs, which are distributed in the entire genome, subgenome, or specific chromosomal regions (Choi et al. 2014; Lim et al. 2007). TRs are organized in a head-to-tail arrangement in distinct chromosomal regions (Coluccia et al. 2011; Sharma et al. 2013).

Most sequenced plant genomes contain a large proportion of Class I TE, mostly of the LTR superfamily (Michael and Jackson 2013). While up to 50% of TRs (Plohl et al. 2008) has been reported; the highest representation are mostly those with centromeric origins (Melters et al. 2013). In this chapter, we focused our survey on the major repeats representing Class I and Class II TEs, structural satellite repeats, and housekeeping ribosomal RNA genes in the *B. napus* genome. The analysis of major repeats in its progenitor genomes provides an enhanced understanding in choosing major repeat elements for the survey of the *B. napus* genome (Waminal et al. 2015, 2016). The major repeats include centromeric satellite repeats (CentBn1 and CentBn2), rDNA tandem repeats (5S and 45S), subtelomeric repeats (BnSTRa and BnSTRb), centromeric retrotransposon in *Brassica* (CRB; Lim et al. 2007), pericentromeric retrotransposon specific to *B. rapa* (pCRBr; Lim et al. 2007), and dispersed LTR and TIR elements specific to the *B. oleracea* genome, BoCopia and BoCACTA, respectively.

Centromeric repeats of *B. napus* are categorized into two groups, CentBn1 and CentBn2, and are homologous to their diploid progenitors (CentBr1/CentBr2 and CentBo1/CentBo2 from *B. rapa* and *B. oleracea*, respectively) (Perumal et al. 2017). Collectively, we refer to them as centromeric repeats of *Brassica* (CentB). The 5S and 45S rDNA sequences are conserved among the A_r, C_o, and A_nC_n. The subtelomeric satellite repeats, BnSTRa and BnSTRb, have orthologs in both A_r (Waminal et al. 2015) and C_o (Waminal et al. 2016) genomes. Collectively, we refer to them as *Brassica* subtelomeric repeats, BSTRa

Table 6.2 Total RE fractions in reference genomes of *Brassica rapa*, *B. oleracea*, and *B. napus*

	Brassica rapa 'Chiifu'[b]				*Brassica oleracea* 'capitata'[c]					*Brassica napus* 'Darmor-bzh'[d]		
Transposon categories[a]	Copy numbers	Coverage (bp)	% in Ref.	GP (%)	Copy numbers	Coverage (bp)	% in Ref.	GP (%)	Copy numbers	Coverage (bp)	% in Ref.	GP (%)
Class I												
LTR/*Copia*	30,349	11,292,047	4.13	2.33	68,842	51,827,340	10.85	8.23	97,152	59,466,274	8.05	5.26
LTR/*Gypsy*	19,229	9,327,740	3.42	1.92	48,169	42,322,494	8.86	6.72	68,354	60,377,324	8.18	5.34
LTR/Unknown	11,358	3,768,473	1.38	0.78	21,405	11,626,899	2.43	1.85	31,172	16,050,602	2.17	1.42
SINE	4,248	549,493	0.20	0.11	8,359	1,122,599	0.23	0.18	12,053	1,638,766	0.22	0.15
LINE total	7,090	3,260,523	1.19	0.67	10,345	6,792,364	1.42	1.08	16,003	10,516,160	1.42	0.93
Subtotal	72,274	28,198,276	10.33	5.81	157,120	113,691,696	23.79	18.05	224,734	148,049,126	20.05	13.10
Class II												
hAT	14,722	2,915,918	1.07	0.60	22,113	5,022,703	1.05	0.80	36,356	7,950,833	1.08	0.70
CACTA	17,742	5,289,213	1.94	1.09	40,258	26,517,864	5.55	4.21	58,323	27,277,074	3.69	2.41
PIF/Harbinger	2,057	581,852	0.21	0.12	3,994	1,375,937	0.29	0.22	5,853	1,790,014	0.24	0.16
Tc1/Mariner	13,307	3,022,686	1.11	0.62	18,119	5,622,275	1.18	0.89	34,488	8,342,337	1.13	0.74
MITE/Tourist	16,867	2,720,339	1.00	0.56	29,593	6,967,515	1.46	1.11	51,770	9,356,495	1.27	0.83
DNA-Unknown	29,919	7,453,903	2.73	1.54	43,933	13,054,872	2.73	2.07	75,309	19,547,930	2.65	1.73
Helitron	46,182	10,206,949	3.74	2.10	65,310	18,932,278	3.96	3.01	107,645	28,297,585	3.83	2.50
Subtotal	140,796	32,190,860	11.79	6.64	223,320	77,493,444	16.22	12.30	369,744	102,562,268	13.89	9.08
Tandem repeats	1,879	565,650.00	0.20	0.12	262,840	53,622	1.61	0.53	331,850	97,713.30	0.01	0.01
Unclassified		2,183,677	0.80	0.45		4,105,469	0.86	0.65		6,383,948	0.86	0.56
Subtotal	1,879	2,749,327	1.00	0.57	262,840	4,159,091	0.86	1.18	331,850	6,481,661	0.88	0.57
Total	214,949	63,138,463	23.11	13.02	643,280	195,344,231	40.87	31.53	926,328	257,093,055.30	34.82	22.75

[a]Adapted from Chalhoub et al. (2014)
[b]Estimated genome size: 485 Mb, assembly size: 283.8 Mb, without N gaps: 273.1 Mb
[c]Estimated genome size: 630 Mb, assembly size: 515.34 Mb, without N gaps: 477.85 Mb
[d]Estimated genome size: 1130 Mb, assembly size: 850.3 Mb, without N gaps: 738.3 Mb

and BSTRb (Perumal et al. 2017). Centromeric retrotransposon of *Brassica* (CRB), a *Ty1/copia* LTR retrotransposon, which is mostly associated with heterochromatic regions and is intermingled with CentB (Lim et al. 2007) in the Oleracea lineage (A_r and C_o) chromosomes, remained conserved in the A_nC_n genome. Sequences of genome-specific transposons such as the pericentromeric retrotransposon of *B. rapa* (pCRBr), *B. oleracea Ty1/Copia* retrotransposon (BoCopia), and *B. oleracea* CACTA transposon (BoCACTA) remained conserved in A_nC_n genome (Lim et al. 2007; Perumal et al. 2017; Waminal et al. 2016).

6.4 Genomic Abundance and Distribution of Major Repeats in *B. napus*

Plant genomes sequenced to date have considerable amount of unassembled fractions (Michael and Jackson 2013). Assembly statistics often only provide a general view of what was efficiently anchored in the genome assembly; hence, it does not provide the proportional genomic abundance of these elements based on the actual genome content (Waminal et al. 2015; Schatz et al. 2012). Reasonably, estimating their abundance can be achieved by read mapping of WGS reads on the repeat unit sequence and by molecular cytogenetic mapping through fluorescence in situ hybridization (FISH). These approaches will elucidate their proportional abundance, genomic distribution, and impact for evolution (Waminal et al. 2015, 2016, 2018; Schatz et al. 2012; Lee et al. 2017).

6.4.1 In Silico WGS Read Mapping and Cytogenetic Mapping

Based on current available data, only about 75% of the 1130 Mbp *B. napus* genome was assembled into scaffolds (Fig. 6.1). Of these scaffolds, only 57% (645 Mbp) was anchored unambiguously into pseudo-chromosomes (Chalhoub et al. 2014). Among the ten repeat families included in this survey (Table 6.3), BSTRb and BSTRa had the highest and second-highest genome proportion (GP) in the pseudo-chromosomes, respectively, followed only, but with considerably lower copy numbers, by CentB1 and CentB2. No rDNA or TE sequences were represented in the anchored assembly, except for the truncated BoCopia element. From the unanchored sequences in the current assembly, several copies of 5S rDNA, and a single copy each of *CACTA* and pCRBr elements were added to the total captured REs (Table 6.3). Nevertheless, still no 45S rDNA and CRB were included in the total assembly. Their long sequence, highly repetitive nature, and pericentromeric location could help explain their exclusion in the current assembly (Pop and Salzberg 2008; Wang et al. 2011a). Altogether, these major repeats covered only about 2.3 Mbp or less than 1% in the current assembly (Table 6.3). As expected, this assembly did not provide robust information about the proportional abundance of major REs in the *B. napus* genome.

Upon read mapping of WGS to representative sequences of these elements, the captured elements increased dramatically to about 124 Mb, which is equivalent to about 11% of the genome (Table 6.3). All elements were well represented, even the 45S rDNA, which was not represented in the assembly, was about double the copies of both BSTRs together. In fact, it had the second-highest genome proportion (2.7%), second only to CentB1 (3.6%). In terms of copy number, the shortest elements, CentB1 and CentB2, had the most numbers (228,030 and 51,093, respectively), and the dispersed BoCopia had the least (284 copies). Overall, based on total accumulated length of all ten repeat elements, the in silico WGS read mapping captured 55 times more than what was present in the current assembly. This corroborates the observed accumulation of major repeats in the unassembled portions.

In the current *B. napus* genome assembly, about 35% comprises TEs alone—excluding TRs (Chalhoub et al. 2014). This represented 97% of the total REs included in the assembly (Fig. 6.1). With the RE families in this work, we captured 11% of the total genome which represented 45%

Table 6.3 Summary of major repeat content in the *Brassica napus* 'Darmor-*bzh*' reference genome assembly and in WGS sequence of nine accessions

Repeat element	Unit length (bp)	Amounts in *B. napus* reference genome							Estimated amounts in 1 × WGS (1130 Mbp)[a]			GP, FISH (%)	GA (%)[c]	Element source
		Pseudo-molecule (645 Mb)		Unanchored scaffolds (205 Mb)		Total (850 Mb)		GP[b] (%)	Copy	Length (Kb)	GP[b] (%)			
		Copy	Kb	Copy	Kb	Copy	Kb							
CentB1	177	307	51.3	29	4.7	336	55.9	0.005	228,030.9 (±60748.1)	40,361.5 (±10752.4)	3.572	8.0	0.14	Liu et al. (2014)
CentB2	177	313	51.1	205	33.7	518	84.8	0.008	51,092.9 (±16975.8)	9043.4 (±3004.7)	0.800	2.0	0.94	Liu et al. (2014)
5S nrDNA	501	–	–	45	22.1	45	22.1	0.002	5146.9 (±1712.7)	2578.6 (±858.1)	0.228	0.9	0.86	Waminal et al (2015)
45S nrDNA	7456	–	–	–	0.0	–	–	0.000	4088.8 (±2353.2)	30,485.1 (±17546.5)	2.698	5.3	0.00	Waminal et al (2015)
BSTRa	350	1217	408.0	300	101.0	1517	509.0	0.045	20,348.8 (±10066.5)	7122 (±3523.3)	0.630	2.7	7.15	This study
BSTRb	351	3716	1251.5	916	317.6	4632	1569.1	0.139	23,141.8 (±12753.7)	8122.7 (±4476.6)	0.719	4.1	19.32	This study
CRB	5908	–	–	–	0.0	–	–	0.000	1168.4 (±303.1)	6901.5 (±1790.1)	0.611	2.8	0.00	Liu et al. (2014)
pCRBr	8395	–	–	1	8.1	1	8.1	0.001	960.4 (±406.9)	8216.6 (±3468.8)	0.727	1.9	0.10	Lim et al. (2007)
BoCopia	6711	1	5.6	–	0.0	1	5.6	0.000	284.4 (±52.6)	1909.8 (±353.3)	0.169	1.0	0.29	This study
BoCACTA	7675	–	–	1	7.6	1	7.6	0.001	1265.6 (±171.6)	9713.1 (±1318.5)	0.860	2.5	0.08	Alix et al. (2008)
Total		5554	1767.5	1496	496.7	7050	2254.1	0.199	335,528.9 (±89796.8)	124,454.5 (±17634.4)	11.014	31.2	1.81	

[a]Repeats were estimated based on the *in silico* read mapping analysis on 9 *B. napus* accessions and the average values were represented to 1 × WGS coverage (850 Mb)
[b]Genome proportion (GP): (total amounts in Kb/1130,000 Kb) × 100
[c]Genome assembly (GA) repeats: percentage of repeat amounts in reference genome compared with the estimated amounts in 1 × WGS
Kb Amounts in kilo base pair, *SD* standard deviation

of the total unassembled fractions. The remaining 55% of this fraction needs to be analyzed further. Considering that only four TEs and six TRs were used in this survey could partially explain why only 11% of the genome was captured. Apparently, it may be necessary to also check other elements, especially other non-LTR retrotransposons, *Ty1/Copia*, other *CACTA* elements, miniature TEs, and *Helitron*s, which are well represented in the current assembly (Chalhoub et al. 2014). Other members may not have been captured in the assembly and not included in our current analysis. However, an underestimation is possible considering several limitations and biases in identifying and estimating REs through computational analyses (Macas et al. 2007; Schatz et al. 2012; Treangen and Salzberg 2011).

To address this concern in quantifying genomic abundance while simultaneously checking genomic distribution of major REs, we carried out molecular cytogenetic analysis using FISH with the ten REs as probes. By signal-to-whole-chromosome area ratio, FISH facilitated estimation of this hidden fraction to about 31% (Table 6.3). CentB1 occupied the largest genomic portion followed by 45S rDNA, a pattern in congruence with that observed through WGS mapping. There is a general proportional increase of repeats estimated through molecular cytogenetics compared with those from in silico analysis. However, it is important to note that it is possible that these results could be an overestimate considering the wider area covered by fluorescence than the actual physical size. Thus, there is room for the development of more accurate RE quantification approaches. However, presently both WGS read mapping and FISH present a plausible approach toward this objective.

6.4.2 Cytogenetic Mapping of Major Repeats Discriminates Subgenomes and Individual Chromosomes

Identification of individual chromosomes is necessary for integrating genetic linkage groups and physical maps (Jiang and Gill 2006), and in understanding the dynamics of genomes in comparative cytogenomic studies (Iourov et al. 2008), particularly in the context of crop improvement. However, this has often been difficult especially among crops with small chromosome sizes, monomorphic chromosome arm ratios, and similar chromosome lengths (Waminal et al. 2012; Pich et al. 1995). Even with the availability of several routinely used cytogenetic markers such as 5S and 45S rDNA, which are localized to only a few chromosomes, other chromosomes are often difficult to identify. This is further aggravated by polyploidy, which increases chromosome number (Vrána et al. 2015). For instance, identifying subgenomes in *B. napus* has proven difficult due to the high homology of the A_n and C_n subgenomes (Snowdon et al. 1997). Although genomic in situ hybridization (GISH) studies have discriminated these two subgenomes, clear distinct discriminating signals are often difficult to obtain (Snowdon et al. 1997; Howell et al. 2008). Meanwhile, genomic distribution of major repetitive DNA has shown potential for identifying individual chromosomes and in resolving subgenomes without GISH (Macas et al. 2007; Hribova et al. 2010; Alix et al. 2008; Choi et al. 2014).

The C_n subgenome-specific hybridization of BoCACTA and BoCopia elements enabled easy and accurate discrimination between A_n and C_n subgenomes without many background signals using a general FISH procedure without the need for block DNA such as needed in GISH (Fig. 6.2; Alix et al. 2008). This was particularly useful in discriminating the overlapping chromosome lengths of the shorter chromosomes of C_n from longer chromosomes of A_n. It is important to note that although C_n chromosomes are generally longer than A_n, the shorter chromosomes of C_n such as C_n09 could be difficult to distinguish from those in A_n such as A_n07.

Another important method to accurately identify chromosomes is multicolor-FISH (Koo et al. 2004; Wang et al. 2012). This allows mapping of several probes (five probes in this case) in one FISH experiment (e.g., Kato et al. 2004), and if chromosomes are in good condition,

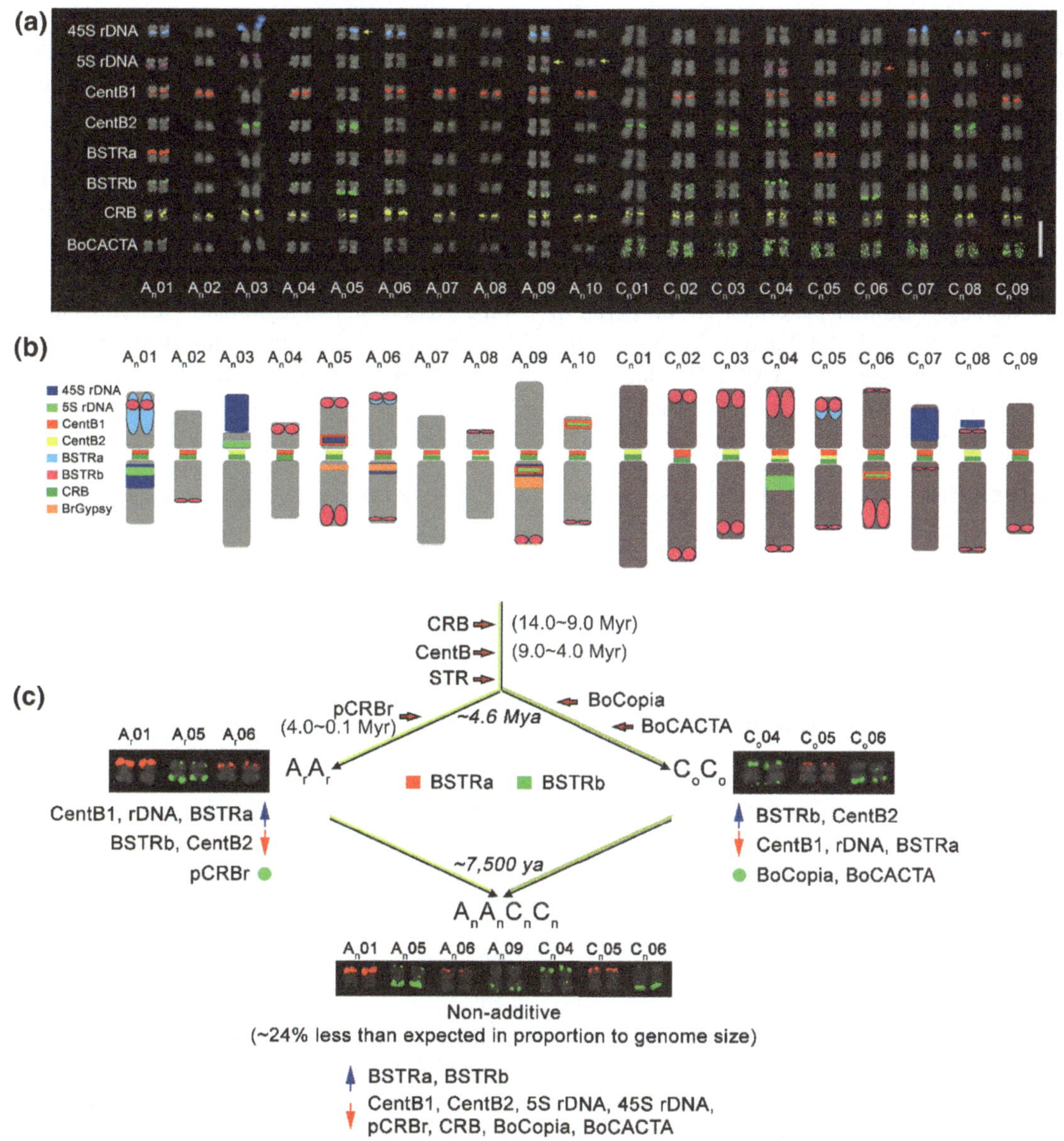

Fig. 6.2 Genomic distribution and evolution of major repeats in *Brassica napus*. **a** Karyogram of *B. napus* based on the distribution of major DNA repeats. Yellow and red arrows indicate major chromosomal rearrangements within the A_nA_n and C_nC_n subgenomes, respectively. Note that the C_nC_n chromosomes, although fewer in number, are generally larger than those of the A_nA_n, reflecting the genomic difference between the two diploid species. CRB is seen in all chromosomes while BoCACTA elements are specific to C_nC_n subgenome. Bar = 10 μm. **b** Karyotype idiogram of *B. napus* rDNA with red border represents hemizygous loci, most likely from homeologous unequal crossover. Darker chromosomes of the C_n subgenome indicate preferential hybridization of BoCACTA and BoCopia transposable elements. **c** Evolutionary dynamics of *Brassica* major repeats. Blue and red arrows indicate GP increase and decrease, respectively. Green circles indicate subgenome specificity of repeats

slides can be reprobed (Jiang and Gill 2006) up to four or five times; thus, increasing the number of probes ($5 \times 4 \sim 5 = 20 \sim 25$) to be analyzed in a shorter period of time while allowing more accurate characterization of individual chromosomes from a single chromosome spread. In this approach, illegitimate recombinations involving REs could also be easily detected by comparing signal patterns from different probes. For example, an apparent loss of a 45S rDNA locus in one A_n05 homolog resulted to a hemizygous 45S rDNA A_n05 locus (Fig. 6.2). Additionally, another 45S rDNA locus at C_n08 had an unbalanced copy number between the two homologs as manifested by a considerably reduced signal in one homolog. Similar patterns were observed in some 5S rDNA loci (A_n09 and A_n10). Compared with diploid *B. rapa*, the A_n06 STRa locus was more reduced (Waminal et al. 2015). These physically observed changes in locus size could be explained by homologous recombination-mediated unequal crossovers, and tandem repeats are hot spots of these events (Kolomietz et al. 2002; Plohl et al. 2012).

Hence, cytogenetic mapping of these ten REs enabled an estimation of their genome abundance, easy discrimination of subgenomes, and identification of individual chromosomes and some associated illegitimate recombinations. A summary of individual chromosome features observed through FISH analysis is listed in Table 6.4.

Table 6.4 Features of individual *Brassica napus* chromosomes based on cytogenetic mapping of major repeats

A_n01	5S and 45S rDNA loci colocalized on the pericentromeric area of the long arm, CentBo1 on centromere, major BoSTRa/b locus at the telomeric area of the short arm, a weak proximal BoSTRc signal on the short arm, CRB
A_n02	CentBo1 on centromere and a weak BoSTRc on the subtelomere of the long arm, CRB
A_n03	NOR at the short arm, 5S rDNA at proximal region of short arm, CentBo2 on centromere, no STR signals, CRB
A_n04	CentBo1 on centromere, BoSTRc on subtelomeric region of both arms with stronger signal on short arm, CRB
A_n05	CentBo2 on centromere, major BoSTRc at telomeric region of long arm, and another weaker BoSTRc on the subtelemeric region of the short arm, hemizygous 45S rDNA translocation on pericentric region of short arm, pCRBr, CRB
A_n06	45S rDNA locus at the pericentromeric area of long arm, CentBo1 on centromere, the major BoSTRa/b locus at telomeric region of short arm is reduced compared to its ortholog in *B. rapa* (A_r), weak colocalized BoSTRc locus on short arm, pCRBr, CRB
A_n07	CentBo1, CRB
A_n08	CentBo1, weak BrSTRb at telomeric region of short arm, CRB
A_n09	Increased 45S rDNA signal at the intercalary region of the long arm when compared with its ortholog in *B. rapa*, CentBo1 on centromere, BoSTc at telomeric region of long arm, pCRBr, CRB
A_n10	Weak 5S rDNA locus at intercalary region of short arm, CentBo1, weak BoSTRc on long arm, CRB
C_n01	Increased CentBo2 signal compared to its paralog in *B. oleracea*, no other readily observable repeat signals, CRB
C_n02	CentBo1 signal, weak centromeric BoSTRc and telomeric BoSTRc on both arms, CRB
C_n03	CentBo2, telomeric BoSTRc at both arms, CRB
C_n04	With 5S rDNA at the pericentromeric area of long arm, CentBo1 and CentBo2, telomeric BoSTRc at both arms with major signal on short arm, CRB
C_n05	With CentBo1 and CentBo2, centromeric and telomeric BoSTRa/b on short arm being the only major BoSTRa/b signal, weak telomeric BoSTRc, CRB
C_n06	CentBo1, major BoSTRc signal on long arm, CRB
C_n07	45S rDNA on short arm, CentBo1, CRB
C_n08	Weak 45S rDNA on short arm, CentBo2, weak BoSTRc on both arms, CRB
C_n09	CentBo1 and weak CentBo2, weak telomeric BoSTRc on long arm, CRB

6.5 Comparative Repeatomics Reveals the Dynamics of Major Repeats in *Brassica* Species

The ancestral karyotype of the family Brassicaceae consisted of eight chromosomes which underwent several rounds of genome duplication and subsequent lineage-specific rearrangements, particularly involving REs, resulting to nine chromosomes of *B. oleracea* being larger than those of the ten chromosomes of *B. rapa* (Liu et al. 2014; Lysak et al. 2006). Comparative analysis of major repeats between *B. oleracea* and *B. rapa* genomes provides two scenarios of RE dynamics. The first suggests a continuous amplification of TEs and TRs in the *B. oleracea* genome over time after its divergence with *B. rapa* about 4.6 million years ago, increasing the genome size of *B. oleracea* (630 Mbp) to more than that of *B. rapa* (540 Mbp) (Liu et al. 2014). The second posits a rapid loss of transposable elements, e.g., BoCACTA, in *B. rapa* during divergence from *B. oleracea*, which was possibly driven by a slower reestablishment of epigenetic control that could have prevented homology-dependent, illegitimate recombination-induced repeat loss in *B. rapa* (Fedoroff 2012; Kelly et al. 2015).

During allopolyploidization, the merging of two genomes often results in genomic shock (Fedoroff and Bennetzen 2013; Fedoroff 2012; Renny-Byfield et al. 2013). This consequently initiates genome reprogramming by altering epigenetic makeup. Although the exact mechanisms and timeframe by which these events happen is not yet fully understood (Fedoroff 2013a), we know that this often leads to genome downsizing through elimination of DNA segments (Renny-Byfield et al. 2013; Renny-Byfield and Wendel 2014), often repetitive DNA fragments, a process aimed at reestablishing stable meiotic pairing and fertility in incipient allopolyploids (Fedoroff 2012; Renny-Byfield and Wendel 2014). In the absence of genome downsizing and element amplification in allopolyploids, an additive number of elements relative to the diploid progenitors can be expected. However, genome downsizing after allopolyploidization seems to be a rule rather than an exemption, although increased genome sizes have been reported (Renny-Byfield et al. 2013). DNA loss could even be biased toward a specific subgenome such as those observed in *Nicotiana tabacum* (Renny-Byfield et al. 2012), and allopolyploid cotton (Paterson et al. 2012). Consequently, the resulting allopolyploid often has a unique genomic make up relative to the diploid progenitors.

Accordingly, eight of the ten *B. napus* repeat elements in this survey showed a non-additive reduction of size; in fact, about a 24% GP reduction than what was expected relative to the genome sizes of the diploid progenitors (Figs. 6.2 and 6.3). The assembled centromeric repeats were the most reduced, followed by 45S rDNA, pCRBr, 5S rDNA, BoCACTA, BoCopia, and CRB (Fig. 6.3). However, BSTRs show non-additive amplification in the *B. napus* genome compared with its diploid progenitors, with BSTRb having more copies than BSTRa (Figs. 6.2 and 6.3). Satellite DNA regions are amplified/contracted in a very short evolutionary time as a result of unequal crossover between homologous sequences (for review on satellite DNA evolution, Plohl et al. 2012). Moreover, 45S rDNA loci are often targets of rapid locus elimination and reorganization among polyploids (Pellicer et al. 2010b, c). An increase of BSTRs in *B. napus* may have added benefits and consequently could have undergone positive selection, whereas other extra elements of other repeat families may not be necessary at all (Plohl et al. 2012).

Aside from interspecific variations with its diploid progenitors, copy number and GR size variation among the nine *B. napus* accessions were also observed (Fig. 6.4). Seven of the nine accessions showed relatively similar amounts of REs. However, two accessions, Bn-1 and Bn-2, generally had much lower TR copies, although they have much more 45S rDNA, compared with the other seven accessions (Fig. 6.4a, c). Additionally, centromeric and pericentromeric retrotransposons were more abundant in these two accessions (Fig. 6.4b, d). A similar intra-species repeat number variation was reported among

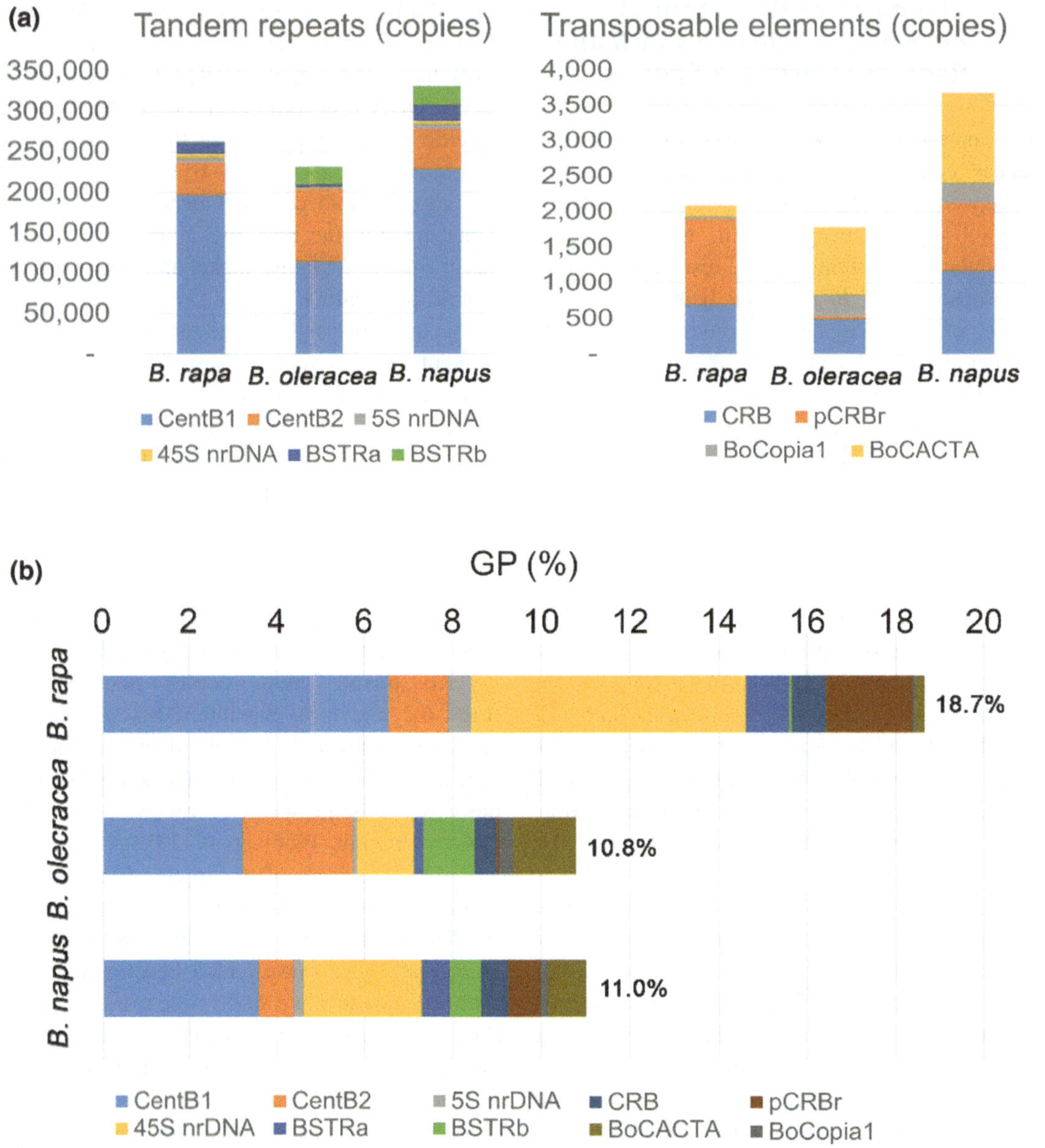

Fig. 6.3 Comparative analysis of major repeats in *B. rapa*, *B. oleracea*, and *B. napus*. **a** Copy number of each repeat family in each genome analyzed. Values for TR and TE are shown in separate charts. **b** Genome proportions of major repeats for each genome analyzed

several *B. oleracea* morphotypes (Perumal et al. 2017). In this previous study, some morphotypes, such as cauliflower and broccoli, had more CentBo1 than CentBo2. Some morphotypes, or accessions, apparently have unique RE compositions. We are aware of the limitations of in silico analysis in quantifying these repeats, which could have contributed to the observed value differences, especially considering the fewer WGS reads used in Bn-1 and Bn-2 (Table 6.1). However, the stark higher abundance of 45S rDNA and total TE in these two *B. napus* accessions indicate RE abundance independent from the amount of random WGS reads used.

Fig. 6.4 Summary of major repeat composition in the nine *B. napus* accessions based on reference mapping. **a**, **b** Copy number-based genomic representation of tandem repeats and transposable elements among nine *B. napus* accessions. **c**, **d** Corresponding length in kilobase from **a**, **b**

Consequently, the impact of variation in RE abundance warrants further analyses, especially when taking into account a previous report that demonstrated a link between TE abundance variation and environmental adaptation (Kalendar et al. 2000).

Meanwhile, genome specificity of some TEs, as observed in the diploid progenitors, has been retained within the *B. napus* genome. BoCopia and BoCACTA retained their C_n specificity, and pCRBr its A_n specificity (Figs. 6.2 and 6.3). How certain elements are retained in different subgenomes, in the context of allopolyploidization, can be explained by epigenetic control mechanisms (Plohl et al. 2012; Fedoroff 2012), which will be discussed further in the following section.

6.6 Evolutionary Implications for Fluctuation of RE Fraction

Genome size variation within and among species, regardless of organism complexity or the number of protein coding genes, is commonly known as the C-value paradox (Pagel and Johnstone 1992; Freeling et al. 2015; Rosbash et al. 1974; Eddy 2012), and is attributed, as mentioned above, mainly to the size fluctuation of REs copy number. For instance, differential accumulation of several TE families, including tandem repeats, has defined the genomes of several *Fritillaria* species in the absence of WGD (Kelly et al. 2015). This phenomenon also caused intraspecies genome size variations in *Helianthus annuus* and other plants (Price and Johnston 1996; Wendel and Wessler 2000). Moreover, growing evidence supports the importance of this genomic fraction in proper genome function and evolution (Wei et al. 2013; Nowak 1994; Shapiro and von Sternberg 2005; Pardue and DeBaryshe 2003; Hall et al. 2005; Biémont and Vieira 2006; Freeling et al. 2015; Kalendar et al. 2000), particularly regarding their significant roles in chromosome segregation, gene expression, and heterochromatin maintenance (Pardue and DeBaryshe 2003; Biemont 2010; Sarilar et al. 2011; Sampath et al. 2013; Wolfgruber et al. 2009; Goodier et al. 2012; Peng and Karpen 2008). An important study by Kalendar et al. (2000) revealed the link between RE content fluctuation among individuals within a species and environmental adaptation. These examples demonstrate the adaptive and evolutionary importance of REs.

Studying the fluctuation of RE fraction is therefore an invaluable approach in understanding phylogenetic relationships, since genomic changes are quantifiable and can reveal variations among accessions and species. For example, the RE composition of *B. rapa* and *B. oleracea* are unique for each species (Waminal et al. 2015), so are the RE compositions among accessions in *B. oleracea* (Waminal et al. 2016) and *B. napus*. However, in *Brassica*, no studies have yet shown the direct phenotypic impact of RE fraction size variation to crop biology.

In *Brassica*, heterochromatins are mostly localized in centromeric and pericentromeric regions (Lim et al. 2007), where most REs are localized (Fig. 1A, B). CRB is a common centromeric component of the *B. rapa* (A_r), *B. nigra* (B_n), and *B. oleracea* (C_o) genomes. However, the absence of CentB hybridization in *B. nigra* supports the earlier divergence of the B_n genome from the A_r and C_o genomes (Lim et al. 2007; Koo et al. 2011; Arias et al. 2014). However, FISH analysis of *Brassica* STR showed genome-specific evolution of these subtelomeric repeats (Waminal et al. 2016) since their divergence. BSTRa seemed to be 'preferentially' selected in the *B. rapa* genome compared with BSTRb, while the opposite was observed in *B. oleracea*. This eventually led to a greater abundance of BSTRb than BSTRa in *B. napus* after the genome merger. Mechanisms that control their retention or elimination are being studied in more detail (Fedoroff 2012; Fablet and Vieira 2011).

Understanding how REs are controlled is necessary to exploit their underlying potential for crop improvement. Studies on sophisticated plant epigenetic control mechanisms, (Haag and Pikaard 2011; Slotkin and Martienssen 2007; Fedoroff 2012; Bennetzen and Wang 2014) for example, have elucidated this objective. DNA and histone modifications, which have a central feedback control mechanism involving siRNAs, are at the core of genome dynamics regulation to ensure genome homeostasis (see Haag and Pikaard 2011; Peng and Karpen 2008; Fedoroff and Bennetzen 2013; Fedoroff 2012). Events such as abiotic stress responses (Petit et al. 2010), polyploidization, or small-scale duplications (Renny-Byfield et al. 2013; De Smet et al. 2013) that disrupt this homeostasis can initiate TE and TR removal or accumulation. The trade-off between removal and accumulation of repeat elements depends on the temporal reestablishment of the epigenetic mechanisms that buffer the adverse effects of TEs and TRs, such as aneuploidy—or worse, sterility (Kelly et al. 2015; Fedoroff 2012). After genomic shock, rapid reestablishment of epigenetic control enables regulation of REs, locking them to recombinationally inert heterochromatin, resulting in larger genomes than when epigenetic mechanisms were reestablished more slowly. The latter provides more opportunities for homologous and illegitimate recombinations that removes DNA fragments and causes genome downsizing to occur (Kelly et al. 2015; Fedoroff 2012).

The same mechanisms (i.e., unequal crossovers of homologous sequences and repeat transposition) that are responsible for DNA segment deletion are also models that explain the homogenization and spread of repeats between sister chromatids, homologous chromosomes, and non-homologous chromosomes (Walsh 1987; Cohen et al. 2003; Hall et al. 2005; Charlesworth et al. 1994; Dover 1982). Unequal crossovers usually result in higher-order repeat units consisting of more than one type of element and variation in lengths of arrays (Hall et al. 2005; Talbert and Henikoff 2010; Plohl et al. 2012). Other mechanisms such as gene conversion, repeat transposition, and rolling circle replication may amplify satellite arrays and cause their spread into non-homologous chromosomes (Hall et al. 2005; Dover 1986; Plohl et al. 2012). Epigenetic control is an active cellular mechanism that controls when recombination and transposition should occur. Nonetheless, clear reasons regarding how and why they happen in response to abiotic stresses are unknown (Fedoroff 2013b).

6.7 Summary and Perspectives

As demonstrated in previous studies in *Pisum sativum* (Macas et al. 2007), *Musa acuminata* (Hribova et al. 2010), and some *Brassica* species (Waminal et al. 2015; Perumal et al. 2017; Waminal et al. 2016), a survey of plant genomes using NGS data and reference-guided mapping (Kim et al. 2015) together with FISH analysis is an excellent approach for quantifying and physically mapping repetitive genomic elements that are mostly omitted during assembly. This approach captured about 11% of the *B. napus* genome and enabled comparative 'repeatomics' analysis with its diploid progenitors. The fluctuating pattern of total RE fraction between *B. napus* and its diploid progenitors, as well as among different *B. napus* accessions further demonstrates that RE dynamics is responsible for the huge genome size variations among accessions of the same species (Wendel and Wessler 2000) or species in the same genus (Kelly et al. 2015; Renny-Byfield et al. 2013). We know that epigenetic control is at the center of this fluctuation; nevertheless, even with the current advances in genomics and epigenetics, accurate reasons for how and why these REs respond to abiotic stresses remain unknown. However, with further research, a robust explanation of the mechanisms that underlie the interconnectedness of the environment, genome, and organisms will be determined.

The empirically demonstrated correlation of TE size variation and environmental adaptation within species (Kalendar et al. 2000) is interesting, but whether repeatomics could have a predictive value in relation to agronomically favorable traits is questionable, but perhaps worth pursuing, particularly in the context of crop improvement, such as the oil content in *B. napus* (Delourme et al. 2006). Additionally, the power of WGS and FISH estimation approaches may further be corroborated by optical mapping (Tang et al. 2015; Lam et al. 2012) to provide accurate, single-molecule resolutions of mega-base tandem repeats, which represented a large portion of the unassembled fractions of the three species in this work. This would be particularly useful in analyzing RE fractions to support sequencing projects of species with large genomes, such as *Allium* species (Jakse et al. 2008), *Fritillaria* species (Kelly et al. 2015), and *Paris japonica* (Pellicer et al. 2010a).

References

Alix K, Joets J, Ryder CD, Moore J, Barker GC, Bailey JP, King GJ, Heslop-Harrison JS (2008) The CACTA transposon Bot1 played a major role in Brassica genome divergence and gene proliferation. Plant J 56(6):1030–1044. https://doi.org/10.1111/j.1365-313X.2008.03660.x

Alkan C, Sajjadian S, Eichler EE (2011) Limitations of next-generation genome sequence assembly. Nat Methods 8(1):61–65. https://doi.org/10.1038/nmeth.1527

Arias T, Beilstein MA, Tang M, McKain MR, Pires JC (2014) Diversification times among Brassica (Brassicaceae) crops suggest hybrid formation after 20 million years of divergence. Am J Bot 101(1):86–91. https://doi.org/10.3732/ajb.1300312

Bennetzen JL, Wang H (2014) The contributions of transposable elements to the structure, function, and evolution of plant genomes. Annu Rev Plant Biol 65:505–530. https://doi.org/10.1146/annurev-arplant-050213-035811

Biemont C (2010) A brief history of the status of transposable elements: from junk DNA to major players in evolution. Genetics 186(4):1085–1093. https://doi.org/10.1534/genetics.110.124180

Biémont C, Vieira C (2006) Genetics: junk DNA as an evolutionary force. Nature 443(7111):521–524

Chadwick BP (2009) Macrosatellite epigenetics: the two faces of DXZ4 and D4Z4. Chromosoma 118(6):675–681. https://doi.org/10.1007/s00412-009-0233-5

Chalhoub B, Denoeud F, Liu S, Parkin IAP, Tang H, Wang X, Chiquet J, Belcram H, Tong C, Samans B, Corréa M, Da Silva C, Just J, Falentin C, Koh CS, Le Clainche I, Bernard M, Bento P, Noel B, Labadie K, Alberti A, Charles M, Arnaud D, Guo H, Daviaud C, Alamery S, Jabbari K, Zhao M, Edger PP, Chelaifa H, Tack D, Lassalle G, Mestiri I, Schnel N, Le Paslier M-C, Fan G, Renault V, Bayer PE, Golicz AA, Manoli S, Lee T-H, Thi VHD, Chalabi S, Hu Q, Fan C, Tollenaere R, Lu Y, Battail C, Shen J, Sidebottom CHD, Wang X, Canaguier A, Chauveau A, Bérard A, Deniot G, Guan M, Liu Z, Sun F, Lim YP, Lyons E, Town CD, Bancroft I, Wang X, Meng J, Ma J, Pires JC, King GJ, Brunel D, Delourme R, Renard M, Aury J-M, Adams KL, Batley J, Snowdon RJ, Tost J, Edwards D, Zhou Y, Hua W, Sharpe AG, Paterson AH, Guan C, Wincker P (2014) Early allopolyploid evolution in the post-Neolithic *Brassica napus* oilseed genome.

Science 345(6199):950–953. https://doi.org/10.1126/science.1253435
Charlesworth B, Sniegowski P, Stephan W (1994) The evolutionary dynamics of repetitive DNA in eukaryotes. Nature 371(6494):215–220. https://doi.org/10.1038/371215a0
Choi HI, Waminal NE, Park HM, Kim NH, Choi BS, Park M, Choi D, Lim YP, Kwon SJ, Park BS, Kim HH, Yang TJ (2014) Major repeat components covering one-third of the ginseng (*Panax ginseng* C. A. Meyer) genome and evidence for allotetraploidy. Plant J (For Cell and Molecular Biology) 77(6):906–916. https://doi.org/10.1111/tpj.12441
Cizkova J, Hribova E, Humplikova L, Christelova P, Suchankova P, Dolezel J (2013) Molecular analysis and genomic organization of major DNA satellites in banana (Musa spp.). PLoS ONE 8(1):e54808. https://doi.org/10.1371/journal.pone.0054808
Claros MG, Bautista R, Guerrero-Fernandez D, Benzerki H, Seoane P, Fernandez-Pozo N (2012) Why assembling plant genome sequences is so challenging. Biology 1 (2):439–459. https://doi.org/10.3390/biology1020439
Cohen S, Yacobi K, Segal D (2003) Extrachromosomal circular DNA of tandemly repeated genomic sequences in Drosophila. Genome Res 13(6A):1133–1145. https://doi.org/10.1101/gr.907603
Coluccia E, Pichiri G, Nieddu M, Coni P, Manconi S, Deiana AM, Salvadori S, Mezzanotte R (2011) Identification of two new repetitive elements and chromosomal mapping of repetitive DNA sequences in the fish Gymnothorax unicolor (Anguilliformes: Muraenidae). Eur J Histochem (EJH) 55(2):e12. https://doi.org/10.4081/ejh.2011.e12
De Smet R, Adams KL, Vandepoele K, Van Montagu MC, Maere S, Van de Peer Y (2013) Convergent gene loss following gene and genome duplications creates single-copy families in flowering plants. Proc Natl Acad Sci 110(8):2898–2903
Delourme R, Falentin C, Huteau V, Clouet V, Horvais R, Gandon B, Specel S, Hanneton L, Dheu J, Deschamps M (2006) Genetic control of oil content in oilseed rape (*Brassica napus* L.). Theoret Appl. Genetics 113(7):1331–1345
Dover G (1982) Molecular drive: a cohesive mode of species evolution. Nature 299(5879):111–117
Dover GA (1986) Molecular drive in multigene families: how biological novelties arise, spread and are assimilated. Trends Genet 2:159–165
Eddy SR (2012) The C-value paradox, junk DNA and ENCODE. Curr Biol 22(21):R898–R899
Fablet M, Vieira C (2011) Evolvability, epigenetics and transposable elements. Biomol Concepts 2(5):333–341. https://doi.org/10.1515/BMC.2011.035
Fedoroff NV (2012) Presidential address. Transposable elements, epigenetics, and genome evolution. Science 338(6108):758–767
Fedoroff NV (2013a) McClintock and epigenetics. In: Plant transposons and genome dynamics in evolution. Wiley, pp 61–70. https://doi.org/10.1002/9781118500156.ch4
Fedoroff NV (2013b) Molecular genetics and epigenetics of CACTA elements. Methods Mol Biol 1057:177–192. https://doi.org/10.1007/978-1-62703-568-2_13
Fedoroff NV, Bennetzen JL (2013) Transposons, genomic shock, and genome evolution. In: Plant transposons and genome dynamics in evolution. Wiley, pp 181–201. https://doi.org/10.1002/9781118500156.ch10
Freeling M, Xu J, Woodhouse M, Lisch D (2015) A solution to the C-value paradox and the function of junk DNA: the genome balance hypothesis. Mol Plant 8(6):899–910. https://doi.org/10.1016/j.molp.2015.02.009
Gao D, Jiang N, Wing RA, Jiang J, Jackson SA (2015) Transposons play an important role in the evolution and diversification of centromeres among closely related species. Front Plant Sci 6
Goodier JL, Cheung LE, Kazazian HH Jr (2012) MOV10 RNA helicase is a potent inhibitor of retrotransposition in cells. PLoS Genet 8(10):e1002941. https://doi.org/10.1371/journal.pgen.1002941
Haag JR, Pikaard CS (2011) Multisubunit RNA polymerases IV and V: purveyors of non-coding RNA for plant gene silencing. Nat Rev Mol Cell Biol 12 (8):483–492. https://doi.org/10.1038/nrm3152
Hall SE, Luo S, Hall AE, Preuss D (2005) Differential rates of local and global homogenization in centromere satellites from arabidopsis relatives. Genetics 170(4):1913–1927. https://doi.org/10.1534/genetics.104.038208
Hamilton JP, Buell CR (2012) Advances in plant genome sequencing. Plant J (For Cell and Molecular Biology) 70(1):177–190. https://doi.org/10.1111/j.1365-313X.2012.04894.x
Heslop-Harrison JS (2000) Comparative genome organization in plants: from sequence and markers to chromatin and chromosomes. Plant Cell 12(5):617–636
Howell EC, Kearsey MJ, Jones GH, King GJ, Armstrong SJ (2008) A and C genome distinction and chromosome identification in *Brassica napus* by sequential fluorescence in situ hybridization and genomic in situ hybridization. Genetics 180(4):1849–1857. https://doi.org/10.1534/genetics.108.095893
Hribova E, Neumann P, Matsumoto T, Roux N, Macas J, Dolezel J (2010) Repetitive part of the banana (*Musa acuminata*) genome investigated by low-depth 454 sequencing. BMC Plant Biol 10:204. https://doi.org/10.1186/1471-2229-10-204
Iourov IY, Vorsanova SG, Yurov YB (2008) Molecular cytogenetics and cytogenomics of brain diseases. Curr Genomics 9(7):452–465. https://doi.org/10.2174/138920208786241216
Jakse J, Meyer JD, Suzuki G, McCallum J, Cheung F, Town CD, Havey MJ (2008) Pilot sequencing of onion genomic DNA reveals fragments of transposable elements, low gene densities, and significant gene enrichment after methyl filtration. Mol Genetics Genomics (MGG) 280(4):287–292. https://doi.org/10.1007/s00438-008-0364-z

Jiang JM, Gill BS (2006) Current status and the future of fluorescence in situ hybridization (FISH) in plant genome research. Genome/Nat Res Counc Can (Genome/Conseil national de recherches Canada) 49 (9):1057–1068. https://doi.org/10.1139/G06-076

Kalendar R, Tanskanen J, Immonen S, Nevo E, Schulman AH (2000) Genome evolution of wild barley (*Hordeum spontaneum*) by BARE-1 retrotransposon dynamics in response to sharp microclimatic divergence. Proc Natl Acad Sci USA 97(12):6603–6607. https://doi.org/10.1073/pnas.110587497

Kato A, Lamb JC, Birchler JA (2004) Chromosome painting using repetitive DNA sequences as probes for somatic chromosome identification in maize. Proc Natl Acad Sci 101(37):13554–13559. https://doi.org/10.1073/pnas.0403659101

Kelly LJ, Renny-Byfield S, Pellicer J, Macas J, Novak P, Neumann P, Lysak MA, Day PD, Berger M, Fay MF, Nichols RA, Leitch AR, Leitch IJ (2015) Analysis of the giant genomes of Fritillaria (Liliaceae) indicates that a lack of DNA removal characterizes extreme expansions in genome size. New Phytol 208(2):596–607. https://doi.org/10.1111/nph.13471

Kim K, Lee S-C, Lee J, Yu Y, Yang K, Choi B-S, Koh H-J, Waminal NE, Choi H-I, Kim N-H, Jang W, Park H-S, Lee J, Lee HO, Joh HJ, Lee HJ, Park JY, Perumal S, Jayakodi M, Lee YS, Kim B, Copetti D, Kim S, Kim S, Lim K-B, Kim Y-D, Lee J, Cho K-S, Park B-S, Wing RA, Yang T-J (2015) Complete chloroplast and ribosomal sequences for 30 accessions elucidate evolution of Oryza AA genome species. Sci Rep 5:15655. https://doi.org/10.1038/srep15655, http://www.nature.com/articles/srep15655#supplementary-information

Kolomietz E, Meyn MS, Pandita A, Squire JA (2002) The role of Alu repeat clusters as mediators of recurrent chromosomal aberrations in tumors. Genes Chromosom Cancer 35(2):97–112

Koo DH, Plaha P, Lim YP, Hur Y, Bang JW (2004) A high-resolution karyotype of *Brassica rapa* ssp. pekinensis revealed by pachytene analysis and multicolor fluorescence in situ hybridization. TAG Theoretical and applied genetics Theoretische und angewandte Genetik 109(7):1346–1352. https://doi.org/10.1007/s00122-004-1771-0

Koo DH, Hong CP, Batley J, Chung YS, Edwards D, Bang JW, Hur Y, Lim YP (2011) Rapid divergence of repetitive DNAs in Brassica relatives. Genomics 97 (3):173–185. https://doi.org/10.1016/j.ygeno.2010.12.002

Lam ET, Hastie A, Lin C, Ehrlich D, Das SK, Austin MD, Deshpande P, Cao H, Nagarajan N, Xiao M (2012) Genome mapping on nanochannel arrays for structural variation analysis and sequence assembly. Nat Biotechnol 30(8):771–776

Lee J, Waminal NE, Choi HI, Perumal S, Lee SC, Nguyen VB, Jang W, Kim NH, Gao LZ, Yang TJ (2017) Rapid amplification of four retrotransposon families promoted speciation and genome size expansion in the genus *Panax*. Sci Rep 7(1):9045. https://doi.org/10.1038/s41598-017-08194-5

Lim KB, Yang TJ, Hwang YJ, Kim JS, Park JY, Kwon SJ, Kim J, Choi BS, Lim MH, Jin M, Kim HI, de Jong H, Bancroft I, Lim Y, Park BS (2007) Characterization of the centromere and peri-centromere retrotransposons in *Brassica rapa* and their distribution in related Brassica species. Plant J (For Cell and Molecular Biology) 49 (2):173–183. https://doi.org/10.1111/j.1365-313X.2006.02952.x

Liu S, Liu Y, Yang X, Tong C, Edwards D, Parkin IAP, Zhao M, Ma J, Yu J, Huang S, Wang X, Wang J, Lu K, Fang Z, Bancroft I, Yang T-J, Hu Q, Wang X, Yue Z, Li H, Yang L, Wu J, Zhou Q, Wang W, King GJ, Pires JC, Lu C, Wu Z, Perumal S, Wang Z, Guo H, Pan S, Yang L, Min J, Zhang D, Jin D, Li W, Belcram H, Tu J, Guan M, Qi C, Du D, Li J, Jiang L, Batley J, Sharpe AG, Park B-S, Ruperao P, Cheng F, Waminal NE, Huang Y, Dong C, Wang L, Li J, Hu Z, Zhuang M, Huang Y, Huang J, Shi J, Mei D, Liu J, Lee T-H, Wang J, Jin H, Li Z, Li X, Zhang J, Xiao L, Zhou Y, Liu Z, Liu X, Qin R, Tang X, Liu W, Wang Y, Zhang Y, Lee J, Kim HH, Denoeud F, Xu X, Liang X, Hua W, Wang X, Wang J, Chalhoub B, Paterson AH (2014) The *Brassica oleracea* genome reveals the asymmetrical evolution of polyploid genomes. Nat Commun 5. https://doi.org/10.1038/ncomms4930

Lysak MA, Berr A, Pecinka A, Schmidt R, McBreen K, Schubert I (2006) Mechanisms of chromosome number reduction in *Arabidopsis thaliana* and related Brassicaceae species. Proc Natl Acad Sci USA 103(13):5224–5229. https://doi.org/10.1073/pnas.0510791103

Macas J, Neumann P, Navratilova A (2007) Repetitive DNA in the pea (*Pisum sativum* L.) genome: comprehensive characterization using 454 sequencing and comparison to soybean and *Medicago truncatula*. BMC Genom 8:427. https://doi.org/10.1186/1471-2164-8-427

Mehrotra S, Goyal V (2014) Repetitive sequences in plant nuclear DNA: types, distribution, evolution and function. Genomics Proteomics Bioinf 12(4):164–171. https://doi.org/10.1016/j.gpb.2014.07.003

Melters DP, Bradnam KR, Young HA, Telis N, May MR, Ruby JG, Sebra R, Peluso P, Eid J, Rank D, Garcia JF, Derisi JL, Smith T, Tobias C, Ross-Ibarra J, Korf I, Chan SW (2013) Comparative analysis of tandem repeats from hundreds of species reveals unique insights into centromere evolution. Genome Biol 14 (1):R10. https://doi.org/10.1186/gb-2013-14-1-r10

Michael TP, Jackson S (2013) The first 50 plant genomes. Plant Genome-Us 6(2). https://doi.org/10.3835/plantgenome2013.03.0001in

Mihai P, Salzberg SL, Shumway M (2002) Genome sequence assembly: algorithms and issues. Computer 35(7):47–54. https://doi.org/10.1109/MC.2002.1016901

Nowak R (1994) Mining treasures from 'junk DNA'. Science 263(5147):608–610

Pagel M, Johnstone RA (1992) Variation across species in the size of the nuclear genome supports the junk-DNA explanation for the C-value paradox. Proc R Soc Lond B 249(1325):119–124

Pardue ML, DeBaryshe PG (2003) Retrotransposons provide an evolutionarily robust non-telomerase mechanism to maintain telomeres. Annu Rev Genet 37:485–511. https://doi.org/10.1146/annurev.genet.38.072902.093115

Paterson AH, Wendel JF (2015) Unraveling the fabric of polyploidy. Nat Biotech 33(5):491–493. https://doi.org/10.1038/nbt.3217

Paterson AH, Wendel JF, Gundlach H, Guo H, Jenkins J, Jin D, Llewellyn D, Showmaker KC, Shu S, Udall J, Yoo MJ, Byers R, Chen W, Doron-Faigenboim A, Duke MV, Gong L, Grimwood J, Grover C, Grupp K, Hu G, Lee TH, Li J, Lin L, Liu T, Marler BS, Page JT, Roberts AW, Romanel E, Sanders WS, Szadkowski E, Tan X, Tang H, Xu C, Wang J, Wang Z, Zhang D, Zhang L, Ashrafi H, Bedon F, Bowers JE, Brubaker CL, Chee PW, Das S, Gingle AR, Haigler CH, Harker D, Hoffmann LV, Hovav R, Jones DC, Lemke C, Mansoor S, ur Rahman M, Rainville LN, Rambani A, Reddy UK, Rong JK, Saranga Y, Scheffler BE, Scheffler JA, Stelly DM, Triplett BA, Van Deynze A, Vaslin MF, Waghmare VN, Walford SA, Wright RJ, Zaki EA, Zhang T, Dennis ES, Mayer KF, Peterson DG, Rokhsar DS, Wang X, Schmutz J (2012) Repeated polyploidization of Gossypium genomes and the evolution of spinnable cotton fibres. Nature 492(7429):423–427. https://doi.org/10.1038/nature11798

Pellicer J, Fay MF, Leitch IJ (2010a) The largest eukaryotic genome of them all? Bot J Linn Soc 164(1):10–15. https://doi.org/10.1111/j.1095-8339.2010.01072.x

Pellicer J, Garcia S, Canela MA, Garnatje T, Korobkov AA, Twibell JD, Valles J (2010b) Genome size dynamics in Artemisia L. (Asteraceae): following the track of polyploidy. Plant Biol (Stuttg) 12(5):820–830. https://doi.org/10.1111/j.1438-8677.2009.00268.x

Pellicer J, Garnatje T, Hidalgo O, Tagashira N, Valles J, Kondo K (2010c) Do polyploids require proportionally less rDNA loci than their corresponding diploids? Examples from *Artemisia* subgenera *Absinthium* and *Artemisia* (Asteraceae, Anthemideae). Plant Biosyst 144(4):841–848. https://doi.org/10.1080/11263504.2010.522783

Peng JC, Karpen GH (2008) Epigenetic regulation of heterochromatic DNA stability. Curr Opin Genet Dev 18(2):204–211. https://doi.org/10.1016/j.gde.2008.01.021

Perumal S, Waminal NE, Lee J, Lee J, Choi BS, Kim HH, Grandbastien MA, Yang TJ (2017) Elucidating the major hidden genomic components of the A, C, and AC genomes and their influence on *Brassica* evolution. Sci Rep 7(1):17986. https://doi.org/10.1038/s41598-017-18048-9

Petit M, Guidat C, Daniel J, Denis E, Montoriol E, Bui QT, Lim KY, Kovarik A, Leitch AR, Grandbastien MA, Mhiri C (2010) Mobilization of retrotransposons in synthetic allotetraploid tobacco. New Phytol 186(1):135–147. https://doi.org/10.1111/j.1469-8137.2009.03140.x

Pich U, Meister A, Macas J, Doležel J, Lucretti S, Schubert I (1995) Primed in situ labelling facilitates flow sorting of similar sized chromosomes. Plant J 7 (6):1039–1044

Plohl M, Luchetti A, Mestrovic N, Mantovani B (2008) Satellite DNAs between selfishness and functionality: structure, genomics and evolution of tandem repeats in centromeric (hetero)chromatin. Gene 409(1–2):72–82. https://doi.org/10.1016/j.gene.2007.11.013

Plohl M, Mestrovic N, Mravinac B (2012) Satellite DNA evolution. Genome Dyn 7:126–152. https://doi.org/10.1159/000337122

Pop M, Salzberg SL (2008) Bioinformatics challenges of new sequencing technology. Trends Genet 24(3):142–149. https://doi.org/10.1016/j.tig.2007.12.006

Price HJ, Johnston JS (1996) Influence of light on DNA content of *Helianthus annuus* Linnaeus. Proc Natl Acad Sci USA 93(20):11264–11267

Renny-Byfield S, Wendel JF (2014) Doubling down on genomes: polyploidy and crop plants. Am J Bot 101 (10):1711–1725

Renny-Byfield S, Kovarik A, Chester M, Nichols RA, Macas J, Novak P, Leitch AR (2012) Independent, rapid and targeted loss of highly repetitive DNA in natural and synthetic allopolyploids of *Nicotiana tabacum*. PloS one 7(5). https://doi.org/10.1371/journal.pone.0036963 (ARTN e36963)

Renny-Byfield S, Kovarik A, Kelly LJ, Macas J, Novak P, Chase MW, Nichols RA, Pancholi MR, Grandbastien MA, Leitch AR (2013) Diploidization and genome size change in allopolyploids is associated with differential dynamics of low- and high-copy sequences. Plant J 74(5):829–839. https://doi.org/10.1111/tpj.12168

Rosbash M, Ford PJ, Bishop JO (1974) Analysis of the C-value paradox by molecular hybridization. Proc Natl Acad Sci USA 71(9):3746–3750

Sampath P, Lee SC, Lee J, Izzah NK, Choi BS, Jin M, Park BS, Yang TJ (2013) Characterization of a new high copy Stowaway family MITE, BRAMI-1 in Brassica genome. BMC Plant Biol 13:56. https://doi.org/10.1186/1471-2229-13-56

Sarilar V, Marmagne A, Brabant P, Joets J, Alix K (2011) BraSto, a Stowaway MITE from Brassica: recently active copies preferentially accumulate in the gene space. Plant Mol Biol 77(1–2):59–75. https://doi.org/10.1007/s11103-011-9794-9

Schadt EE, Turner S, Kasarskis A (2010) A window into third-generation sequencing. Hum Mol Genet 19(R2): R227–R240. https://doi.org/10.1093/hmg/ddq416

Schatz MC, Witkowski J, McCombie WR (2012) Current challenges in de novo plant genome sequencing and assembly. Genome Biol 13(4):243. https://doi.org/10.1186/gb4015

Shapiro JA, von Sternberg R (2005) Why repetitive DNA is essential to genome function. Biol Rev Camb Philos Soc 80(2):227–250

Sharma A, Wolfgruber TK, Presting GG (2013) Tandem repeats derived from centromeric retrotransposons. BMC Genom 14(1):142. https://doi.org/10.1186/1471-2164-14-142

Slotkin RK, Martienssen R (2007) Transposable elements and the epigenetic regulation of the genome. Nat Rev Genet 8(4):272–285. https://doi.org/10.1038/nrg2072

Snowdon RJ, Köhler W, Friedt W, Köhler A (1997) Genomic in situ hybridization in Brassica amphidiploids and interspecific hybrids. Theoret Appl Genetics 95(8):1320–1324. https://doi.org/10.1007/s001220050699

Soltis DE, Gitzendanner MA, Stull G, Chester M, Chanderbali A, Chamala S, Jordon-Thaden I, Soltis PS, Schnable PS, Barbazuk WB (2013) The potential of genomics in plant systematics. Taxon 62(5):886–898

Talbert PB, Henikoff S (2010) Centromeres convert but don't cross. PLoS Biol 8(3):e1000326. https://doi.org/10.1371/journal.pbio.1000326

Tang H, Lyons E, Town CD (2015) Optical mapping in plant comparative genomics. Genome/Nat Res Counc Can (Genome/Conseil national de recherches Canada) 1:8

Treangen TJ, Salzberg SL (2011) Repetitive DNA and next-generation sequencing: computational challenges and solutions. Nat Rev Genet 13(1):36–46. https://doi.org/10.1038/nrg3117

Vrána J, Kubaláková M, Číhalíková J, Valárik M, Doležel J (2015) Preparation of sub-genomic fractions enriched for particular chromosomes in polyploid wheat. Biol Plant 59(3):445–455. https://doi.org/10.1007/s10535-015-0522-1

Walsh JB (1987) Persistence of tandem arrays: implications for satellite and simple-sequence DNAs. Genetics 115(3):553–567

Waminal N, Park HM, Ryu KB, Kim JH, Yang TJ, Kim HH (2012) Karyotype analysis of Panax ginseng C.A. Meyer, 1843 (Araliaceae) based on rDNA loci and DAPI band distribution. Comp Cytogenet 6 (4):425–441. https://doi.org/10.3897/compcytogen.v6i4.3740

Waminal NE, Perumal S, Lim K-B, Park B-S, Kim HH, Yang T-J (2015) Genomic survey of the hidden components of the *B. rapa* Genome. In: *The Brassica rapa* genome. Springer, pp 83–96

Waminal NE, Perumal S, Lee J, Kim HH, Yang T-J (2016) Repeat evolution in *Brassica rapa* (AA), *B. oleracea* (CC), and *B. napus* (AACC) genomes. Plant Breeding Biotechnol 4(2):107–122

Waminal NE, Pellerin RJ, Jang W, Kim HH, Yang T-J (2018) Characterization of chromosome-specific microsatellite repeats and telomere repeats based on low-coverage whole-genome sequence reads. Plant Breeding Biotechnol 6(1):1–8

Wang X, Torres MJ, Pierce G, Lemke C, Nelson LK, Yuksel B, Bowers JE, Marler B, Xiao Y, Lin L, Epps E, Sarazen H, Rogers C, Karunakaran S, Ingles J, Giattina E, Mun JH, Seol YJ, Park BS, Amasino RM, Quiros CF, Osborn TC, Pires JC, Town C, Paterson AH (2011a) A physical map of *Brassica oleracea* shows complexity of chromosomal changes following recursive paleopolyploidizations. BMC Genom 12:470. https://doi.org/10.1186/1471-2164-12-470

Wang X, Wang H, Wang J, Sun R, Wu J, Liu S, Bai Y, Mun JH, Bancroft I, Cheng F, Huang S, Li X, Hua W, Wang J, Wang X, Freeling M, Pires JC, Paterson AH, Chalhoub B, Wang B, Hayward A, Sharpe AG, Park BS, Weisshaar B, Liu B, Li B, Liu B, Tong C, Song C, Duran C, Peng C, Geng C, Koh C, Lin C, Edwards D, Mu D, Shen D, Soumpourou E, Li F, Fraser F, Conant G, Lassalle G, King GJ, Bonnema G, Tang H, Wang H, Belcram H, Zhou H, Hirakawa H, Abe H, Guo H, Wang H, Jin H, Parkin IA, Batley J, Kim JS, Just J, Li J, Xu J, Deng J, Kim JA, Li J, Yu J, Meng J, Wang J, Min J, Poulain J, Wang J, Hatakeyama K, Wu K, Wang L, Fang L, Trick M, Links MG, Zhao M, Jin M, Ramchiary N, Drou N, Berkman PJ, Cai Q, Huang Q, Li R, Tabata S, Cheng S, Zhang S, Zhang S, Huang S, Sato S, Sun S, Kwon SJ, Choi SR, Lee TH, Fan W, Zhao X, Tan X, Xu X, Wang Y, Qiu Y, Yin Y, Li Y, Du Y, Liao Y, Lim Y, Narusaka Y, Wang Y, Wang Z, Li Z, Wang Z, Xiong Z, Zhang Z, *Brassica rapa* Genome Sequencing Project C (2011b) The genome of the mesopolyploid crop species *Brassica rapa*. Nature genetics 43 (10):1035–1039. https://doi.org/10.1038/ng.919

Wang DO, Matsuno H, Ikeda S, Nakamura A, Yanagisawa H, Hayashi Y, Okamoto A (2012) A quick and simple FISH protocol with hybridization-sensitive fluorescent linear oligodeoxynucleotide probes. RNA 18(1):166–175. https://doi.org/10.1261/rna.028431.111

Wei L, Xiao M, An Z, Ma B, Mason AS, Qian W, Li J, Fu D (2013) New insights into nested long terminal repeat retrotransposons in Brassica species. Mol Plant 6(2):470–482. https://doi.org/10.1093/mp/sss081

Wendel JF, Wessler SR (2000) Retrotransposon-mediated genome evolution on a local ecological scale. Proc Natl Acad Sci 97(12):6250–6252. https://doi.org/10.1073/pnas.97.12.6250

Wolfgruber TK, Sharma A, Schneider KL, Albert PS, Koo D-H, Shi J, Gao Z, Han F, Lee H, Xu R, Allison J, Birchler JA, Jiang J, Dawe RK, Presting GG (2009) Maize centromere structure and evolution: sequence analysis of centromeres 2 and 5 reveals dynamic loci shaped primarily by retrotransposons. PLoS Genet 5(11):e1000743. https://doi.org/10.1371/journal.pgen.1000743

Zhou X, Ren L, Meng Q, Li Y, Yu Y, Yu J (2010) The next-generation sequencing technology and application. Protein Cell 1(6):520–536. https://doi.org/10.1007/s13238-010-0065-3

7 Homoeologous Exchanges and Gene Losses Generate Diversity and Differentiate the *B. napus* Genome from that of Its Ancestors

Birgit Samans, Rod Snowdon and Annaliese S. Mason

Abstract

Brassica napus (rapeseed) is a young allopolyploid species, formed from the union between progenitor species *B. rapa* (2*n* = AA) and *B. oleracea* (2*n* = CC) less than 7500 years ago during human agricultural history. Hence, *B. napus* provides a useful model for investigation of the structural changes that can occur following allopolyploidization, as well as how these changes can affect phenotypic variation in this important crop species. In this chapter, we describe the types and patterns of structural variation that can occur and have been observed in both natural and resynthesized *B. napus*, major methodologies for detecting these events and the subsequent effects on phenotype so far observed. In the future, increasingly high-quality genome sequences and other technological advances are expected to more thoroughly elucidate the role of phenomena such as homoeologous exchanges and gene loss on genome evolution and crop phenotypes.

B. Samans · R. Snowdon (✉) · A. S. Mason
Plant Breeding Department, IFZ Research Centre for Biosystems, Land Use and Nutrition, Justus Liebig University, Heinrich-Buff-Ring 26-32, 35392 Giessen, Germany
e-mail: Rod.Snowdon@agrar.uni-giessen.de

7.1 *Brassica napus* as a Model Crop Allopolyploid

The young allopolyploid species *Brassica napus* (2*n* = 4*x* = 38) formed from interspecific hybridization between *B. rapa* (2*n* = AA) and *B. oleracea* (2*n* = CC), most probably just a few hundred to a few thousand years ago. Due to its very recent origin, *B. napus* provides a useful model for investigation of the structural genome changes that can occur following allopolyploidization, as well as how these changes can affect phenotypic variation and human selection in an important crop species.

Polyploidization is a common phenomenon that is widespread in the plant kingdom. It arises frequently either by genome duplication within a species (autopolyploidization) or by hybridization between two species contributing two or more divergent chromosome sets (allopolyploidization) (Stebbins 1947; Grant 1981; Briggs and Walters 1997). Although allopolyploid plant species were originally thought to be more common than autopolyploid plant species (Hegarty and Hiscock 2008) due to advantages such as hybrid vigour, recent research suggests that autopolyploidy may actually be more common (Barker et al. 2016). Regardless, all flowering plants are now known to have undergone at least one polyploidy event in their evolutionary history (Jiao et al. 2011), with many species undergoing several (Wendel 2000; Comai 2005;

S. Liu et al. (eds.), *The Brassica napus Genome*, Compendium of Plant Genomes,
https://doi.org/10.1007/978-3-319-43694-4_7

Leitch and Leitch 2008; Soltis et al. 2014, 2016). Hence, many clearly diploid species, such as *Arabidopsis thaliana* and *Zea mays*, are actually what we call 'paleopolyploid' species that have undergone ancient polyploidization events followed by diploidization, an evolutionary process where the polyploid genome reverts to a functionally diploid one (Shoemaker et al. 2006; Woodhouse et al. 2010). Polyploidization contributes to plant evolution by creating new species which have the potential to adapt to changing environments (Wendel 2000; Ramsey and Schemske 2002; Soltis and Soltis 2009; Jackson and Chen 2010). It is also of high importance for crop plant domestication and improvement, as many modern crop species, including rapeseed, wheat, cotton, oats and coffee, are polyploids which outperform their corresponding diploid progenitors, e.g. by increased biomass, vigour and fertility, higher yield and better growth (Paterson 2005; Dubcovsky and Dvorak 2007; Soltis et al. 2009).

7.2 Genetic Consequences of Polyploidization

Before a new polyploid plant can be successfully established in the plant kingdom, it has to overcome a bottleneck of high genetic instability. The collision of two genomes in the nucleus of an allopolyploid, also described as 'genomic shock' (McClintock 1984), is a major challenge for the plant and often leads to an evolutionary 'dead end', as it goes along with reduced fertility (Comai et al. 2000; Comai 2005; Madlung et al. 2005). Different studies have documented that genomes of young allopolyploids are often highly dynamic, with extensive genetic and epigenetic changes (Wolfe 2001; Song and Messing 2003). The most common forms of novel genetic variation produced in allopolyploid genomes are chromosome rearrangements, duplications and deletions. These are caused by recombination events between chromosomes belonging to different subgenomes in the allopolyploid, i.e. by recombination between 'homoeologues' (ancestrally related chromosomes or chromosome segments), rather than strictly enforced pairing between homologous chromosomes, as occurs in stable, diploid species. Novel polyploid plants that survive to reproductive age are faced with the need to develop diploid-like chromosome pairing behaviour, which is necessary for stable genome transmission and subsequent species establishment (Comai 2005). However, this process of cytological diploidization is often accompanied by selection of favourable structural and functional genome variants that arise frequently from interchanges between homoeologous chromosomes (Soltis and Soltis 2000; Wendel 2000).

What happens after the hybridization of two genomes and how newly formed polyploids are stabilized are key questions in the evolution of polyploid plants. Diploidization often leads to extensive genomic rearrangements, gene losses and homoeologous exchanges (HE). Homoeologous exchanges include reciprocal and non-reciprocal translocations. Reciprocal translocations occur when two subgenomes 'swap' chromosome segments with each other, whereas non-reciprocal translocations occur when a region belonging to one subgenome is duplicated and the corresponding homoeologous region in the other subgenome is deleted. Both reciprocal and non-reciprocal translocations (traditionally called homoeologous reciprocal translocations and homoeologous non-reciprocal translocations, or HRTs and HNRTs) occur as a result of homoeologous recombination events during meiosis (Nicolas et al. 2007; Gaeta and Chris Pires 2010). Many of the genomic changes observed in recent allopolyploids, such as in *Tragopogon miscellus* (Chester et al. 2012), *B. napus* (Gaeta et al. 2007; Szadkowski et al. 2010; Xiong et al. 2011) and cotton (*Gossypium hirsutum*) (Flagel et al. 2012) are consistent with homoeologous exchanges.

As a consequence of diploidization, allopolyploid plant genomes often do not reflect the sum of the genomes of both parental species. Genome size and the number of genes are often much lower than one would predict from a doubled set. Clear genome size reduction has been described for tobacco (Leitch and Leitch 2008), cotton

(Paterson et al. 2012) and some synthetic wheat lines (Ozkan 2003). Most variations in genome size are due to the removal of repetitive sequences rather than as a consequence of reduced gene content (Han et al. 2005; Lim et al. 2007; Renny-Byfield et al. 2011, 2012, 2013 Buggs et al. 2012). Genome size reduction can also be biased, e.g. occur preferentially or at an increased rate in one subgenome. In tobacco, the paternal subgenome shows a size reduction, whereas the maternal subgenome remains intact (Renny-Byfield et al. 2011, 2012).

An important strategy to analyse the impact of polyploidization on genetic variability is to compare the genomes of the polyploid species with the genomes of the parental progenitors, if these are known. This approach to assess the impact of polyploidization on the genome sequence has been conducted on both old and young polyploid species. As well, it is sometimes possible to recreate ancestral allopolyploidization events by artificially crossing between the extent progenitor species, in order to investigate genome variation in the newly resynthesized allopolyploids. Polyploid systems which have been studied quite intensively are *Gossypium, Brassica, Nicotiana, Glycine*, *Arabidopsis*, *Senecio, Spartina, Triticum* and *Tragopogon*. Studies in these crop and wild genera have focused on a range of different aspects of polyploidization, including karyotype evolution, genomic changes, gene expression, tissue-specific gene expression, epigenetic changes, repetitive elements, alternative splicing, eco-geography, population genetics, reproductive ecology and physiology (reviewed in Soltis et al. 2016). Of particular importance is young, recently formed polyploids like *Spartina*, *Tragopogon* and *Senecio*, and species that can be recreated synthetically such as *B. napus* (Song et al. 1995; Gaeta et al. 2007; Chalhoub et al. 2014), wheat (Yang et al. 2009; Li et al. 2014b), cotton (Liu et al. 2001), *Tragopogon* (Tate et al. 2009; Malinska et al. 2010) and *Arabidopsis* (Madlung et al. 2005). Synthetic polyploids are good models with which to monitor the immediate genetic and genomic consequences of polyploidization while excluding effects introduced by genome shaping due to further selection processes. Additionally, the exact progenitor genotypes of synthetic lines are known, which is usually not the case for natural allopolyploids such as *B. napus* (rapeseed) or *Triticum aestivum* (wheat).

Comparative analyses in different species have revealed that the consequences of polyploidization observed so far are highly complex and variable. Common mechanisms have been described, but completely different patterns are often seen, such that there is no ubiquitous pattern across all species (Soltis et al. 2016). Genetic changes, often quite extensive, have been reported in the allopolyploids *B. napus* (Parkin et al. 1995; Sharpe et al. 1995; Udall et al. 2005; Chalhoub et al. 2014; Samans et al. 2017), *Tragopogon mirus* and *T. miscellus* (Buggs et al. 2010, 2011; Chester et al. 2012, 2013, 2015; Buggs et al. 2014), *Arabidopsis suecica* (Comai 2000; Madlung and Comai 2004; Pontes et al. 2004; Madlung et al. 2005; Wang et al. 2006), *Coffea arabica* (Lashermes et al. 2014) and newly resynthesized allohexaploid bread wheat (Feldman et al. 1997; Ozkan et al. 2001; Feldman and Levy 2009; Liu et al. 2009). In comparison, newly synthesized allopolyploid cotton behaves quite differently, with no observable genetic changes (Liu et al. 2001). Genetic changes have all been described at various genomic levels (Table 7.1)

7.3 Diploidization-Induced Gene Loss

Doubling of the genome and the gene content is associated with a range of potential advantages for the resulting polyploids (Comai 2005; Doyle et al. 2008). A number of different mechanisms have been described as causes for this advantageous situation, including increased gene dosage, heterosis, the development of new gene functions (neofunctionalization) through relaxation of selection pressure for existing gene functions and enabling the partitioning of gene expression (subfunctionalization). However, most of these advantages are only short term, as they rely on gene duplicates being present. Over evolutionary

Table 7.1 Genetic changes in different organisms

Type	Organisms	References
Chromosome rearrangements, deletions and duplications	Wheat *Brassica napus* *Gossypium hirsutum* *Nicotiana tabacum* *Coffea arabica* L. Arabidopsis	Ozkan et al. (2001) Chalhoub et al. (2014) Samans et al. (2017) Parkin et al. (1995) Pires et al. (2004a) Nicolas et al. (2007) Li et al. (2015) Wendel et al. (1995) Sierro et al. (2014) Kenton et al. (1993) Lashermes et al. (2014) Lashermes et al. (2016) Madlung et al. (2002) Kenton et al. (1993)
Changes in rDNA loci	Soybean	Joly et al. (2004)
Transposon activation	Wheat Arabidopsis *Brassica napus*	Kashkush et al. (2002) Madlung et al. (2005) Comai et al. (2000) Alix and Heslop-Harrison (2004)
Epigenetic changes	Cotton Senecio Arabidopsis Spartina *Brassica napus*	Adams et al. (2003) Hegarty et al. (2005) Madlung et al. (2005) Wang et al. (2004) Salmon et al. (2005) Lukens et al. (2006)

time, another scenario rapidly presents itself: that of diploidization and gene loss. In fact, despite having undergone multiple rounds of whole-genome duplication, paleopolyploids vary only a little in their total gene number. In many ancient polyploids, following polyploidization the subsequent diploidization process resulted in a high number of single-copy genes via deletion of a large fraction of the duplicated genes. Over time, smaller and smaller fractions of genes remain duplicated (Thomas et al. 2006; Woodhouse et al. 2010; Freeling et al. 2012; Albalat and Canestro 2016). Comparative analyses of syntenic regions in *A. thaliana and maize resulting from their last whole-genome duplication events (WGD) showed large differences in gene content, with many duplicated genes returned to single-copy status* (Thomas et al. 2006).

Gene loss can occur either on the structural (physical loss of DNA sequence) or the functional (loss of gene expression or function) level. Mechanisms leading to gene loss on the structural level include deletions via non-homologous chromosome interactions during meiosis which lead to non-reciprocal exchanges, ectopic sequence loss, transposon insertions and the accumulation of indel mutations (Woodhouse et al. 2010; Tang et al. 2012). On the functional level, gene losses can take place via mutations that result in pseudogenization, neofunctionalization and subfunctionalization, often preceded or detectable by gene expression changes (Ohno 1970; Wendel 2000; Freeling 2008; Edger and Pires 2009; Freeling 2009; Salse et al. 2009; Woodhouse et al. 2010; Freeling et al. 2012).

Gene loss rarely appears randomly. Genes which remain as single-copy often belong to specific functional categories that influence gene family size or which are dependent on genomic position (Albalat and Canestro 2016). Genes preferentially retained as single-copy are overrepresented in the functional categories of DNA-repair, recombination, enzyme activity, kinase activity, transport, tRNA ligation, defence and categories associated with domestication processes (Samans et al. submitted; Blanc and Wolfe 2004; Duarte et al. 2010; de Smet et al. 2013; Samans et al. 2017). In extreme cases, entire gene families can be deleted from the genomes of certain lineages (Aravind et al. 2000; Demuth and Hahn 2009). In comparison, genes preferentially preserved in multiple copies commonly belong to functional groups involved in basic cellular machinery, nucleotide-sugar metabolism and regulatory functions (Blanc and Wolfe 2004; Thomas et al. 2006; Sankoff et al. 2010; Pont et al. 2011). Another factor frequently affecting gene loss is biased fractionation, whereby one subgenome tends to be preferentially lost relative to another, more 'dominant' subgenome in an allopolyploid over time (Gaeta et al. 2007; Chang et al. 2010; Woodhouse et al. 2010; Schnable and Freeling 2011; Schnable et al. 2012; Samans et al. 2017). Biased fractionation has been observed in *Brassica rapa* (Chen et al. 2013), *A. thaliana* (Thomas et al. 2006; Wang et al. 2006), *Z. mays* (Schnable et al. 2011), *Coffee arabica* (Combes et al. 2013) and *T. aestivum* (Wang et al. 2006; Pont et al. 2013; Li et al. 2014a).

7.4 Methods for Investigating Variation in Genome Structure

Several different methods can be used to detect genomic structural variations. Strong evidence supporting the mechanisms of genomic modification in polyploid plant genomes has come from cytogenetic methods and genome-wide molecular marker analyses (Song et al. 1995; Osborn et al. 2003; Szadkowski et al. 2010; Mason et al. 2014). Classical and molecular cytogenetic techniques (e.g. in situ hybridization) have enabled the identification of large chromosomal changes such as aneuploidy, translocations and loss of repeats (Pontes et al. 2004; Skalicka et al. 2005; Lim et al. 2008; Xiong et al. 2011; Tang et al. 2014). Fluorescent in situ hybridization (FISH) uses fluorescently labelled DNA (either specific DNA sequences, or whole genomic DNA, the latter also known as genomic in situ hybridization or GISH) as probes to hybridize to chromosome spreads, allowing labelling of foreign or subgenomes and chromosomes (GISH) or enabling the localization of the position of repetitive DNA and unique sequences along the chromosomes using a fluorescence microscope. Although relatively inexpensive, the major limitations of cytogenetic methods are how labour-intensive generation of quantitative data is, as well as a limited resolution that often leads to an underestimation of changes.

Genetic linkage maps based on molecular markers have also proven to be important tools for studying genomic rearrangements over the last two decades. The first molecular marker studies used restriction fragment length polymorphism (RFLP) in *B. napus* (Song et al. 1995; Gaeta et al. 2007; Rygulla et al. 2008), *T. miscellus* (Leitch and Bennett 1997) and wheat (Liu et al. 1998) and amplified fragment length polymorphism (AFLP) in *Arabidopsis* (Madlung et al. 2005) and *Spartina* (Salmon et al. 2005). Nowadays, high-throughput genotyping arrays based on single-nucleotide polymorphisms (SNPs) allow assessment of many markers simultaneously (Gore et al. 2009), providing more sophisticated methods for producing high-density genetic maps and enabling the detection of genomic rearrangements at much higher resolution. Additionally, a large number of samples can be analysed in parallel using genotyping array technology (Gore et al. 2009; Ganal et al. 2012). This is a useful feature for investigation of large segregating mapping populations, which are usually needed to fully resolve translocation and chromosome rearrangement events via methods such as linkage mapping and marker segregation ratios (reviewed in Mason et al. 2017 for review). Beside the identification of genomic changes based on inheritance of allelic variation, SNP-based genotyping arrays can also be applied to detect copy number variation (deletion and duplication events) using comparative and consecutive hybridization intensities based on their physical position in the genome (Grandke et al.; Mason et al. 2015; Grandke et al. 2017). This method was successfully applied to detect copy number variation of A and C genome SNP alleles in an allohexaploid *Brassica* ((*B. napus* × *B. carinata*) × *B. juncea*) microspore-derived population (Mason et al. 2015) using the *B. napus* Illumina Infinium 60K Brassica array (Clarke et al. 2016).

Next-generation sequencing technologies have developed rapidly over the last decade. The use of second-generation sequencing technologies like Roche/454 pyrosequencing and Illumina/Solexa sequencing provided enormous power and new opportunities to access the complex polyploid genomes of many major crops (Varshney et al. 2009; Duran et al. 2010; Edwards and Batley 2010). Dropping costs and improved technologies in the 'genomic era' have subsequently increased the number of whole-genome sequences and re-sequenced accessions available for comparative studies (Mardis 2008; Abecasis et al. 2010). Although second-generation sequencing facilitated detection of SNPs and small variants (e.g. indels; insertion/deletion events), detecting large genomic changes remains a challenge. The assembly of a de novo genome using short reads is difficult as the reads are often fragmented, miss important genes and contain collapsed repeat regions that mess up the detection of rearrangements (Li et al. 2010; Schatz et al. 2010). Sequencing polyploid species is even more challenging due to the increased genome complexity offered by high levels of sequence redundancy (Fu et al. 2016). As a result, progenitor genomes (when progenitor species are known) are often sequenced first, in order to use their sequences as a 'template' for the allopolyploid genomes. This approach was also used for *B. napus*, by first sequencing the diploid progenitors *B. rapa* and *B. oleracea* (Wang et al. 2011; Chalhoub et al. 2014; Liu et al. 2014). Resequencing approaches, where individual genotypes are aligned to the genome sequence of a closely related genotype or reference genome, allow detection of different structural variants: regions with increased or decreased read depth after mapping of sequenced reads to the reference genome usually indicate the presence of duplication and deletion events, respectively. Compared to a de novo assembly, it is cheaper to perform resequencing, as lower sequence coverage is required, and the data are less complex to analyse (Hormozdiari et al. 2009; Barrick et al. 2014; Samans et al. 2017).

Meanwhile, new single molecule sequencing technologies like Pacific Biosciences (PacBio), Single Molecule Real-Time (SMRT) sequencing, the Illumina Tru-seq Synthetic Long-Read technology and the Oxford Nanopore Technologies sequencing platform, which can produce average read lengths of up to 100,000 bp have greatly improved our ability to analyse genome structure (Rhoads and Au 2015; Sakai et al. 2015). The longer read length achieved with these technologies is a critical factor for obtaining high-quality genome assemblies, as these long reads can span otherwise difficult to assemble duplicated regions and regions with a high density of repetitive elements (Lee et al. 2016).

Asides from sequencing technologies, new techniques such as optical mapping, whereby fluorescent tags are used to label DNA sequences and map regions from 50 to 250 kbp are available and/or under development in many species (Zhou et al. 2007; Chamala et al. 2013; Tang et al. 2015). Combining single-molecule sequencing technologies with new mapping technologies may allow the production of scaffolds spanning entire chromosome arms, providing an optimized environment for structural analysis and the identification of genomic variation (Burton et al. 2013; Cao et al. 2014; Pendleton et al. 2015).

7.5 The *Brassica napus* Genome and the Consequences of Allopolyploidization

The allopolyploid species *B. napus* (AACC, $2n = 4x = 38$) is an important oilseed crop that originated around 7500 years ago from interspecific hybridizations between the diploid parental species *B. rapa* (AA, $2n = 20$) and *B. oleracea* (CC, $2n = 18$) (Gaeta et al. 2007; Xiong et al. 2011; Chalhoub et al. 2014). *B. napus* has become an important model for studies of de novo allopolyploidization, as synthetic *B. napus* can be readily generated (with the help of tissue culture), and the reference genome sequences are now available for both *B. napus* and its parental progenitor species (Wang et al. 2011; Chalhoub et al. 2014; Liu et al. 2014). The first study showing genomic rearrangements in *B. napus* was published by Song et al. (1995). The authors observed genomic changes in the F2 to F5 generations of newly resynthesized *B. napus* accessions using molecular markers. Further studies on synthetic *B. napus* accessions demonstrated that genomic rearrangements in different synthetic accessions, either from the same parental background or originating from different progenitor genotypes can vary greatly in the frequency and type of genomic rearrangements, from only a few small rearrangements to whole-chromosome substitutions (Gaeta et al. 2007; Szadkowski et al. 2010; Xiong et al. 2011; Chalhoub et al. 2014; Samans et al. 2017). In newly synthesized accessions, genomic rearrangements start immediately in the first generation after hybridization, with the highest meiotic disruption in the first meiosis of resynthesized *B. napus* (Szadkowski et al. 2010). The frequency of rearrangements is positively correlated with the genetic distance between the progenitor genomes in most species and additionally negatively correlated with fertility (Gaeta et al. 2007). A study comprising whole-genome DNA resequencing data of 31 diverse adapted and 20 synthetic *B. napus* accessions found significantly less and smaller genomic changes in the adapted compared to the synthetic accessions (Samans et al. 2017), although this is possibly a result of selection for smaller rearranged regions over time, as large rearrangements have also been identified in natural *B. napus* (Mason et al. 2017). Comparative analyses of the *B. napus* ('Darmor-bzh') A_n and C_n subgenomes revealed large collinearity to the corresponding diploid *B. rapa* 'Chiifu' (A_r) and *B. oleracea* 'TO1000' (C_o) genomes (Chalhoub et al. 2014).

Gene loss is a common event following whole-genome duplication or allopolyploidization (Langham et al. 2004; Thomas et al. 2006) and is considered part of the diploidization processes whereby the genomes of polyploids reduce in size and lose duplicated genetic information over time. Comparing the gene content of the *B. napus* ('Darmor-bzh v4.1') A_n and C_n subgenomes with the gene content of *B. rapa* 'Chiifu' v1.5 and *B. oleracea* 'TO1000' genomes, respectively, 38,661 orthologous gene pairs were identified. Most orthologous gene pairs in *B. rapa* and *B. oleracea* remained as homoeologous pairs in *B. napus* 'Darmor-*bzh*', with only 112 genes on the A subgenome and 91 genes on the C subgenome identified as lost (Chalhoub et al. 2014). Genomic changes in synthetic and adapted *B. napus* accessions are not random. For adapted *B. napus* accessions, an asymmetric fractionation with predominant gene loss in the C subgenome was described. Whole-genome DNA resequencing data of 31 diverse adapted *B. napus* accessions described common deletions at distinct chromosomal regions on chromosomes C1, C2, C4 and C9 in nearly all investigated natural *B. napus* accessions, suggesting strong selection against homoeologous gene copies in the C subgenome (Samans et al. 2017). Similarly, a higher number of genetic changes in the C subgenome than in the A genome were found by Gaeta et al. (2007) in 50 synthetic *B. napus* allopolyploids of a population. As well as biased gene loss by subgenome location, biased gene loss by function has also been described (Samans et al. 2017). Genes that were deleted within the panel of 31 natural *B. napus* accessions belonged to gene families with the ontologies 'cytosol to endoplasmic reticulum (ER) transport', 'seed development', 'reproductive processes', 'vernalization response' and pectin catabolism', suggesting that loss of these genes may have played a role in differential environmental adaptations.

As well as biased gene loss by subgenome location, HEs are also known to be unequally distributed across chromosomes in both adapted and resynthesized accessions. HEs show clearly more A_n to C_n events where the A subgenome fragment is duplicated and the corresponding homoeologous C subgenome region is deleted in comparison with C_n to A_n translocations (Chalhoub et al. 2014; Samans et al. 2017).

This was demonstrated for large HEs but also for gene conversions and exchanges at the single-nucleotide level (Chalhoub et al. 2014). HEs appear between all homoeologous linkage groups, but more frequently between the chromosomes A1/C1, A2/C2 A3/C3 and A9/C9, chromosome pairs which share large synteny blocks with high similarity (Gaeta et al. 2007; Chalhoub et al. 2014; Samans et al. 2017). This is because homoeologous exchanges and most deletion/duplication events in *B. napus* are a consequence of homoeologous recombination between A and C chromosomes. Multivalent pairing between the homoeologous A and C genome has been frequently observed in different synthetic *B. napus*, AC allohaploids and various interspecific hybrid types (Jenczewski et al. 2003; Osborn et al. 2003; Leflon et al. 2006; Nicolas et al. 2007, 2008, 2009; Mason et al. 2010). As common areas of rearrangement are increasingly described for adapted and resynthesized accessions, the presence of hot spots for homoeologous crossing over and recombination will hopefully soon be identified (Samans et al. 2017).

In *B. napus* 'Darmor-*bzh*' on the whole-genome level, non-reciprocal exchanges between subgenomes account for ~86% of the mutations differentiating *B. napus* from its progenitors *B. rapa* and *B. oleracea* (Chalhoub et al. 2014). As well, these genomic rearrangements may tend to lead to overall size reduction of the C_n subgenome relative to the A_n subgenome, further differentiating the C_n subgenome from the progenitor C_o genome. This occurs due to the higher number of deletions and HEs in the C subgenome, and the relatively larger sequence size of these deletions and HEs, relative to those in the A_n genome (Samans et al. 2017). Although the A and C genomes have similar gene numbers, gene density in the C genome is much less, as this genome has a higher fraction of repetitive elements (Alix and Heslop-Harrison 2004; Howell et al. 2008). As rearrangement events occur based on homoeologous gene locations rather than on absolute sequence position along chromosomes, this means that when an A genome segment is duplicated on the C genome in a HE, some sequence is actually lost as the larger C genome fragment is replaced by the smaller A genome fragment (Samans et al. 2017).

7.6 Homoeologous Exchanges and Karyotype Change in Other Recently Formed Allopolyploid Species

Similar to in young allopolyploid *B. napus*, genomic changes have been reported in many other synthesized and adapted allopolyploids. The recently formed (<90 yrs) allopolyploid species *T. mirus* and *T. miscellus* show frequent homoeologous losses and large chromosomal changes including translocations; interestingly similar patterns of rearrangements were observed both in synthetic and natural accessions (Buggs et al. 2010, 2011, 2014; Chester et al. 2012, 2013, 2015; Soltis and Soltis 2012). Synthetic and natural allotetraploid *A. suecica* (12,000–300,000 yrs ago) (Comai 2000; Madlung and Comai 2004; Pontes et al. 2004; Madlung et al. 2005; Wang et al. 2006), the recent allopolyploid *C. arabica* (<100,000 yrs) (Lashermes et al. 2014) and newly resynthesized allohexaploid bread wheat (Feldman et al. 1997; Ozkan et al. 2001; Feldman and Levy 2009; Liu et al. 2009) also show extensive rearrangements, whereas synthetic allopolyploid cotton behaves quite differently, with a predominantly additive genome composition with only a few epigenetic changes (Liu et al. 2001). Similar to in *B. napus,* rearrangement patterns in wheat are not random, and similarities observed between different synthetic lines suggest a directed process (Ozkan et al. 2001; Feldman and Levy 2012).

Only a few studies have been conducted on the other crop allopolyploid *Brassica* species *B. juncea* ($2n$ = AABB) and *B. carinata* ($2n$ = BBCC). Although the study (Axelsson et al. 2000) found no evidence for genetic changes using RFLP marker, the study by Song et al. (1995) observed extensive genetic changes between the progenitor A and B genomes (in *B. rapa* and *B. nigra,* respectively) and the A and B subgenomes of synthetic *B. juncea*.

Biased, genome-wide increased gene expression of A subgenome over B subgenome homoeologues has been described in both adapted and synthetic *B. juncea* accessions (Yang et al. 2016). However, the greater divergence between the B and A/C subgenomes (Lagercrantz and Lydiate 1996) and the fact that resynthesized *B. juncea* and *B. carinata* tend to demonstrate far more stable meiosis with only little homoeologous pairing; relative to resynthesized *B. napus* (Prakash and Chopra 1988) suggests that the frequency and size of HEs may be low between the B and A/C genomes under natural conditions.

7.7 Gene Losses Induce Phenotypic Variation

Recently, increasing attention has been paid to the role of gene duplications and deletions and HEs on phenotypic variation in polyploids (Mason and Snowdon 2016; Stein et al. 2017). The increase of genomic data provided by new sequencing technologies has elevated our ability to detect gene losses and may lead to increased exploitation of these as a potential source for phenotypic diversity (Albalat and Canestro 2016). A recent study by Schiessl et al. (2014) analysed differences between morphotypes in relation to copy number variation in flowering-time genes. They sequenced 29 regulatory flowering-time genes in four genetically and phenotypically diverse *B. napus* accessions. The genotype set included a winter-type oilseed rape, a winter fodder rape, a spring-type oilseed rape (all *B. napus* ssp. *napus*) and a swede (*B. napus* ssp. *napobrassica*), exhibiting differences in winter-hardiness, vernalization requirement and flowering behaviour. In comparison with the early flowering fodder rape accession and the winter-sensitive swede, they found a reduced copy number for the transcription factor *CONSTANS* in the late flowering, winter hardy accession. *CO* is a central day length regulator necessary for flowering transition. Hence, the lower gene copy number was proposed to relate to the late-flowering behaviour of this accession, providing some of the first concrete evidence in *B. napus* for gene loss affecting phenotype. A further example in *B. napus* for the influence of gene loss on phenotype was described for the auxin response factor 18 (*ARF18*) gene. A 165-bp deletion within the gene was found to affect both the seed weight and the seed length: this deletion prevents ARF18 from forming homodimers, which results in loss of its binding activity (Liu et al. 2015).

The link between phenotype and gene loss has also been demonstrated in other species. In wheat, for example, copy number variation has been identified as a significant regulator of flowering time. Wheat genotypes with only one copy of *Ppd1* on the B genome (*Ppd-B1*) are photoperiod sensitive, whereas a higher copy number (2–4 copies) results in a day-neutral, early flowering phenotype (Diaz et al. 2012; Langer et al. 2014; Wurschum et al. 2015). Plants with an increased copy number of *Vrn-A1* are more dependent on vernalization, such that longer periods of cold are required to potentiate flowering (Diaz et al. 2012).

7.8 Homoeologous Exchanges Influence Phenotypic Variation

The impact of homoeologous exchanges on important phenotypic traits such as regulation of flowering time, flowering size, plant height, leaf morphology and size, glucosinolate content and yield is increasingly being demonstrated (Schranz and Osborn 2000; Pires et al. 2004a; Schranz and Osborn 2004; Gaeta et al. 2007).

FLOWERING LOCUS C (FLC) is a key adaptive gene controlling vernalisation and photoperiod responses (Turck and Coupland 2011). In *B. napus* homoeologous gene copies of FLC on A_n2/C_n2, A_n3/C_n3, A_n10/C_n9 and a C_n9 locus are preserved in comparison with progenitor species *B. rapa* and *B. oleracea* (Chalhoub et al. 2014). Different homoeologous exchanges of the FLC gene have been assumed to be the mechanism by which novel phenotypic variation in flowering time was created in synthetic *B. napus* lines (Pires et al. 2004b; Gaeta et al. 2007). Chalhoub et al. (2014) compared the semi-winter oilseed accessions 'Yudal' and 'Aburamasari' and the late-flowering swede/rutabaga accession 'Sensation NZ' for the presence of FLC gene copies. For the homoeologous A2/C2 FLC gene, they found an A_n2 to C_n2 translocation in both semi-winter accessions and a homoeologous exchange between C_n9 and A_n10, (including 2 FLC genes) in the late-flowering swede/rutabaga (Chalhoub et al. 2014). Additionally, for both regions and for the A_n3/C_n3 FLC locus, a quantitative trait locus (QTL) for flowering time was described (Zou et al. 2012).

Following intensive breeding efforts several decades ago, significant reductions in glucosinolate (GSL) and erucic acid content in rapeseed oil were achieved, allowing rapeseed (newly christened 'canola' to differentiate varieties on the basis of oil quality) to be used for human and animal nutrition (Dupont et al. 1989). Hence, it might be reasonable to expect either a reduction of the number of GSL biosynthesis genes in the *B. napus* genome of modern cultivars or an increased number of genes involved in GSL catabolism as the reason for reduced GSL content. However, with 101 GSL biosynthesis genes in the A_n subgenome and 97 in the C_n subgenome, the progenitor GSL biosynthesis genes from *B. rapa* and *B. oleracea* are widely conserved (Chalhoub et al. 2014). Genes involved in the GLS catabolism are all retained, and only 3 *B. oleracea* and 1 *B. rapa* GSL biosynthesis genes are lost (Chalhoub et al. 2014). Interestingly, a putative impact on the GSL content was in fact assigned to a deletion of both homoeologous GSL genes on A_n2/C_n2, where the A_n2 segment with the missing GSL gene has replaced the C_n2 homoeologue (Chalhoub et al. 2014). This genomic area is also associated with two QTL for total aliphatic GSL content (Delourme et al. 2013). A deleted GSL gene on A_n9 and its retained homoeologous gene on C_n9 can be co-localized with two further QTL for aliphatic GSL content (Delourme et al. 2013). An impact of an A_n9 to C_n8 translocation on seed acid detergent lignin was also described (Liu et al. 2012). Although other examples of the effects of homeologous exchanges on phenotype in *B. napus* are sparse, QTLs for yield and *Sclerotinia* resistance could also be associated with HEs from previous research (Osborn et al. 2003; Zhao et al. 2006), and with increasing genomic information many more are expected to be found in future (Mason and Snowdon 2016).

7.9 Conclusions

Allopolyploidization, whereby two different species come together to make a new species, very commonly involves structural genomic rearrangements. Increasingly, we are able to detect and map phenomena such as homoeologous exchanges (HEs), duplication/deletion events and gene losses using sophisticated sequencing approaches and associated technologies such as optical mapping. In *B. napus*, genomic resources such as good quality reference genomes and increasingly large quantities of resequencing data for different accessions have identified numerous structural rearrangements and variations in both natural and synthetic lines (Schmutzer et al. 2015; Samans et al. 2017). These resources have revealed that while the *B. napus* subgenomes are fundamentally unchanged for the most part from those of the respective diploid progenitor species, this may be due to the relative youth of this <7500-year-old allopolyploid. Closer inspection of natural germplasm has revealed the start of the diploidization process, with preferential replacement of the C subgenome with A subgenome gene copies, and variation for particular HEs and duplication/deletion events within the wider germplasm pool of this species. Several studies have also clearly demonstrated that structural variation can also play a major role in phenotypic variation, both in the evolution and targeted breeding of *B. napus* and in synthetic lines. In future, the role of homoeologous exchanges in the evolutionary and ongoing differentiation of the *B. napus* genome from that of its progenitor species *B. rapa* and *B. oleracea* may be fully elucidated, enhancing our understanding and utilization of this phenomenon in rapeseed breeding.

References

Abecasis GR, Altshuler D, Auton A, Brooks LD, Durbin RM, Gibbs RA, Hurles ME, McVean GA (2010) A map of human genome variation from population-scale sequencing. Nature 467:1061–1073. https://doi.org/10.1038/nature09534

Adams KL, Cronn R, Percifield R, Wendel JF (2003) Genes duplicated by polyploidy show unequal contributions to the transcriptome and organ-specific reciprocal silencing. Proc Nat Acad Sci 100(8):4649–4654

Albalat R, Canestro C (2016) Evolution by gene loss. Nat Rev Genet 17:379–391. https://doi.org/10.1038/nrg.2016.39

Alix K, Heslop-Harrison JS (2004) The diversity of retroelements in diploid and allotetraploid Brassica species. Plant Mol Biol 54:895–909

Aravind L, Watanabe H, Lipman DJ, Koonin EV (2000) Lineage-specific loss and divergence of functionally linked genes in eukaryotes. Proc Natl Acad Sci U S A 97:11319–11324. https://doi.org/10.1073/pnas.200346997

Axelsson T, Bowman CM, Sharpe AG, Lydiate DJ, Lagercrantz U (2000) Amphidiploid *Brassica juncea* contains conserved progenitor genomes. Genome 43:679–688

Barker MS, Arrigo N, Baniaga AE, Li Z, Levin DA (2016) On the relative abundance of autopolyploids and allopolyploids. New Phytol 210:391–398. https://doi.org/10.1111/nph.13698

Barrick JE, Colburn G, Deatherage DE, Traverse CC, Strand MD, Borges JJ, Knoester DB, Reba A, Meyer AG (2014) Identifying structural variation in haploid microbial genomes from short-read resequencing data using breseq. BMC Genomics 15:1039. https://doi.org/10.1186/1471-2164-15-1039

Blanc G, Wolfe KH (2004) Functional divergence of duplicated genes formed by polyploidy during Arabidopsis evolution. Plant Cell 16:1679–1691. https://doi.org/10.1105/tpc.021410

Briggs D, Walters SM (1997) Plant variation and evolution. Cambridge University Press

Buggs RJA, Elliott NM, Zhang L, Koh J, Viccini LF, Soltis DE, Soltis PS (2010) Tissue-specific silencing of homoeologs in natural populations of the recent allopolyploid *Tragopogon mirus*. New Phytol 186:175–183. https://doi.org/10.1111/j.1469-8137.2010.03205.x

Buggs RJA, Zhang L, Miles N, Tate JA, Gao L, Wei W, Schnable PS, Barbazuk WB, Soltis PS, Soltis DE (2011) Transcriptomic shock generates evolutionary novelty in a newly formed, natural allopolyploid plant. Curr Biol 21:551–556. https://doi.org/10.1016/j.cub.2011.02.016

Buggs RJA, Renny-Byfield S, Chester M, Jordon-Thaden IE, Viccini LF, Chamala S, Leitch AR, Schnable PS, Barbazuk WB, Soltis PS, Soltis DE (2012) Next-generation sequencing and genome evolution in allopolyploids. Am J Bot 99:372–382. https://doi.org/10.3732/ajb.1100395

Buggs RJA, Wendel JF, Doyle JJ, Soltis DE, Soltis PS, Coate JE (2014) The legacy of diploid progenitors in allopolyploid gene expression patterns. Philos Trans R Soc Lond B Biol Sci 369. https://doi.org/10.1098/rstb.2013.0354

Burton JN, Adey A, Patwardhan RP, Qiu R, Kitzman JO, Shendure J (2013) Chromosome-scale scaffolding of de novo genome assemblies based on chromatin interactions. Nat Biotechnol 31:1119–1125. https://doi.org/10.1038/nbt.2727

Cao H, Hastie AR, Cao D, Lam ET, Sun Y, Huang H, Liu X, Lin L, Andrews W, Chan S, Huang S, Tong X, Requa M, Anantharaman T, Krogh A, Yang H, Cao H, Xu X (2014) Rapid detection of structural variation in a human genome using nanochannel-based genome mapping technology. Gigascience 3:34. https://doi.org/10.1186/2047-217X-3-34

Chalhoub B, Denoeud F, Liu S, Parkin IA, Tang H, Wang X, Chiquet J, Belcram H, Tong C, Samans B, Correa M, Da SC, Just J, Falentin C, Koh CS, Le Clainche I, Bernard M, Bento P, Noel B, Labadie K, Alberti A, Charles M, Arnaud D, Guo H, Daviaud C, Alamery S, Jabbari K, Zhao M, Edger PP, Chelaifa H, Tack D, Lassalle G, Mestiri I, Schnel N, Le Paslier MC, Fan G, Renault V, Bayer PE, Golicz AA, Manoli S, Lee TH, Thi VH, Chalabi S, Hu Q, Fan C, Tollenaere R, Lu Y, Battail C, Shen J, Sidebottom CH, Canaguier A, Chauveau A, Berard A, Deniot G, Guan M, Liu Z, Sun F, Lim YP, Lyons E, Town CD, Bancroft I, Meng J, Ma J, Pires JC, King GJ, Brunel D, Delourme R, Renard M, Aury JM, Adams KL, Batley J, Snowdon RJ, Tost J, Edwards D, Zhou Y, Hua W, Sharpe AG, Paterson AH, Guan C, Wincker P (2014) Plant genetics. Early allopolyploid evolution in the post-Neolithic *Brassica napus* oilseed genome. Science 345:950–953. https://doi.org/10.1126/science.1253435

Chamala S, Chanderbali AS, Der JP, Lan T, Walts B, Albert VA, Depamphilis CW, Leebens-Mack J, Rounsley S, Schuster SC, Wing RA, Xiao N, Moore R, Soltis PS, Soltis DE, Barbazuk WB (2013) Assembly and validation of the genome of the nonmodel basal angiosperm Amborella. Science 342:1516–1517. https://doi.org/10.1126/science.1241130

Chang PL, Dilkes BP, McMahon M, Comai L, Nuzhdin SV (2010) Homoeolog-specific retention and use in allotetraploid *Arabidopsis suecica* depends on parent of origin and network partners. Genome Biol 11:R125. https://doi.org/10.1186/gb-2010-11-12-r125

Chen ECH, Buen Abad Najar C, Zheng C, Brandts A, Lyons E, Tang H, Carretero-Paulet L, Albert VA, Sankoff D (2013) The dynamics of functional classes of plant genes in rediploidized ancient polyploids. BMC Bioinf 14(Suppl 15):S19. https://doi.org/10.1186/1471-2105-14-S15-S19

Chester M, Gallagher JP, Symonds VV, da Silva Cruz, Veruska Ana, Mavrodiev EV, Leitch AR, Soltis PS,

Soltis DE (2012) Extensive chromosomal variation in a recently formed natural allopolyploid species, *Tragopogon miscellus* (Asteraceae). Proc Natl Acad Sci U S A 109:1176–1181. https://doi.org/10.1073/pnas.1112041109

Chester M, Lipman MJ, Gallagher JP, Soltis PS, Soltis DE (2013) An assessment of karyotype restructuring in the neoallotetraploid *Tragopogon miscellus* (Asteraceae). Chromosome Res 21:75–85. https://doi.org/10.1007/s10577-013-9339-y

Chester M, Riley RK, Soltis PS, Soltis DE (2015) Patterns of chromosomal variation in natural populations of the neoallotetraploid *Tragopogon mirus* (Asteraceae). Heredity (Edinb) 114:309–317. https://doi.org/10.1038/hdy.2014.101

Clarke WE, Higgins EE, Plieske J, Wieseke R, Sidebottom C, Khedikar Y, Batley J, Edwards D, Meng J, Li R, Lawley CT, Pauquet J, Laga B, Cheung W, Iniguez-Luy F, Dyrszka E, Rae S, Stich B, Snowdon RJ, Sharpe AG, Ganal MW, Parkin IAP (2016) A high-density SNP genotyping array for *Brassica napus* and its ancestral diploid species based on optimised selection of single-locus markers in the allotetraploid genome. Theor Appl Genet 129:1887–1899. https://doi.org/10.1007/s00122-016-2746-7

Comai L (2000) Genetic and epigenetic interactions in allopolyploid plants. Plant Mol Biol 43:387–399

Comai L, Tyagi AP, Winter K, Holmes-Davis R, Reynolds SH, Stevens Y, Byers B (2000) Phenotypic instability and rapid gene silencing in newly formed arabidopsis allotetraploids. Plant Cell 12:1551–1568

Comai L (2005) The advantages and disadvantages of being polyploid. Nat Rev Genet 6:836–846. https://doi.org/10.1038/nrg1711

Combes M-C, Dereeper A, Severac D, Bertrand B, Lashermes P (2013) Contribution of subgenomes to the transcriptome and their intertwined regulation in the allopolyploid *Coffea arabica* grown at contrasted temperatures. New Phytol 200:251–260. https://doi.org/10.1111/nph.12371

Delourme R, Falentin C, Fomeju BF, Boillot M, Lassalle G, Andre I, Duarte J, Gauthier V, Lucante N, Marty A, Pauchon M, Pichon JP, Ribiere N, Trotoux G, Blanchard P, Riviere N, Martinant JP, Pauquet J (2013) High-density SNP-based genetic map development and linkage disequilibrium assessment in *Brassica napus* L. BMC Genomics 14:120. https://doi.org/10.1186/1471-2164-14-120

Demuth JP, Hahn MW (2009) The life and death of gene families. BioEssays 31:29–39. https://doi.org/10.1002/bies.080085

Diaz A, Zikhali M, Turner AS, Isaac P, Laurie DA (2012) Copy number variation affecting the Photoperiod-B1 and Vernalization-A1 genes is associated with altered flowering time in wheat (*Triticum aestivum*). PLoS ONE 7:e33234. https://doi.org/10.1371/journal.pone.0033234

Doyle JJ, Flagel LE, Paterson AH, Rapp RA, Soltis DE, Soltis PS, Wendel JF (2008) Evolutionary genetics of genome merger and doubling in plants. Annu Rev Genet 42:443–461. https://doi.org/10.1146/annurev.genet.42.110807.091524

Duarte JM, Wall PK, Edger PP, Landherr LL, Ma H, Pires JC, Leebens-Mack J, Depamphilis CW (2010) Identification of shared single copy nuclear genes in Arabidopsis, Populus, Vitis and Oryza and their phylogenetic utility across various taxonomic levels. BMC Evol Biol 10:61. https://doi.org/10.1186/1471-2148-10-61

Dubcovsky J, Dvorak J (2007) Genome plasticity a key factor in the success of polyploid wheat under domestication. Science 316:1862–1866. https://doi.org/10.1126/science.1143986

Dupont J, White PJ, Johnston KM, Heggtveit HA, McDonald BE, Grundy SM, Bonanome A (1989) Food safety and health effects of canola oil. J Am Coll Nutr 8:360–375

Duran C, Eales D, Marshall D, Imelfort M, Stiller J, Berkman PJ, Clark T, McKenzie M, Appleby N, Batley J, Basford K, Edwards D (2010) Future tools for association mapping in crop plants. Genome 53:1017–1023. https://doi.org/10.1139/G10-057

Edger PP, Pires JC (2009) Gene and genome duplications: the impact of dosage-sensitivity on the fate of nuclear genes. Chromosome Res 17:699–717. https://doi.org/10.1007/s10577-009-9055-9

Edwards D, Batley J (2010) Plant genome sequencing: applications for crop improvement. Plant Biotechnol J 8:2–9. https://doi.org/10.1111/j.1467-7652.2009.00459.x

Feldman M, Liu B, Segal G, Abbo S, Levy AA, Vega JM (1997) Rapid elimination of low-copy DNA sequences in polyploid wheat: a possible mechanism for differentiation of homoeologous chromosomes. Genetics 147:1381–1387

Feldman M, Levy AA (2009) Genome evolution in allopolyploid wheat—a revolutionary reprogramming followed by gradual changes. J Genet Genomics 36:511–518. https://doi.org/10.1016/S1673-8527(08)60142-3

Feldman M, Levy AA (2012) Genome evolution due to allopolyploidization in wheat. Genetics 192:763–774. https://doi.org/10.1534/genetics.112.146316

Flagel LE, Wendel JF, Udall JA (2012) Duplicate gene evolution, homoeologous recombination, and transcriptome characterization in allopolyploid cotton. BMC Genomics 13:302. https://doi.org/10.1186/1471-2164-13-302

Freeling M (2008) The evolutionary position of subfunctionalization, downgraded. Genome Dyn 4:25–40. https://doi.org/10.1159/000126004

Freeling M (2009) Bias in plant gene content following different sorts of duplication: tandem, whole-genome, segmental, or by transposition. Annu Rev Plant Biol 60:433–453. https://doi.org/10.1146/annurev.arplant.043008.092122

Freeling M, Woodhouse MR, Subramaniam S, Turco G, Lisch D, Schnable JC (2012) Fractionation mutagenesis and similar consequences of mechanisms removing dispensable or less-expressed DNA in plants. Curr Opin Plant Biol 15:131–139. https://doi.org/10.1016/j.pbi.2012.01.015

Fu Y, Xiao M, Yu H, Mason AS, Yin J, Li J, Zhang D, Fu D (2016) Small RNA changes in synthetic *Brassica napus*. Planta 244:607–622. https://doi.org/10.1007/s00425-016-2529-z

Gaeta RT, Chris Pires J (2010) Homoeologous recombination in allopolyploids: the polyploid ratchet. New Phytol 186:18–28. https://doi.org/10.1111/j.1469-8137.2009.03089.x

Gaeta RT, Pires JC, Iniguez-Luy F, Leon E, Osborn TC (2007) Genomic changes in resynthesized *Brassica napus* and their effect on gene expression and phenotype. Plant Cell 19:3403–3417. https://doi.org/10.1105/tpc.107.054346

Ganal MW, Polley A, Graner E-M, Plieske J, Wieseke R, Luerssen H, Durstewitz G (2012) Large SNP arrays for genotyping in crop plants. J Biosci 37:821–828

Gore MA, Chia J-M, Elshire RJ, Sun Q, Ersoz ES, Hurwitz BL, Peiffer JA, McMullen MD, Grills GS, Ross-Ibarra J, Ware DH, Buckler ES (2009) A first-generation haplotype map of maize. Science 326:1115–1117. https://doi.org/10.1126/science.1177837

Grandke F, Snowdon R, Samans B (2017) gsrc: an R package for genome structure rearrangement calling. Bioinformatics 33:545–546. https://doi.org/10.1093/bioinformatics/btw648

Grandke F, Snowdon RJ, Samans B (submitted) gsrc—an R package for genome structure rearrangement calling

Grant V (1981) Pant Speciation. Columbia University Press, New York

Han F, Fedak G, Guo W, Liu B (2005) Rapid and repeatable elimination of a parental genome-specific DNA repeat (pGc1R-1a) in newly synthesized wheat allopolyploids. Genetics 170:1239–1245. https://doi.org/10.1534/genetics.104.039263

Hegarty MJ, Hiscock SJ (2008) Genomic clues to the evolutionary success of polyploid plants. Curr Biol 18:44. https://doi.org/10.1016/j.cub.2008.03.043

Hegarty MJ et al. (2005) Development of anonymous cDNA microarrays to study changes to the Senecio floral transcriptome during hybrid speciation. Mol Ecol 14:2493–2510

Hormozdiari F, Alkan C, Eichler EE, Sahinalp SC (2009) Combinatorial algorithms for structural variation detection in high-throughput sequenced genomes. Genome Res 19:1270–1278. https://doi.org/10.1101/gr.088633.108

Howell EC, Kearsey MJ, Jones GH, King GJ, Armstrong SJ (2008) A and C genome distinction and chromosome identification in *Brassica napus* by sequential fluorescence in situ hybridization and genomic in situ hybridization. Genetics 180:1849–1857. https://doi.org/10.1534/genetics.108.095893

Jackson S, Chen ZJ (2010) Genomic and expression plasticity of polyploidy. Curr Opin Plant Biol 13:153–159. https://doi.org/10.1016/j.pbi.2009.11.004

Jenczewski E, Eber F, Grimaud A, Huet S, Lucas MO, Monod H, Chevre AM (2003) PrBn, a major gene controlling homeologous pairing in oilseed rape (*Brassica napus*) haploids. Genetics 164:645–653

Jiao Y, Wickett NJ, Ayyampalayam S, Chanderbali AS, Landherr L, Ralph PE, Tomsho LP, Hu Y, Liang H, Soltis PS, Soltis DE, Clifton SW, Schlarbaum SE, Schuster SC, Ma H, Leebens-Mack J, Depamphilis CW (2011) Ancestral polyploidy in seed plants and angiosperms. Nature 473:97–100. https://doi.org/10.1038/nature09916

Joly S, Rauscher JT, Sherman-Broyles SL, Brown AHD, Doyle JJ (2004) Evolutionary dynamics and preferential expression of homoeologous 18S-5.8S-26S nuclear ribosomal genes in natural and artificial Glycine allopolyploids. Mol Biol Evol

Kashkush K, Feldman M, Levy AA (2002) Gene loss, silencing and activation in a newly synthesized wheat allotetraploid. Genetics 160:1651–1659

Kenton A, Parokonny AS, Gleba YY, Bennett MD (1993) Characterization of the Nicotiana tabacum L. genome by molecular cytogenetics. Mol Gen Genet 240:159–169.

Lagercrantz U, Lydiate DJ (1996) Comparative genome mapping in Brassica. Genetics 144:1903–1910

Langer SM, Longin CFH, Wurschum T (2014) Flowering time control in European winter wheat. Front Plant Sci 5:537. https://doi.org/10.3389/fpls.2014.00537

Langham RJ, Walsh J, Dunn M, Ko C, Goff SA, Freeling M (2004) Genomic duplication, fractionation and the origin of regulatory novelty. Genetics 166:935–945

Lashermes P, Combes M-C, Hueber Y, Severac D, Dereeper A (2014) Genome rearrangements derived from homoeologous recombination following allopolyploidy speciation in coffee. Plant J 78:674–685. https://doi.org/10.1111/tpj.12505

Lee H, Gurtowski J, Yoo S, Nattestad M, Marcus S, Goodwin S, McCombie WR, Schatz M (2016) Third-generation sequencing and the future of genomics

Leflon M, Eber F, Letanneur JC, Chelysheva L, Coriton O, Huteau V, Ryder CD, Barker G, Jenczewski E, Chevre AM (2006) Pairing and recombination at meiosis of *Brassica rapa* (AA) × *Brassica napus* (AACC) hybrids. Theor Appl Genet 113:1467–1480. https://doi.org/10.1007/s00122-006-0393-0

Leitch IJ, Bennett MD (1997) Polyploidy in angiosperms. Trends Plant Sci 2:470–476. https://doi.org/10.1016/S1360-1385(97)01154-0

Leitch AR, Leitch IJ (2008) Genomic plasticity and the diversity of polyploid plants. Science 320:481–483. https://doi.org/10.1126/science.1153585

Li R, Fan W, Tian G, Zhu H, He L, Cai J, Huang Q, Cai Q, Li B, Bai Y, Zhang Z, Zhang Y, Wang W, Li J, Wei F, Li H, Jian M, Li J, Zhang Z, Nielsen R, Li D, Gu W, Yang Z, Xuan Z, Ryder OA, Leung FC-C, Zhou Y, Cao J, Sun X, Fu Y, Fang X, Guo X, Wang B, Hou R, Shen F, Mu B, Ni P, Lin R, Qian W, Wang G, Yu C, Nie W, Wang J, Wu Z, Liang H, Min J, Wu Q, Cheng S, Ruan J, Wang M, Shi Z, Wen M, Liu B, Ren X, Zheng H, Dong D, Cook K, Shan G, Zhang H, Kosiol C, Xie X, Lu Z, Zheng H, Li Y, Steiner CC, Lam TT-Y, Lin S, Zhang Q, Li G,

Tian J, Gong T, Liu H, Zhang D, Fang L, Ye C, Zhang J, Hu W, Xu A, Ren Y, Zhang G, Bruford MW, Li Q, Ma L, Guo Y, An N, Hu Y, Zheng Y, Shi Y, Li Z, Liu Q, Chen Y, Zhao J, Qu N, Zhao S, Tian F, Wang X, Wang H, Xu L, Liu X, Vinar T, Wang Y, Lam T-W, Yiu S-M, Liu S, Zhang H, Li D, Huang Y, Wang X, Yang G, Jiang Z, Wang J, Qin N, Li L, Li J, Bolund L, Kristiansen K, Wong GK-S, Olson M, Zhang X, Li S, Yang H, Wang J, Wang J (2010) The sequence and de novo assembly of the giant panda genome. Nature 463:311–317. https://doi.org/10.1038/nature08696

Li A, Liu D, Wu J, Zhao X, Hao M, Geng S, Yan J, Jiang X, Zhang L, Wu J, Yin L, Zhang R, Wu L, Zheng Y, Mao L (2014a) mRNA and small RNA transcriptomes reveal insights into dynamic homoeolog regulation of allopolyploid heterosis in nascent hexaploid wheat. Plant Cell 26:1878–1900. https://doi.org/10.1105/tpc.114.124388

Li J, Wan H-S, Yang W-Y (2014b) Synthetic hexaploid wheat enhances variation and adaptive evolution of bread wheat in breeding processes. J Syst Evol 52:735–742. https://doi.org/10.1111/jse.12110

Lim KY, Kovarik A, Matyasek R, Chase MW, Clarkson JJ, Grandbastien MA, Leitch AR (2007) Sequence of events leading to near-complete genome turnover in allopolyploid Nicotiana within five million years. New Phytol 175:756–763. https://doi.org/10.1111/j.1469-8137.2007.02121.x

Lim KY, Soltis DE, Soltis PS, Tate J, Matyasek R, Srubarova H, Kovarik A, Pires JC, Xiong Z, Leitch AR (2008) Rapid chromosome evolution in recently formed polyploids in Tragopogon (Asteraceae). PLoS ONE 3:e3353. https://doi.org/10.1371/journal.pone.0003353

Liu B, Vega JM, Feldman M (1998) Rapid genomic changes in newly synthesized amphiploids of Triticum and Aegilops. II. Changes in low-copy coding DNA sequences. Genome 41:535–542

Liu B, Brubaker CL, Mergeai G, Cronn RC, Wendel JF (2001) Polyploid formation in cotton is not accompanied by rapid genomic changes. Genome 44:321–330

Liu B, Xu C, Zhao N, Qi B, Kimatu JN, Pang J, Han F (2009) Rapid genomic changes in polyploid wheat and related species: implications for genome evolution and genetic improvement. J Genet Genomics 36:519–528. https://doi.org/10.1016/S1673-8527(08)60143-5

Liu L, Stein A, Wittkop B, Sarvari P, Li J, Yan X, Dreyer F, Frauen M, Friedt W, Snowdon RJ (2012) A knockout mutation in the lignin biosynthesis gene CCR1 explains a major QTL for acid detergent lignin content in *Brassica napus* seeds. Theor Appl Genet 124:1573–1586. https://doi.org/10.1007/s00122-012-1811-0

Liu S, Yang X, Tong C, Edwards D, Parkin IAP, Zhao M, Ma J, Yu J, Huang S, Wang X, Wang J, Lu K, Fang Z, Bancroft I, Yang T-J, Hu Q, Wang X, Yue Z, Li H, Yang L, Zhou Q, Wang W, King GJ, Pires JC, Lu C, Wu Z, Sampath P, Wang Z, Guo H, Pan S, Yang L, Min J, Zhang D, Jin D, Li W, Belcram H, Tu J, Guan M, Qi C, Du D, Li J, Jiang L, Batley J, Sharpe AG, Park B-S, Ruperao P, Waminal NE, Huang Y, Dong C, Wang L, Li J, Hu Z, Huang Y, Huang J, Shi J, Mei D, Liu J, Lee T-H, Wang J, Jin H, Li Z, Li X, Zhang J, Xiao L, Zhou Y, Liu Z, Liu X, Qin R, Tang X, Liu W, Wang Y, Lee J, Kim HH, Denoeud F, Xu X, Liang X, Hua W, Wang J, Chalhoub B, Paterson AH (2014) The *Brassica oleracea* genome reveals the asymmetrical evolution of polyploid genomes. Nat Commun 5:3930. https://doi.org/10.1038/ncomms4930

Liu J, Hua W, Hu Z, Yang H, Zhang L, Li R, Deng L, Sun X, Wang X, Wang H (2015) Natural variation in ARF18 gene simultaneously affects seed weight and silique length in polyploid rapeseed. Proc Natl Acad Sci U S A 112:32. https://doi.org/10.1073/pnas.1502160112

Lukens LN, Pires JC, Leon E, Vogelzang R, Oslach L, Osborn T (2006) Patterns of sequence loss and cytosine methylation within a population of newly resynthesized Brassica napus allopolyploids. Plant Physiol 140:336–348

Madlung A, Comai L (2004) The effect of stress on genome regulation and structure. Ann Bot 94:481–495. https://doi.org/10.1093/aob/mch172

Madlung A, Tyagi AP, Watson B, Jiang H, Kagochi T, Doerge RW, Martienssen R, Comai L (2005) Genomic changes in synthetic Arabidopsis polyploids. Plant J 41:221–230. https://doi.org/10.1111/j.1365-313X.2004.02297.x

Malinska H, Tate JA, Matyasek R, Leitch AR, Soltis DE, Soltis PS, Kovarik A (2010) Similar patterns of rDNA evolution in synthetic and recently formed natural populations of Tragopogon (Asteraceae) allotetraploids. BMC Evol Biol 10:291. https://doi.org/10.1186/1471-2148-10-291

Mardis ER (2008) The impact of next-generation sequencing technology on genetics. Trends Genet 24:133–141. https://doi.org/10.1016/j.tig.2007.12.007

Mason AS, Huteau V, Eber F, Coriton O, Yan G, Nelson MN, Cowling WA, Chèvre A-M (2010) Genome structure affects the rate of autosyndesis and allosyndesis in AABC, BBAC and CCAB Brassica interspecific hybrids. Chromosome Res 18:655–666. https://doi.org/10.1007/s10577-010-9140-0

Mason AS, Batley J, Bayer PE, Hayward A, Cowling WA, Nelson MN (2014) High-resolution molecular karyotyping uncovers pairing between ancestrally related Brassica chromosomes. New Phytol 202:964–974. https://doi.org/10.1111/nph.12706

Mason AS, Takahira J, Atri C, Samans B, Hayward A, Cowling WA, Batley J, Nelson MN (2015) Microspore culture reveals complex meiotic behaviour in a trigenomic Brassica hybrid. BMC Plant Biol 15:173. https://doi.org/10.1186/s12870-015-0555-9

Mason AS, Snowdon RJ (2016) Oilseed rape: learning about ancient and recent polyploid evolution from a recent crop species. Plant Biol. https://doi.org/10.1111/plb.12462

Mason AS, Higgins EE, Snowdon RJ, Batley J, Stein A, Werner C, Parkin IAP (2017) A user guide to the

Brassica 60K Illumina Infinium™ SNP genotyping array. Theor Appl Genet 130:621–633. https://doi.org/10.1007/s00122-016-2849-1
McClintock B (1984) The significance of responses of the genome to challenge. Science 226:792–801
Nicolas SD, Le Mignon G, Eber F, Coriton O, Monod H, Clouet V, Huteau V, Lostanlen A, Delourme R, Chalhoub B, Ryder CD, Chevre AM, Jenczewski E (2007) Homeologous recombination plays a major role in chromosome rearrangements that occur during meiosis of *Brassica napus* haploids. Genetics 175:487–503. https://doi.org/10.1534/genetics.106.062968
Nicolas SD, Leflon M, Liu Z, Eber F, Chelysheva L, Coriton O, Chevre AM, Jenczewski E (2008) Chromosome 'speed dating' during meiosis of polyploid Brassica hybrids and haploids. Cytogenet Genome Res 120:331–338. https://doi.org/10.1159/000121082
Nicolas SD, Leflon M, Monod H, Eber F, Coriton O, Huteau V, Chevre A-M, Jenczewski E (2009) Genetic regulation of meiotic cross-overs between related genomes in *Brassica napus* haploids and hybrids. Plant Cell 21:373–385. https://doi.org/10.1105/tpc.108.062273
Ohno S (1970) Evolution by gene duplication. Springer-Verlag, Berlin, New York
Osborn TC, Butrulle DV, Sharpe AG, Pickering KJ, Parkin IAP, Parker JS, Lydiate DJ (2003) Detection and effects of a homeologous reciprocal transposition in *Brassica napus*. Genetics 165:1569–1577
Ozkan H (2003) Nonadditive changes in genome size during allopolyploidization in the wheat (Aegilops-Triticum) group. J Hered 94:260–264. https://doi.org/10.1093/jhered/esg053
Ozkan H, Levy AA, Feldman M (2001) Allopolyploidy-induced rapid genome evolution in the wheat (Aegilops-Triticum) group. Plant Cell 13:1735–1747
Parkin IA, Sharpe AG, Keith DJ, Lydiate DJ (1995) Identification of the A and C genomes of amphidiploid *Brassica napus* (oilseed rape). Genome 38:1122–1131
Paterson AH (2005) Polyploidy, evolutionary opportunity, and crop adaptation. Genetica 123:191–196
Paterson AH, Wendel JF, Gundlach H, Guo H, Jenkins J, Jin D, Llewellyn D, Showmaker KC, Shu S, Udall J, Yoo M, Byers R, Chen W, Doron-Faigenboim A, Duke MV, Gong L, Grimwood J, Grover C, Grupp K, Hu G, Lee T-H, Li J, Lin L, Liu T, Marler BS, Page JT, Roberts AW, Romanel E, Sanders WS, Szadkowski E, Tan X, Tang H, Xu C, Wang J, Wang Z, Zhang D, Zhang L, Ashrafi H, Bedon F, Bowers JE, Brubaker CL, Chee PW, Das S, Gingle AR, Haigler CH, Harker D, Hoffmann LV, Hovav R, Jones DC, Lemke C, Mansoor S, ur Rahman M, Rainville LN, Rambani A, Reddy UK, Rong J, Saranga Y, Scheffler BE, Scheffler JA, Stelly DM, Triplett BA, van Deynze A, Vaslin MFS, Waghmare VN, Walford SA, Wright RJ, Zaki EA, Zhang T, Dennis ES, Mayer KFX, Peterson DG, Rokhsar DS, Wang X, Schmutz J (2012) Repeated polyploidization of Gossypium genomes and the evolution of spinnable cotton fibres. Nature 492:423–427. https://doi.org/10.1038/nature11798
Pendleton M, Sebra R, Pang AWC, Ummat A, Franzen O, Rausch T, Stutz AM, Stedman W, Anantharaman T, Hastie A, Dai H, Fritz MH-Y, Cao H, Cohain A, Deikus G, Durrett RE, Blanchard SC, Altman R, Chin C-S, Guo Y, Paxinos EE, Korbel JO, Darnell RB, McCombie WR, Kwok P-Y, Mason CE, Schadt EE, Bashir A (2015) Assembly and diploid architecture of an individual human genome via single-molecule technologies. Nat Meth 12:780–786. https://doi.org/10.1038/nmeth.3454
Pires JC, Zhao J, Schranz ME, Leon EJ, Quijada PA, Lukens LN, Osborn TC (2004a) Flowering time divergence and genomic rearrangements in resynthesized Brassica polyploids (Brassicaceae). Biol J Lin Soc 82:675–688. https://doi.org/10.1111/j.1095-8312.2004.00350.x
Pires JC, Lim KY, Kovarik A, Matyasek R, Boyd A, Leitch AR, Leitch IJ, Bennett MD, Soltis PS, Soltis DE (2004b) Molecular cytogenetic analysis of recently evolved Tragopogon (Asteraceae) allopolyploids reveal a karyotype that is additive of the diploid progenitors. Am J Bot 91:1022–1035. https://doi.org/10.3732/ajb.91.7.1022
Pont C, Murat F, Confolent C, Balzergue S, Salse J (2011) RNA-seq in grain unveils fate of neo- and paleopolyploidization events in bread wheat (*Triticum aestivum* L.). Genome Biol 12:R119. https://doi.org/10.1186/gb-2011-12-12-r119
Pont C, Murat F, Guizard S, Flores R, Foucrier S, Bidet Y, Quraishi UM, Alaux M, Dolezel J, Fahima T, Budak H, Keller B, Salvi S, Maccaferri M, Steinbach D, Feuillet C, Quesneville H, Salse J (2013) Wheat syntenome unveils new evidences of contrasted evolutionary plasticity between paleo- and neoduplicated subgenomes. Plant J 76:1030–1044. https://doi.org/10.1111/tpj.12366
Pontes O, Neves N, Silva M, Lewis MS, Madlung A, Comai L, Viegas W, Pikaard CS (2004) Chromosomal locus rearrangements are a rapid response to formation of the allotetraploid *Arabidopsis suecica* genome. Proc Natl Acad Sci U S A 101:18240–18245. https://doi.org/10.1073/pnas.0407258102
Prakash S, Chopra VL (1988) Introgression of resistance to shattering in *Brassica napus* from *Brassica juncea* through non-homologous recombination. Plant Breed 101:167–168. https://doi.org/10.1111/j.1439-0523.1988.tb00283.x
Ramsey J, Schemske DW (2002) Neopolyploidy in flowering plants. Annu Rev Ecol Syst 33:589–639. https://doi.org/10.1146/annurev.ecolsys.33.010802.150437
Renny-Byfield S, Chester M, Kovarik A, Le Comber SC, Grandbastien M-A, Deloger M, Nichols RA, Macas J, Novak P, Chase MW, Leitch AR (2011) Next generation sequencing reveals genome downsizing in allotetraploid *Nicotiana tabacum*, predominantly through the elimination of paternally derived repetitive

DNAs. Mol Biol Evol 28:2843–2854. https://doi.org/10.1093/molbev/msr112

Renny-Byfield S, Kovarik A, Chester M, Nichols RA, Macas J, Novak P, Leitch AR (2012) Independent, rapid and targeted loss of highly repetitive DNA in natural and synthetic allopolyploids of *Nicotiana tabacum*. PLoS ONE 7:e36963. https://doi.org/10.1371/journal.pone.0036963

Renny-Byfield S, Kovarik A, Kelly LJ, Macas J, Novak P, Chase MW, Nichols RA, Pancholi MR, Grandbastien M-A, Leitch AR (2013) Diploidization and genome size change in allopolyploids is associated with differential dynamics of low- and high-copy sequences. Plant J 74:829–839. https://doi.org/10.1111/tpj.12168

Rhoads A, Au KF (2015) PacBio sequencing and its applications. Genomics Proteomics Bioinf 13:278–289. https://doi.org/10.1016/j.gpb.2015.08.002

Rygulla W, Snowdon RJ, Friedt W, Happstadius I, Cheung WY, Chen D (2008) Identification of quantitative trait loci for resistance against *Verticillium longisporum* in oilseed rape (*Brassica napus*). Phytopathology 98:215–221. https://doi.org/10.1094/PHYTO-98-2-0215

Sakai H, Naito K, Ogiso-Tanaka E, Takahashi Y, Iseki K, Muto C, Satou K, Teruya K, Shiroma A, Shimoji M, Hirano T, Itoh T, Kaga A, Tomooka N (2015) The power of single molecule real-time sequencing technology in the de novo assembly of a eukaryotic genome. Sci Rep 5:16780. https://doi.org/10.1038/srep16780

Salmon A, Ainouche ML, Wendel JF (2005) Genetic and epigenetic consequences of recent hybridization and polyploidy in Spartina (Poaceae). Mol Ecol 14:1163–1175. https://doi.org/10.1111/j.1365-294X.2005.02488.x

Salse J, Abrouk M, Murat F, Quraishi UM, Feuillet C (2009) Improved criteria and comparative genomics tool provide new insights into grass paleogenomics. Brief Bioinform 10:619–630. https://doi.org/10.1093/bib/bbp037

Samans B, Chalhoub B, Snowdon RJ (submitted) Surviving a genome collision: genomic signatures of allopolyploidization

Samans B, Chalhoub B, Snowdon RJ (2017) Surviving a genome collision: genomic signatures of allopolyploidization in the recent crop species. Plant Genome 10:0. https://doi.org/10.3835/plantgenome2017.02.0013

Sankoff D, Zheng C, Zhu Q (2010) The collapse of gene complement following whole genome duplication. BMC Genomics 11:313. https://doi.org/10.1186/1471-2164-11-313

Schatz MC, Delcher AL, Salzberg SL (2010) Assembly of large genomes using second-generation sequencing. Genome Res 20:1165–1173. https://doi.org/10.1101/gr.101360.109

Schmutzer T, Samans B, Dyrszka E, Ulpinnis C, Weise S, Stengel D, Colmsee C, Lespinasse D, Micic Z, Abel S, Duchscherer P, Breuer F, Abbadi A, Leckband G, Snowdon R, Scholz U (2015) Species-wide genome sequence and nucleotide polymorphisms from the model allopolyploid plant *Brassica napus*. Sci Data 2:150072. https://doi.org/10.1038/sdata.2015.72

Schnable JC, Freeling M (2011) Genes identified by visible mutant phenotypes show increased bias toward one of two subgenomes of maize. PLoS ONE 6: e17855. https://doi.org/10.1371/journal.pone.0017855

Schnable JC, Springer NM, Freeling M (2011) Differentiation of the maize subgenomes by genome dominance and both ancient and ongoing gene loss. Proc Natl Acad Sci U S A 108:4069–4074. https://doi.org/10.1073/pnas.1101368108

Schnable JC, Wang X, Pires JC, Freeling M (2012) Escape from preferential retention following repeated whole genome duplications in plants. Front Plant Sci 3:94. https://doi.org/10.3389/fpls.2012.00094

Schranz ME, Osborn TC (2000) Novel flowering time variation in the resynthesized polyploid *Brassica napus*. J Hered 91:242–246

Schranz ME, Osborn TC (2004) De novo variation in life-history traits and responses to growth conditions of resynthesized polyploid *Brassica napus* (Brassicaceae). Am J Bot 91:174–183. https://doi.org/10.3732/ajb.91.2.174

Sharpe AG, Parkin IA, Keith DJ, Lydiate DJ (1995) Frequent nonreciprocal translocations in the amphidiploid genome of oilseed rape (*Brassica napus*). Genome 38:1112–1121

Shoemaker RC, Schlueter J, Doyle JJ (2006) Paleopolyploidy and gene duplication in soybean and other legumes. Curr Opin Plant Biol 9:104–109. https://doi.org/10.1016/j.pbi.2006.01.007

Sierro N, Battey JND, Ouadi S, Bakaher N, Bovet L, Willig A, Goepfert S, Peitsch MC, Ivanov NV (2014) The tobacco genome sequence and its comparison with those of tomato and potato. Nat Commun 5(1)

Skalicka K, Lim KY, Matyasek R, Matzke M, Leitch AR, Kovarik A (2005) Preferential elimination of repeated DNA sequences from the paternal, *Nicotiana tomentosiformis* genome donor of a synthetic, allotetraploid tobacco. New Phytol 166:291–303. https://doi.org/10.1111/j.1469-8137.2004.01297.x

de Smet R, Adams KL, Vandepoele K, Montagu Van, Marc CE, Maere S, van de Peer Y (2013) Convergent gene loss following gene and genome duplications creates single-copy families in flowering plants. Proc Natl Acad Sci U S A 110:2898–2903. https://doi.org/10.1073/pnas.1300127110

Soltis DE, Albert VA, Leebens-Mack J, Bell CD, Paterson AH, Zheng C, Sankoff D, Depamphilis CW, Wall PK, Soltis PS (2009) Polyploidy and angiosperm diversification. Am J Bot 96:336–348. https://doi.org/10.3732/ajb.0800079

Soltis DE, Visger CJ, Soltis PS (2014) The polyploidy revolution then… and now: Stebbins revisited. Am J Bot 101:1057–1078. https://doi.org/10.3732/ajb.1400178

Soltis DE, Visger CJ, Marchant DB, Soltis PS (2016) Polyploidy: pitfalls and paths to a paradigm. Am J Bot 103:1146–1166. https://doi.org/10.3732/ajb.1500501

Soltis PS, Soltis DE (2000) The role of genetic and genomic attributes in the success of polyploids. Proc Natl Acad Sci 97:7051–7057. https://doi.org/10.1073/pnas.97.13.7051

Soltis PS, Soltis DE (2009) The role of hybridization in plant speciation. Annu Rev Plant Biol 60:561–588. https://doi.org/10.1146/annurev.arplant.043008.092039

Soltis PS, Soltis DE (2012) Polyploidy and genome evolution. Springer, Berlin, New York

Song K, Lu P, Tang K, Osborn TC (1995) Rapid genome change in synthetic polyploids of Brassica and its implications for polyploid evolution. Proc Natl Acad Sci U S A 92:7719–7723

Song R, Messing J (2003) Gene expression of a gene family in maize based on noncollinear haplotypes. Proc Natl Acad Sci U S A 100:9055–9060. https://doi.org/10.1073/pnas.1032999100

Stebbins GL Jr (1947) Types of polyploids; their classification and significance. Adv Genet 1:403–429

Stein A, Coriton O, Rousseau-Gueutin M, Samans B, Schiessl SV, Obermeier C, Parkin IAP, Chèvre A-M, Snowdon RJ (2017) Mapping of homoeologous chromosome exchanges influencing quantitative trait variation in *Brassica napus*. Plant Biotechnol J 15:1478–1489. https://doi.org/10.1111/pbi.12732

Szadkowski E, Eber F, Huteau V, Lode M, Huneau C, Belcram H, Coriton O, Manzanares-Dauleux MJ, Delourme R, King GJ, Chalhoub B, Jenczewski E, Chevre A-M (2010) The first meiosis of resynthesized *Brassica napus*, a genome blender. New Phytol 186:102–112. https://doi.org/10.1111/j.1469-8137.2010.03182.x

Tang H, Woodhouse MR, Cheng F, Schnable JC, Pedersen BS, Conant G, Wang X, Freeling M, Pires JC (2012) Altered patterns of fractionation and exon deletions in Brassica rapa support a two-step model of paleohexaploidy. Genetics 190:1563–1574. https://doi.org/10.1534/genetics.111.137349

Tang Z, Li M, Chen L, Wang Y, Ren Z, Fu S (2014) New types of wheat chromosomal structural variations in derivatives of wheat-rye hybrids. PLoS ONE 9:e110282. https://doi.org/10.1371/journal.pone.0110282

Tang H, Lyons E, Town CD (2015) Optical mapping in plant comparative genomics. Gigascience 4:3. https://doi.org/10.1186/s13742-015-0044-y

Tate JA, Symonds VV, Doust AN, Buggs RJA, Mavrodiev E, Majure LC, Soltis PS, Soltis DE (2009) Synthetic polyploids of *Tragopogon miscellus* and *T. mirus* (Asteraceae): 60 Years after Ownbey's discovery. Am J Bot 96:979–988. https://doi.org/10.3732/ajb.0800299

Thomas BC, Pedersen B, Freeling M (2006) Following tetraploidy in an Arabidopsis ancestor, genes were removed preferentially from one homeolog leaving clusters enriched in dose-sensitive genes. Genome Res 16:934–946. https://doi.org/10.1101/gr.4708406

Turck F, Coupland G (2011) Plant science. When vernalization makes sense. Science 331:36–37. https://doi.org/10.1126/science.1200786

Udall JA, Quijada PA, Osborn TC (2005) Detection of chromosomal rearrangements derived from homologous recombination in four mapping populations of *Brassica napus* L. Genetics 169:967–979. https://doi.org/10.1534/genetics.104.033209

Varshney RK, Nayak SN, May GD, Jackson SA (2009) Next-generation sequencing technologies and their implications for crop genetics and breeding. Trends Biotechnol 27:522–530. https://doi.org/10.1016/j.tibtech.2009.05.006

Wang J (2004) Stochastic and epigenetic changes of gene expression in arabidopsis polyploids. Genetics 167(4):1961–1973

Wang J, Tian L, Lee H-S, Wei NE, Jiang H, Watson B, Madlung A, Osborn TC, Doerge RW, Comai L, Chen ZJ (2006) Genomewide nonadditive gene regulation in Arabidopsis allotetraploids. Genetics 172:507–517. https://doi.org/10.1534/genetics.105.047894

Wang X, Wang H, Wang J, Liu S, Bai Y, Mun J-H, Bancroft I, Huang S, Li X, Hua W, Wang J, Wang X, Freeling M, Pires JC, Paterson AH, Chalhoub B, Wang B, Hayward A, Sharpe AG, Park B-S, Weisshaar B, Liu B, Li B, Tong C, Song C, Duran C, Peng C, Geng C, Koh C, Lin C, Edwards D, Mu D, Di Shen, Soumpourou E, Fraser F, Conant G, Lassalle G, King GJ, Wang H, Belcram H, Zhou H, Hirakawa H, Abe H, Guo H, Jin H, Parkin IAP, Batley J, Kim J-S, Just J, Li J, Xu J, Kim JA, Li J, Yu J, Meng J, Wang J, Min J, Poulain J, Hatakeyama K, Wu K, Wang L, Fang L, Trick M, Links MG, Zhao M, Jin M, Ramchiary N, Drou N, Berkman PJ, Cai Q, Huang Q, Li R, Tabata S, Cheng S, Zhang S, Huang S, Sato S, Sun S, Kwon S-J, Choi S-R, Lee T-H, Fan W, Zhao X, Tan X, Xu X, Qiu Y, Yin Y, Li Y, Du Y, Lim Y, Narusaka Y, Wang Y, Wang Z, Li Z, Wang Z, Xiong Z, Zhang Z (2011) The genome of the mesopolyploid crop species *Brassica rapa*. Nature Genetics 43:1035–1039. https://doi.org/10.1038/ng.919

Wendel JF (2000) Genome evolution in polyploids. Plant Mol Biol 42:225–249. https://doi.org/10.1023/A:1006392424384

Wendel JF, Schnabel A, Seelanan T (1995) Bidirectional interlocus concerted evolution following allopolyploid speciation in cotton (Gossypium). Proc Nat Acad Sci USA 92:280–284

Wolfe KH (2001) Yesterday's polyploids and the mystery of diploidization. Nat Rev Genet 2:333–341. https://doi.org/10.1038/35072009

Woodhouse MR, Schnable JC, Pedersen BS, Lyons E, Lisch D, Subramaniam S, Freeling M (2010) Following tetraploidy in maize, a short deletion mechanism removed genes preferentially from one of the two homologs. PLoS Biol 8:e1000409. https://doi.org/10.1371/journal.pbio.1000409

Wurschum T, Boeven PHG, Langer SM, Longin CFH, Leiser WL (2015) Multiply to conquer: copy number variations at Ppd-B1 and Vrn-A1 facilitate global adaptation in wheat. BMC Genet 16:96. https://doi.org/10.1186/s12863-015-0258-0

Xiong Z, Gaeta RT, Pires JC (2011) Homoeologous shuffling and chromosome compensation maintain genome balance in resynthesized allopolyploid *Brassica napus*. Proc Natl Acad Sci U S A 108:7908–7913. https://doi.org/10.1073/pnas.1014138108

Yang W, Liu D, Li J, Zhang L, Wei H, Hu X, Zheng Y, He Z, Zou Y (2009) Synthetic hexaploid wheat and its utilization for wheat genetic improvement in China. J Genetics Genomics 36:539–546. https://doi.org/10.1016/S1673-8527(08)60145-9

Yang J, Liu D, Wang X, Ji C, Cheng F, Liu B, Hu Z, Chen S, Pental D, Ju Y, Yao P, Li X, Xie K, Zhang J, Wang J, Liu F, Ma W, Shopan J, Zheng H, Mackenzie SA, Zhang M (2016) The genome sequence of allopolyploid *Brassica juncea* and analysis of differential homoeolog gene expression influencing selection. Nat Genet. https://doi.org/10.1038/ng.3657

Zhao J, Udall JA, Quijada PA, Grau CR, Meng J, Osborn TC (2006) Quantitative trait loci for resistance to *Sclerotinia sclerotiorum* and its association with a homeologous non-reciprocal transposition in *Brassica napus* L. Theor Appl Genet 112:509–516. https://doi.org/10.1007/s00122-005-0154-5

Zhou S, Bechner MC, Place M, Churas CP, Pape L, Leong SA, Runnheim R, Forrest DK, Goldstein S, Livny M, Schwartz DC (2007) Validation of rice genome sequence by optical mapping. BMC Genomics 8:278. https://doi.org/10.1186/1471-2164-8-278

Zou X, Suppanz I, Raman H, Hou J, Wang J, Long Y, Jung C, Meng J (2012) Comparative analysis of FLC homologues in Brassicaceae provides insight into their role in the evolution of oilseed rape. PLoS ONE 7: e45751. https://doi.org/10.1371/journal.pone.0045751

8 Fractionization of Polyploid Duplicated Genes: Gene Loss, Expression Divergence, and Epigenetic Regulation in *Brassica napus*

Chaobo Tong, Rafaqat Ali Gill, Yang Xiang, Lixin Ma, Xiaohui Cheng, Junyan Huang and Shengyi Liu

Abstract

Plant genome polyploidization and subsequent evolution is a crucial process for trait innovation and new species formation. Recurrent whole-genome duplication (WGD) existing in *Brassica napus* created abundant genetic variation, which make it an ideal model for studying the patterns and mechanism of polyploid genome evolution. Based on the availability of reference genomes and high-throughput 'omics' data, the duplicated genes derived from different WGD events were identified, enabling investigation of their expression divergence and epigenetic regulation. This chapter introduced generation and loss of duplicated genes in multiple cycles of whole-genome duplication and described expression divergence of duplicates, particularly those homeologous genes. Alternative splicing events may play an important role in expression divergence of the duplicate genes. Expression divergence of duplicated genes may be regulated by epigenetic mechanisms, especially subgenome interaction-related small RNA produced from transposable elements in the case that the two subgenomes have asymmetrical transposable elements and their abundance. In the future, the new 'omics' technologies could be used to accurately quantify expression divergence and its regulatory mechanisms of duplicated genes in relation to phenotypic changes to uncover polyploid genome evolution.

C. Tong · R. A. Gill · X. Cheng · J. Huang
S. Liu (✉)
Key Laboratory of Biology and Genetic Improvement of Oil Crops, the Ministry of Agriculture, Oil Crops Research Institute of the Chinese Academy of Agricultural Sciences, Wuhan 430062, China
e-mail: liusy@oilcrops.cn

C. Tong · L. Ma · X. Cheng · J. Huang · S. Liu
Hubei Collaborative Innovation Center for Green Transformation of Bio-Resources, Hubei University, Wuhan, China

Y. Xiang
Guizhou Rapeseed Institute, Guizhou Academy of Agricultural Sciences, Guiyang, China

8.1 The Duplicated Genes and Their Loss Rate in *B. napus*

In this chapter, whole-genome duplication (WGD) refers to two or more sets of parental genomes of a plant whose parents are either a same species or two close relatives. If WGD was very old, its extant descendant genomes in its lineage plants may just retain a portion syntenic to the original one. Plant polyploids can be divided into allopolyploids and autopolyploid, and ploidy levels can be triple, quadruple, sextuple, octuple, and even more. In nature, allopolyploids occupy a large proportion (Van de

S. Liu et al. (eds.), *The Brassica napus Genome*, Compendium of Plant Genomes,
https://doi.org/10.1007/978-3-319-43694-4_8

Peer et al. 2017) including important crops such as wheat, *Brassica napus* oilseed rape, and cotton. Here, our discussions are restricted to allopolyploids represented by *B. napus*.

Allopolyploids were generally formed from two close relative progenitors which share a common farther progenitor, and thus a large number of orthologous genes in the two relatives were created and they share similar DNA sequences and/or same functions. Based on inter- or intra-species syntenic analyses, 32,699 orthologous gene pairs were identified between the two diploid progenitors *B. rapa* and *B. oleracea*. Further, most (27,360) of them were conserved among the 31,526 homeologous pairs in *B. napus*. After detailed checking, most of the non-shared parental orthologs and homeologous gene pairs were actually caused by the absence of annotation or collapsed regions in one of four genomes compared. After validating by PCR, only 119 A_n and 83 C_n gene losses were confirmed after polyploidization, and the A_n subgenome showed a slightly higher frequency of gene losses than the C_n subgenome. Among them, 36% of losses are small segmental deletions of two to four adjacent genes. It was concluded that there was no abundant gene loss during the short evolutionary time after the recent allopolyploidization (7,500 years ago).

Furthermore, many diploids also have been revealed to be paleopolyploidy as evidenced by presence of many clearly detectable ancient WGD-derived syntenic blocks (subgenomes) (Lee et al. 2013). A whole-genome triplication (WGT) event occurred (about 15 million years ago, Mya) in the ancestor of the Brassiceae lineage after its divergence from *Arabidopsis* (24–29 Mya) and are shared in *Brassica* species including two subgenomes of *B. napus* (Liu et al. 2014; Lysak et al. 2005; Parkin et al. 2005; Wang et al. 2011). Based on *Arabidopsis–Brassica* syntenic analysis, significant gene losses were observed in the triplicated blocks with an overall retention rate of 1.2-fold of *Arabidopsis* genes in the corresponding syntenic regions of *Brassica* species (Fig. 8.1). In the syntenic regions, about 53% of ancestor genes just retained one copy, about 35% retained two copies and only about 11% retained triplicates in *Brassica* species. The WGT-derived three subgenomes showed different rates of gene loss/retention and were defined as the least fractionated blocks (LF), the medium fractionated blocks (MF1), and the most fractionated blocks (MF2), respectively. The LF subgenome retains ~70% of the genes found in *A. thaliana*, whereas the MF1 and MF2 subgenomes retain substantially lower proportions of retained genes (~46 and ~36%, respectively). Overall, 11,448, 11,493, and 21,348 duplicated gene pairs derived from 15 Mya-Brassiceae-specific WGT events were identified in *B. rapa*, *B. oleracea,* and *B. napus*, respectively. This pattern supported the hypothesis that two steps were responsible for hexaploidization: MF1 and MF2 combined firstly and experienced substantial fractionation in a tetraploid nucleus, and then LF joined and underwent relatively slight gene losses.

In addition, three older polyploidization events, called α, β, and γ events, happened in different evolutionary time points predating the divergence of *Arabidopsis* and *Brassica* species (Bowers et al. 2003; Lyons et al. 2008; Vision et al. 2000). Furthermore, Jiao et al. (2011) identified two more ancient WGD events (called ε and ζ) occurred in the ancestor of angiosperms, which suggested all the angiosperm plants are paleopolyploid. Therefore, *B. napus* contained the legacy of at least five rounds of WGD events before Brassiceae-lineage-specific WGT and recent allopolyploid (A and C genome merging) events (Fig. 8.1). By syntenic analysis between the basal angiosperm *Amborella trichopoda*, the basal eudicot *Vitis vinifera*, the model crucifer *Arabidopsis thaliana,* and *Brassica* species (*B. rapa*, *B. oleracea,* and *B. napus*), the 72× genome multiplication in *B. napus* was confirmed, including 69 specific regions from *Amborella* were detected to be matched up to 72 regions in *B. napus* (Chalhoub et al. 2014) (Fig. 8.1). In summary, recurrent WGD duplication created abundant genetic variation in *B. napus*, which makes it an ideal model for studying the patterns and mechanism of polyploid evolution. The identified WGD-related duplicated genes provided solid basis for further study on

(a)

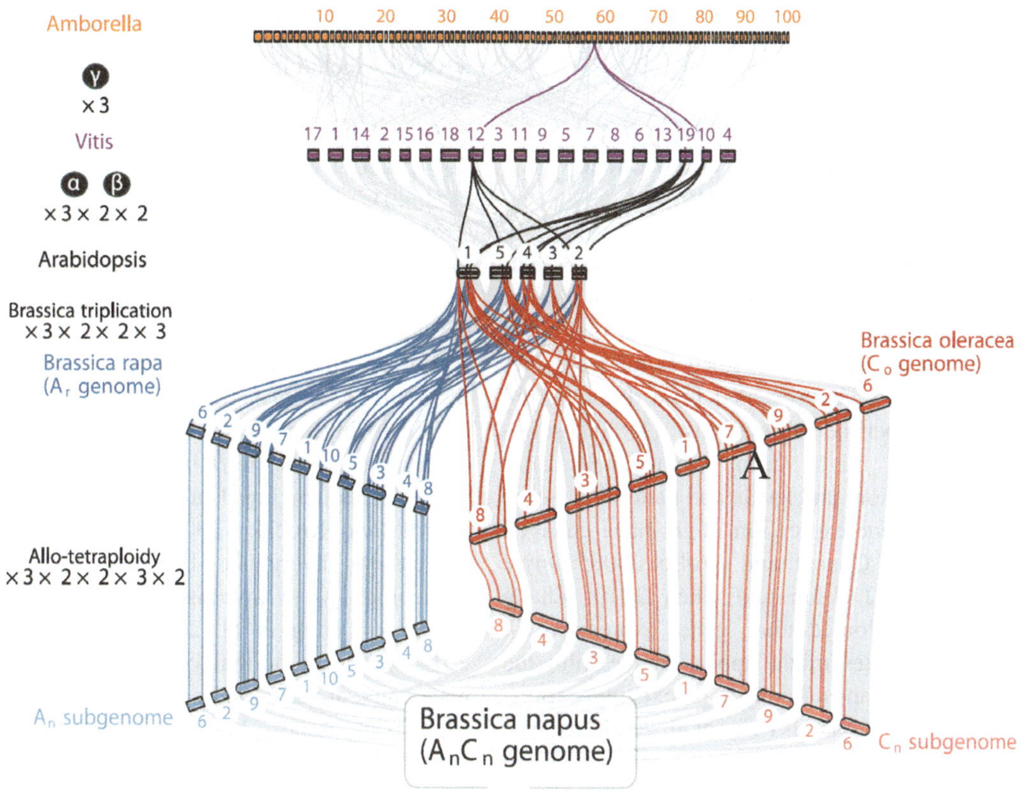

(b) The number of duplicated gene pairs derived from different WGD events

Species	The number of annotated genes in genome	The number of duplicated gene pairs		
		Recent allopolyploidy (A and C genome merging) event (5000-10000 years ago)	*Brassiceae*-specific WGT event (~13 million years ago)	*Brassicaceae* alpha event (~ 35 million years ago)
B. napus	101040	31526	21348	8875
B. rapa	41174	—	11448	5778
B. oleracea	45758	—	11493	5692
A. thaliana	27379	—	—	3742

Fig. 8.1 Syntenic analysis for identifying duplicated genes derived from different WGD events, **a** the syntenic comparison between *A. trichopoda*, *V. vinifera*, *Arabidopsis thaliana*, and *Brassica* species *B. rapa*, *B. oleracea*, and *B. napus*. Gray wedges in the background highlight conserved syntenic blocks each with more than ten gene pairs. A typical ancestral region in *Amborella* is expected to match up to 72 regions in *B. napus*. Adopted from Chalhoub et al. (2014) (Copyright 2014 by the American Association for the Advancement of Science), **b** the number of duplicated gene pairs derived from different WGD events in *B. rapa*, *B. oleracea*, *B. napus*

their expression divergence and epigenetic regulation in *B. napus*.

8.2 The Expression Divergence of WGD-Derived Duplicated Genes in *Brassica napus*

After allopolyploidization from two close relative progenitors, levels and patterns of duplicated gene expression from the progenitors may have reshuffled and can be described as the following categories or layers revealed from different and separate experimental systems: (1) Transcriptome shock might occur immediately after polyploidization, and levels and patterns of some duplicated gene expression might have changed (Chen 2007; Conant and Wolfe 2008; Doyle et al. 2008; Jackson and Chen 2010; Zhang 2003). (2) Non-additive gene expression of duplicated genes might appear and exhibit the non-vertical transmission of preexisting expression patterns of parents (Yoo et al. 2014). (3) The patterns of expression divergence between duplicated genes could be described as expression-level dominance, transgressive expression, and homeolog expression bias (Buggs et al. 2014; Grover et al. 2012). (4) Probably, these unequal expression patterns of homeologs in allopolyploids have led to the non-functionalization, subfunctionalization, and neo-functionalization in long evolutionary terms (Chaudhary et al. 2009; Chen and Pikaard 1997; Combes et al. 2013; Edger et al. 2016; Flagel and Wendel 2010; Force et al. 1999). However, all duplicated gene pairs can simply be divided into the inherited and the non-inherited expression levels/patterns—the former is the same as those in two parents and the latter changed from those of parents. Regardless of these classifications, as long as expression divergence of duplicated gene pairs, they have profound impact on genome differentiation toward diploidization and corresponding phenotypic changes while highly co-expressed gene pairs may generally have certain role in maintaining genome stability and polyploid adaptation. Thus, these differentiating processes could contribute to the polyploid plasticity, e.g., creating novel traits and increasing adaption to environments. Technological advances including the availability of reference genomes and RNA-seq data from different tissues in *B. napus* and its parents (Chalhoub et al. 2014; Liu et al. 2014; Parkin et al. 2014; Sun et al. 2017; Wang et al. 2011) and their relatives *Arabidopsis* have made *B. napus* be a more important model for systematical investigation of the pattern of expression divergence between duplicated genes.

Before the whole-genome sequence of *B. napus* was finished, studies have been conducted to investigate the expression divergence of duplicated genes in *B. napus* using microarray and EST sequencing data (Albertin et al. 2007; Whittle and Krochko 2009). Subsequently, the expression divergence between A_n and C_n homeologs was investigated using Illumina RNA-seq (Tong et al. 2013; Chalhoub et al. 2014). It was found that, of 29,736 homeologous gene pairs, at least one homeolog expressed in at least one tissue. According to comparisons between leaves and roots and between the subgenomes, 45 expression patterns were grouped (Chalhoub et al. 2014). 17,326 (58.3%) gene pairs of A_n and C_n homeologs contributed similarly to gene expression in both contrasting tissues ($P > 0.01$). However, for 4,665 (15.7%) and 5,137 (17.3%) gene pairs, homeolog C_n or A_n contributed more, showing biased expression in both tissues, but no evidence for pronounced genome dominance was observed. In root, leaf, flower, and silique tissues from the other reference genome of the cultivar 'ZS11,' 32–40% of the homeologous genes was diverged in terms of expression ratios (>2 or <1/2, FDR < 0.001 and P-value $\leq$ 0.05), but ANOVA analysis still showed that the homeologous A_n and C_n genomes exhibited no expression bias in the whole-subgenome scale ($P > 0.05$) (Sun et al. 2017). In addition, in a parallel comparison of homeologous gene pairs among synthetic allopolyploids, 26–31% of A_n and C_n duplicated genes displayed homeolog bias in expression levels toward A or C genome, and nearly half of them were shared among different hybrids (Zhang et al. 2016). Nevertheless, the overall homeolog expressions bias in *B. napus* hybrids were also balanced, i.e., expression bias toward A or C subgenome is equal. The absence of significant genome dominance in the recent *B. napus* allopolyploid concurs with old polyploids (Garsmeur et al. 2014), but contrasts with many other polyploids (Cheng et al. 2012; Schnable et al. 2011; Tang et al. 2012; Grover et al. 2012).

Divergent patterns of tissue-specific gene expression among homeologous gene pairs were also revealed. Among 1,062 gene pairs (3.7%), the homeolog A_n was more expressed in leaves than in roots whereas the homeolog C_n was more expressed in roots than in leaves (Chalhoub et al. 2014). In 12 different tissues, a substantial part of them showed tissue-specific expression divergence of homeologous gene pairs and only 30% of them shared the same pattern ($A_n > C_n$ or $A_n < C_n$) among all tissues (Sun et al. 2017). These patterns suggested that some homeolog gene pairs are already evolving toward subfunctionalization in the different tissues, which may provide potential phenotypic plasticity in young allopolyploids. GO analysis of the gene pairs with homeolog expression bias revealed that the genes involved in structural molecule activity, generation of precursor metabolites and energy and ribosome were enriched in A_n-bias patterns, whereas those involved in oxidoreductase activity, translation, and cytoplasm were enriched in C_n-bias patterns.

However, among different subgenomes derived from the Brassiceae-lineage-specific WGT occurred ~15 Mya, there are significantly 'genome dominance' in all of *B. rapa*, *B. oleracea,* and *B. napus* (Cheng et al. 2012; Liu et al. 2014; Wang et al. 2011), in contrast to 'genome equivalence' between A_n and C_n. The genes in the less fractioned subgenome (LF) were dominantly expressed over their paralogs/orthologs in more fractioned subgenomes (MF1 and MF2) by analysis of RNA-seq data generated from callus, root, leaf, stem, flower, and silique of *B. oleracea*, *B. rapa,* and *B. napus.* By detailed checking, ~40% of WGT paralogous gene pairs are differentially expressed in *Brassica* species, suggesting potential subfunctionalization of these genes. ~38% of duplicated gene pairs belonging to transcription factors showed differentiated expression, and paralogs with GO categories related to membrane, catalytic activity, and defense response exhibited a higher ratio of differentiated expression among WGT-derived duplicates.

Based on the contrasting pattern between recent and older polyploidization events, we speculated that *B. napus* is just at the beginning of the process of gene loss and diploidization. Investigation in multiple accessions of *B. rapa* revealed that genes in the LF subgenome had less non-synonymous or frameshift mutations than genes in MF1 and MF2, indicating that the subgenome LF was under significantly more selection pressure to sweep the functional mutations in comparison to MFs (Cheng et al. 2012, 2016). On another hand, the epigenetic mechanisms may play a role in their functional diversification, evidenced by differential patterns of DNA methylation and small RNAs among different subgenomes in *B. napus* (Chalhoub et al. 2014).

8.3 The Alternative Splicing Divergence of Duplicated Genes in *B. napus*

During the last seven years, many plant species have been shown each producing a large number of alternative splicing (AS) transcripts (isoforms or variants) from individual genes, and more and more studies have indicated alternative splicing changes in duplicated genes after polyploidization (Chalhoub et al. 2014; Liu et al. 2014; Su et al. 2006; Talavera et al. 2007; Zhou et al. 2011). Polyploid homeologous duplicates may retain the same AS events as those in the progenitors, or one or both lose or gain novel AS events after polyploidization, leading to different AS patterns between homeologs. However, few studies have analyzed the impact of either gene duplication or polyploidy on AS. Because there are more and more evidences indicating that AS transcripts from a single gene may have different key domains and they conduct different (not similar) functions (Santos et al. 2011; Zhang et al. 2009, 2010), we here use the term 'variant' for these transcripts where necessary.

Based on RNA-seq data from different tissues, four main types of AS events as intron retention (IR), exon skipping (ES), alternative 5' splice site donor (A5SS), and alternative 3' splice site acceptor (A3SS) were identified in *B. rapa*, *B. oleracea*, and *B. napus* (Table 8.1) (Chalhoub et al. 2014; Liu et al. 2014; Tong et al. 2013). Totally, 25–40% of *Brassica* genes was detected to be alternatively spliced with frequent intron retention and rare exon skipping. It was interestingly that genes with small introns were preferentially to produce intron retention events, whereas large introns use an exon definition mechanism in exon shipping events. In addition, the GO enrichment analysis showed that AS genes were associated with signal transduction, regulation, response, binding and catalytic activity, etc. Meanwhile, about 30% of transcription factors genes underwent AS, especially in ARF-, AP2-, MIKC-, C3H-, and MYB-related families. All these evidences suggested that alternative splicing may serve as an important and prevalent mechanism for these 'AS-preferred' genes to function in genome.

The comparison of AS difference between all A_n and C_n homeologs pairs showed that AS patterns can change rapidly after polyploidy, and AS changes after allopolyploidy were much more common than homeolog silencing (Zhou et al. 2011). Based on AS identified from RNA-seq data, 20% of them exhibited different AS patterns that the specific AS events occurred in only A_n or C_n genes within homeologs pairs. Importantly, the AS differentiation in homeologs largely appears to be tissue- or stress-specific manner as many AS was organ specific or induced by abiotic stress treatments. Zhou et al. compared AS events in 82 duplicated gene pairs (homeologs) of *B. napus* using RT-PCR and sequencing assays, 26–30% of the duplicated genes showed changes in AS compared with the parents, including many cases of AS event loss or gain after polyploidization (Zhou et al. 2011). Interestingly, most of the changes were homeolog-specific losses of AS in one homeolog, and only one case there was a homeolog-specific gain of an AS event. In addition, many AS events after allopolyploidy were detected to be parallel losses

Table 8.1 Alternative splicing events identified in different tissues of *Brassica* species

Species	Tissues	Intron retention (events/genes)	Alternative 3′ splicing (events/genes)	Alternative 5′ splicing (events/genes)	Exon skipping (events/genes)
B. napus ('ZS11')	Callus	34,878/11,340	5,614/4,600	2,338/2,059	326/294
	Leaf	18,225/6,733	5,955/4,529	2,633/2,227	308/262
	Root	45,869/13,949	4,796/3,854	2,531/2,162	456/400
	Flower	33,855/11,187	3,454/3,054	1,405/1,290	385/346
	Silique	38,722/11,153	4,800/3,983	1,973/1,780	561/507
B. rapa ('Chiifu-401')	Root	3,776/1,630	1,111/1,010	449/433	281/260
	Stem	2,869/1,290	1,123/1,032	518/494	324/278
	Leaf	3,325/1,313	910/833	353/86	242/215
	Flower	3,427/1,401	1,228/1,087	538/494	336/285
	Silique	6,876/2,379	783/718	295/284	294/260
B. oleracea ('02-12')	Root	4,530/2,230	1,356/1,242	569/489	350/305
	Stem	3,268/1,863	1,365/1,265	718/560	425/369
	Leaf	3,895/1,635	1,025/911	416/105	316/283
	Flower	3,895/1,695	1,296/1,154	947/569	406/385
	Silique	7,563/2,596	896/819	369/305	425/363

in the two independently resynthesized lines, suggesting that changes in AS after allopolyploidy is not entirely random, and many of them were repeatable.

AS is able not only to remove functional domains to produce non-functional transcripts, thus regulating gene dosage (Kalsotra and Cooper 2011; Nilsen and Graveley 2010), but also cause subfunctionalization between WGT-derived paralogous genes (Zhang et al. 2010; Zhou et al. 2011). For WGT-derived paralogs, they have higher ratio of genes with AS than singletons. AS transcripts may be required for gene balance (Birchler et al. 2005; Thomas et al. 2006). These 'AS-preferred' genes were enriched in transcription factors and signal transducer and also were more likely to have been retained as duplicates after WGT events (Tong et al. 2013). By comparing the AS pattern identified from RNA-seq data, larger AS differentiation was displayed among WGT-derived paralogs. In *B. oleracea* and *B. rapa*, 35.5% (8,467) of orthologous gene pairs showed differential expression due to AS variation. When only counting intron retention and exon skipping, 9.3% (2,215) of gene pairs differs. Analysis of AS variants of paralogous gene pairs that have identical numbers of exons demonstrated that these variants with either different variants or differential expression of the same variants caused >20 and >44% of such paralogous genes to be differentially expressed in *B. oleracea* and *B. rapa*, respectively. In conclusion, divergence in AS variants of gene pairs may present an important mechanism for subfunctionalization of duplicated genes and an important layer of gene regulation, and thus provides a genetic basis for polyploid evolution and new species formation.

8.4 Expression Divergence of WGD-Derived Duplicated Genes via Epigenetic Regulation

What are regulatory mechanisms for the above divergence of expression among homeologous gene pairs remains unclear. Some hypotheses were proposed for the mechanisms (Chen 2007; Guan et al. 2014; Hollister et al. 2011; Ng et al. 2012; Pang et al. 2009; Shen et al. 2015). For example, for immediate changes after polyploidization, the copy number and expression of regulatory genes were doubled such as transcription factors and DNA methylation-related enzymes and cofactors if their expression does not proportionally decease after two close relatives genome merger, and thus corresponding target duplicated genes may have increased or decreased in expression, whereas the others have reverse impact on target gene expression because their copy number increases but less than double and thus have less competitive power for shared elements for working complexes; the dosage balance of the network may impose expression of some duplicated genes (but the reason or exact regulation is poorly understood). Among these, DNA methylation, particularly small RNA-directed DNA methylation, may play a critical role.

For studying the DNA methylation regulation to gene expression of duplicated genes, genome-wide bisulfite sequencing has been performed in 'Darmor-*Bzh*' and 'ZS11,' two reference cultivars of *B. napus*. The correlation analysis showed that gene expression is generally inversely related to DNA methylation levels (CpG, CHG, and CHH) both in 'Darmor-*Bzh*' and 'ZS11' (Chalhoub et al. 2014; Sun et al. 2017). As expected, repetitive elements were highly methylated, CDS and gene body sequences had intermediate levels of methylation, and promoters (UTRs) were the least methylated sequence types (Chalhoub et al. 2014). However, the C_n subgenome was found to be more methylated than the A_n subgenome in all sequence types in terms of CpG, CHG, and CHH cytosine contexts. ~10% of total homeologous gene pairs were differentially methylated between A_n and C_n homeologs in both roots and leaves across the gene body sequence and/or UTRs. Among them, ~34% showed expected higher expression for the less methylated homeologs and could be explained by corresponding methylation variation in UTRs or gene body. Meanwhile, 12% of differentially methylated

A_n-C_n homeologous gene pairs was not conserved among tissues, suggesting tissue-specific methylation regulation. All these results showed evidence for the important role of DNA methylation regulating expression divergence of duplicated genes. However, no transcriptomic dominance between A_n and C_n is still a mystery although C_n genes were more often flanked by highly methylated transposons (Chalhoub et al. 2014; Sun et al. 2017).

The striking result for WGT-derived subgenomes in *Brassica* (including A_n, A_r, C_n and C_o) is that there is a good relationship between transcriptomic dominance and bias methylation pattern (Parkin et al. 2014). The levels of methylation in all contexts were lowest for the least fractionated subgenome (LF) in *Brassica* species while its expression is dominant over more fractionated subgenomes (MF1 and MF2), suggesting that cytosine methylation possibly played a significant role in establishing relative expression differences and diversification among the three subgenomes.

Because allopolyploid subgenomes carry different types and contents of transposable elements (TE) producing 24-nt small RNA which can direct DNA methylation (RdDM), it is reasonable to infer that dosage and complementation of 24-nt small RNA from the two subgenomes may have different impact on genes and regulatory elements on the subgenomes (Hollister et al. 2011). Genome-wide investigation of the relationships of gene expression with transposable element (TE) distribution and small RNA targeting showed that the 24-nt small RNAs target TEs were less inserted in LF subgenome and are negatively correlated to the dominant expression of individual paralogous gene pairs (Cheng et al. 2016; Woodhouse et al. 2014). The biased distribution of TEs among the subgenomes and the targeting of 24-nt small RNAs together produce the dominant expression phenomenon at a subgenome scale. From this view, epigenetic modification of TEs via the RdDM pathway provided one potential mechanism that explains dominance at the subgenomic level in *Brassica* paleopolyploids. More importantly, a recent hypothesis further suggested that when two suites of distinct siRNAs and TEs suddenly merged within a cell, siRNAs might target the sequences other than its original ones resulting in genome-wide inter-genomic interaction (Wendel et al. 2016). This type of inter-genomic interaction via siRNAs from each other may contribute to the genome stability and the expression changes between the subgenomes. Unfortunately, this hypothesis has not been thoroughly tested, and there are no related reports about the degree and way how siRNAs could regulate the different subgenomes each other in *trans*.

8.5 Perspectives

As long as expression divergence of duplicated gene pairs exists, they have profound impact on genome differentiation toward diploidization and corresponding phenotypic changes. Therefore, maintenance and divergence of duplicated gene expression is one of most important topics in polyploids and plant evolution as well as in crop improvement.

The genome sequencing and identification of duplicated genes have set a resource foundation and expression studies of these duplicated genes have added new insights into the evolution of polyploidy *B. napus*. However, up to now, investigations are just at the beginning. There are many to be exploited. First, genome and transcript resource need to be significantly improved by updating the reference genome and directly identifying full length RNA by new sequencing technologies such as the PacBio platform. The data available represents only a few tissues and cultivars. Given the importance of tissue-specific expression divergence and epigenetic regulation on duplicated genes, we need more data from distinct tissues under specific developmental stage or stress, enabling the search for broader patterns from different tissues or individuals. The interplay between different subgenomes after polyploidization should be addressed based on combination analysis of sets of mRNA and sRNA sequencing data, but precise identification and function studies of small RNA are challenges.

References

Albertin W, Alix K, Balliau T, Brabant P, Davanture M, Malosse C, Valot B, Thiellement H (2007) Differential regulation of gene products in newly synthesized *Brassica napus* allotetraploids is not related to protein function nor subcellular localization. BMC Genom 8:56

Birchler JA, Riddle NC, Auger DL, Veitia RA (2005) Dosage balance in gene regulation: biological implications. Trends Genet 21:219–226

Bowers JE, Chapman BA, Rong J, Paterson AH (2003) Unravelling angiosperm genome evolution by phylogenetic analysis of chromosomal duplication events. Nature 422:433–438

Buggs RJA, Wendel JF, Doyle JJ, Soltis DE, Soltis PS, Coate JE (2014) The legacy of diploid progenitors in allopolyploid gene expression patterns. Philos Trans R Soc Lond B 369

Chalhoub B, Denoeud F, Liu S, Parkin IA, Tang H, Wang X, Chiquet J, Belcram H, Tong C, Samans B et al (2014) Plant genetics. Early allopolyploid evolution in the post-neolithic *Brassica napus* oilseed genome. Science 345:950–953

Chaudhary B, Flagel L, Stupar RM, Udall JA, Verma N, Springer NM, Wendel JF (2009) Reciprocal silencing, transcriptional bias and functional divergence of homeologs in polyploid cotton (gossypium). Genetics 182:503–517

Chen ZJ (2007) Genetic and epigenetic mechanisms for gene expression and phenotypic variation in plant polyploids. Annu Rev Plant Biol 58:377–406

Chen ZJ, Pikaard CS (1997) Transcriptional analysis of nucleolar dominance in polyploid plants: biased expression/silencing of progenitor rRNA genes is developmentally regulated in Brassica. Proc Natl Acad Sci USA 94:3442–3447

Cheng F, Sun C, Wu J, Schnable J, Woodhouse MR, Liang J, Cai C, Freeling M, Wang X (2016) Epigenetic regulation of subgenome dominance following whole genome triplication in *Brassica rapa*. New Phytol 211:288–299

Cheng F, Wu J, Fang L, Sun S, Liu B, Lin K, Bonnema G, Wang X (2012) Biased gene fractionation and dominant gene expression among the subgenomes of *Brassica rapa*. PLoS ONE 7:e36442

Combes MC, Dereeper A, Severac D, Bertrand B, Lashermes P (2013) Contribution of subgenomes to the transcriptome and their intertwined regulation in the allopolyploid *Coffea arabica* grown at contrasted temperatures. New Phytol 200:251–260

Conant GC, Wolfe KH (2008) Turning a hobby into a job: how duplicated genes find new functions. Nat Rev Genet 9:938–950

Doyle JJ, Flagel LE, Paterson AH, Rapp RA, Soltis DE, Soltis PS, Wendel JF (2008) Evolutionary genetics of genome merger and doubling in plants. Annu Rev Genet 42:443–461

Edger PP, Smith RD, McKain MR, Cooley AM, Vallejo-Marin M, Yuan Y, Bewick AJ, Ji L, Platts AE, Bowman MJ (2016) Subgenome dominance in an interspecific hybrid, synthetic allopolyploid, and a 140 year old naturally established neo-allopolyploid monkeyflower. bioRxiv: 094797

Flagel LE, Wendel JF (2010) Evolutionary rate variation, genomic dominance and duplicate gene expression evolution during allotetraploid cotton speciation. New Phytol 186:184–193

Force A, Lynch M, Pickett FB, Amores A, Yan YL, Postlethwait J (1999) Preservation of duplicate genes by complementary, degenerative mutations. Genetics 151:1531–1545

Garsmeur O, Schnable JC, Almeida A, Jourda C, D'Hont A, Freeling M (2014) Two evolutionarily distinct classes of paleopolyploidy. Mol Biol Evol 31:448–454

Grover CE, Gallagher JP, Szadkowski EP, Yoo MJ, Flagel LE, Wendel JF (2012) Homoeolog expression bias and expression level dominance in allopolyploids. New Phytol 196:966–971

Guan XY, Song QX, Chen ZJ (2014) Polyploidy and small RNA regulation of cotton fiber development. Trends Plant Sci 19:516–528

Hollister JD, Smith LM, Guo YL, Ott F, Weigel D, Gaut BS (2011) Transposable elements and small RNAs contribute to gene expression divergence between *Arabidopsis thaliana* and *Arabidopsis lyrata*. Proc Natl Acad Sci USA 108:2322–2327

Jackson S, Chen ZJ (2010) Genomic and expression plasticity of polyploidy. Curr Opin Plant Biol 13:153–159

Jiao Y, Wickett NJ, Ayyampalayam S, Chanderbali AS, Landherr L, Ralph PE, Tomsho LP, Hu Y, Liang H, Soltis PS et al (2011) Ancestral polyploidy in seed plants and angiosperms. Nature 473:97–100

Kalsotra A, Cooper TA (2011) Functional consequences of developmentally regulated alternative splicing. Nat Rev Genet 12:715–729

Lee TH, Tang H, Wang X, Paterson AH (2013) PGDD: a database of gene and genome duplication in plants. Nucleic Acids Res 41:D1152–D1158

Liu S, Liu Y, Yang X, Tong C, Edwards D, Parkin IA, Zhao M, Ma J, Yu J, Huang S et al (2014) The *Brassica oleracea* genome reveals the asymmetrical evolution of polyploid genomes. Nat Commun 5:3930

Lyons E, Pedersen B, Kane J, Alam M, Ming R, Tang H, Wang X, Bowers J, Paterson A, Lisch D et al (2008) Finding and comparing syntenic regions among Arabidopsis and the outgroups papaya, poplar, and grape: CoGe with rosids. Plant Physiol 148:1772–1781

Lysak MA, Koch MA, Pecinka A, Schubert I (2005) Chromosome triplication found across the tribe Brassiceae. Genome Res 15:516–525

Ng DWK, Lu J, Chen ZJ (2012) Big roles for small RNAs in polyploidy, hybrid vigor, and hybrid incompatibility. Curr Opin Plant Biol 15:154–161

Nilsen TW, Graveley BR (2010) Expansion of the eukaryotic proteome by alternative splicing. Nature 463:457–463

Pang MX, Woodward AW, Agarwal V, Guan XY, Ha M, Ramachandran V, Chen XM, Triplett BA, Stelly DM,

Chen ZJ (2009) Genome-wide analysis reveals rapid and dynamic changes in miRNA and siRNA sequence and expression during ovule and fiber development in allotetraploid cotton (*Gossypium hirsutum* L.). Genome Biol 10

Parkin IA, Gulden SM, Sharpe AG, Lukens L, Trick M, Osborn TC, Lydiate DJ (2005) Segmental structure of the *Brassica napus* genome based on comparative analysis with *Arabidopsis thaliana*. Genetics 171:765–781

Parkin IA, Koh C, Tang H, Robinson SJ, Kagale S, Clarke WE, Town CD, Nixon J, Krishnakumar V, Bidwell SL et al (2014) Transcriptome and methylome profiling reveals relics of genome dominance in the mesopolyploid *Brassica oleracea*. Genome Biol 15:R77

Santos ME, Athanasiadis A, Leitao AB, DuPasquier L, Sucena E (2011) Alternative splicing and gene duplication in the evolution of the FoxP gene subfamily. Mol Biol Evol 28:237–247

Schnable JC, Springer NM, Freeling M (2011) Differentiation of the maize subgenomes by genome dominance and both ancient and ongoing gene loss. Proc Natl Acad Sci USA 108:4069–4074

Shen EH, Zou J, Behrens FH, Chen L, Ye CY, Dai ST, Li RY, Ni M, Jiang XX, Qiu J et al (2015) Identification, evolution, and expression partitioning of miRNAs in allopolyploid *Brassica napus*. J Exp Bot 66:7241–7253

Su Z, Wang J, Yu J, Huang X, Gu X (2006) Evolution of alternative splicing after gene duplication. Genome Res 16:182–189

Sun F, Fan G, Hu Q, Zhou Y, Guan M, Tong C, Li J, Du D, Qi C, Jiang L et al (2017) The high-quality genome of *Brassica napus* cultivar 'ZS11' reveals the introgression history in semi-winter morphotype. Plant J Cell Mol Biol

Talavera D, Vogel C, Orozco M, Teichmann SA, de la Cruz X (2007) The (in)dependence of alternative splicing and gene duplication. PLoS Comput Biol 3:e33

Tang H, Woodhouse MR, Cheng F, Schnable JC, Pedersen BS, Conant G, Wang X, Freeling M, Pires JC (2012) Altered patterns of fractionation and exon deletions in *Brassica rapa* support a two-step model of paleohexaploidy. Genetics 190:1563–1574

Thomas BC, Pedersen B, Freeling M (2006) Following tetraploidy in an Arabidopsis ancestor, genes were removed preferentially from one homeolog leaving clusters enriched in dose-sensitive genes. Genome Res 16:934–946

Tong C, Wang X, Yu J, Wu J, Li W, Huang J, Dong C, Hua W, Liu S (2013) Comprehensive analysis of RNA-seq data reveals the complexity of the transcriptome in *Brassica rapa*. BMC Genom 14:689

Van de Peer Y, Mizrachi E, Marchal K (2017) The evolutionary significance of polyploidy. Nat Rev Genet 18:411–424

Vision TJ, Brown DG, Tanksley SD (2000) The origins of genomic duplications in Arabidopsis. Science 290:2114–2117

Wang X, Wang H, Wang J, Sun R, Wu J, Liu S, Bai Y, Mun JH, Bancroft I, Cheng F et al (2011) The genome of the mesopolyploid crop species *Brassica rapa*. Nat Genet 43:1035–1039

Wendel JF, Jackson SA, Meyers BC, Wing RA (2016) Evolution of plant genome architecture. Genome Biol 17

Whittle CA, Krochko JE (2009) Transcript profiling provides evidence of functional divergence and expression networks among ribosomal protein gene paralogs in *Brassica napus*. Plant Cell 21:2203–2219

Woodhouse MR, Cheng F, Pires JC, Lisch D, Freeling M, Wang XW (2014) Origin, inheritance, and gene regulatory consequences of genome dominance in polyploids. Proc Natl Acad Sci USA 111:5283–5288

Yoo MJ, Liu X, Pires JC, Soltis PS, Soltis DE (2014) Nonadditive gene expression in polyploids. Annu Rev Genet 48:485–517

Zhang D, Pan Q, Tan C, Zhu B, Ge X, Shao Y, Li Z (2016) Genome-Wide gene expressions respond differently to A-subgenome origins in *Brassica napus* synthetic hybrids and natural allotetraploid. Front Plant Sci 7:1508

Zhang JZ (2003) Evolution by gene duplication: an update. Trends Ecol Evol 18:292–298

Zhang PG, Huang SZ, Pin AL, Adams KL (2010) Extensive divergence in alternative splicing patterns after gene and genome duplication during the evolutionary history of Arabidopsis. Mol Biol Evol 27:1686–1697

Zhang Z, Zhou L, Wang P, Liu Y, Chen X, Hu L, Kong X (2009) Divergence of exonic splicing elements after gene duplication and the impact on gene structures. Genome Biol 10:R120

Zhou R, Moshgabadi N, Adams KL (2011) Extensive changes to alternative splicing patterns following allopolyploidy in natural and resynthesized polyploids. Proc Natl Acad Sci USA 108:16122–16127

Brassica Mitochondrial and Chloroplast Genomes

9

Pu Chu, Jianmei Chen and Rongzhan Guan

Abstract

The energy-converting organelles, mitochondria and chloroplasts, play important roles in plant growth and development. The coordination between nuclear and organellar genomes has been reported extensively in plants, and studies on organellar genomes are important supplements to nuclear genome research. Benefitting from high-throughput sequencing technology, mitochondrial genome sequences have been reported in many plants, especially in *Brassica* species. Here, the size, composition, and structure variation of *Brassica* mitochondria genomes are introduced. There are fewer reports of complete chloroplast genome sequences in *Brassica* compared to those of mitochondrial genomes. The involvements of mitochondrial and chloroplast genomes in evolution and *Brassica* breeding are discussed.

P. Chu · R. Guan (✉)
State Key Laboratory of Crop Genetics and Germplasm Enhancement, College of Agriculture, Nanjing Agricultural University, Nanjing 210095, China
e-mail: guanrzh@njau.edu.cn

J. Chen
National Navel Orange Engineering Research Center, College of Navel Orange, Gannan Normal University, Ganzhou 341000, China

9.1 Introduction

Mitochondria and chloroplasts control energy conversion in plant cells (Alberts et al. 2002) and synthesize amino acids, lipids, nucleotides, vitamins, and porphyrins to sustain the functional metabolism in plants and thus play important roles in plant growth and development (Inoue 2007). Plant mitochondria are associated with the determination of cytoplasmic male sterility (Sandhu et al. 2007), stress responses (Huang et al. 2011), regulation of programmed cell death (Diamond and McCabe 2011), nitrate sensing and GA-mediated pathways for growth and flowering (Pellny et al. 2008). Chloroplasts are the organelles that define plants and contain highly conserved genes fundamental to plant life, such as those involved in chlorophyll biosynthesis, photosynthesis, and retrograde signaling (Jarvis and López-Juez 2013; Jensen and Leister 2014).

Previous analyses suggested that ancient eubacterial invasions gave rise to mitochondria and chloroplasts (Gray et al. 1999). In this process, endosymbionts lost the bulk of their genomes, necessitating the evolution of elaborate mechanisms for organelle biogenesis and metabolite exchange (Dyall et al. 2004). Gene flow between the two organelle genomes and the nuclear genome has led to increased genetic diversity (Petit et al. 2005), a phenomenon that has been widely reported in the plant kingdom.

S. Liu et al. (eds.), *The Brassica napus Genome*, Compendium of Plant Genomes,
https://doi.org/10.1007/978-3-319-43694-4_9

Therefore, coordination between the nucleus and organelles is crucial for plants' survival (Woodson and Chory 2008), and studies on the mitochondrial and chloroplast genome will facilitate our understanding of the whole genome in plants.

Plant mitochondrial genomes have unique features compared to their counterparts in animals and fungi, such as a dynamic structure (Ogihara et al. 2005), incorporation of foreign DNA (Tanaka et al. 2012), and large and dramatic variations in size (Kubo and Newton 2008), which range from 208 kb for *Brassica hirta* (Palmer and Herbon 1987) to over 11.3 Mb for *Silene conica* (Sloan et al. 2012). Active recombination via repeat sequences is thought to be responsible for the dynamic nature and multipartite organization of the mitochondrial genome in all angiosperms investigated (Woloszynska 2010; Wang et al. 2014), which may produce dramatic variation within closely related species (Alverson et al. 2010; Palmer and Herbon 1988). Compared with the plant mitochondrial genomes, chloroplast genomes are relatively small, ranging from 120 to 220 kb, encoding 120–130 genes. Genes for the organelle's own genetic system and photosynthesis-related genes are found in the chloroplast genome in flowering plants (Repkova 2010; Bock 2014). In contrast to plant mitochondria, sequence analysis has revealed that the genome organization and coding capacity of chloroplasts are highly conserved in higher plants (Grevich and Daniel 2005).

The genus *Brassica* (family *Brassicaceae*) contains six cultivated species, including three elementary diploid species: *Brassica rapa* (nuclear genome constitution AA), *Brassica nigra* (BB), and *Brassica oleracea* (CC); and three amphidiploids: *Brassica juncea* (AABB), *Brassica napus* (AACC), and *Brassica carinata* (BBCC) (Nagaharu 1935). *Brassica* species served as early models for understanding the structure, function, and content of plant mitochondrial genomes (Grewe et al. 2014). Considering that *Brassica* species have the smallest mitochondrial genomes among higher plants, investigation of their mitochondrial genomes may reveal the minimum sequence requirement for a mitochondrial genome of higher plants (Handa 2003). The first complete restriction map for a higher plant mitochondrial genome was that of *Brassica campestris* (Palmer and Shields 1984). Since then, taking advantage of high-throughput sequencing technology, the complete mitochondrial genome sequences of all six cultivated species and some related wild species in the genus *Brassica* have been reported. For the chloroplast genome, the origin and evolution of *Brassica* species were first explored using mutational analysis of the maternally inherited chloroplast genome about 30 years ago (Palmer et al. 1983). Recently, the complete chloroplast genome sequences of *B. rapa* and *B. napus* have been reported and their evolutionary implications have been discussed (Hu et al. 2011).

In this chapter, we offer a brief introduction to *Brassica* mitochondria and chloroplast genomes, including data from recent investigations, as well as implications for breeding and evolutionary studies.

9.2 *Brassica* Mitochondria Genome

9.2.1 Mitochondria Genome Size

The mitochondrial genome of *B. hirta* (white mustard), a related wild species in the genus *Brassica*, is only 208 kb, which is the smallest mitochondrial genome in plants reported to date (Palmer and Herbon 1987). The sizes of whole single circular mitochondrial genomes of the *B. rapa* (*cam*), *B. juncea* (*jun*), *B. napus* (Samuels et al. 2013), *B. oleracea* (*ole*, accession "08C717"), and *B. carinata* (*car*) are 219,747, 219,766, 221,853, 360,271, and 232,241 bp, respectively (Chang et al. 2011; Handa 2003). Previous studies have reported physical maps of the *cam* (Palmer and Shields 1984) and *ole* (Chetritl et al. 1984) mitochondrial genomes. The length of the *cam* sequence was almost the same as that obtained from the physical map (219.7 kb vs. 218 kb). However, the length of the *ole* mitotype is much larger than that previously reported from physical mapping (219 kb), because of a duplication of a 141.8 kb segment in its genome. Variation of the mitochondrial genomes within *B. oleracea* has been reported. The mitotype of cv. "Fujiwase" (denoted as *ole*-F in this chapter) is a 219,952 bp circular mitochondrial genome (Tanaka et al. 2014), which is similar to that of *B. oleracea* var botrytis (Grewe et al. 2014) and smaller than the previously reported mitochondrial genome of *B. oleracea* accession "08C717" (360 kb). Cytoplasmic male sterility (CMS) lines and their iso-nuclear maintainer lines may have mitotypes of different sizes. For example, two mitotypes of *B. napu*, *pol* and *nap*, are 22,412 bp (Handa 2003) and 221,853 bp, respectively (Chen et al. 2011). In addition, a heterogeneous *B. napus* mitochondrial genome was sequenced whose size (258,473 bp) is different from those of *pol* and *nap* mitotypes. The *B. juncea hau* CMS mitochondrial genome is 247,903 bp, which is obviously larger than the maintainer line and another normal type line "J163-4" (both 219,863 bp). Recently, the mitochondrial genome of the sixth species, *B. nigra* (Clare et al. 2008), was sequenced and its size is 232,145 bp (Yamagishi et al. 2014), which is slightly smaller than that of *B. carinata*.

9.2.2 Mitochondria Genome Composition

Despite the size variation of the mitochondria genomes in these six cultivated *Brassica* species (220–360 kb), the compositions of the mtDNAs are similar. These *Brassica* mitochondrial mitotypes share 36 protein-coding genes, three ribosomal genes (*rrn*5, *rrn*18, and *rrn*26), and 15 tRNA genes (Table 9.1). The total number of genes varies with mitotypes, ranging from 53 in *car* to 95 in *ole*: The number of protein-coding genes varies from 33 to 56, and the number of tRNA varies from 17 to 35. The numbers of open reading frames (ORFs) without known function are also different among mitotypes, ranging from 29 in *hau* to 46 in *nap*.

The numbers of genes with known functions are almost the same in all mitotypes (Table 9.2);

Table 9.1 Total length and gene numbers of 11 *Brassica* mitotypes[a]

Feature	*cam*	*jun*	*hau*	*hau M*	*nig*	*car*	*nap*	*pol*	*oguC*	*ole*	*ole*-F
Genome size (bp)	219,747	219,766	247,903	219,863	232,145	232,241	221,853	223,412	258,473	360,271	219,952
Total gene	55	55	63	61	53	53	55	56	59	95	54
Protein-coding genes	34	34	35	36	33	33	35	35	33	56	34
rRNA	3	3	3	3	3	3	3	3	3	4	3
tRNA	18	18	25	22	17	17	17	18	23	35	17
Unidentified ORFs	44	44	29	31	–	36	46	45	39	44	40

cam, *jun*, *ole*, and *car* denote the mitotype of *B. rapa*, *B. juncea*, *B. oleracea*, and *B. carinata*, respectively (Chen et al. 2011). *ole*-F denotes the mitotype of *B. oleracea* from Fujiwase (Tanaka et al. 2014). *nig* denotes the mitotype of *B. nigra* (Yamagishi et al. 2014). *pol* and *nap* denote two mitotypes of *B. napus* (Handa 2003; Chen et al. 2011), while *oguC* denotes the mitotype of *Ogura-cms-cybrid* (*oguC*) *B. napus* (Wang et al. 2012a). *hau* and *hau M* denote the mitotype of *hau* CMS *B. juncea* line and its maintainer line, respectively (Heng et al. 2014)

[a]Adapted from Chang et al. (2011)

however, the *ole*, *car*, and *nig* mitotypes lack the complex IV-related *cox2-2* gene found in the other mitotypes. The CMS-related genes (homologous to *orf224* and *orf222*) found in the *pol* and *nap* mitotypes are not found in other mitotypes. Thirty-eight duplicated genes with known functions are present in *ole*. The differences in the number of functional genes between *ole* and the other mitotypes have been attributed to the 141.8 kb segment duplication in the *ole* mitotype.

9.2.3 Mitochondrial Genome Structure

Repeats, including short repeats (30–99 bp), intermediate repeats (50–500 bp), and large repeats (>1 kb), are closely related to the structure variations of plant mitochondrial genomes. Short repeats associated with an irreversible reorganization (Andre et al. 1992) are uniformly distributed in the six *Brassica* mitochondrial genomes, which may imply that there are prerequisites for such an irreversible rearrangement (Chang et al. 2011). Intermediate repeats mediate low-frequency, asymmetric DNA exchange (Arrieta-Montiel et al. 2009) which is associated with rapid stoichiometric changes in genome configuration (Shedge et al. 2007), referred to as substoichiometric shifting (Small et al. 1987).

Large repeats usually account for most of the observed genome structural complexity in plants (Arrieta-Montiel and Mackenzie 2011). Large repeats mediate high-frequency reciprocal DNA exchanges that can give rise to the subdivision of the genome into a multipartite configuration (Palmer and Shields 1984; Lonsdale et al. 1981; Chang et al. 2013). In *Brassica*, this phenomenon was first reported in the *B. campestris* mitochondrial genome (Palmer and Shields 1984). The 218 kb master chromosome has been postulated to interconvert with the two smaller circles (135 and 83 kb) via a co-integration-resolution pathway mediated by reciprocal recombination within the 2 kb repeat, which is shared by all three circles.

Four kinds of large repeats have been observed in the mitochondrial genomes of the six species of *Brassica* (Fig. 9.1). The R1 repeat (3605 bp) carries two exons of the *nad5* gene, and the R2, RB, and R repeats are mtDNA fragments of 141.8 kb, 2427 bp, and 6580 bp, respectively (Chang et al. 2011). A pair of RB repeats and only one copy of the R1 repeat are found in the *pol*, *nap*, *ole*-F, *jun*, and *cam* mitotypes. The overall structure of the *B. nigra* mt genome is identical to that of the *car* mitotype, which contains one copy of the RB repeat, two copies of the R repeats, and one copy of the R1 repeat (Yamagishi et al. 2014).

The multipartite structures of *cam*, *jun*, *ole*-F, *pol*, and *nap* may result from the same large RB repeats, and the multipartite structure of *car* may result from R repeats. The *ole* mitotype (Chang et al. 2011) contains three pairs of large repeats, R1, R2, and RB, which makes its multipartite structure too complex to predict. The sizes of the predicted multipartite circles for the five species' mitotypes, except that of *nig*, are listed in Table 9.3.

9.2.4 *Brassica napus* Mitochondria Genomes

In spite of the differences found in genome size and total gene number, the compositions of the mtDNAs of the six crop species of *Brassica* are similar, and multipartite structures of the mitochondria genome resulting from the large repeats are commonly reported in *Brassica* crops.

Brassica napus is the most important oilseed crop in the genus *Brassica*. The complete mitochondrial genome of *B. napus* was first sequenced in 2003 and compared with that of *Arabidopsis thaliana* (Handa 2003). In that study, the complete nucleotide sequence of the *B. napus* (cv. Wester) mitochondrial genome was determined. The genome is 221,853 bp in size, containing 34 protein-coding genes, three rRNA genes, 17 tRNA genes, and 45 other ORFs larger than 100 codons in size. To better understand mitochondrial genome evolution in higher plants, comparative analyses have been performed. The mtDNAs of *B. napus* and *A. thaliana* share nearly the same set of functional genes. The protein-coding regions are

Table 9.2 Gene contents of *Brassica* mitotypes[a]

Product group	Gene	*cam*	*jun*	*nig*	*car*	*nap*	*pol*	*ole*	*ole*-F
Complex I	*nad1*	+	+	+	+	+	+	+	+
	nad2	+	+	+	+	+	+	+	+
	nad3	+	+	+	+	+	+	+2	+
	nad4	+	+	+	+	+	+	+2	+
	nad4L	+	+	+	+	+	+	+2	+
	nad5	+	+	+	+	+	+	+2	+
	nad6	+	+	+	+	+	+	+2	+
	nad7	+	+	+	+	+	+	+	+
	nad9	+	+	+	+	+	+	+2	+
Complex III	*cob*	+	+	+	+	+	+	+2	+
Complex IV	*cox1*	+	+	+	+	+	+	+2	+
	cox2-1	+	+	+	+	+	+	+2	+
	cox2-2	+	+	−	−	+	+	−	+
	cox3	+	+	+	+	+	+	+	+
Complex V	*atp1*	+	+	+	+	+	+	+2	+
	atp4	+	+	+	+	+	+	+2	+
	atp6	+	+	+	+	+	+	2	+
	atp8	+	+	+	+	+	+	+	+
	atp9	+	+	+	+	+	+	+	+
Cytochrome c	*ccmB*	+	+	+	+	+	+	+2	+
	ccmC	+	+	+	+	+	+	+	+
	ccmFN1	+	+	+	+	+	+	+	+
	ccmFN2	+	+	+	+	+	+	+2	+
	ccmFC	+	+	+	+	+	+	+	+
Other ORF	*tatC*	+	+	+	+	+	+	+2	+
	matR	+	+	+	+	+	+	+2	+
	orf222/4	−	−	−	−	+	+	−	−
Ribosome	*rps3*	+	+	+	+	+	+	+2	+
	rps4	+	+	+	+	+	+	+	+
	rps7	+	+	+	+	+	+	+2	+
	rps12	+	+	+	+	+	+	+2	+
	rps14	+	+	+	+	+	+	+2	+
	rpl2	+	+	+	+	+	+	+2	+
	rpl5	+	+	+	+	+	+	+2	+
	rpl16	+	+	+	+	+	+	+2	+
tRNA									
Asparagine	*trnN*	+	+	+	+	+	+	+2	+
Aspartic	*trnD*	+	+	+	+	+	+	+2	+
Cysteine	*trnC*	+	+	+	+	+	+	+2	+
Glutamic	*trnE*	+	+	+	+	+	+	+2	+

(continued)

Table 9.2 (continued)

Product group	Gene	*cam*	*jun*	*nig*	*car*	*nap*	*pol*	*ole*	*ole*-F
Glutamine	*trnQ*	+	+	+	+	+	+	+2	+
	trnG	+	+	+	+	+	+	+	+
Histidine	*trnH*	+2	+2	+	+	+	+2	+4	+2
Isoleucine	*trnI*	+	+	+	+	+	+	+2	+
Lysine	*trnK*	+	+	+	+	+	+	+2	+
Methionine	*trnM*	+	+	+	+	+	+	+2	+
fMethionine	*trnfM*	+	+	+	+	+	+	+2	+
Proline	*trnP*	+	+	+	+	+	+	+2	+
Serine	*trnS*	+3	+3	+3	+3	+3	+3	+6	+3
Tryptophan	*trnW*	+	+	+	+	+	+	2	+
Tyrosine	*trnY*	+	+	+	+	+	+	+2	+
rRNA	*rrn5*	+	+	+	+	+	+	+	+
	rrn18	+	+	+	+	+	+	+	+
	rrn26	+	+	+	+	+	+	+2	+

+ Denotes present, − denotes absent. Gene copy number is shown after +. *cam*, *jun*, *ole*, and *car* denote the mitotype of *B. rapa*, *B. juncea*, *B. oleracea*, and *B. carinata*, respectively (Chen et al. 2011). *ole*-F denotes the mitotype of *B. oleracea* from Fujiwase (Tanaka et al. 2014). *nig* denotes the mitotype of *B. nigra* (Yamagishi et al. 2014). *pol* and *nap* denote two mitotypes of *B. napus* (Handa 2003; Chen et al. 2011)
[a]Adapted from Chang et al. 2011

extremely conserved and the intron numbers and positions are completely identical. However, non-coding parts of the mitochondrial DNA are very dynamic with respect to structural changes, sequence acquisition, and/or sequence loss, and no additional ORFs are shared between these two closely related plants. The rapeseed mitochondrial genome could recombine into two subgenomic circles via the 2427 bp repeats, which is different from the DNA sequences involved in the intramolecular recombination in *Arabidopsis*.

Mitochondria play an important role in the determination of CMS in plants. The *B. napus* lines can be classified naturally into two cytoplasmic groups, the *pol* and *nap* mitotypes (Shiga and Bata 1973). The *B. napus pol* mitotype is 223,412 bp, which is larger than the previously reported *nap* mitotype of the fertile *B. napus* variety Westar (Fig. 9.2). The *pol* mitotype encodes 34 proteins, three ribosomal RNAs, and 18 tRNAs, among which 48 and five are identical to or differ only marginally from their counterparts in the *nap* cytoplasm, respectively. Two near-identical copies of *trnH* are found in the *pol* mitotype, while only one copy is found in the *nap* mitotype. Forty-four ORFs, including *orf122* and *orf132*, are unique to the *pol* mitotype. Except for the CMS-related genes *orf222* and *orf224*, the functions of most ORFs remain unknown and need further exploration. The structural differences between the *pol* and *nap* sequences have been analyzed, and at least five rearrangement events were detected. The *pol* mitotype is presumed to contain one master circle accompanied by two smaller circles of 86.2 and 137.1 kb, which is different from the *nap* mitotype. Genome sequencing and PCR assays suggested that the *pol* and *nap* mitotypes probably coexist within one *B. napus* plant, and large variation in the copy number ratio of mitotypes has been found, even among cultivars sharing the same cytoplasm.

Recently, the complete sequence of a heterogeneously mitochondrial genome of *Ogura-cms-*

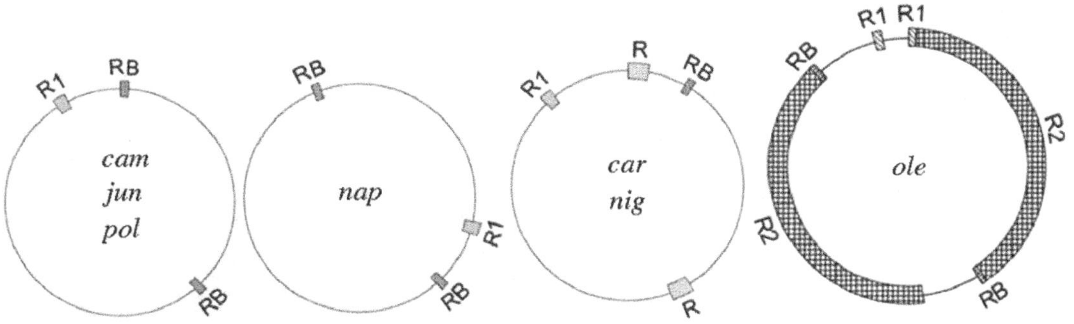

Fig. 9.1 Large repeats exist in the *Brassica* mitotypes (Adapted from Chang et al. 2011)

Table 9.3 Predicted multipartite sizes (bp) in *Brassica* mitotypes

Mitotype	Master circle	Big circle	Small circle
cam	219,747	137,111	82,636
jun	219,766	137,123	82,643
car	232,241	136,493	95,748
nap	221,853	124,908	96,945
pol	223,412	137,132	86,280
oguc	258,473	201,863	56,610
ole-F	219,952	170,039	49,913

hybrid (*oguC*) *B. napus*, which was derived from somatic fusion between *B. napus* and a sterile radish, was sequenced (Wang et al. 2012a). The circular mitochondrial genome is 258,473 bp in length, encoding 33 proteins, three ribosomal RNA sequences, and 23 tRNA sequences. The authors compared the oguC mitochondrial genome with the *nap* mitochondrial genome and detected 40 point mutations in the 33 protein-coding genes. Moreover, two copies of *atp9* were found in the *oguC* mitochondrial genome, while the *cox2-2* gene in *nap* is missing. The presence of a pair of large repeat sequences (9731 bp) was suggested to be responsible for the multipartite structure of *oguC*, which comprises a master circle and two smaller subgenomic circles (56,610 and 20,1863 bp). After comparisons with other reported *Brassica* mitochondrial genomes, the authors speculated that the *tatC* gene and the unique regions U3 and U7 in *oguC* must be introgressed from radish.

RNA editing events in the *B. napus* mitochondria genome were investigated, and 427 C to U conversions were identified in genes and ORFs of *B. napus* mitochondrial transcripts. Nine ORFs longer than 150 amino acids were subjected to editing analysis, and only one partial editing event was detected in the *orf188* transcript. Only 358 editing sites are shared by both *B. napus* and *A. thaliana* mitochondria, while 99.2% of the protein-coding regions are similar in their primary DNA sequences between *B. napus* and *Arabidopsis*, suggesting that diversification of RNA editing is more rapid than that of coding information.

9.3 *Brassica* Chloroplast Genomes

9.3.1 Chloroplast DNA Variation in *Brassica*

The earliest mutational analysis of the chloroplast genome was performed in 1983 to explore the origin and evolution of a hybrid species complex in the genus *Brassica* (Palmer et al. 1983). *Brassica* chloroplast DNAs from 22 accessions were used, and comparative restriction analysis revealed that one small inversion may have occurred during the evolution of the *Brassica* chloroplast genome, in addition to numerous small deletions–insertions and point mutations. The results indicated that the chloroplast DNAs from all assigned amphidiploids (*B. carinata*, *B. juncea*, and *B. napus*) were identical to those of their respective

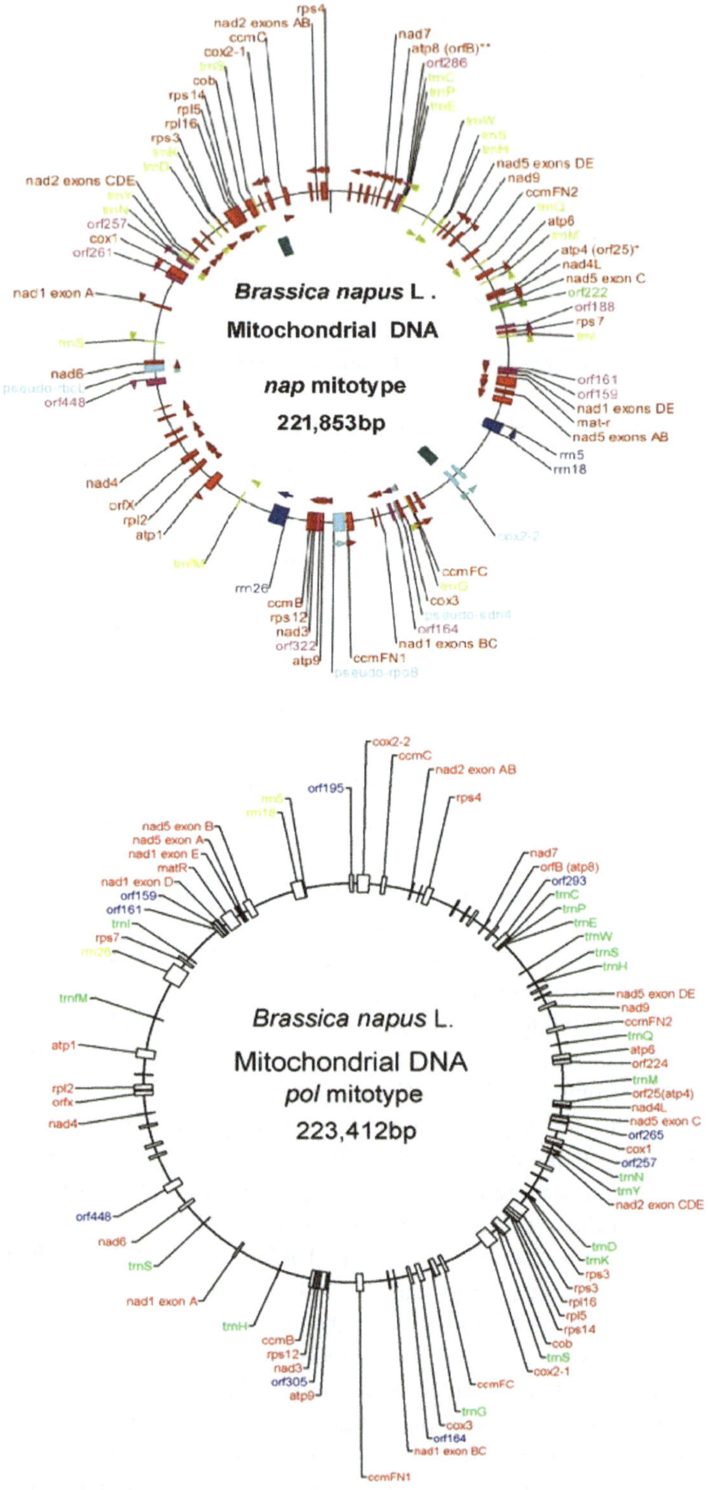

Fig. 9.2 Gene organization of the *Brassica napus* mitochondrial genome (Adapted from Handa 2003, with permission from Oxford University Press; Chen et al. 2011)

maternal parents, suggesting that all these hybridizations occurred very recently. Chloroplast DNA (cpDNA) diversity in 14 wild *Brassicas* (including 31 accessions) was recently evaluated using the PCR-restriction fragment length polymorphism (RFLP) technique and revealed 219 intergeneric/interspecific or intraspecific polymorphic fragments. The authors provided an efficient method for characterizing or confirming the maternal lineage of natural hybrids and alloplasmic lines developed by cross-breeding between wild and crop *Brassicas* (Sarin et al. 2015).

9.3.2 *Brassica rapa* Chloroplast Genomes

The chloroplast genomes of three *B. rapa* accessions were determined using Solexa sequencing technology (Wu et al. 2012). Using the *B. rapa* sequence as the reference, more than 99.96% of the cp genome in the three tested accessions is covered. The *B. rapa* cp genome is 153,482 bp, with an 83,282 bp large single copy (LSC) region and a 26,212 bp small single copy (SSC) region. The two IR copies in the cp genome are both 26,212 bp. The *B. rapa* cp genes were annotated and 89 potential protein-coding genes (including eight genes duplicated in the inverted repeat), eight rRNA genes, and 37 tRNA genes were assigned to the *B. rapa* cp genome (Fig. 9.3). Analysis of sequence polymorphisms of cp genomes within *B. rapa* species reveals 31 and eight single nucleotide polymorphisms (SNPs) in the Z16 and FT cp genome sequences, respectively, compared with the reference cp sequence of Chiifu-402-41.

9.3.3 *Brassica napus* Chloroplast Genomes

The chloroplast genome of *B. napus* (cv. zy036) was de novo sequenced and reported (Hu et al. 2011). The complete *B. napus* cpDNA sequence is 152,850 bp (Fig. 9.4), comprising 74 protein-coding genes, 30 tRNA genes, four rRNA genes, and five conserved ORFs (*ycf*). A pair of 26,035 bp inverted repeat regions (IRa and IRb), which are separated by an SSC (17,760 bp) and an LSC (83,030 bp), have been found in the *B. napus* cpDNA. Eighty-six simple sequence repeats were identified in the *B. napus* cpDNA. The *B. napus* genome organization and gene contents were compared with other chloroplast genomes available in the NCBI public database. The gene content and organization, and the number of genes and introns of the *B. napus* cpDNA, are identical to *B. rapa*, and similar to other typical land plant species (Table 9.4). The cpDNA of *B. napus* and *B. rapa* shows very low sequence divergence of 0.133% in the coding regions and 0.275% in the intron regions, suggesting that the mutation rate in the intron region is twice that of the gene coding region. The total length of the spacer region is 402 bp longer in *B. rapa* compared with *B. napus*, which may be attributed to small deletions in some spacer regions. In addition, 0.348% sequence divergence in the intergenic spacer regions was observed between the cpDNA of *B. napus* and *B. rapa*.

9.4 Evolutionary Implications

The small, relatively constant size and conserved evolution of the chloroplast genome make it an ideal molecule for phylogenetic studies of different plant species (Palmer et al. 1983). U's triangle, which comprises six crop species in the genus *Brassica*, provides a typical example of the evolution of plant species through interspecific hybridization and polyploidization. The mitochondrial and chloroplast genomes have been used to determine the origin of genomes in amphiploids. Phylogenetic analysis based on chloroplast DNA indicated that the *Brassica* species could be divided into two ancient evolutionary lineages: the "*nigra*" lineage and the "*rapa/oleracea*" lineage (Warwick and Black 1991). Chloroplast non-coding regions in seven species were sequenced (Fig. 9.5), and the results showed that the rate of nucleotide substitution in the *rapa/oleracea* lineage is at least 1.5 times that

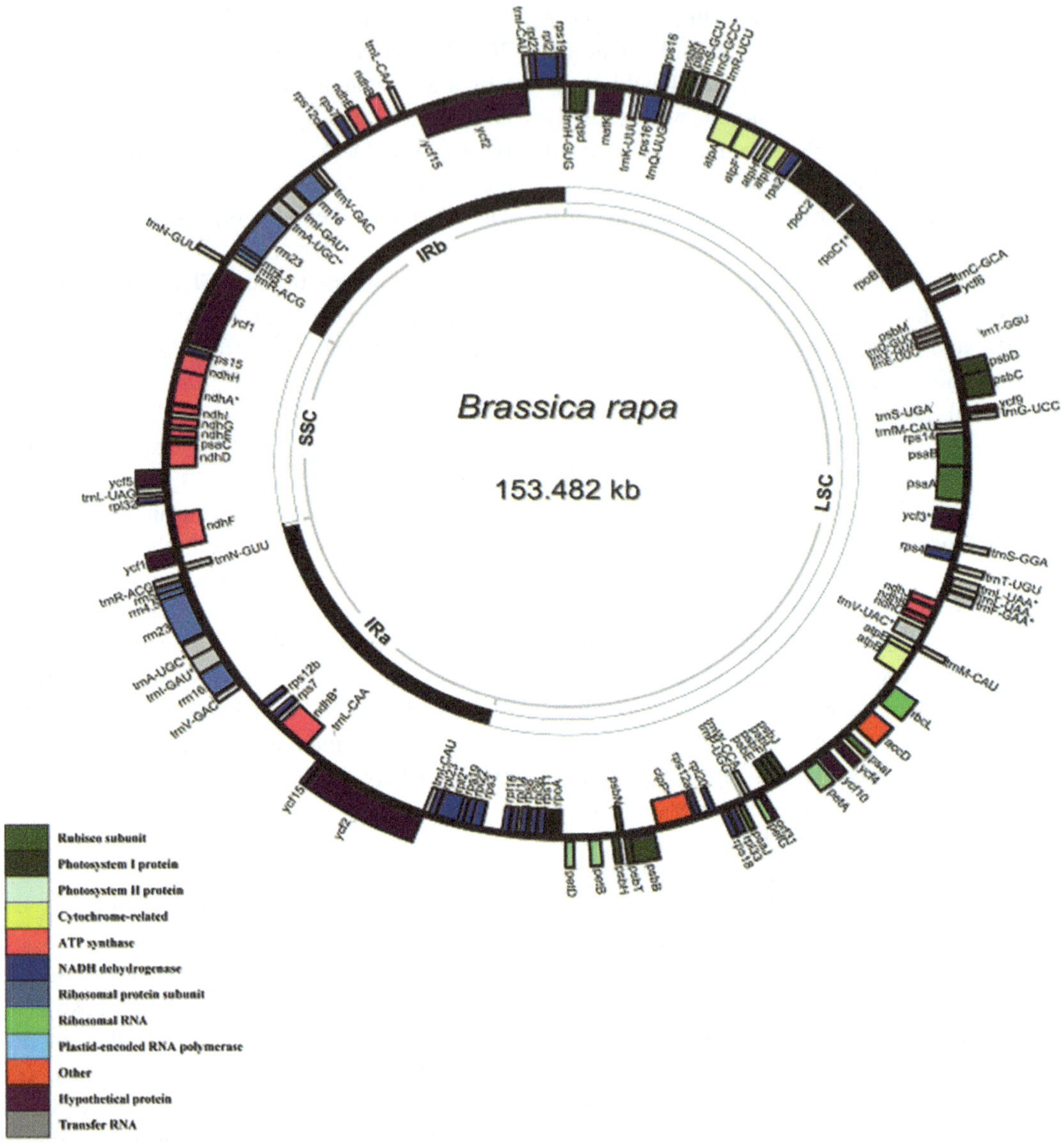

Fig. 9.3 Circular gene map of *Brassica rapa* chloroplast genome (Wu et al. 2012)

in the *nigra* lineage (Yang et al. 2002), which is consistent with the conclusion from RFLP analysis that *B. rapa/oleracea* DNA sequences vary more than those of *B. nigra* (Song et al. 1988). Moreover, the results of that study supported the view that *Raphanus* may be derived from hybridization between the *rapa/oleracea* and *nigra* lineages and estimated the date of this hybridization event to be approximately 0.4 T years after the divergence between the two *Brassica* lineages. Phylogenetic analysis of 61 protein-coding genes from 48 taxa supported the hypothesis that *B. rapa* may be the maternal parent of *B. napus* cv. zy036 (Hu et al. 2011).

Mitochondrial genomes were also used to elucidate the evolutionary mechanism in *Brassica* (Fig. 9.6). The cytoplasm donors of *B. juncea* and *B. carinata* were *B. rapa* and *B. nigra*, respectively (Uchimiya and Wildman 1978), a deduction that was supported by mitochondrial genome

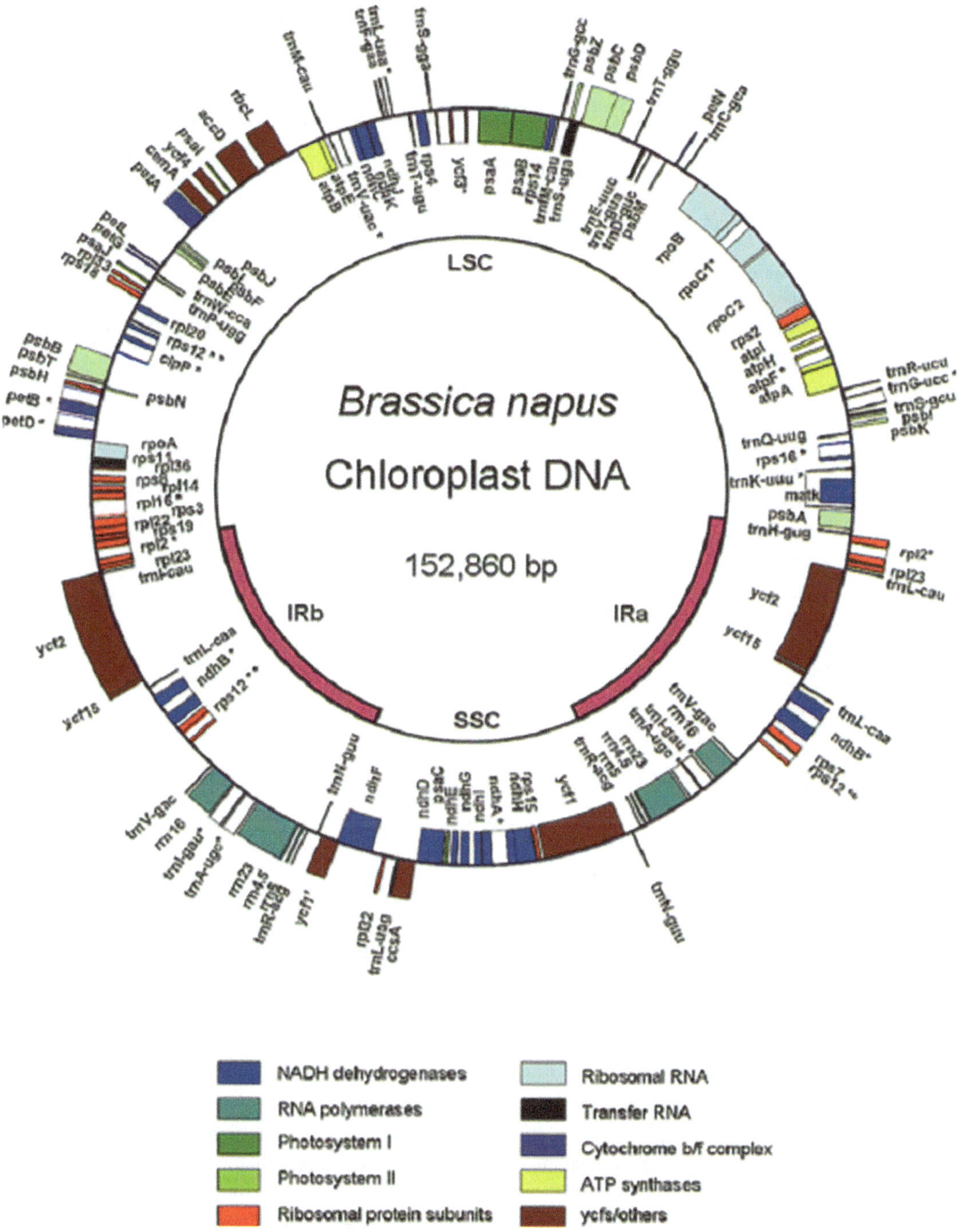

Fig. 9.4 Gene map of *B. napus* cpDNA (Reprinted from Hu et al. 2011 with permission of Springer Nature)

Table 9.4 Genes present in *B. napus* cpDNA (Adapted from Hu et al. 2011 with permission of Springer Nature)

DNA numbers	Group of genes	Name of genes[a]
Self-replication	rRNA genes	*rrn4.5*(×2), *rrn5*(×2), *rrn16*(×2), *rrn23*(×2)
	tRNA genes	*trnA-UGC*(×2)*, *trnC-GCA*, *trnD-GUC*, *trnE-UUC*, *trnF-GAA*, *trnfM-CAU*, *trnG-GCC*, *trnG-UCC**, *trnH-GUC*, *trnI-CAU*(×2), *trnI-GAU*(×2)*, *trnK-UUU**, *trnL-UAA**, *trnL-CAA*(×2), *trnL-UAG*, *trnM-CAU*, *trnN-GUU* (×2), *trnP-UGG*, *trnQ-UUG*, *trnR-UCU*, *trnR-ACG*(×2), *trnS-GCU*, *trnS-UGA*, *trnS-GGA*, *trnT-GGU*, *trnT-UGU*, *trnV-UAC**, *trnV-GAC*(×2), *trnW-CCA*, *trnY-GUA*
	Small subunit of ribosome	*rps2*, *rps3*, *rps4*, *rps7*(×2), *rps8*, *rps11*, *rps12*(×2)**, *rps14*, *rps15*, *rps16**, *rps18*, *rps19*(×2, *part*)
	Large subunit of ribosome	*rpl2*(×2)*, *rpl14*, *rpl16**, *rpl20*, *rpl22*, *rpl23*(×2), *rpl32*, *rpl33*, *rpl36*
	DNA-dependent RNA polymerase	*rpoA*, *rpoB*, *rpoC1**, *rpoC2*
Genes for photosynthesis	Subunits of NADH-dehydrogenase	*ndhA**, *ndhB*(×2)*, *ndhC*, *ndhD*, *ndhE*, *ndhF*, *ndhG*, *ndhH*, *ndhI*, *ndhJ*, *ndhK*
	Subunits of photosystem I	*psaA*, *psaB*, *psaC*, *psaI*, *psaJ*
	Subunits of photosystem II	*psbA*, *psbB*, *psbC*, *psbD*, *psbE*, *psbF*, *psbH*, *psbI*, *psbJ*, *psbK*, *psbL*, *psbM*, *psbN*, *psbT*, *psbZ*
	Subunits of cytochrome b/f complex	*petA*, *petB**, *petD**, *petG*, *petL*, *petN*
	Subunits of ATP synthase	*AtpA*, *AtpB*, *AtpE*, *AtpF**, *AtpH*, *AtpI*
	Large subunit of rubisco	*rbcL*
Miscellaneous proteins	Maturase	*matK*
	Protease	*clpP***
	Envelop membrane protein	*cemA*
	Subunit of Acetyl-CoA-carboxylase	*accD*
	c-type cytochrom synthesis gene	*ccsA*
Genes of unknown function	Conserved open reading frames (ORF, *ycf*)	*ycf1*(×2, *part*), *ycf2*(×2), *ycf3***, *ycf4*, *ycf15*(×2)

[a]One and two asterisks reflect one- and two-intron containing genes, respectively. Genes located in the IR regions are indicated by the (×2) symbol after the gene name

sequencing (Yamagishi et al. 2014). Cluster analysis of the six mitotypes based on indels and SNPs showed that *cam* and *pol* are closely related, and *ole* slightly diverges from the *cam–pol* class, while *nap* and *car* have diverged the furthest from the *cam–pol* group (Chang et al. 2011). Evolutionary events such as inheritance, duplication, rearrangement, genome compaction, and mutation may have been involved in the mechanism of mitochondrial genome formation in *Brassica*, and the importance of the roles of the three evolutionary factors in genome formation could be arranged in the following order: mutation > segment indel > genome rearrangement. Gene evolution of mitochondrial genomes within the *Brassicaceae* family was further analyzed using the *Raphanus sativus* (*sat*) mitotype, together with six other reported *Brassica* mitotypes (Chang et al. 2013). In this study, a synonymous substitution (*dS*) tree constructed using SNPs in the exons of 32 conserved genes indicated that the relationships among the seven *Brassicaceae* mitotypes could be classified into three types. Type I describes relatively low evolutionary divergence, such as that shown among *pol, cam, nap,* and *ole*. Type II classification, which involves slightly higher divergence than type I, suggested that *car* is more closely related to *sat* than to the other *Brassica* mitotypes. Type III describes the highest evolutionary divergence, such as between *tha* and the *Brassica* mitotypes or *sat*.

9.5 Cytoplasm Application in *Brassica* Breeding

The maternal inheritance of chloroplast and mitochondrial genomes is important for breeding programs, because the direction of cross- and successful fertilization in wide hybridization depends on the type of cytoplasm/maternal lineage in *Brassica* (Chapman and Goring 2010; Sarin et al. 2015). CMS is one of the most important applications of organelle genomes and has been developed and adopted in the practical breeding of *Brassica* crops (Yamagishi and Bhat 2014). The molecular and genetic basis of CMS in plants has been studied extensively and is associated with aberrant recombination in the mitochondrial genome. Several chimeric genes in mitochondrial genomes responsible for CMS have been identified. Moreover, the importance of the chloroplast genome has been assessed, and the implications of chloroplast genomes in crop breeding for abiotic stress resistance and high oil content have been explored.

CMS may occur spontaneously or arise from intergeneric crosses, interspecific crosses, or intraspecific crosses (Kiang et al. 1993). The CMS-associated mitochondrial genomes of crop species in *Brassica* were reported, including *B. napus*, *B. oleracea,* and *B. juncea*. Ogura CMS was discovered in an unidentified variety of wild Japanese radish (*R. sativus*) by Ogura in 1968 (Ogura 1968) and has been studied extensively and used in F_1 breeding of *B. napus*, *B. juncea*, *B. oleracea,* and *R. sativus* (Yamagishi and Bhat 2014). Other spontaneous CMS variants include *hau* CMS in *B. juncea* (Wan et al. 2008), Polima (*pol*) CMS (Fu 1981), and 681A in *B. napus* (Liu et al. 2005). The *nap* CMS lines, Shan2A CMS and MI CMS, arose from intraspecific crosses in *B. napus* and have been applied widely in F_1 hybrid breeding.

Alloplasmic CMS originating by interspecific or intergeneric hybridizations has also been reported in *Brassica* (Yamagishi and Bhat 2014). CMS lines of *B. napus* with the cytoplasm of *Brassica tournefortii*, *Diplotaxis muralis*, *Diplotaxis siifolia*, and *Enarthrocarpus lyratus* have been produced. *Brassica oxyrrhina*, *D. muralis*, *E. lyratus*, *Eruca sativa,* and *Moricandia arvensis* induce CMS in *B. rapa*. The cytoplasms of *D. muralis*, *Erucastrum canariense*, and *M. arvensis* induce sterility in *B. oleracea*, while *B. oxyrrhina*, *B. tournefortii*, *D. berthautii*, *D. catholica*, *D. erucoides*, *D. siifolia*, *E. lyratus,* and *E. canariense* confer CMS in *B. juncea*. In addition to sexual

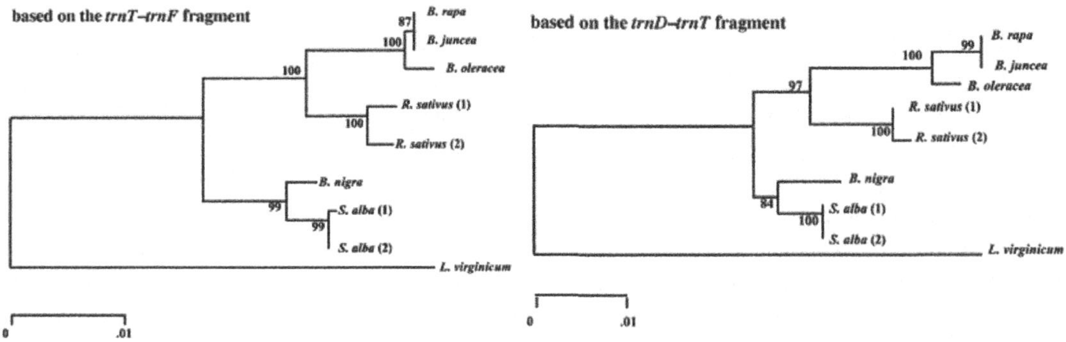

Fig. 9.5 Neighbor-joining trees based on the sequence data of chloroplast non-coding regions (Adapt from Yang et al. 2002) with permission of Elsevier

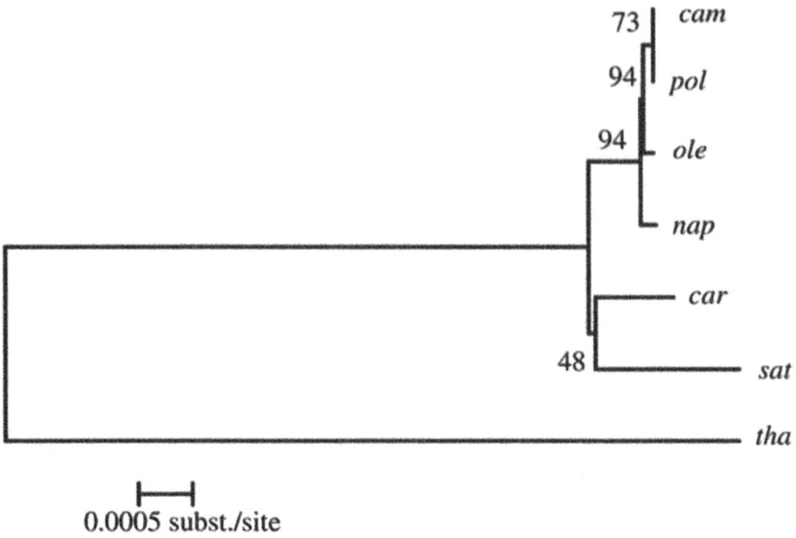

Fig. 9.6 Phylogeny of seven Brassicaceae mitotypes (Chang et al. 2013)

hybridization, wild species have been used for cell fusion to produce male sterile lines. *A. thaliana*, *Orychophragmus violaceus*, *Sinapis arvensis,* and *Trachystoma ballii* cytoplasms were transferred into *B. napus*, *B. juncea,* and *B. oleracea* by somatic cell fusion to obtain CMS (reviewed by Yamagishi and Bhat 2014).

The mechanism underlying CMS in plants has been studied extensively, and chimeric genes responsible for CMS were identified (Fig. 9.7). A chimeric gene is generated through rearrangement and recombination, and such genes might disrupt plant mitochondrial respiratory genes, resulting in severely deleterious phenotypes, including stunting, stripping, and female sterility, in addition to male sterility (Chase 2007). Structural analysis of the *atp6* gene regions of *pol* and *nap* mitotypes showed that rearrangements in the *pol* mitochondrial genome occurring upstream of *atp6* have generated *orf224*, which is co-transcribed with *atp6*, and induced altered organization and expression of the *afp6* gene (Singh and Brown 1991). The structure of the protein encoded by *orf222* is similar to that of the protein product of *orf224* and was found to be responsible for *nap* CMS in *B. napus* (L' Homme et al. 1997). In addition, research in both *B. napus* and *B. juncea* found that different mitotypes coexist substoichiometrically in CMS lines and their maintainer lines. The coexistence of mitochondrial mitotypes and substoichiometric shifting is thought to explain the emergence of CMS in *B. napus* (Chen et al. 2011). The *orf220* isolated from CMS stem mustard shares 79% homology with *orf222* (*nap*) and 81% with *orf224* (*pol*) of *B. napus* and is associated with male sterility in *B. juncea* (Yang et al. 2005). Comparative analysis indicated that the *hau* CMS mitochondrial genome is highly rearranged compared with that of its iso-nuclear maintainer line and further confirmed that *orf288* was a cytoplasmic male sterility-associated gene in *B. juncea*. The formation of the CMS-associated gene in the *hau* CMS line is associated with three large repeats downstream of *orf288* (Heng et al. 2014), and these findings might provide new insights into the mechanism of natural CMS. An ancient origin ORF, *orf108*, is widely distributed among wild allies of *Brassica*, is co-transcribed with *atpA,* and is associated with male sterility in all three CMS *B. juncea* lines carrying *Diplotaxis* sp. cytoplasm (*D. berthautii*, *D. catholica* and *D. erucoides*), suggesting that CMS lines of different origin and morphology share a common molecular basis (Kumar et al. 2012). In addition to mitochondrial genome sequences, single sequence repeats (SSRs) derived from the chloroplast genome

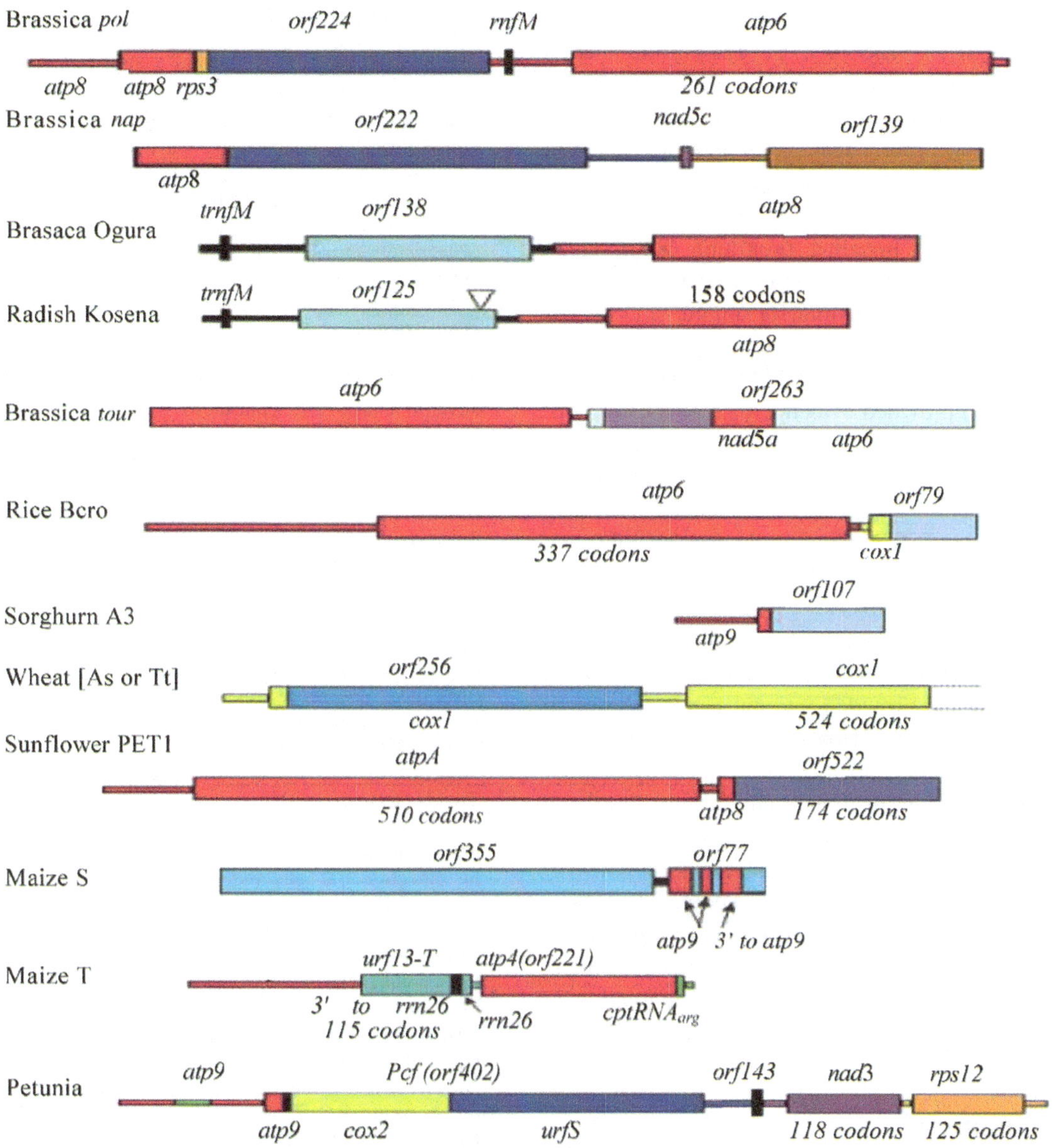

Fig. 9.7 Chimeric genes associated with CMS (Reprinted from Hanson and Bentolila 2004 with permission of American Society Of Plant Physiologists). Shades of blue indicate unknown reading frames within CMS-associated regions. Shades of yellow indicate genes for subunits of cytochrome oxidase. Orange indicates ribosomal protein genes. Shades of brown indicate conserved unidentified reading frames found in multiple vascular plant mtDNAs. Green indicates chloroplast-derived sequences

have also been used for CMS-type identification in *B. oleracea* (Wang et al. 2012b). In that study, f11 cpSSR primers revealed polymorphism among six *B. oleracea* CMS types (*nig* CMS, *Ogu* CMSR1, *Ogu* CMSR2, *Ogu* CMSR3, *Ogu* CMSHY, and *pol* CMS).

The wild relatives of crops often exhibit resistance to biotic and abiotic stresses (Rieseberg and Willis 2007) and are also a source of male sterility-inducing cytoplasm in cultivars (Bang et al. 2011). Therefore, interspecific hybridizations between wild species and their related crops could promote beneficial gene transfer between species and play an important role in crop improvement (Chu et al. 2014). Maternal inheritance of chloroplast genomes has been observed in the genus *Brassica* (Zhang et al. 2012). The type of cytoplasm/maternal lineage in Brassicas can influence the direction of a cross and the extent of success achieved in wide hybridization (Chapman and Goring 2010; Yamane et al. 2005). Thus, evaluation of chloroplast genome diversity in *Brassica* is important for breeding programs (Sarin et al. 2015). Chloroplast genomes could also be applied in crop breeding for abiotic stress resistance. For example, deep sequencing of the chloroplast genome of Chinese cabbage (*B. rapa*) provided evidence of a novel subset of small RNAs derived from the chloroplast (csRNAs), and many members of these csRNA families are highly sensitive to heat stress (Wang et al. 2011). This finding might provide a fast and efficient way to improve the heat resistance of important crops.

Recently, the potential role of the chloroplast genome in seed oil content regulation has been reported and provided rational targets for future oilseed breeding (Hua et al. 2012). Seed lipid synthesis is dependent upon the supply of photosynthate from maternal plant tissues (Baud and Lepiniec 2010). A significant positive correlation between silique (a maternal tissue) photosynthetic ability and seed oil content has been detected. The expression levels of chloroplast genes correlate closely with the oil content, indicating their importance in the regulation of the silique wall and seed oil content.

References

Alberts B, Jonhson A, Lewis J, Raff M, Roberts K, Walter P (2002) Energy conversion: mitochondria and chloroplasts

Alverson AJ, Wei X, Rice DW, Stern DB, Barry K, Palmer JD (2010) Insights into the evolution of mitochondrial genome size from complete sequences of *Citrullus lanatus* and *Cucurbita pepo* (Cucurbitaceae). Mol Biol Evol 27:1436–1448

Andre C, Levy A, Walbot V (1992) Small repeated sequences and the structure of plant mitochondrial genomes. Trends Genet 8:128–132

Arrieta-Montiel MP, Mackenzie SA (2011) Plant mitochondrial genomes and recombination. Plant Mitochondria (Springer)

Arrieta-Montiel MP, Shedge V, Davila J, Christensen AC, Mackenzie SA (2009) Diversity of the Arabidopsis mitochondrial genome occurs via nuclear-controlled recombination activity. Genetics 183:1261–1268

Bang S, Tsutsui K, Shim S, Kaneko Y (2011) Production and characterization of the novel CMS line of radish (*Raphanus sativus*) carrying *Brassica maurorum* cytoplasm. Plant Breed 130:410–412

Baud S, Lepeniec L (2010) Physiological and developmental regulation of seed oil production. Prog Lipid Res 49:235–249

Bock R (2014) Genetic engineering of the chloroplast: novel tools and new applications. Curr Opin Biotechnol 26:7–13

Chang S, Chen J, Wang Y, Gu B, He J, Chu P, Guan R (2013) The mitochondrial genome of *Raphanus sativus* and gene evolution of cruciferous mitochondrial types. J Genet Genomics 40:117–126

Chang S, Yang T, Du T, Huang Y, Chen J, Yan J, He J, Guan R (2011) Mitochondrial genome sequencing helps show the evolutionary mechanism of mitochondrial genome formation in Brassica. BMC Genom 12:497

Chapman LA, Goring DR (2010) Pollen–pistil interactions regulating successful fertilization in the Brassicaceae. J Exp Bot 61:1987–1999

Chase CD (2007) Cytoplasmic male sterility: a window to the world of plant mitochondrial-nuclear interactions. Trends Genet 23:81–90

Chen J, Guan R, Chang S, Du T, Zhang H, Xing H (2011) Substoichiometrically different mitotypes coexist in mitochondrial genomes of *Brassica napus* L. PLoS ONE 6:e17662

Chetritl P, Mathieu C, Muller JP, Vedel F (1984) Physical and gene mapping of cauliflower (*Brassica oleracea*) mitochondrial DNA. Curr Genet 8:413–421

Chu P, Liu HJ, Yang Q, Wang YK, Yan GX, Guan RZ (2014) An RNA-seq transcriptome analysis of floral buds of an interspecific Brassica hybrid between B-carinata and B-napus. Plant Reprod 27:225–237

Clare EL, Kerr KC, Von Konigslow TE, Wilson JJ, Hebert PD (2008) Diagnosing mitochondrial DNA diversity: applications of a sentinel gene approach. J Mol Evol 66:362–367

Diamond M, Mccabe PF (2011) Mitochondrial regulation of plant programmed cell death. Plant Mitochondria (Springer)

Dyall SD, Brown MT, Johnson PJ (2004) Ancient invasions: from endosymbionts to organelles. Science 304:253–257

Fu TD (1981) Production and research of rapeseed in the People's Republic of China. Cruciferae Newsl. 6:6–7

Gray MW, Burger G, Lang BF (1999) Mitochondrial evolution. Science 283:1476–1481

Grevich JJ, Daniel H (2005) Chloroplast genetic engineering: recent advances and future perspectives. Crit Rev Plant Sci 24:83–107

Grewe F, Edger PP, Keren I, Sultan L, Pires JC, Ostersetzer-Biran O, Mower JP (2014) Comparative analysis of 11 Brassicales mitochondrial genomes and the mitochondrial transcriptome of *Brassica oleracea*. Mitochondrion 19:135–143

Handa H (2003) The complete nucleotide sequence and RNA editing content of the mitochondrial genome of rapeseed (*Brassica napus* L.): comparative analysis of the mitochondrial genomes of rapeseed and *Arabidopsis thaliana*. Nucleic Acids Res 31:5907–5916

Hanson MR, Bentolila S (2004) Interactions of mitochondrial and nuclear genes that affect male gametophyte development. Plant Cell 16:S154–S169

Heng SP, Wei C, Jing B, Wan ZJ, Wen J, Yi B, Ma CZ, Tu JX, Fu TD, Shen JX (2014) Comparative analysis of mitochondrial genomes between the hau cytoplasmic male sterility (CMS) line and its iso-nuclear maintainer line in *Brassica juncea* to reveal the origin of the CMS-associated gene orf288. BMC Genom 15

Hu Z-Y, Hua W, Huang SM, Wang H-Z (2011) Complete chloroplast genome sequence of rapeseed (*Brassica napus* L.) and its evolutionary implications. Genet Resour Crop Evol 58:875–887

Hua W, Li RJ, Zhan GM, Liu J, Li J, Wang XF, Liu GH, Wang HZ (2012) Maternal control of seed oil content in *Brassica napus*: the role of silique wall photosynthesis. Plant J 69:432–444

Huang S, Millar AH, Taylor NL (2011) The plant mitochondrial proteome composition and stress response: conservation and divergence between monocots and dicots. Plant Mitochondria 207–239

Inoue K (2007) The chloroplast outer envelope membrane: the edge of light and excitement. J Integr Plant Biol 49:1100–1111

Jarvis P, López-Juez E (2013) Biogenesis and homeostasis of chloroplasts and other plastids. Nat Rev Mol Cell Biol 14:787–802

Jensen PE, Leister D (2014) Chloroplast evolution, structure and functions. F1000Prime Rep 6:40

Kiang AS, Connolly V, Mcconnell D, Kavanagh T (1993) Cytoplasmic male sterility (CMS) in *Lolium perenne* L.: 1. Development of a diagnostic probe for the male-sterile cytoplasm. Theor Appl Genet 86:781–787

Kubo T, Newton KJ (2008) Angiosperm mitochondrial genomes and mutations. Mitochondrion 8:5–14

Kumar P, Vasupalli N, Srinivasan R, Bhat SR (2012) An evolutionarily conserved mitochondrial orf108 is associated with cytoplasmic male sterility in different alloplasmic lines of *Brassica juncea* and induces male sterility in transgenic *Arabidopsis thaliana*. J Exp Bot 63:2921–2932

L' Homme Y, Stahl RJ, LI XQ, Hameed A, Brown GG (1997) Brassica nap cytoplasmic male sterility is associated with expression of a mtDNA region containing a chimeric gene similar to the pol CMS-associated orf224 gene. Curr Genet 31:325–335

Liu Z, Guan C, Zhao F, Chen S (2005) Inheritance and mapping of a restorer gene for the rapeseed cytoplasmic male sterile line 681A. Plant Breed 124:5–8

Lonsdale DM, Thompson RD, Hodge TP (1981) The integrated forms of the S1 and S2 DNA elements of maize male sterile mitochondrial DNA are flanked by a large repeated sequence. Nucleic Acids Res 9:3657–3669

Ogihara Y, Yamazaki Y, Murai K, Kanno A, Terachi T, Shiina T, Miyashita N, Nasuda S, Nakamura C, Mori N (2005) Structural dynamics of cereal mitochondrial genomes as revealed by complete nucleotide sequencing of the wheat mitochondrial genome. Nucleic Acids Res 33:6235–6250

Ogura H (1968) Studies on the new male-sterility in Japanese radish, with special reference to the utilization of this sterility towards the practical raising of hybrid seeds. Mem Fac Agric Kagoshima Univ 6:39–78

Palmer JD, Herbon LA (1987) Unicircular structure of the *Brassica hirta* mitochondrial genome. Curr Genet 11:565–570

Palmer JD, Herbon LA (1988) Plant mitochondrial DNA evolved rapidly in structure, but slowly in sequence. J Mol Evol 28:87–97

Palmer JD, Shields CR (1984) Tripartite structure of the *Brassica campestris* mitochondrial genome. Nature 307:437–440

Palmer JD, Shields CR, Cohen DB, Orton TJ (1983) Chloroplast DNA evolution and the origin of amphidiploid Brassica species. Theor Appl Genet 65:181–189

Pellny TK, Van Aken O, Dutilleul C, Wolff T, Groten K, Bor M, De Paepe R, Reyss A, Van Breusegem F, Noctor G (2008) Mitochondrial respiratory pathways modulate nitrate sensing and nitrogen-dependent regulation of plant architecture in *Nicotiana sylvestris*. Plant J 54:976–992

Petit RJ, Duminil J, Fineschi S, Hampe A, Salvini D, Vendramin GG (2005) Invited review: comparative organization of chloroplast, mitochondrial and nuclear diversity in plant populations. Mol Ecol 14:689–701

Repkova J (2010) Potential of chloroplast genome in plant breeding. Czech J Genet Plant Breed 46:103–113

Rieseberg LH, Willis JH (2007) Plant speciation. Science 317:910–914

Samuels DC, Wonnapinij P, Chinnery PF (2013) Preventing the transmission of pathogenic mitochondrial DNA mutations: Can we achieve long-term benefits from germ-line gene transfer? Hum Reprod 28:554–559

Sandhu APS, Abdelnoor RV, Mackanzie SA (2007) Transgenic induction of mitochondrial rearrangements for cytoplasmic male sterility in crop plants. Proc Natl Acad Sci 104:1766–1770

Sarin B, Martín JP, Kaula BC, Mohanty A (2015) Chloroplast DNA variations in wild Brassicas and their implication in breeding and population genetics studies. Scientifica 2015

Shedge V, Arrieta-Montiel M, Christensen AC, Mackanzie SA (2007) Plant mitochondrial recombination surveillance requires unusual RecA and MutS homologs. Plant Cell 19:1251–1264
Shiga T, Bata S (1973) Cytoplasmic male sterility in oil seed rape. *Brassica napus* L., and its utilization to breeding. Jpn J Breed 23:187–197
Singh M, Brown GG (1991) Suppression of cytoplasmic male-sterility by nuclear genes alters expression of a novel mitochondrial gene region. Plant Cell 3:1349–1362
Sloan DB, Alverson AJ, Chuchalovcak JP, Wu M, Mccauley DE, Palmer JD, Taylor DR (2012) Rapid evolution of enormous, multichromosomal genomes in flowering plant mitochondria with exceptionally high mutation rates. PLoS Biol 10:e1001241
Small ID, Isaac PG, Leaver CJ (1987) Stoichiometric differences in DNA molecules containing the atpA gene suggest mechanisms for the generation of mitochondrial genome diversity in maize. EMBO J 6:865
Song KM, Osborn TC, Williams PH (1988) Brassica taxonomy based on nuclear restriction fragment length polymorphisms (RFLPs): 2. Preliminary analysis of subspecies within *B. rapa* (syn. campestris) and *B. oleracea*. Theor Appl Genet 76:593–600
Tanaka Y, Tsuda M, Yasumoto K, Terachi T, Yamagishi H (2014) The complete mitochondrial genome sequence of *Brassica oleracea* and analysis of coexisting mitotypes. Curr Genet 60:277–284
Tanaka Y, Tsuda M, Yasumoto K, Yamagishi H, Terachi T (2012) A complete mitochondrial genome sequence of Ogura-type male-sterile cytoplasm and its comparative analysis with that of normal cytoplasm in radish (*Raphanus sativus* L.). BMC Genom 13:352
Nagaharu U (1935) Genome analysis in Brassica with special reference to the experimental formation of *B. napus* and peculiar mode of fertilization. Jpn J Bot 7:389–452
Uchimiya H, Wildman S (1978) Evolutin of fraction I protein in relation to origin of amphididiploid Brassica species and other members of the Cruciferae. J Hered 69:299–303
Wan ZJ, Jing B, Tu JX, Ma CZ, Shen JX, Yi B, Wen J, Huang T, Wang XJ, Fu TD (2008) Genetic characterization of a new cytoplasmic male sterility system (hau) in *Brassica juncea* and its transfer to *B-napus*. Theor Appl Genet 116:355–362
Wang J, Jiang J, Li X, Li A, Zhang Y, Guan R, Wang Y (2012a) Complete sequence of heterogenous-composition mitochondrial genome (*Brassica napus*) and its exogenous source. BMC Genom 13:675
Wang L, Yu X, Wang H, Lu YZ, De Ruiter M, Prins M, He YK (2011) A novel class of heat-responsive small RNAs derived from the chloroplast genome of Chinese cabbage (*Brassica rapa*). BMC Genom 12:289
Wang Q, Zhang Y, Fang Z, Liu Y, Yang L, Zhuang M (2012b) Chloroplast and mitochondrial SSR help to distinguish allo-cytoplasmic male sterile types in cabbage (*Brassica oleracea* L. var. capitata). Mol Breed 30:709–716
Wang Y, Chu P, Yang Q, Chang S, Chen J, Hu M, Guan R (2014) Complete mitochondrial genome of *Eruca sativa* Mill. (Garden rocket). PLoS ONE 9: e105748
Warwick SI, Black LD (1991) Molecular systematics of Brassica and allied genera (Subtribe Brassicinae, Brassiceae)—chloroplast genome and cytodeme congruence. Theor Appl Genet 82:81–92
Woloszynska M (2010) Heteroplasmy and stoichiometric complexity of plant mitochondrial genomes—though this be madness, yet there's method in't. J Exp Bot 61:657–671
Woodson JD, Chory J (2008) Coordination of gene expression between organellar and nuclear genomes. Nat Rev Genet 9:383–395
Wu J, Liu B, Cheng F, Ramchiary N, Choi SR, Lim YP, Wang XW (2012) Sequencing of chloroplast genome using whole cellular DNA and Solexa sequencing technology. Front Plant Sci 3
Yamagishi H, Bhat SR (2014) Cytoplasmic male sterility in Brassicaceae crops. Breed Sci 64:38–47
Yamagishi H, Tanaka Y, Terachi T (2014) Complete mitochondrial genome sequence of black mustard (*Brassica nigra*; BB) and comparison with *Brassica oleracea* (CC) and *Brassica carinata* (BBCC). Genome 57:577–582
Yamane K, Lü N, Ohnishi O (2005) Chloroplast DNA variations of cultivated radish and its wild relatives. Plant Sci 168:627–634
Yang JH, Zhang MF, Yu JQ, Zhang S, Wang T, Chen ZJ (2005) Identification of alloplasmic cytoplasmic male-sterile line of leaf mustard synthesized by intra-specific hybridization. Plant Sci 168:865–871
Yang YW, Tai PY, Chen Y, Li WH (2002) A study of the phylogeny of *Brassica rapa*, *B. nigra*, *Raphanus sativus*, and their related genera using noncoding regions of chloroplast DNA. Mol Phylogenet Evol 23:268–275
Zhang Y, Fang Z, Wang Q, Liu Y, Yang L, Zhuang M, Sun P (2012) Chloroplast subspecies-specific SNP detection and its maternal inheritance in *Brassica oleracea* L. by using a dCAPS marker. J Hered 103:606–611

Diversity and Evolution of *B. napus* Chloroplast Genome

10

Sampath Perumal, Jonghoon Lee,
Nomar Espinosa Waminal, Shengyi Liu and Tae-Jin Yang

Abstract

Chloroplast genomes (cpDNAs) are a vital resource for studying plant genome diversity, origin and evolution. *B. napus*, an important oilseed crop, is a recently formed allotetraploid between *B. rapa* and *B. oleracea*. In this chapter, we explored the genetic diversity and evolutionary origin of the three types of *B. napus* cpDNA. We exclusively assembled and characterized the complete cpDNAs of nine *B. napus* accessions using Illumina whole-genome sequence data for this study. Based on the genetic diversity and phylogenetic analysis of three cytotypes with its progenitor species, we provide a possible explanation for the origin of the most common nap-type cpDNA in *B. napus* genome. Overall, this study discusses the diversity, evolution and origin of the *B. napus* chloroplast genome and also provides new resources for *Brassica* breeding and evolutionary studies.

10.1 Introduction

The cpDNAs are cytoplasmic genomes, which are conservatively inherited uniparentally mostly via maternal inheritance and play various roles other than photosynthesis. For example, biochemical processes such as fatty acid synthesis, nitrogen metabolism and immune response in plants are associated with cpDNA function (Mullet 1988; Birky 1995; Jansen and Ruhlman 2012). The cpDNA is a circular genome with a size about 59–218 kb and contains a typical quadripartite structure, with a pair of inverted repeats (IRs) flanked by large and small single-copy regions (Chumley et al. 2006; Jansen and Ruhlman 2012; Delannoy et al. 2011). The IRs play important role in intermolecular homologous recombination in order to produce isomeric structure of cpDNA (Palmer et al. 1983). Highly conserved nature of the cpDNA makes them vital tool in studying genetic and

S. Perumal · J. Lee · N. E. Waminal · T.-J. Yang (✉)
Department of Plant Science, Plant Genomics and Breeding Institute, and Research Institute of Agriculture and Life Sciences, College of Agriculture and Life Sciences, Seoul National University, Seoul 151-921, Republic of Korea
e-mail: tjyang@snu.ac.kr

S. Perumal
Agriculture and Agri-Food Canada, 107 Science Place, Saskatoon, SK S7N 0X2, Canada

N. E. Waminal
Department of Life Science, Plant Biotechnology Institute, Sahmyook University, Seoul 139-742, Republic of Korea

S. Liu
Key Laboratory of Biology and Genetic Improvement of Oil Crops, the Ministry of Agriculture, Oil Crops Research Institute of the Chinese Academy of Agricultural Sciences, Wuhan 430062, China

S. Liu et al. (eds.), *The Brassica napus Genome*, Compendium of Plant Genomes,
https://doi.org/10.1007/978-3-319-43694-4_10

genome diversity and phylogenetic and systematic evolutionary analyses (Shu et al. 2015). Development of cpDNA-based markers for species authentication and barcoding has been comparatively easier than the nuclear genome (Nock et al. 2011; Kim et al. 2013). Due to the mostly maternal inheritance, the cpDNA has high advantage in tracking down the parental origin or parentage in interspecific hybrid (Allender and King 2010).

B. napus (AACC, $2n = 4x = 38$) belongs to genus *Brassica* and is an economically important oilseed crop yielding food, biofuels, and lubricants (Bonnema 2011). It is an allotetraploid plant with recent evolutionary ($\sim$7500 years) and domestication history (<500 years) (Röbbelen et al. 1989). As a natural allopolyploid, *B. napus* originated from hybridization between two diploid species, *B. rapa* (AA, $2n = 2x = 20$) and *B. oleracea* (CC, $2n = 2x = 18$) (Parkin et al. 1995). Depending on the artificial or natural cross direction, *B. napus* cytoplasm may be derived from either of its progenitors. A recent study about exploration of *B. napus* and its progenitor genome exposes the high-level genome rearrangement caused by non-homeologous exchanges between the parental sub-genome in *B. napus* (Cheung et al. 2009; Chalhoub et al. 2014). However, due to extensive homeologous recombination, high-level rearrangements were observed which hinder control of chromosome pairing that leads to unstable hybrid formation (Chang et al. 2011; Leflon et al. 2006; Chalhoub et al. 2014; Sharpe et al. 1995). Unlike *B. napus*, *B. juncea* (AABB, $2n = 4x = 36$), which is also an important allotetraploid from the genus *Brassica*, has remained considerably unchanged since its polyploidization from progenitor species, *B. rapa* (AA, $2n = 2x = 20$) and *B. nigra* (BB, $2n = 2x = 16$). The genetic map developed from the natural and synthetic parents of *B. juncea* exhibited disomic inheritance and comparison of its A and B subgenomes revealed collinearity with their respective progenitor diploid genomes (Axelsson et al. 2000).

Primary results based on the *B. napus* cpDNAs suggest that *B. napus* has three kinds of cytotypes which may have derived from *B. oleracea* (ole-type), *B. rapa* (rap-type) and its own (nap-type) (Hu et al. 2011; Allender and King 2010; Qiao et al. 2015). Various studies based on partial or complete cpDNA of *B. napus* could not bring a clear conclusion about the maternal origin of the *B. napus* cpDNA (Mei et al. 2011; Song and Osborn 1992; Qiao et al. 2015). To date, numerous controversies have arisen from attempts to decipher the molecular mechanisms underlying the origin and evolution of the *B. napus* genome (Zamani-Nour et al. 2013). In this investigation, we explore the diversity and evolutionary origin of the *B. napus* cpDNA genome based on complete chloroplast genome sequence of 11 *B. napus* accessions (Table 10.1).

10.2 Chloroplast Genome Assembly and Characterization of Nine *B. napus* Accessions

Advancement of next-generation sequencing technology (NGS) has offered remarkable advantage in understanding the genomes. Low-coverage (1× haploid equivalent) whole-genome shotgun (WGS) sequences from the nine *B. napus* accessions were used to assemble complete and error-free chloroplast genome by dnaLCW method (Kim et al. 2015b) (Table 10.1; Fig. 10.1). All nine accessions used in this study are inbred lines. It is important to note that the accessions M083 (Bn-2) and H165 (Bn-7) are derived from multi-parental and synthetic origin, respectively. Accession Bn-2 is a Asian semi-winter type oilseed rape, which was derived from double haploid (DH) line of various multiple crossing with inbred lines (Liu et al. 2005). Likewise, accession Bn-7 was obtained by embryo rescue and chromosome doubling of an interspecific haploid from the cross between *B. oleracea* ssp. *capitata* var. *sabauda* and *B. rapa* ssp. *chinensis* (Jesske et al. 2013).

Each cpDNA of the nine assembled *B. napus* consists of a typical quadripartite structure with size range from 152,833 to 153,502 bp. Unlike cpDNA, mitochondrial genome (mtDNA), another organelle genome, exhibited high

Table 10.1 Summary statistics of *B. napus* accessions and chloroplast genome assembly

ID	Chloroplast genome						NGS read summary		Accession name	Genotype description	Genbank No
	Size (bp)	LSC (bp)	SSC (bp)	IR (bp)	Mean coverage (x)	GC%	Amounts (Mbp)	Genome coverage (x)			
Bn-1	153,453	83,227	17,794	26,216	456	36.32	2,126.6	1.9	Zhongsuang11	Asian winter type	KU324625
Bn-2	153,502	83,301	17,775	26,213	1,287	36.35	1,273.8	1.1	M083	Asian semi-winter type	KU324626
Bn-3	153,453	83,227	17,794	26,216	260	36.36	13,900.7	12.3	Aburamasari	South asian origin	KU324627
Bn-4	153,452	83,226	17,794	26,216	264	36.35	8,679.2	7.7	Avisol	European winter type	KU324628
Bn-5	153,472	83,247	17,793	26,216	196	36.35	6,029.7	5.3	Darmor-Bzh	European winter type	KU324629
Bn-6	153,471	83,225	17,814	26,216	1,949	36.35	13,323.1	11.8	Grüner Schnittkohl	Siberian Kale type	KU324630
Bn-7	153,368	83,137	17,837	26,197	2,205	36.35	15,928.4	14.1	H165	Resynthesized oilseed	KU324631
Bn-8	153,452	83,227	17,793	26,216	2,255	36.36	14,923.5	13.2	Swede Sensation NZ	Rutabaga type	KU324632
Bn-9	153,450	83,228	17,790	26,216	633	36.35	14,105.3	12.5	Yudal	South asian origin	KU324633
Bn-NCBI	152,833	83,003	17,760	26,035		36.35			ZY036	Asian semi-winter type	GQ861354
BnCp-1 (pan)	152,831	83,002	17,759	26,035		36.32			Pangenome		Qiao et al. (2015)

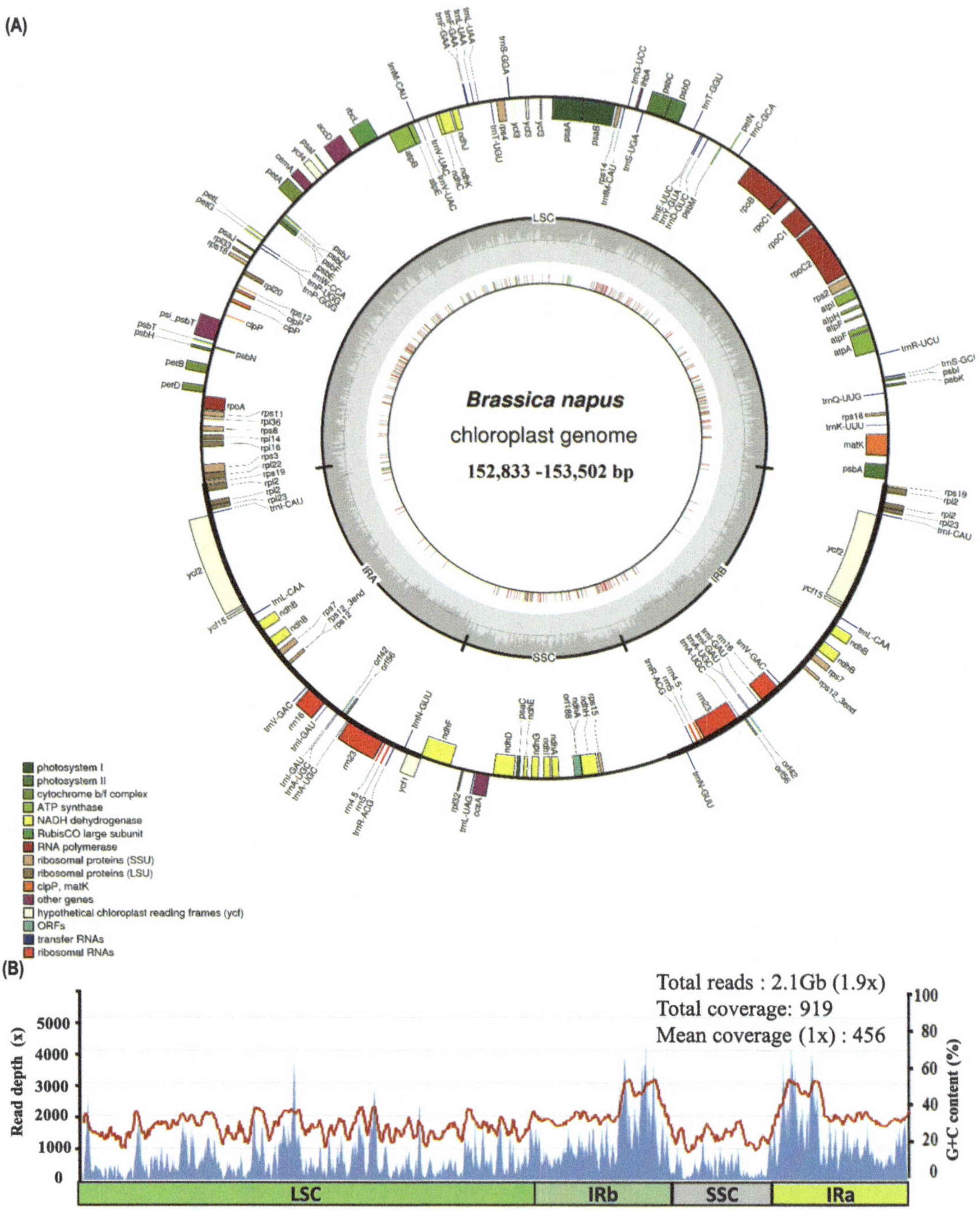

Fig. 10.1 Chloroplast genome structure of *Brassica napus*. **a** Gene map of *B. napus* chloroplast genome sequence was created using OGDRAW (Lohse et al. 2013). Genes transcribed clockwise and counterclockwise are indicated on the outside and inside of the large circle, respectively. Genes associated with different functional groups are color coded. Four parts of chloroplast genome and GC content are indicated on the middle circle. The innermost circle represents the sequence variation as SNP (red bar), INDEL (green bar) and copy number variation (blue bar). **b** Estimation of coverage of chloroplast genome by mapping of raw reads on Bn-1 cpDNA with GC distribution (red graph)

diversity and evolution in *Brassica.* Around 140 kb (219,747–360,271 bp) size variation was observed among the mtDNAs of *B. rapa* and *B. oleracea* and allopolyploids (Yang et al. 2015; Chang et al. 2011), suggesting that cpDNAs are vastly more conserved structure than mitochondrial genomes. The genome annotation based on DOGMA tool and manual curation has revealed 113 individual genes including 74 protein-coding genes, 30 tRNA, 4 rRNA and five open reading frames, which is similar to the reported *B. napus* (Bn-NCBI) cpDNA (Hu et al. 2011; Wyman et al. 2004). The overall GC content is 36.3%, which is parallel to its close relatives such as *B. rapa* (36.3), *B. oleracea* (36.3), *B. nigra* (36.3), *Raphanus sativus* (36.3) and *A. thaliana* (36.2). We also observed the differences in terms of copy numbers; the mean cpDNA coverage for a haploid genome was found to have 11-fold variation (196–2255 copies) based on clc_reference mapping approach. The newly developed cpDNAs of the nine accessions with complete annotation can be accessed from the Genbank with accession numbers listed in Table 10.1.

10.3 Diversity and Phylogenetic Relationship of *B. napus* cpDNA

Though the cpDNAs are considered to be evolving slowly and are generally highly conserved, considerable variations were observed in the coding and non-coding regions especially in rapidly evolving regions such as intergenic spacers and intronic regions (Zeng et al. 2012). cpDNA markers derived from disparity sites were widely accepted for numerous applications including genetic differentiation, cytoplasmic diversity, molecular barcoding, monitoring transgene introgression and population and phylogenetic studies (Flannery et al. 2006; Woo et al. 2013; Shu et al. 2015; Kundu et al. 2013; Wang et al. 2012). Owing to artificial breeding programs and intentional introgression, increases in genetic diversity in *B. napus* have been achieved. In addition, exploring the allelic variation responsible for the genetic changes will provide a way for crop enhancement and dissecting complex agronomic traits (Qian et al. 2006; Song et al. 1994; Wang et al. 2014; Szadkowski et al. 2010). However, using a limited set of cpDNA markers may cause inaccurate results which raise concerns on drawing the right conclusion (Allender et al. 2007; Allender and King 2010; Flannery et al. 2006).

We have obtained complete chloroplast genome sequences of nine *B. napus* accessions which includes the three types of cpDNAs such as, rap-type, ole-type and nap-type based on homology with cp genomes from *B. rapa*, *B. oleracea* and *B. napus*–unique, respectively (Qiao et al. 2015; Allender and King 2010) (Table 10.1). Genome-wide cpDNA nucleotide similarity search for 13 *Brassica* accessions including 11 *B. napus* including two previously reported *B. napus* cpDNA (Bn-NCBI and BnCp-1) and its progenitors (*B. rapa* and *B. oleracea*) has revealed high homology within *B. napus* (98.9–100%) (Table 10.2). Despite the conserved gene content and gene order in those of 11 accessions, more than 450 genetic variations were observed including 332 single nucleotide polymorphisms (SNPs), 118 insertions and deletions (INDELs) and 4 copy number variation (CNVs) (Fig. 10.1a). The differential nucleotide count analysis showed 0–1554 and 7–1549 variations sites as intra- and interspecies diversity, respectively (Table 10.2).

B. napus cpDNAs were highly diverged with *B. oleracea* and *B. rapa*. Alignment by mVISTA tool showed high genome conservation in the genic regions than intronic and intergenic spacer regions. Similar to other angiosperms, the non-coding regions of the cpDNA show high sequence divergence than the coding regions (Li et al. 2015). Out of 113 genes, *rpoC1*, *rpob*, *rbs12*, *psbB*, *rpL16* and *ycf1* are potential hotspot regions for development of barcoding markers in the 11 *B. napus* accessions (Hollingsworth et al. 2011; Kim et al. 2015a). Among the 11 *B. napus* accessions, Bn-2, (multi inter-crossed synthetic *B. napus*) and Bn-7 (resynthesized origin: synthetic *B. napus*) are highly diverged with other *B. napus* cpDNA (Fig. 10.2). High amount of genetic differentiation compared with the

Table 10.2 Inter- and intra-species similarity and diversity of *B. napus* chloroplast genome with its progenitors

Similarity (%)/variation sites (n)[a]	Bn-NCBI	Bn-1	Bn-2	Bn-3	Bn-4	Bn-5	Bn-6	Bn-7	Bn-8	Bn-9	BnCp-1 (pan)	Bo-NCBI	Br-NCBI
Bn-NCBI	ID	99.5	99	99.5	99.5	99.5	99.5	98.9	99.5	99.5	99.9	98.9	99.1
Bn-1	680	ID	99.4	100	99.9	99.9	99.9	99.3	99.9	99.9	99.5	99.3	99.4
Bn-2	1,402	811	ID	99.4	99.4	99.4	99.4	99.3	99.4	99.4	99	99.3	99.9
Bn-3	680	0	811	ID	99.9	99.9	99.9	99.3	99.9	99.9	99.5	99.3	99.4
Bn-4	675	7	816	7	ID	99.9	99.9	99.3	99.9	99.9	99.5	99.3	99.4
Bn-5	696	28	835	28	23	ID	99.9	99.3	99.9	99.9	99.5	99.3	99.4
Bn-6	698	30	839	30	27	48	ID	99.3	99.9	99.9	99.5	99.3	99.4
Bn-7	1,554	958	955	958	952	973	978	ID	99.3	99.3	99	99.9	99.4
Bn-8	681	1	812	1	8	29	31	959	ID	99.9	99.5	99.3	99.4
Bn-9	686	6	817	6	13	34	36	964	5	ID	99.5	99.3	99.4
BnCp-1 (pan)	37	644	1,392	644	651	672	672	1,533	645	650	ID	99	99.1
Bo-NCBI	1,549	953	936	953	947	968	971	7	954	959	1,528	ID	99.4
Br-NCBI	1,381	790	39	790	795	814	818	913	791	796	1,371	910	ID

[a]Upper and lower diagonal shows the percentage of similarities and number of sequence variations among the chloroplast genomes analyzed

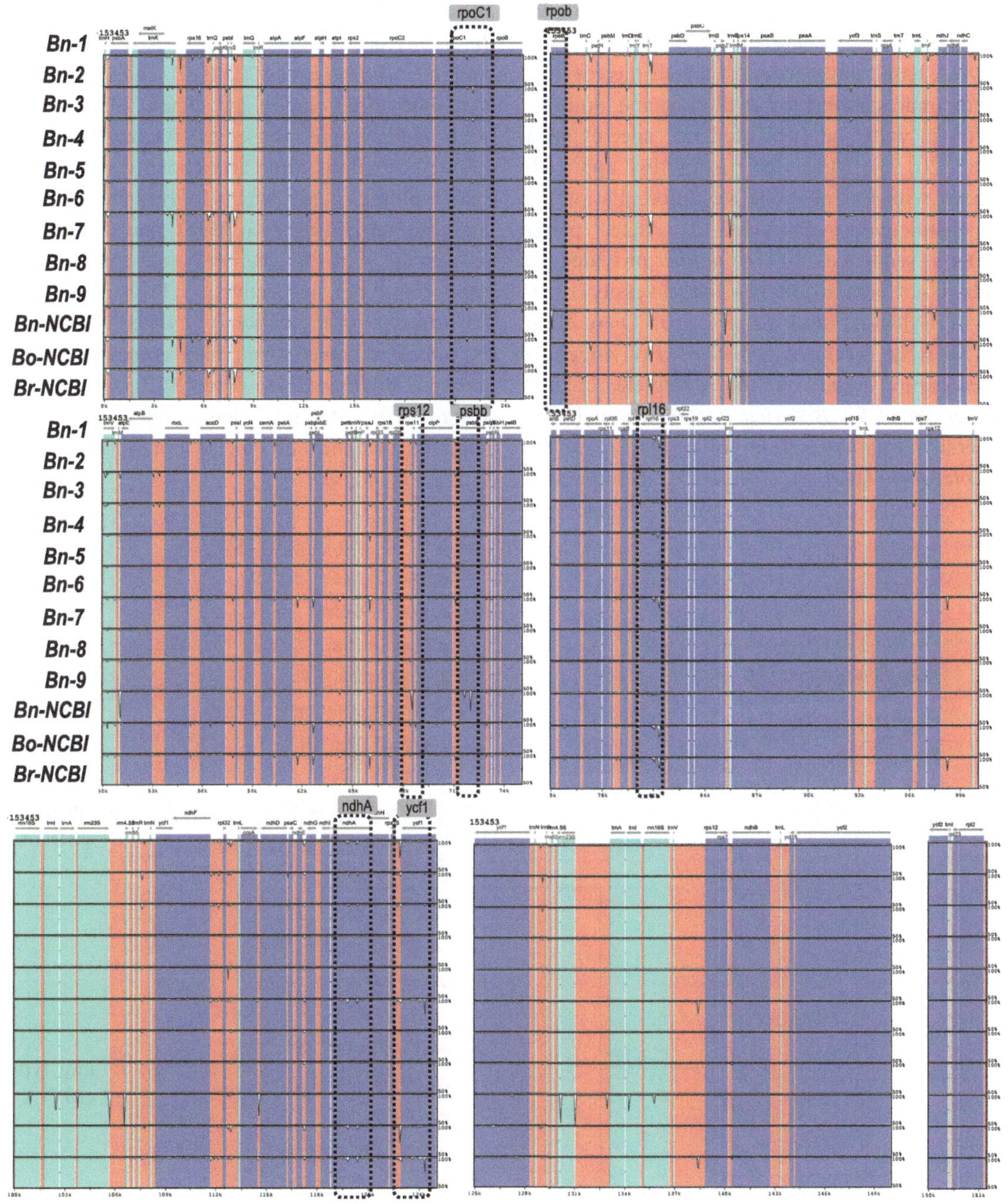

Fig. 10.2 Sequence comparison and visualization of 10 *B. napus* cpDNA with its progenitor genomes. Complete cpDNA sequence-based identity plot was developed by mVISTA. Genome regions are color coded; blue block, conserved gene; sky-blue block, tRNA and rRNA; red block, intergenic region. Prominent genic regions for molecular validation were marked as dotted box

recently developed accession (Bn-2 and Bn-7) suggest that the established accessions have undergone rapid evolutionary changes (Fig. 10.2). Furthermore, understanding the diversity in the gene pool will help for breeding improvement and hybrid formation.

The cpDNA provides high-resolution data, thus widely accepted for population and phylogenetic analyses (Bailey et al. 2006; Panda et al. 2003; Wang et al. 2005). The cpDNA-based phylogenetic analysis has revealed better understanding of evolution and domestication in wild and cultivated rice species (Kim et al. 2015b). Moreover, using this approach, species with less or moderate differentiation can be distinctively classified. To date, partial cpDNA sequences of *B. napus* provided an ambiguous conclusion for the diversity of *B. napus* (Allender and King 2010; Qiao et al. 2015). Here, we have generated a phylogenetic relationship based on the complete cpDNA sequences of 23 *Brassica* accessions including several morphotypes from *B. rapa* (Br1-5) and *B. oleracea* (Bo1-5) (Fig. 10.3). Comparative phylogenetic analysis of complete cpDNA of 11 *B. napus* accessions with five *B. rapa* (Br1-5) and five *B. oleracea* (Bo1-5) categorized the taxa into three clades and clearly distinguishes each species and subspecies. Among the 11 *B. napus* accessions, nine accessions were grouped into a unique clade which follows the nap-type and the remaining two Bn-2 and Bn-7 were associated with *B. oleracea* (ole-type) and *B. rapa* (rap-type), respectively. This suggests that the nap-type is the major type of cpDNA in the *B. napus* genome which also corresponds with previous findings (Qiao et al. 2015; Allender and King 2010) cytotypes. Hence, our analysis also supports that the nap-type cytoplasm is a major type in *B. napus*.

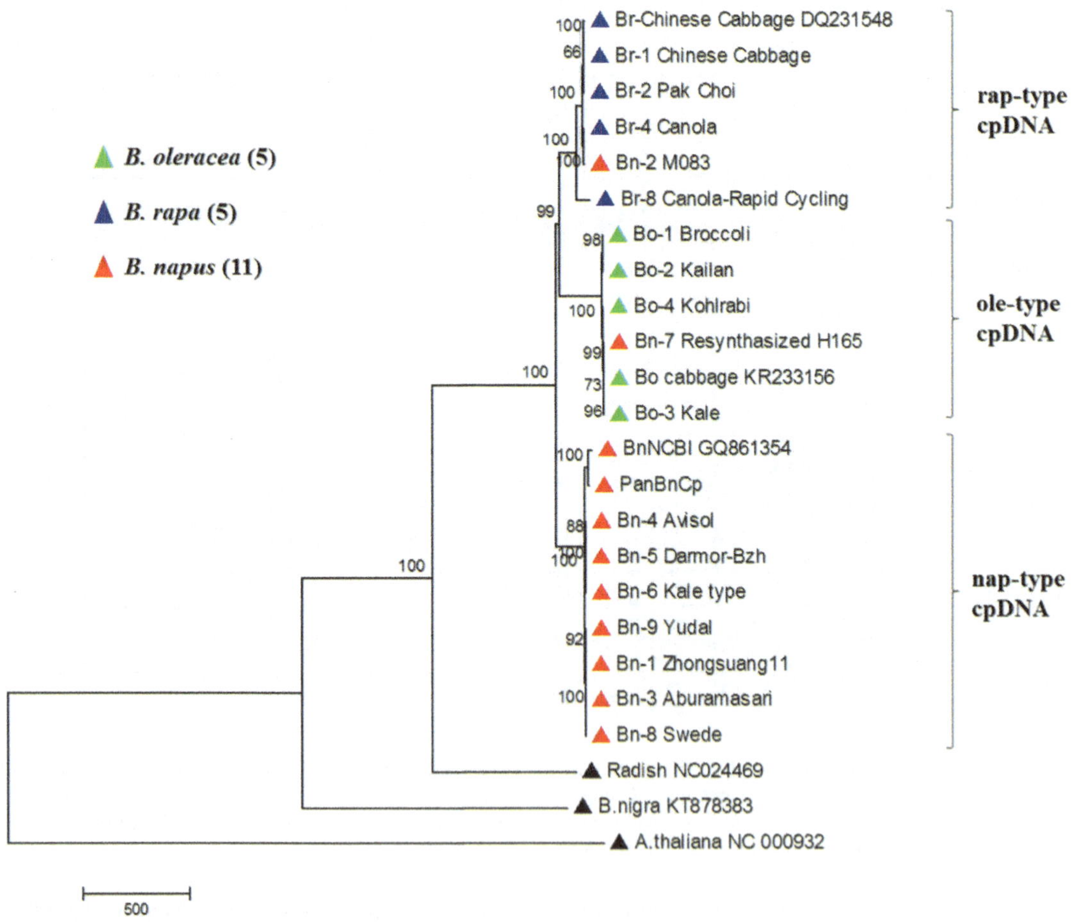

Fig. 10.3 Phylogenetic analysis based on complete chloroplast genome sequences of *B. napus* accessions and its relative species. Complete cpDNA of 11 *B. napus*, five *B. rapa* (Br1-5), five *B. oleracea* (Bol-5), *R. sativus*, *B. nigra* and *A. thaliana* used to develop neighbor-joining tree with 1000 bootstrap replications by MEGA6

10.4 Origin and Evolution of the *B. napus* Chloroplast Genome

Polyploidization is a major evolutionary force in the evolution of *Brassica* species. *Brassica* diploids, *B. rapa* (AA), *B. nigra* (BB), *B. oleracea* (CC) evolved from a common hexaploid ancestor (Sharma et al. 2014) around 13 million years ago (mya) (Gupta 2013; Yang et al. 2006). Two distinct lineage of *Brassica* diploids (rapa/oleracea and nigra lineage) that have formed around 9–13 mya with hexaploid ancestor were clearly explained by plastid genome analysis (Sharma et al. 2014; Kaur et al. 2014). The allotetraploid *B. napus* (AC) was formed quite recently around 7500 years ago by hybridization and polyploidization of the diploid progenitor *B. rapa* and *B. oleracea* (Chalhoub et al. 2014). The cpDNA is one of the important tools in identifying parental origins and in clearly elucidating the origin for many species including rice, wheat and apple (Zou et al. 2015; Haider 2012; Nikiforova et al. 2013). Lack of wild relatives and various cytogenetic and genomic studies on *B. napus* support its polyphyletic origin (Warwick et al. 2003).

Unlike other two tetraploids, *B. juncea* (AB) and *B. carinata* (BC), *B. napus* chloroplast did not follow with either of the parental genomes (A or C genome) (Li et al. 2017). Studies have been performed to clarify the genetic relationships of the major diploid and tetraploid *Brassica* species, but the origin of the chloroplast in the AC genome is still unclear (Qiao et al. 2015; Allender and King 2010). Furthermore, the cpDNA of the *B. napus* has unique origin (nap-type) and its genetic relationship with its diploid ancestors remains controversial (Qiao et al. 2015). cpDNA analysis based on *rpo* regions of 488 *B. napus* accessions revealed that more than 92% were associated with nap-type and differentiated with their ancestor (Qiao et al. 2015). Identification of exact origin of the *B. napus* will help to understand genome for stable hybrid formation, overcome self-incompatibility and creation of fertile plants required for crop improvement.

B. napus has three cytotypes including two diploid progenitors type and nap-type, which is also well supported by previous analysis (Qiao et al. 2015; Allender and King 2010). Since the origin of the Bn-2 (multi inter-crossed origin) and Bn-7 (synthetic origin with *B. rapa* as a maternal parent) were obvious, it is possible that the cytoplasm of the Bn-2 and Bn-7 could be grouped into *B. oleracea* and *B. rapa* genotypes, respectively. In addition, because the *B. napus* genome has high sexual compatibility with close relatives such as *B. rapa*, *R. sativus* and *Sinapis alba*, it is possible that *B. napus* cytoplasm may have derived from close relatives by natural or artificial crossing (Wang et al. 2005; Warwick et al. 2003). For example, Polima and Ogura cytoplasmic male sterility (CMS) lines achieved through introgression of cytotypes derived from polish winter oilseed rape and radish (Witt et al. 1991; Pellan-Delourme and Renard 1988). However, we could not identify any off types or CMS types among the 11 accessions since all *B. napus* accessions have clearly grouped into three cytotypes. Recently, cpDNA analysis of more diverse *B. rapa* genotypes revealed two types of chloroplast genomes rapa-type1 (= rap-type) and rapa-type2 (= nap-type). Though the rap-type chloroplast genome is found to be common to *B. rapa*, rapa-type2 is unique for some Italian Broccoletto genotypes of *B. rapa* (Li et al. 2017). Further analysis indicated that the Italian Broccoletto genotype is expected to be the donor for the nap-type chloroplast genome of *B. napus*. Moreover, nap-type chloroplast genome was maintained in the Italian Broccoletto genotype by geographical isolation or maternal dominance since its divergence (4.7 mya), and the Italian Broccoletto genome was utilized as the maternal parent to generate the AC genome 7500 years ago (Li et al. 2017).

10.5 Conclusion and Perspectives

Chloroplast genome has been applied to decode the plant evolution and systematics (Jansen and Ruhlman 2012). Studies on *B. napus* chloroplast genome has revealed three cytotypes in which the nap-type (>92%) was of unknown origin and discrete to both parental progenitors, *B. rapa* and *B. oleracea* (Qiao et al. 2015). Artificial *B. napus* lines which were developed by interspecies hybridization has widened its genetic diversity which helps increase its environmental adaptability, improved production and quality (Qian et al. 2006). However, genetic factors such as self-incompatibility, unbalanced gametes and environmental causes such as biotic and abiotic stress hinder further improvement of *B. napus* (Leflon et al. 2006; Cifuentes et al. 2010). Identification of the original parents does not only clarify the evolutionary history but also enables the closer investigation of chromosome pairing mechanisms to produce stable artificial *B. napus* hybrids. Furthermore, agronomically important elite alleles that are present in the progenitors will help to improve the crop management and production.

Using the reconstructed chloroplast genome sequences of various *B. napus* accessions, we investigated the genetic diversity and evolution. The comparative genomics studies revealed that cpDNAs were well diversified among the *B. napus* and with its progenitors. Inter- and intra-cytotype variations including the recently developed synthetic *B. napus* will serve as important resources for *Brassica* breeding and evolutionary analysis. Phylogenetic analysis revealed that *B. napus* carry three kinds of cytotypes, rap-type, ole-type and nap-type, and comparative analysis with its progenitors revealed that the Italian Broccoletto genotype is the possible source for the origin of nap-type cp genome. Our study provides further evidence to clarify the phylogenetic origin and evolution of the three cytotypes of *B. napus* chloroplast genome. However, it is still not clear how the rapa-type2/nap-type chloroplast genome became the common maternal parent for most (92%) of AC genomes, although the Italian Broccoletto genotype is not prevalent in the A genome.

References

Allender C, Allainguillaume J, Lynn J, King GJ (2007) Simple sequence repeats reveal uneven distribution of genetic diversity in chloroplast genomes of *Brassica oleracea* L. and (n = 9) wild relatives. Theor Appl Genet 114:609–618

Allender CJ, King GJ (2010) Origins of the amphiploid species *Brassica napus* L. investigated by chloroplast and nuclear molecular markers. BMC Plant Biol 10:54

Axelsson T, Bowman C, Sharpe A, Lydiate D, Lagercrantz U (2000) Amphidiploid *Brassica juncea* contains conserved progenitor genomes. Genome 43:679–688

Bailey CD, Koch MA, Mayer M, Mummenhoff K, O'Kane SL, Warwick SI, Windham MD, Al-Shehbaz IA (2006) Toward a global phylogeny of the Brassicaceae. Mol Biol Evol 23:2142–2160

Birky CW (1995) Uniparental inheritance of mitochondrial and chloroplast genes: mechanisms and evolution. Proc Natl Acad Sci 92:11331–11338

Bonnema A (2011) Diversity and taxonomy of Brassica oil crops. In: Genetics, genomics and breeding of oilseed Brassicas, 47

Chalhoub B, Denoeud F, Liu S, Parkin IA, Tang H, Wang X, Chiquet J, Belcram H, Tong C, Samans B (2014) Early allopolyploid evolution in the post-neolithic *Brassica napus* oilseed genome. Science 345:950–953

Chang C, Kakihara F, Hondo K, Kato M (2011) The cytoplasm effect comparison between *Brassica napus* and *Brassica carinata* on floral characteristics of *Brassica oleracea*. Plant Breed 130:73–79

Cheung F, Trick M, Drou N, Lim YP, Park J-Y, Kwon S-J, Kim J-A, Scott R, Pires JC, Paterson AH (2009) Comparative analysis between homoeologous genome segments of *Brassica napus* and its progenitor species reveals extensive sequence-level divergence. Plant Cell 21:1912–1928

Chumley TW, Palmer JD, Mower JP, Fourcade HM, Calie PJ, Boore JL, Jansen RK (2006) The complete chloroplast genome sequence of Pelargonium× hortorum: organization and evolution of the largest and most highly rearranged chloroplast genome of land plants. Mol Biol Evol 23:2175–2190

Cifuentes M, Eber F, Lucas M-O, Lode M, Chèvre A-M, Jenczewski E (2010) Repeated polyploidy drove different levels of crossover suppression between homoeologous chromosomes in *Brassica napus* allohaploids. Plant Cell 22:2265–2276

Delannoy E, Fujii S, Des Francs-Small CC, Brundrett M, Small I (2011) Rampant gene loss in the underground orchid *Rhizanthella gardneri* highlights evolutionary constraints on plastid genomes. Mol Biol Evol 28:2077–2086

Flannery M, Mitchell F, Coyne S, Kavanagh T, Burke J, Salamin N, Dowding P, Hodkinson T (2006) Plastid genome characterisation in Brassica and Brassicaceae using a new set of nine SSRs. Theor Appl Genet 113:1221–1231

Gupta SK (2013) The importance, origin, and evolution. In: Biotechnology of crucifers, Springer

Haider N (2012) Evidence for the origin of the B genome of bread wheat based on chloroplast DNA. Turk J Agric For 36:13–25

Hollingsworth PM, Graham SW, Little DP (2011) Choosing and using a plant DNA barcode. PLoS ONE 6:e19254

Hu Z-Y, Hua W, Huang S-M, Wang H-Z (2011) Complete chloroplast genome sequence of rapeseed (*Brassica napus* L.) and its evolutionary implications. Genet Resour Crop Evol 58:875–887

Jansen RK, Ruhlman TA (2012) Plastid genomes of seed plants. In: Genomics of chloroplasts and mitochondria. Springer

Jesske T, Olberg B, Schierholt A, Becker H (2013) Resynthesized lines from domesticated and wild *Brassica taxa* and their hybrids with *B. napus* L.: genetic diversity and hybrid yield. Theor Appl Genet 126:1053–1065

Kaur P, Banga S, Kumar N, Gupta S, Akhatar J, Banga SS (2014) Polyphyletic origin of *Brassica juncea* with *B. rapa* and *B. nigra* (Brassicaceae) participating as cytoplasm donor parents in independent hybridization events. Am J Bot 101:1157–1166

Kim JH, Jung J-Y, Choi H-I, Kim N-H, Park JY, Lee Y, Yang T-J (2013) Diversity and evolution of major Panax species revealed by scanning the entire chloroplast intergenic spacer sequences. Genet Resour Crop Evol 60:413–425

Kim K, Lee S-C, Lee J, Lee HO, Joh HJ, Kim N-H, Park H-S, Yang T-J (2015a) Comprehensive survey of genetic diversity in chloroplast genomes and 45S nrDNAs within *Panax ginseng* species. PLoS ONE 10:e0117159

Kim K, Lee S-C, Lee J, Yu Y, Yang K, Choi B-S, Koh H-J, Waminal NE, Choi H-I, Kim N-H (2015b) Complete chloroplast and ribosomal sequences for 30 accessions elucidate evolution of Oryza AA genome species. Sci Rep, 5

Kundu A, Topdar N, Sarkar D, Sinha MK, Ghosh A, Banerjee S, Das M, Balyan HS, Mahapatra B, Gupta PK (2013) Origins of white (*Corchorus capsularis* L.) and dark (*C. olitorius* L.) jute: a reevaluation based on nuclear and chloroplast microsatellites. J Plant Biochem Biotechnol 22:372–381

Leflon M, Eber F, Letanneur J, Chelysheva L, Coriton O, Huteau V, Ryder C, Barker G, Jenczewski E, Chevre A (2006) Pairing and recombination at meiosis of *Brassica rapa* (AA)× *Brassica napus* (AACC) hybrids. Theor Appl Genet 113:1467–1480

Li P, Zhang S, Li F, Zhang S, Zhang H, Wang X, Sun R, Bonnema G, Borm TJA (2017) A phylogenetic analysis of chloroplast genomes elucidates the relationships of the six economically important Brassica species comprising the triangle of U. Front Plant Sci 8:111

Li X, Yang Y, Henry RJ, Rossetto M, Wang Y, Chen S (2015) Plant DNA barcoding: from gene to genome. Biol Rev 90:157–166

Liu S, Wang H, Zhang J, Fitt BDL, Xu Z, Evans N, Liu Y, Yang W, Guo X (2005) In vitro mutation and selection of doubled-haploid *Brassica napus* lines with improved resistance to *Sclerotinia sclerotiorum*. Plant Cell Rep 24:133–144

Lohse M, Drechsel O, Kahlau S, Bock R (2013) OrganellarGenomeDRAW—a suite of tools for generating physical maps of plastid and mitochondrial genomes and visualizing expression data sets. Nucleic Acids Res 41:W575–W581

Mei J, Li Q, Qian L, Fu Y, Li J, Frauen M, Qian W (2011) Genetic investigation of the origination of allopolyploid with virtually synthesized lines: application to the C subgenome of *Brassica napus*. Heredity 106:955–961

Mullet JE (1988) Chloroplast development and gene expression. Annu Rev Plant Physiol Plant Mol Biol 39:475–502

Nikiforova SV, Cavalieri D, Velasco R, Goremykin V (2013) Phylogenetic analysis of 47 chloroplast genomes clarifies the contribution of wild species to the domesticated apple maternal line. Mol Biol Evol 30:1751–1760

Nock CJ, Waters DL, Edwards MA, Bowen SG, Rice N, Cordeiro GM, Henry RJ (2011) Chloroplast genome sequences from total DNA for plant identification. Plant Biotechnol J 9:328–333

Palmer JD, Shields CR, Cohen DB, Orton TJ (1983) Chloroplast DNA evolution and the origin of amphidiploid Brassica species. Theor Appl Genet 65:181–189

Panda S, Martin J, Aguinagalde I (2003) Chloroplast and nuclear DNA studies in a few members of the *Brassica oleracea* L. group using PCR-RFLP and ISSR-PCR markers: a population genetic analysis. Theor Appl Genet 106:1122–1128

Parkin I, Sharpe A, Keith D, Lydiate D (1995) Identification of the A and C genomes of amphidiploid Brassica napus (oilseed rape). Genome 38:1122–1131

Pellan-Delourme R, Renard M (1988) Cytoplasmic male sterility in rapeseed (*Brassica napus* L.): female fertility of restored rapeseed with "Ogura" and cybrids cytoplasms. Genome 30:234–238

Qian W, Meng J, Li M, Frauen M, Sass O, Noack J, Jung C (2006) Introgression of genomic components from Chinese *Brassica rapa* contributes to widening the genetic diversity in rapeseed (*B. napus* L.), with emphasis on the evolution of Chinese rapeseed. Theor Appl Genet 113:49–54

Qiao J, Cai M, Yan G, Wang N, Li F, Chen B, Gao G, Xu K, Li J, Wu, X (2015) High-throughput multiplex cpDNA resequencing clarifies the genetic diversity and genetic relationships among *Brassica napus*, *Brassica rapa* and *Brassica oleracea*. Plant Biotechnol J

Röbbelen S, Downey RK, Ashri A (1989) Oilcrops of the world: their breeding and utilization. McGraw-Hill Publishing Company

Sharma S, Padmaja KL, Gupta V, Paritosh K, Pradhan AK, Pental D (2014) Two plastid DNA lineages—Rapa/Oleracea and Nigra—within the tribe Brassiceae can be best explained by reciprocal crosses at hexaploidy: evidence from divergence times of the plastid genomes and R-block genes of the A and B genomes of *Brassica juncea*. PloS ONE 9

Sharpe A, Parkin I, Keith D, Lydiate D (1995) Frequent nonreciprocal translocations in the amphidiploid genome of oilseed rape (*Brassica napus*). Genome 38:1112–1121

Shu J, Liu Y, Li Z, Zhang L, Fang Z, Yang L, Zhuang M, Zhang Y, Lv H (2015) Organelle simple sequence repeat markers help to distinguish carpelloid stamen and normal cytoplasmic male sterile sources in Broccoli. PLoS ONE 10:e0138750

Song K, Osborn TC (1992) Polyphyletic origins of *Brassica napus*: new evidence based on organelle and nuclear RFLP analyses. Genome 35:992–1001

Song K, Tang K, Osborn TC, Lu P (1994) Genome variation and evolution of *Brassica amphidiploids*. In: ISHS Brassica symposium-IX crucifer genetics workshop, vol 407, pp 35–44

Szadkowski E, Eber F, Huteau V, Lodé M, Huneau C, Belcram H, Coriton O, Manzanares-Dauleux M, Delourme R, King GJ (2010) The first meiosis of resynthesized *Brassica napus*, a genome blender. New Phytol 186:102–112

Wang A, Yang M, Liu J (2005) Molecular phylogeny, recent radiation and evolution of gross morphology of the rhubarb genus Rheum (Polygonaceae) inferred from chloroplast DNA trnL-F sequences. Ann Bot 96:489–498

Wang N, Li F, Chen B, Xu K, Yan G, Qiao J, Li J, Gao G, Bancroft I, Meng J (2014) Genome-wide investigation of genetic changes during modern breeding of Brassica napus. Theor Appl Genet 127:1817–1829

Wang Q, Zhang Y, Fang Z, Liu Y, Yang L, Zhuang M (2012) Chloroplast and mitochondrial SSR help to distinguish allo-cytoplasmic male sterile types in cabbage (*Brassica oleracea* L. var. capitata). Mol Breed 30:709–716

Warwick S, Simard M-J, Légère A, Beckie H, Braun L, Zhu B, Mason P, Séguin-Swartz G, Stewart Jr C (2003) Hybridization between transgenic *Brassica napus* L. and its wild relatives: *Brassica rapa* L., *Raphanus raphanistrum* L., *Sinapis arvensis* L., and *Erucastrum gallicum* (Willd.) OE Schulz. Theor Appl Genet 107:528–539

Witt U, Hansen S, Albaum M, Abel W (1991) Molecular analyses of the CMS-inducing 'Polima' cytoplasm in *Brassica napus* L. Curr Genet 19:323–327

Woo H-J, Lim M-H, Shin K-S, Martins B, Lee B-K, Cho H-S, Mallory-Smith CA (2013) Development of a chloroplast DNA marker for monitoring of transgene introgression in *Brassica napus* L. Biotech Lett 35:1533–1539

Wyman SK, Jansen RK, Boore JL (2004) Automatic annotation of organellar genomes with DOGMA. Bioinformatics 20:3252–3255

Yang J, Liu G, Zhao N, Chen S, Liu D, Ma W, Hu Z, Zhang M (2015) Comparative mitochondrial genome analysis reveals the evolutionary rearrangement mechanism in Brassica. Plant Biol

Yang T-J, Kim JS, Kwon S-J, Lim K-B, Choi B-S, Kim J-A, Jin M, Park JY, Lim M-H, Kim H-I (2006) Sequence-level analysis of the diploidization process in the triplicated FLOWERING LOCUS C region of *Brassica rapa*. Plant Cell 18:1339–1347

Zamani-Nour S, Clemens R, Möllers C (2013) Cytoplasmic diversity of *Brassica napus* L., *Brassica oleracea* L. and *Brassica rapa* L. as determined by chloroplast microsatellite markers. Genet Resour Crop Evol 60:953–965

Zeng C-L, Wang G-Y, Wang J-B, Yan G-X, Chen B-Y, Xu K, Li J, Gao G-Z, Wu X-M, Zhao B (2012) High-throughput discovery of chloroplast and mitochondrial DNA polymorphisms in Brassicaceae species by ORG-EcoTILLING

Zou X-H, Du Y-S, Tang L, Xu X-W, Doyle JJ, Sang T, Ge S (2015) Multiple origins of BBCC allopolyploid species in the rice genus (Oryza). Sci Rep 5

11 Case Study for Trait-Related Gene Evolution: Oil Biosynthesis Genes

Zhiyong Hu and Wei Hua

Abstract

Brassica napus was formed through recent hybridization between *Brassica oleracea* and *Brassica rapa* and is an important source of edible oil. The genomic characterization of the families of genes involved in oil biosynthesis was undertaken in the *B. napus* genome assembly to assess the potential impact of selection breeding on gene content and function. We compared oil biosynthesis genes and found that the genes number has a huge difference in 14 different species. There are 2482 homologs in *B. napus* cv. ZS11 and only 120 homologs in *Jatropha curcas*. There is a 4.1 fold expansion over *Arabidopsis thaliana* and >20 fold expansion over *J. curcas*. However, the distributions of the gene number in the acyl metabolism pathway are highly similar in all the 14 species. The fatty acid elongation and wax biosynthesis pathway, the phospholipid signaling pathway, and the galactolipid, sulfolipid, and phospholipid synthesis pathway are the top pathways in terms of the gene number. A total of 19 positive selection genes were identified in *B. napus*. Among them, 5 genes are in the phospholipid signaling pathway and 5 genes in the triacylglycerol and fatty acid degradation pathway. These results will help better understand the mechanism and evolution of oil biosynthesis genes.

Z. Hu · W. Hua (✉)
The Key Laboratory of Biology and Genetic Improvement of Oil Crops, the Ministry of Agriculture, Oil Crops Research Institute, Chinese Academy of Agricultural Sciences, Wuhan, Hubei, China
e-mail: huawei@oilcrops.cn

11.1 Introduction

Flowering plants propagate generation after generation through production of seeds. Carbon and nitrogen are deposited into seed oil, protein, and starch to support the establishment of the young seedlings (Wang et al. 2007). In the past centuries, the importance of these storage reserves as food and feed brought in continued breeding efforts and successful improvements of crop seed composition and yield. Today, crops are not only essential for food supplies to the growing population but also demanded by expanding industrial markets beyond human nutrition. For example, vegetable oil is the important lipid source for the production of lubricants, inks, paints, and biodiesels. Although genetic engineering offers great potential to speed up the process of crop improvement, such an effort relies on a good understanding of the molecular mechanisms underlying substance storage. Thus, regulation of oil biosynthesis in plant seeds has been extensively studied.

S. Liu et al. (eds.), *The Brassica napus Genome*, Compendium of Plant Genomes,
https://doi.org/10.1007/978-3-319-43694-4_11

The pathways involved in plant storage lipid biosynthesis have recently been reviewed in detail elsewhere (Snyder et al. 2009). Extensive research has led to an unprecedented level of knowledge regarding the pathways and genes involved in lipid biosynthesis, especially in the model plant *A. thaliana* (Baud and Lepiniec 2009; Hua et al. 2012). The intricate transcription regulatory system that controls Arabidopsis seed development has been determined. In this network, the B3 transcription factors FUS3, ABI3, and LEC2, as well as the transcriptional activator LEC1, are master regulators of seed development and reserve accumulation (Santos-Mendoza et al. 2008).

The acyltransferases involved in the two pathways leading to triacylglycerol (TAG) play an important role in the utilization of unusual fatty acids (Snyder et al. 2009). The classical sn-glycerol-3-phosphate or the Kennedy pathway involves the sequential acyl-CoA-dependent acylation of sn-glycerol-3-phosphate catalyzed by sn-glycerol-3-phosphate acyltransferase (GPAT), lysophosphatidic acid acyltransferase (LPAAT) and diacylglycerol acyltransferase (DGAT), respectively. And phosphatidic acid phosphatase catalyzes the dephosphorylation of phosphatidic acid (PA) before the final acylation. In the second pathway, acyl groups can be channeled into phosphatidylcholine (PC) via the activity of lysophosphatidylcholine acyltransferase (LPCAT), and then subsequently transferred from PC to TAG via the activity of phospholipid: diacylglycerol acyltransferase (PDAT) (Dahlqvist et al. 2000; Stahl et al. 2004).

Acyl-CoA-binding protein (ACBP) plays an important housekeeping role in lipid metabolism by maintaining the intracellular acyl-CoA pool (Yurchenko and Weselake 2011). ACBP is involved in lipid biosynthesis and transport, gene expression, and membrane biogenesis. WRINKLED1, a key regulator of seed oil biosynthesis in *Arabidopsis* (*A. thaliana*), has been implicated in the control of genes encoding enzymes for plastidial glycolysis and fatty acid biosynthesis (Cernac and Benning 2004; Pouvreau et al. 2011). WRINKLED1 specifies the regulatory activities of LEC1 and LEC2 with respect to production of storage lipids (Baud et al. 2007; Mu et al. 2008). In addition, application of stable isotope labeling methods has enabled in vivo examination of the metabolic fluxes in developing *B. napus* seeds, and these studies have determined the relative contributions of plastidial, mitochondrial and cytosolic pathways to fatty acid biosynthesis (Schwender et al. 2003, 2004, 2006; Ruuska et al. 2004; Goffman et al. 2005). However, despite the great advances mentioned above, we still know relatively little about the overall regulation of lipid synthesis and accumulation, or the partitioning of carbon between oil and other storage products, and these deficiencies continue to hinder progress in breeding seeds with high oil contents (Weselake et al. 2009; Baud and Lepiniec 2010).

B. napus was formed by recent allopolyploidization after hybridization between *B. oleracea* and *B. rapa* (Hasan et al. 2008; Chalhoub et al. 2014; Sun et al. 2017) and is primarily grown as one of the most important oilseed crops. Together with soybean and oil palm, they provide a major part of edible oil in the world. Moreover, *B. napus* seed oil is with high-quality nutritional composition, (Hu et al. 2013). In *B. napus*, recent breeding has focused on the selection of lines with high oil content and optimized oil composition in the seed (Sakhno 2010). Thus, a genomic characterization of the families of genes involved in oil biosynthesis was undertaken in the *B. napus* genome assembly to assess the potential impact of selection breeding on gene content and function (Chalhoub et al. 2014; Sun et al. 2017).

11.2 Variation of the Gene Number of Acyl-Lipid Metabolism

The acyl-lipid metabolism pathway is crucial for oil biosynthesis and plant defense. Chalhoub et al. (2014) made an extensive comparative analysis of *B. napus* with the model crucifer species *A. thaliana*, identified more than 120 different enzymatic reactions, and found 606 genes that played a role in acyl-lipid metabolism

(Chalhoub et al. 2014). We further compared these oil biosynthesis genes in 14 different species and found that there was a huge difference in the copy number of acyl-lipid metabolism genes between these species (Table 11.1, Sun et al. 2017). There are 2482 homologs in *B. napus* cv. ZS11, 2010 homologs in *B. napus* cv. Darmor, and 926 homologs in *Glycine max* which represents the highest number of acyl-lipid metabolism genes annotated in an oilseed plant species genome to date, but only 120 homologs in oilseed plant *J. curcas* (Table 11.1, Sun et al. 2017). This represents a 4.1 fold expansion over *A. thaliana* and >20 fold expansion over *J.curcas*.

Although the total gene number is various, the distributions of genes in acyl metabolism pathways are highly similar in all the 14 species. In terms of gene number, the fatty acid elongation and wax biosynthesis pathway, the phospholipid signaling pathway and the galactolipid, sulfolipid, and phospholipid synthesis pathway are the top three pathways almost in all the 14 species (Table 11.1, Sun et al. 2017).

Table 11.1 Copy number of acyl-lipid metabolism genes in 14 species

Acyl metabolism pathway	Species[a]													
	At	Bn	Bo	Br	Cc	Gm	Jc	Lu	Os	Rc	Sin	Sit	Vv	Zm
Fatty acid synthesis	29	50	70	65	30	55	9	44	9	27	54	30	25	36
Fatty acid elongation, desaturation, and export from plastid	9	40	20	16	10	16	2	16	1	8	15	10	11	14
Galactolipid, sulfolipid, and phospholipid synthesis	68	246	126	115	64	109	1	105	9	51	17	63	55	66
Triacylglycerol biosynthesis	53	193	91	100	34	65	1	59	1	23	49	22	20	26
Triacylglycerol and fatty acid degradation	48	216	114	102	40	77	9	84	9	35	55	44	34	36
Fatty acid elongation and wax biosynthesis	148	653	333	298	89	155	2	115	7	62	1	50	62	53
Sphingolipid biosynthesis	23	86	46	37	25	46	2	43	1	19	33	18	18	23
Mitochondrial fatty acid and lipoic acid synthesis	13	43	21	19	12	21	1	18	0	12	20	11	10	9
Mitochondrial phospholipid synthesis	6	24	11	10	5	10	2	9	2	5	8	2	4	2
Lipid trafficking	6	20	5	7	8	14	2	14	5	5	15	6	6	8
Cutin synthesis and transport	16	67	40	34	20	45	2	14	10	14	21	11	16	10
Suberin synthesis and transport	18	111	57	50	21	40	8	33	15	22	26	14	13	18
Oxylipin metabolism	56	192	92	87	55	94	9	67	22	40	69	31	48	25
Phospholipid signaling	78	293	149	139	64	125	21	104	42	57	110	39	50	43
Pathway, function, or subcellular location uncertain	35	148	81	69	30	54	9	41	19	22	45	21	21	19
Total	606	2482	1256	1148	507	926	120	766	352	402	748	372	393	388

[a]At, Bn, Bo, Br, Cc, Gm, Jc, Lu, Os, Rc, Sin, Sit, Vv, Zm represent *Arabidopsis thaliana*, *Brassica napus* cv. ZS11, *Brassica oleracea*, *Brassica rapa*, *Cajanus cajan*, *Glycine max*, *Jatropha curcas*, *Linum usitatissimum*, *Oryza sativa*, *Ricinus communis*, *Sesamum indicum*, *Setaria italic*, *Vitis vinifera* and *Zea mays*, respectively

11.3 The Expansion and Loss of the Acyl-lipid Metabolism Genes in *B. napus*

Chalhoub et al. (2014) reported that in comparison to the diploids generally there appeared to be no targeted loss of gene copies in the tetraploid genomes. However, 18 orthologs of acyl genes were no longer found in the *B. napus* genome—five appeared to be completely lost, three in the An subgenome and two in the Cn subgenome, and all other (13) missing orthologs had been replaced either through large or short homeologous exchanges (HEs). As observed for most regions of the genome, there was a bias in HES, with the Cn genome copies being preferentially replaced by An genome homeologs; nine Cn genes and four An genes were replaced by counterpart homeologs through larger homeologous recombination events. One Cn gene and one An gene were replaced by An and Cn homeologs, respectively, through individual gene conversion. There did not appear to be any targeted classes of lost genes; the two classes with the largest number of duplicated genes (Fatty Acid Elongation and Wax Biosynthesis and Phospholipid Signaling) also represented the classes with the highest number of lost genes; five and three, respectively.

The expansion of the acyl-lipid metabolism genes could have an impact on the versatility of the trait in *B. napus*, yet the phenotypic differences between the non-oilseed diploids and the oilseed *B. napus* would appear to be largely the result of allelic variation. Chalhoub et al. (2014) confirmed that the *B. napus* orthologs of fatty acid elongase 1 (FAE1) from *B. rapa* (Bra034635) and *B. oleracea* (Bo7g116890) exhibit the respective SNP (C to T) and two-base deletion (AA) alleles associated with low levels of erucic acid in the seeds. This is a key trait that has been selected for human nutrition.

11.4 Identification of Positive Selection Acyl-Lipid Metabolism Genes from *B. napus* and Its Progenitor Species

The *Brassicaceae* is a large eudicot family and includes the model plant *A. thaliana* (Chalhoub et al. 2014; Sun et al. 2017). Brassicas have a propensity for genome duplications and genome mergers (Chalhoub et al. 2014; Sun et al. 2017). *B. napus* was formed ~7500 years ago by hybridization between *B. rapa* and *B. oleracea*, followed by chromosome doubling, a process known as allopolyploidization. Together with more ancient polyploidizations, they conferred an aggregate 72 × genome multiplication detectable since the origin of angiosperms and resulting in high gene content. Chalhoub et al. (2014) examined the *B. napus* genome and the consequences of its recent duplication. The constituent An and Cn subgenomes are engaged in subtle structural, functional, and epigenetic cross talk, with abundant homeologous exchanges. Incipient gene loss and expression divergence have begun. Selection in *B. napus* oilseed types has accelerated the loss of glucosinolate genes, while preserving expansion of oil biosynthesis genes (Chalhoub et al. 2014).

In the study of Chalhoub et al. (2014), they identified orthologous and paralogous sequences within the *B. napus*, *B. oleracea* and *B. rapa* reference genomes. Then they found that an assessment of conserved syntenic putative acyl metabolism genes between all three Brassica species could suggest selection pressures on specific pathways or gene families in response to the breeding strategies applied in the development of low erucic acid, high oil content *B. napus*, as compared to its two diploid progenitors. Conserved syntenic positions within each genome will identify the most likely functional orthologs in the Brassica species. For the 606 *A. thaliana* genes, they identified a total of 986 and 1030 homologs within the Ar and Co genomes, respectively, and an almost identical number of 985 and 1025 homologs within the amphidiploid An and Cn genomes, respectively.

In order to identify the positive selection genes in *B. napus*, we analyzed and compared orthologous and paralogous sequences within the *B. napus* cv. ZS11, *B. napus* cv. Darmor, *B. oleracea*, *B. rapa* reference genomes and the *A. thaliana* one. Based on the analysis of positive selection mutation site, a total of 19 positive selection genes were identified (Table 11.2). There are 8 genes located in An subgenomes and 11 genes located in Cn subgenomes. Among the 19 positive selection genes, 5 genes are in the phospholipid signaling pathway and 5 genes are in the triacylglycerol and fatty acid degradation pathway.

Table 11.2 Positive selection acyl-lipid metabolism genes in *B. napus*

Co	AN-Z	AN-D	CN-Z	CN-D	*A. thaliana*	Mutant site	First acyl biosynthesis pathway	Second acyl biosynthesis pathway
Bo1g126040.1	**BnA01g0035780.1**	**BnaA01g28540D**	BnC03g0552120.1	BnaC01g35830D	AT3G15730	687-R-M-0.997**-AGA-ATG	Phospholipid signaling	Phospholipase D alpha
Bo1g134540.1	**BnA01g0037150.1**	**BnaA01g29510D**	BnC01g0466760.1	BnaC01g37150D	AT3G14205	513-N-K-0.638-AAT-AAA; 540-A-G-0.648GCG-GGG; 547-R-G-0.621-GA-GGA	Phospholipid signaling	Sac domain-containing phosphoinositide phosphatase
Bo2g129970.1	**BnA02g0075260.1**	**BnaA02g24400D**	BnC02g0509430.1	BnaC02g32230D	AT5G46290	147-G-N-0.973*-GGG-AAT	Fatty acid synthesis	Ketoacyl-ACP synthase I
Bo7g118300.1	**BnA03g0157180.1**	**BnaA03g53730D**	BnC07g0834380.1	BnaC07g46420D	AT4G37050	270-I-S-0.883-ATC-AGC	Oxylipin metabolism 1	Acyl-hydrolase (patatin-like)
Bo4g140370.1	**BnA04g0169560.1**	**BnaA04g10460D**	BnC04g0660430.1	BnaC04g32730D	AT5G40610	208-T-S-0.830-ACC-TCC	Prokaryotic galactolipid, sulfolipid, and phospholipid synthesis 1	NAD-dependent glycerol-3-phosphate dehydrogenase
Bo4g188160.1	**BnA04g0182010.1**	**BnaA04g23070D**	BnC04g0678350.1	BnaC04g46320D	AT2G39420	251-F-L-0.823-TTC-CTC	Triacylglycerol and fatty acid degradation	Monoacylglycerol lipase (MAGL)
Bo5g116900.1	**BnA05g0211540.1**	**BnaA05g21240D**	BnUnng0943350.1	BnaC05g34160D	AT3G18860		Oxylipin metabolism 1; oxylipin metabolism 2; phospholipid signaling	Phospholipase A2
Bo8g004200.1	**BnA08g0335160.1**	**BnaA08g28040D**	BnC04g0667150.1	BnaC08g00920D	AT1G04710	130-V-I-0.528-GTT-ATT; 183-Q-K-0.609-CAG-AAG; 312-E-G-0.549-GAG-GGG	Triacylglycerol and fatty acid degradation	Ketoacyl-CoA thiolase
Bo1g105160.1	BnUnng0974260.1	BnaA01g36630D	**BnC04g0626060.1**	**BnaC01g31590D**	AT3G22400	649-M-K-0.865-ATG-AAG	Oxylipin metabolism 1; oxylipin metabolism 2	Lipoxygenase
Bo1g123670.1	BnA01g0035440.1	BnaA01g28170D	**BnC01g0465810.1**	**BnaC01g35460D**	AT3G16170	170-K-E-0.731-AAA-GAA; 306-E-G-0.740-GAA-GGA	Mitochondrial fatty acid and lipoic acid synthesis	Malonyl-CoA synthase
Bo1g134540.1	BnA01g0037150.1	BnaA01g29510D	**BnC01g0466760.1**	**BnaC01g37150D**	AT3G14205	669-I-M-0.912-ATA-ATG	Phospholipid signaling	Sac domain-containing phosphoinositide phosphatase

(continued)

Table 11.2 (continued)

Co	AN-Z	AN-D	CN-Z	CN-D	*A. thaliana*	Mutant site	First acyl biosynthesis pathway	Second acyl biosynthesis pathway
Bo2g094640.1	BnA03g0144790.1	BnaA02g19010D	**BnC02g0500200.1**	**BnaC02g25190D**	AT1G78690	67-T-M-0.937-ACA-ATG	Eukaryotic phospholipid synthesis and editing	Lysophospholipid acyltransferase
Bo5g144910.1	BnA09g0358910.1	BnaA05g31340D	**BnC05g0731200.1**	**BnaC05g45860D**	AT3G05970	136-I-V-0.705-ATT-GTT	Triacylglycerol and fatty acid degradation	Long-chain acyl-CoA synthetase
Bo4g137870.1	BnA04g0168090.1	BnaA04g09220D	**BnC02g0505140.1**	**BnaC04g31430D**	AT5G39400	140-L-F-0.984*-TTG-TTT;165-H-P-0.983*-CAT-CCT;185-D-Y-0.983*-GAT-TAT;341-K-R-0.983*-AAA-AGA	Triacylglycerol and fatty acid degradation	Long-chain Acyl-CoA synthetase
Bo4g196050.1	BnA03g0128710.1	BnaA04g26670D	**BnC04g0684100.1**	**BnaC04g50730D**	AT2G46090	17-H-Q-0.759-CAC-CAA	Phospholipid signaling	Phosphoinositide 3-phosphatase
Bo4g071970.1	BnA07g0282820.1	BnaA07g13590D	**BnC04g0640730.1**	**BnaC04g16280D**	AT2G28070	469-K-N-0.953*-AAA-AAC	Fatty acid elongation and wax biosynthesis	ABC transporter
Bo3g149400.1	BnA08g0323810.1	BnaA08g17030D	**BnC03g0609720.1**	**BnaC03g60040D**	AT4G38570	241-F-I-0.989*-TTC-ATC	Eukaryotic phospholipid synthesis and editing	Phosphatidylinositol synthase
Bo8g058380.1	BnA08g0333420.1	BnaA08g26450D	**BnC08g0849560.1**	**BnaCnng06840D**	AT1G09400	154-I-R-0.892-ATA-AGG	Phospholipid signaling	Phosphoinositide 3-phosphatase
Bo9g181470.1	BnA10g0422410.1	BnaA10g26130D	**BnC09g0934550.1**	**BnaCnng10420D**	AT5G04040	111-N-Y-0.552 AAC-TAC	Triacylglycerol and fatty acid degradation	Triacylglycerol lipase (TAGL)

Ar, *B. rapa*; Co, *B. oleracea*; AN-Z/CN-Z, *B. napus* cv. ZS11(An and Cn subgenomes); AN-D/BN-D, *B. napus* cv. Darmor (An and Cn subgenomes); bold shows the positive selection genes

Acknowledgements This work was supported by the National 863 Plan projects (2013AA102602 and 2012AA101107), National Key Basic Research Program of China (2015CB150200), the Industry Technology System of Rapeseed in China (CARS-12), and the Hubei Agricultural Science and Technology Innovation Center of China.

References

Baud S, Lepiniec L (2009) Regulation of de novo fatty acid synthesis in maturing oilseeds of Arabidopsis. Plant Physiol Biochem. (PPB)/Societe francaise de physiologie vegetale 47:448–455

Baud S, Lepiniec L (2010) Physiological and developmental regulation of seed oil production. Prog Lipid Res 49:235–249

Baud S, Mendoza MS, To A, Harscoet E, Lepiniec L, Dubreucq B (2007) WRINKLED1 specifies the regulatory action of LEAFY COTYLEDON2 towards fatty acid metabolism during seed maturation in *Arabidopsis*. Plant J Cell Mol Biol 50:825–838

Cernac A, Benning C (2004) WRINKLED1 encodes an AP2/EREB domain protein involved in the control of storage compound biosynthesis in *Arabidopsis*. Plant J Cell Mol Biol 40:575–585

Chalhoub B, Denoeud F, Liu S, Parkin IA, Tang H, Wang X, Chiquet J, Belcram H, Tong C, Samans B, Correa M, Da Silva C, Just J, Falentin C, Koh CS, Le Clainche I, Bernard M, Bento P, Noel B, Labadie K, Alberti A, Charles M, Arnaud D, Guo H, Daviaud C, Alamery S, Jabbari K, Zhao M, Edger PP, Chelaifa H, Tack D, Lassalle G, Mestiri I, Schnel N, Le Paslier MC, Fan G, Renault V, Bayer PE, Golicz AA, Manoli S, Lee TH, Thi VH, Chalabi S, Hu Q, Fan C, Tollenaere R, Lu Y, Battail C, Shen J, Sidebottom CH, Wang X, Canaguier A, Chauveau A, Berard A, Deniot G, Guan M, Liu Z, Sun F, Lim YP, Lyons E, Town CD, Bancroft I, Wang X, Meng J, Ma J, Pires JC, King GJ, Brunel D, Delourme R, Renard M, Aury JM, Adams KL, Batley J, Snowdon RJ, Tost J, Edwards D, Zhou Y, Hua W, Sharpe AG, Paterson AH, Guan C, Wincker P (2014) Plant genetics. Early allopolyploid evolution in the post-neolithic *Brassica napus* oilseed genome. Science 345:950–953

Dahlqvist A, Stahl U, Lenman M, Banas A, Lee M, Sandager L, Ronne H, Stymne S (2000) Phospholipid: diacylglycerol acyltransferase: an enzyme that catalyzes the acyl-CoA-independent formation of triacylglycerol in yeast and plants. Proc Natl Acad Sci USA 97:6487–6492

Goffman FD, Alonso AP, Schwender J, Shachar-Hill Y, Ohlrogge JB (2005) Light enables a very high efficiency of carbon storage in developing embryos of rapeseed. Plant Physiol 138:2269–2279

Hasan M, Friedt W, Pons-Kuhnemann J, Freitag NM, Link K, Snowdon RJ (2008) Association of gene-linked SSR markers to seed glucosinolate content in oilseed rape (*Brassica napus* ssp. *napus*). TAG. Theoretical and applied genetics. Theoretische und angewandte Genetik 116:1035–1049

Hu ZY, Hua W, Zhang L, Deng LB, Wang XF, Liu GH, Hao WJ, Wang HZ (2013) Seed structure characteristics to form ultrahigh oil content in rapeseed. PLoS ONE 8:e62099

Hua W, Li RJ, Zhan GM, Liu J, Li J, Wang XF, Liu GH, Wang HZ (2012) Maternal control of seed oil content in Brassica napus: the role of silique wall photosynthesis. Plant J Cell Mol Biol 69:432–444

Mu J, Tan H, Zheng Q, Fu F, Liang Y, Zhang J, Yang X, Wang T, Chong K, Wang XJ, Zuo J (2008) LEAFY COTYLEDON1 is a key regulator of fatty acid biosynthesis in *Arabidopsis*. Plant Physiol 148:1042–1054

Pouvreau B, Baud S, Vernoud V, Morin V, Py C, Gendrot G, Pichon JP, Rouster J, Paul W, Rogowsky PM (2011) Duplicate maize Wrinkled1 transcription factors activate target genes involved in seed oil biosynthesis. Plant Physiol 156:674–686

Ruuska SA, Schwender J, Ohlrogge JB (2004) The capacity of green oilseeds to utilize photosynthesis to drive biosynthetic processes. Plant Physiol 136:2700–2709

Sakhno LA (2010) Fatty acid composition variability of rapeseed oil: classical selection and biotechnology. Tsitol Genet 44:70–80

Santos-Mendoza M, Dubreucq B, Baud S, Parcy F, Caboche M, Lepiniec L (2008) Deciphering gene regulatory networks that control seed development and maturation in *Arabidopsis*. Plant J Cell Mol Biol 54:608–620

Schwender J, Ohlrogge JB, Shachar-Hill Y (2003) A flux model of glycolysis and the oxidative pentosephosphate pathway in developing *Brassica napus* embryos. J Biol Chem 278:29442–29453

Schwender J, Shachar-Hill Y, Ohlrogge JB (2006) Mitochondrial metabolism in developing embryos of *Brassica napus*. J Biol Chem 281:34040–34047

Schwender J, Goffman F, Ohlrogge JB, Shachar-Hill Y (2004) Rubisco without the Calvin cycle improves the carbon efficiency of developing green seeds. Nature 432:779–782

Snyder CL, Yurchenko OP, Siloto RM, Chen X, Liu Q, Mietkiewska E, Weselake RJ (2009) Acyltransferase action in the modification of seed oil biosynthesis. New Biotechnol 26:11–16

Stahl U, Carlsson AS, Lenman M, Dahlqvist A, Huang B, Banas W, Banas A, Stymne S (2004) Cloning and functional characterization of a phospholipid:diacylglycerol acyltransferase from *Arabidopsis*. Plant Physiol 135:1324–1335

Sun F, Fan G, Hu Q, Zhou Y, Guan M, Tong C, Li J, Du D, Qi C, Jiang L, Liu W, Huang S, Chen W, Yu J, Mei D, Meng J, Zeng P, Shi J, Liu K, Wang X,

Wang X, Long Y, Liang X, Hu Z, Huang G, Dong C, Zhang H, Li J, Zhang Y, Li L, Shi C, Wang J, Lee SM, Guan C, Xu X, Liu S, Liu X, Chalhoub B, Hua W, Wang H (2017) The high-quality genome of *Brassica napus* cultivar 'ZS11' reveals the introgression history in semi-winter morphotype. Plant J Cell Mol Biology 92:452–468

Wang H, Guo J, Lambert KN, Lin Y (2007) Developmental control of *Arabidopsis* seed oil biosynthesis. Planta 226:773–783

Weselake RJ, Taylor DC, Rahman MH, Shah S, Laroche A, McVetty PB, Harwood JL (2009) Increasing the flow of carbon into seed oil. Biotechnol Adv 27:866–878

Yurchenko OP, Weselake RJ (2011) Involvement of low molecular mass soluble acyl-CoA-binding protein in seed oil biosynthesis. New Biotechnol 28:97–109

Case Study for Trait-Related Gene Evolution: Glucosinolates

12

Kun Lu, Rod Snowdon and Jiana Li

Abstract

Glucosinolates are a group of secondary metabolites involved in plant defense and found mainly in the Brassicales order. While the breakdown products of some glucosinolates are beneficial to human health, many glucosinolates are toxic. The recently sequenced genomes of *Brassica napus* and its parental species *Brassica rapa* and *Brassica oleracea* provided the *Brassica* scientific community with a valuable tool for systematically investigating glucosinolate biosynthesis, transport, and breakdown genes, elucidating the relationship between variation of glucosinolate profiles and the evolution of glucosinolate-related genes in *Brassica* crops. In this chapter, we summarized the variation in glucosinolate composition and content in *Brassica* crops and identified 166, 167, 191, 333 genes in *B. rapa, B. oleracea* var. *capitata, B. oleracea* var. *italica,* and *B. napus,* respectively, as orthologs of 78 glucosinolate biosynthetic, transport, and breakdown genes in *Arabidopsis thaliana*. Among these glucosinolate-related genes, transcription factor, side-chain modification, and breakdown genes experienced significant expansion in the four *Brassica* crops. Moreover, phylogenetic and expression pattern analyses of the glucosinolate-related genes *HAG1*, *MAM*, *AOP,* and *GTR* correspond with the glucosinolate profiles and total seed glucosinolate contents in *B. napus* and its parental species. These results, together with those published previously, provide a valuable resource for understanding the genetic mechanism underlying glucosinolate metabolism and transport and suggest novel approaches for improving the nutritional quality of *Brassica* crops through breeding cultivars with lower glucosinolate contents.

12.1 Introduction

Glucosinolates (GSLs) are a group of sulfur-rich, nitrogen-containing plant secondary metabolites mainly found in the Brassicales order (Fahey et al. 2001), which includes many economically and nutritionally important crops and condiments, such as oilseed rape (*Brassica napus*), broccoli (*Brassica oleracea* var. *italica*), cabbage (*B. oleracea* var. *capitata*), turnip (*Brassica rapa*), mustard (*Brassica juncea* L. *Czern*), and wasabi (*Wasabia japonica*), as well as the model

K. Lu · J. Li (✉)
Chongqing Rapeseed Engineering Research Center, College of Agronomy and Biotechnology, Academy of Agricultural Sciences, Southwest University, Beibei, Chongqing 400715, PR China
e-mail: ljn1950@swu.edu.cn

R. Snowdon
Department of Plant Breeding, Justus Liebig University, Giessen, Germany

S. Liu et al. (eds.), *The Brassica napus Genome*, Compendium of Plant Genomes,
https://doi.org/10.1007/978-3-319-43694-4_12

plant *Arabidopsis thaliana*. In addition, GSLs have also been identified in the genus *Drypetes* of the family *Euphorbiaceae* (Fahey et al. 2001). GSLs share an identical core structure containing a *β*-thioglucose group linked to a sulfonated aldoxime moiety, plus a variable aglycone side chain (R) derived from one of eight amino acids (Halkier and Du 1997). Based on the amino acid precursors and the type of modification to the R group, GSLs can be divided into three major classes: aliphatic, indole, and aromatic GSLs (Halkier and Gershenzon 2006). Aliphatic GSLs are derived from alanine, leucine, isoleucine, valine, and methionine, while indole GSLs and aromatic GSLs are derived from tryptophan and phenylalanine or tyrosine, respectively. By 2000, at least 120 different GSLs were reported in 16 families of the order *Capparales*, and the Brassicaceae family alone was found to contain at least 96 of these. A more recent review elucidated and documented the discovery of additional natural GSL structures, citing around 132 unique GSLs from nature (Agerbirk and Olsen 2012). The structural diversity of these compounds is mainly caused by extensive modification of the variable side chain by elongation of the amino acid precursors and from a wide variety of side-chain modifications, including hydroxylation, oxidation, methylation, glucosylation, desaturation, and sulfation (Halkier and Gershenzon 2006).

Though most GSLs are not bioactive in their intact form, they are rapidly hydrolyzed by an endogenous family of plant enzymes called myrosinases (thioglucoside glucohydrolases (TGGs); EC 3.2.1.147), *β*-glucosidases that are compartmentalized in the vacuoles of myrosin cells, a location separate from that of GSLs. Once plant tissues are damaged by wounding, herbivore or pathogen attack, freezing, or grazing (Bones and Rossiter 2006; Fahey et al. 2001), the myrosinases are mixed with GSLs, resulting in hydrolysis of the thioglycoside bond to yield glucose and an unstable aglucone. The latter compound is either spontaneously rearranged into bioactive isothiocyanate or is converted into alternative hydrolysis products such as simple nitriles, epithionitriles, or organic thiocyanates (Wittstock and Burow 2010). The types of breakdown products of the GSL–myrosinase system depend mainly on the chemical nature of the side chain of the parent GSL, the reaction conditions, and the cofactors that are present (Fahey et al. 2001; Halkier and Gershenzon 2006).

GSLs and their degradation products have been recognized for their roles in plant defense and their distinctive effects on human health and on the flavor of cruciferous vegetables. Glucoraphanin (4-methylsulfinylbutyl, GRA), which is known to reduce the risk of aggressive prostate cancer (Halkier and Gershenzon 2006), is the most widely studied GSL. Despite the importance of certain GSLs and their metabolites to human health, most GSLs are also undesirable substances in *Brassica* crops for animal feed, due to the deleterious effects of their breakdown products on animal growth and reproductive performance. To reduce the levels of GSLs in *Brassica* crops, oilseed rape breeders have devoted much effort to developing genetically improved varieties with lower amounts of GSLs. Significant progress has been made toward this goal through classical breeding approaches, and several varieties with low levels of seed GSLs (less than 30 μmol/g in defatted meal) and erucic acid (less than 2% of the total fatty acids present in the oil) have been released in Canada and marketed under the name "canola." While processed canola meal has been widely accepted in the feed industry as a high-quality feedstuff for livestock and poultry, a number of reports have documented reduced performance in farm animals fed diets containing significant amounts of canola meal (Khajali and Slominski 2012; Leeson et al. 1987).

The completion of genome sequencing of *B. napus* and its parental species *B. rapa* and *B. oleracea* provided the *Brassica* scientific community with a valuable tool for further improving seed quality through regulating and controlling secondary metabolism pathways (Chalhoub et al. 2014; Liu et al. 2014; Parkin et al. 2014; Wang et al. 2011b). In *Arabidopsis*, the close relative of *Brassica* crops, most genes responsible for GSL biosynthesis, breakdown, and transport have been characterized using biochemical and reverse genetics approaches (Halkier and Gershenzon 2006). Based on this research in *Arabidopsis*, orthologous genes involved in GSL metabolism and transport in *Brassica* crops have been identified, allowing for the manipulation of these genes and the development of *Brassica* vegetables with high levels of anticancer GSLs and *B. napus* varieties containing much lower levels of undesirable GSLs.

This chapter presents an overview of the genes responsible for GSL biosynthesis, transport, and breakdown in *Brassica* crops, with special emphasis on elucidating the evolutionary processes that resulted in the variation in GSL profiles of *Brassica* crops. Based on this information, we present perspectives for further research aimed at modifying and reducing different kinds of GSLs in *B. napus*.

12.2 Variation in GSL Composition and Content in *Brassica* Crops

The chemical composition of many Brassicaceae genera has been studied, with a focus on identifying variations in oil content and seed fatty acid and GSL composition (Warwick 2011). Comparative studies of GSL profiles indicate that the type of GSLs present and their concentrations vary considerably between species in the Brassicaceae family, as well as between cultivars of the same species, and within different organs or developmental stages of the same plant (Daxenbichler et al. 1991; Fahey et al. 2001). In addition, the total GSL content and the relative proportion of individual GSLs are also influenced by the genotype and by agronomic and environmental factors (such as growth stage, harvest time, soil moisture, and temperature) (Gu et al. 2012; Padilla et al. 2007; Wang et al. 2012; Yang and Quiros 2010).

Numerous studies have described the GSL contents and composition in representative *Brassica* species, and these data have been compiled in several reviews (Daxenbichler et al. 1991; Fahey et al. 2001; Ishida et al. 2014; Jeffery et al. 2003; Padilla et al. 2007). As many as 20 kinds of GSLs have been identified in commercial *Brassica* crops (Table 12.1), which possess substantially different GSL profiles, and usually only 3 or 4 predominant kinds of GSLs occur in the same plant (Rosa 1997). Comparison of the GSL profiles and concentrations in

Table 12.1 Major GSLs present in *Brassica* crops

GSL name	Trivial name	Systematic name	Abbreviation
Aliphatic 3C	Sinigrin	2-propenyl	SIN
	Glucoiberverin	3-methylthiopropyl	GIV
	Glucoiberin	3-methylsulfinylpropyl	GIB
Aliphatic 4C	Progoitrin	(2R)-2-hydroxy-3-butenyl	PRO
	Gluconapin	3-butenyl	NAP
	Glucoerucin	4-methylthiobutyl	GER
	Glucoraphanin	4-methylsulphinylbutyl	GRA
Aliphatic 5C	Gluconapoleiferin	2-hydroxy-4-pentenyl	GNL
	Glucobrassicanapin	4-pentenyl	GBN
	Glucoberteroin	5-methylthiopentyl	GBE
	Glucoalyssin	5-methylsulphinylpentyl	GAL
Indole	Glucobrassicin	3-indolylmethyl	GBS
	Neoglucobrassicin	1-methoxy-3-indolylmethyl	NGBS
	4-Hydroxyglucobrassicin	4-hydroxy-3-indolylmethyl	4HGBS
	4-Methoxyglucobrassicin	4-methoxy-3-indolylmethyl	4MGBS
Aromatic	Gluconasturtiin	2-phenylethyl	GST

different tissues during different growth stages from four *Brassica* crops of the "triangle of U" (*Brassica carinata*, *Brassica nigra*, *B. juncea*, and *B. rapa*) revealed that sinigrin (2-propenyl, SIN) is the dominant GSL in three mustards (*B. carinata*, *B. nigra*, and *B. juncea*) (Table 12.2), where it represents more than 90% of the total GSL concentration in ripe seeds and over 50% of the total GSL concentration in green tissues (Bellostas et al. 2007). *B. carinata* contains other GSLs, including gluconapin (3-butenyl, NAP), 4-hydroxyglucobrassicin (4-hydroxyindol-3-ylmethyl, 4HGBS), gluconasturtiin (2-phenylethyl, GST), and progoitrin (2-hydroxy-3-butenyl, PRO), the last of which is ultimately decomposed into oxazolidine-2-thiones, which are considered to be goitrogenic compounds in monogastric animals (Bellostas et al. 2007; Fahey et al. 2001). The GSL profile of *B. rapa* is quite distinct from that of the aforementioned three mustards (Table 12.2). In *B. rapa*, 16 GSLs have been identified. Among these, the aliphatic GSLs, NAP, and glucobrassicanapin (4-pentenyl, GBN), and their hydroxylated forms, PRO and gluconapoleiferin (2-hydroxy-4-pentenyl, GNL), were found to be the most abundant, while the concentrations of indolic and aromatic GSLs were low and showed the fewest differences among the different varieties (Cartea and Velasco 2008). Most *B. rapa* varieties had a proportion of NAP of between 70 and 95% of the total GSL content and a proportion of GBN of below 20% of the total GSL content, while other minor GSLs, such as glucoiberin (3-methylsulfinilpropyl, GIV), PRO, glucoalyssin (5-methylsulphinylpentyl, GAL), and GST, accounted for less than 20% of the total GSL content (Padilla et al. 2007).

Diversity in the concentration and type of GSLs is much higher in *B. oleracea* than in *B. rapa* species (Ishida et al. 2014). All *B. oleracea* types and cultivars contain high concentrations of glucobrassicin (3-indolymethyl, GBS) and GIV and most contain substantial amounts of SIN. For example, SIN accounts for most of the GSLs in kale (*B. oleracea* var. *acephala*), while GBS and GIB account for most of those in cabbage (*B. oleracea* var. *capitata*) leaves (Cartea et al. 2008). The most common GSLs found in broccoli (*B. oleracea* var. *italica*) are GRA, SIN, PRO, NAP, and the indole GSLs GBS and neoglucobrassicin (1-methoxy-3-indolylmethyl, NGBS) (Kushad et al. 1999). The predominant GSL GRA (accounting for more than 50% of the total GSLs and the precursor of sulforaphane) is the most important health-promoting compound in broccoli, but only trace amounts of GRA are present in most *B. rapa*, *B. napus*, and *B. juncea* vegetables and oilseeds (Liu et al. 2012; Tian et al. 2005). GRA was not detected in several *B. oleracea* crops, including cabbage, Brussels sprouts (*B. oleracea* var. *gemmifera*), and cauliflower (*B. oleracea* var. *botrytis*). In cauliflower, SIN and GIB are the major aliphatic GSLs present (together occurring at a concentration of 0.42 μmol/g FW), and GBS (1.5 μmol/g FW) and 4-methoxyglucobrassicin (4-methoxy-3-indolylmethyl, 4MGBS, 0.4 μmol/g FW) are the major indole GSLs (Tian et al. 2005). Broccoli sprouts and Brussels sprouts contain higher amounts of total GSLs than do broccoli and cauliflower. The major GSLs detected in broccoli sprouts are 4MBGS, GRA, GER, and GIB (0.385, 1.33, 1.02, and 0.599 μmol/g FW, respectively; Tian et al. 2005; West et al. 2002). GBS (3.74 μmol/g FW) is the most abundant GSL in Brussels sprouts, while the concentration of SIN, PRO, and NAP (1.55, 1.33, and 1.08 μmol/g FW, respectively) is also relatively higher than that of other GSLs (Tian et al. 2005). In Chinese kale (*B. oleracea* var. *alboglabra*), the total and individual GSL contents varied extensively among the different edible parts, and NAP was the most abundant GSL in the edible plant parts (Sun et al. 2011).

Due to their toxic and antinutritive effect on animals, GSLs have long since been regarded as unfavorable components of *B. napus* seeds. Hence, developing a double-low *B. napus* variety with seeds lacking erucic acid and containing only low levels of GSL has been an important objective of rapeseed breeding programs, and much research examining variation in GSL composition and content in *B. napus* has been conducted (Font et al. 2005; Sang et al. 1984).

Table 12.2 Distribution of GSLs among the six *Brassica* crops in the triangle of U

Crop	Aliphatic										Indole				Aromatic
	3-carbon			4-carbon				5-carbon							
	GIV	GIB	SIN	GER	GRA	NAP	PRO	GBN	GAL	GNL	GBS	NGBS	4HGBS	4MGBS	GST
Brassica carinata[a]	–	–	*	–	–	+	+	–	–	–	+	+	+	+	+
Brassica junce [b]	–	–	*	–	–	*	+	–	–	–	+	+	+	+	+
Brassica napus															
Swede[c]	–	–	–	–	+	–	+	+	–	+	+	*	+	*	+
Leaf rape[c]	+	–	–	–	–	+	+	*	+	+	+	+	–	+	+
High GSL variety[d]		–	–	–	+	*	*	+	+	+	+	+	+	+	+
Low-GSL variety[d]	–	–	–	–	+	+	+	+	+	+	+	+	*	+	+
Brassica nigra[a]	–	–	*	–	–	+	–	–	–	–	+	–	+	+	+
Brassica oleracea[c]															
White cabbage	+	*	*	+	+	+	+	+	+	–	*	+	+	+	+
Savoy cabbage	+	*	*	–	+	+	+	+	–	–	*	+	–	+	+
Red cabbage	+	*	*	–	+	+	+	–	–	–	*	+	–	–	–
Kale	+	*	*	–	+	+	+	–	–	–	*	+	+	+	+
Collard	+	+	*	+	–	–	*	–	–	–	*	–	–	–	–
Tronchuda cabbage	+	*	*	–	+	+	+	+	+	–	*	+	+	+	+
Broccoli	–	+	+	+	*	*	*	+	+	+	*	+	+	+	+
Brussels sprouts	+	+	*	–	+	+	*	–	–	–	*	+	–	–	–
Cauliflower	+	*	*	–	+		*	–	–	–	*	+	–	–	–
Kohlrabi	+	+	*	–	+	+	+	–	+	–	*	+	+	+	–
Brassica rapa[b, c]															
Turnip	–	+	–	+	–	*	*	*	–	+	+	+	+	+	*
Turnip greens	+	+	–	+	+	*	+	*	+	+	+	+	+	–	+
Turnip tops	+	+	–	–	–	*	+	*	–	–	+	+	+	–	+
Chinese cabbage	–	+	–	–	–	*	+	*	–	*	+	+	–	+	+
Bok choy	–	–	–	–	–	*	*	+	–	+	+	+	+	–	+

GIV Glucoiberverin (3-methylthiopropyl); *GIB* glucoiberin (3-methylsulfinylpropyl); *SIN* sinigrin (2-propenyl); *GER* glucoerucin (4-methylthiobutyl); *GRA* glucoraphanin (4-methylsulphinylbutyl); *NAP* gluconapin (3-butenyl); *PRO* progoitrin (2-hydroxy-3-butenyl); *GBN* glucobrassicanapin (4-pentenyl); *GAL* glucoalyssin (5-methylsulphinylpentyl); *GNL* gluconapoleiferin (2-hydroxy-4-pentenyl); GBS: glucobrassicin (3-indolylmethyl); *NGBS* neoglucobrassicin (1-methoxy-3-indolylmethyl); *4HGBS* 4-hydroxyglucobrassicin (4-hydroxy-3-indolylmethyl); *4MGBS* 4-methoxyglucobrassicin (4-methoxy-3-indolylmethyl), *GST* Gluconasturtiin (2-phenylethyl). Major GSLs found in each crop are indicated with the * symbol (data sources: [a]Ishida et al. Breed Sci. 2014, 64: 48–59. [b]Yang et al. BMC Genomics 2014, 15:107. [c]Cartea and Velasco Phytochem Rev. 2008, 7:213–229. [d]Li et al. Sci Agric Sin 2005, 38: 1346–1352)

Based on the GSL content, the seeds of 499 *B. napus* accessions were divided into three types, containing high, medium, and low levels of GSLs, and the GSL components of each of these types were systemically analyzed by high-performance liquid chromatography (Li et al. 2005). In *B. napus* varieties containing high and medium levels of GSLs, but not in those containing low levels, the dominant and stable components are PRO and NAP. Although GST and 4HGBS are minor components of *B. napus* varieties containing high levels of GSL, they are major components of varieties containing low GSL levels (Li et al. 2005). Accurately measuring the GSL profiles and identifying the corresponding GSL biosynthetic, breakdown, and transport genes in different *Brassica* crops are of great importance for further improving the GSL profiles in given tissues and organs. For instance, ideal *B. napus* varieties would have high levels of GSLs in the vegetative tissues, but lack GSLs in the seeds.

12.3 Genes Involved in the Metabolism, Transport, and Regulation of GSLs

Substantial advances have recently been made in our understanding of the metabolism and regulation of GSLs in plants, particularly in *Arabidopsis*, where structural and regulatory genes involved in GSL biosynthesis, transport, and degradation pathways have been identified through in vitro biochemical assays and mutant studies (Burow et al. 2010; Radojčić Redovniković et al. 2008; Sønderby et al. 2010).

12.3.1 GSL Biosynthetic Genes

GSL biosynthesis is comprised of three independent stages: (i) amino acid chain elongation, in which additional methylene groups are inserted into the side chain of certain aliphatic and aromatic amino acids, (ii) conversion of the amino acid moiety to form the core structure of GSLs, and (iii) subsequent secondary modifications of side chains to generate chemical diversity (Grubb and Abel 2006; Halkier and Gershenzon 2006). Methionine undergoes a series of chain elongation cycles in which one methylene group is added per time prior to entering the core structure pathway. These chain elongation reactions include deamination by a branched-chain amino acid aminotransferase (BCAT), condensation with acetyl-CoA by a methylthioalkylmalate synthase (MAM), isomerization by an isopropylmalate isomerase (IPMI), and oxidative decarboxylation by an isopropylmalate dehydrogenase (IPM-DH) (Sønderby et al. 2010). The newly formed 2-oxo acid can either be transformed into the corresponding methionine derivative and enter the core GSL structure pathway or undergo another round of chain elongation (Radojčić Redovniković et al. 2008). In *A. thaliana*, three tandemly duplicated and functionally diverse MAM members were identified as being responsible for the condensation step of the chain elongation. Functional analysis demonstrated that AtMAM2 (absent in ecotype Columbia) and AtMAM1 catalyze the condensation reaction of the first and the first two elongation cycles, respectively, for the synthesis of aliphatic GSLs with short carbon chains (3C and 4C, respectively) (Benderoth et al. 2006; Kroymann et al. 2003; Textor et al. 2004), while AtMAM3 catalyzes all six additions of methylene groups and the formation of all aliphatic GSLs, especially long-chain GSLs (6C, 7C, and 8C) (Textor et al. 2007). Hence, the number and expression patterns of *MAM* genes in a plant determine variations in aliphatic GSLs during the earliest stages of GSL biosynthesis and have a fundamental impact on GSL composition and diversity in plant tissues.

The GSL core structure is formed from precursor amino acids via a series of reactions catalyzed by various cytochrome P450 (CYP) monooxygenases (Halkier and Gershenzon 2006). Briefly, the five characterized CYP79 homologs in *Arabidopsis* catalyze the conversion of amino acids to their corresponding aldoximes. *CYP79F1* and *CYP79F2* encode the enzymes that catalyze aldoxime production in the biosynthesis of the major GSLs derived from chain-elongated methionine derivatives. *CYP79B2* and *CYP79B3* catalyze the biosynthesis of indole-3-acetaldoxime from tryptophan, whereas *CYP79A2* converts phenylalanine to phenylacetaldoxime, the precursor of benzyl GSL (Radojčić Redovniković et al. 2008). Biochemical studies identified differences in the substrate specificity of CYP79F1 and CYP79F2, showing that CYP79F1 metabolizes homomethionine and di-, tri-, tetra-, penta-, and hexahomomethionines, resulting in both short- and long-chain methionine derivatives, whereas CYP79F2 only catalyzes the production of long-chain penta- and hexahomomethionines (Chen et al. 2003; Radojčić Redovniković et al. 2008). The aldoximes are further metabolized to form *S*-alkylthiohydroximates by CYP83A1 and CYP83B1, cytochrome P450 of the CYP83 family (Bak and Feyereisen 2001). Both biochemical and transgenic lines of evidence show that CYP83A1 mainly metabolizes the aliphatic aldoximes to form aliphatic GSLs, whereas CYP83B1 mostly metabolizes indole-3-acetaldoxime and aromatic oximes to synthesize the corresponding substrates for indolic and aromatic GSLs, respectively (Bak and Feyereisen 2001; Naur et al. 2003). In a subsequent step, the resulting *S*-alkylthiohydroximates are cleaved to yield thihydroximates by a *C-S* lyase SUR1 (Mikkelsen et al. 2004). The second to last step in the formation of GSLs is the *S*-glycosylation of thihydroximates, a reaction that is catalyzed by glucosyltransferases of the UGT74 family. This reaction appears to be unique and catalyzes the formation of an *S*-glycosidic bond between glucose and the acceptor thiohydroximate, leading to the production of the corresponding desulfo-GSL (Grubb et al. 2004). The results of biochemical and genetic analyses demonstrated that UGT74C1 plays a key role in the biosynthesis of aliphatic GSLs and that UGT74B1 catalyzes the formation of aromatic GSLs (Grubb et al. 2004, 2014). Three sulfotransferase (SOT) proteins perform the final step of GSL biosynthesis. Biochemical characterization showed that SOT16 metabolizes tryptophan- and phenylalanine-derived desulfo-GSLs, whereas SOT17 and SOT18 metabolize long-chained aliphatic desulfo-GSLs (Piotrowski et al. 2004).

After parent GSL formation, a wide range of further modifications can occur on the methionine side chain and occasionally on the glucose moiety (Mikkelsen et al. 2002; Neal et al. 2010), giving rise to an enormous variety of GSL structures. These secondary modifications, which take place in an organ- and developmental stage-specific manner (Radojčić Redovniković et al. 2008; Sønderby et al. 2010), are particularly important as the structure of the side chain largely determines the nature of the products formed following GSL hydrolysis by myrosinases (Sønderby et al. 2010; Wittstock and Halkier 2002). For aliphatic GSLs, these modifications include oxidations, hydroxylations, alkenylations, and benzoylations, while for indole GSLs, they include hydroxylations and methoxylations.

The *S*-oxygenation of aliphatic GSLs is a common modification catalyzed by five flavin-monooxygenases, designated $FMO_{GS\text{-}OX1}$ to $FMO_{GS\text{-}OX5}$ (Li et al. 2008). $FMO_{GS\text{-}OX5}$ shows substrate specificity for the long-chain 8-methylthiooctyl GSLs (8MTOs), whereas $FMO_{GS\text{-}OX1}$ to $FMO_{GS\text{-}OX4}$ exhibit broad chain length specificity and catalyze the conversion from methylthioalkyl (MT) GSL to the corresponding methylsulfinylalkyl (MS) independently of chain length (Li et al. 2008), resulting in the production of the potent cancer-preventive substances sulforaphane (4-methylsulfinylbutyl isothiocyanate, 4MSB ITC), which is derived from GRA, and the 7-methylsulfinylheptyl (7MSOH) and 8-methylsulfinyloctyl (8MSOO) isothiocyanates, derived from 7-methylthioheptyl GSL (7MTH) and 8MTO, respectively (Li et al. 2008). Hence, the five $FMO_{GS\text{-}OX}$ genes could potentially be used in genetic engineering strategies to optimize the GSL profiles of *Brassica* crops. Substantial variation in *Arabidopsis* GSL profiles between different genotypes has expedited the identification of the GS-AOP locus, which encodes the two tandemly duplicated 2-oxoglutarate-dependent dioxygenases, AOP2 and AOP3 (Kliebenstein et al. 2001). AOP2 directly catalyzes the conversion of methylsulfinylalkyl GSLs to the alkenyl GSLs NAP or GBN (n = 2–3), and the GS-OH locus can further convert NAP to PRO (Hansen et al. 2008). AOP3 controls the production of hydroxyalkyl GSLs ($n = 2$) from methylsulfinylalkyl GSLs. When both AOPs are non-functional, the plant accumulates the precursor methylsulfinyl alkyl GSLs (Liu et al. 2014). Secondary modifications of indole GSLs mainly include hydroxylation by CYP81F2, which is essential for the 4-hydroxylation of unmodified indolyl-3-methyl (I3M), and catalyzes the formation of 4-hydroxy I3M (4OH-I3M) and 4-methoxy I3M (4M-I3M) from I3M (Bednarek et al. 2009; Pfalz et al. 2009; Sønderby et al. 2010).

12.3.2 Regulatory Genes of GSL Biosynthesis

Biosynthesis of GSLs is tightly regulated by six R2R3-MYB transcription factors (TFs) belonging to subgroup 12 of the R2R3 MYB family, which has a conserved "[L/F]LN[K/R]VA" motif (Dubos et al. 2010). In *Arabidopsis*, *MYB28*, *MYB29*, and *MYB76* positively regulate the biosynthesis of aliphatic GSLs with partial functional redundancy (Hirai et al. 2007). During aliphatic GSL biosynthesis, *AtMYB28* acts as the major positive regulator and *AtMYB29* as an accessory factor in the response to methyl jasmonate signaling in the *trans*-activation of the aforementioned aliphatic GSL biosynthetic genes, i.e., *AtMAM1*, *AtMAM3*, *AtCYP79F1*, *AtCYP79F2*, *AtCYP83A1*, *AtAOP2*, *AtSOT17*, and *AtSOT18* (Gigolashvili et al. 2008a; Hirai et al. 2007). *Arabidopsis* mutants defective in MYB28 function had decreased amounts of both long- and short-chain aliphatic GSLs, whereas the *myb29* or *myb76* mutant contained significantly reduced levels of short-chained aliphatic GSLs, indicating that MYB28 regulates the biosynthesis of all methylsulfinyl GSLs, whereas MYB29 and MYB76 regulate the biosynthesis of short-chained GSLs (Gigolashvili et al. 2008b). The total aliphatic GSLs but not indolic GSLs were significantly increased in the leaves of plants overexpressing *AtMYB28*, *AtMYB29*, or *AtMYB76* (Gigolashvili et al. 2008b; Hirai et al. 2007). Overexpression of both *AtMYB28* and *AtMYB29* significantly repressed the expression of the indolic GSL pathway genes, indicating that a reciprocal antagonistic relationship exists between the aliphatic and indolic GSL biosynthetic pathways (Gigolashvili et al. 2008a).

Conversely, *AtMYB34*, *AtMYB51*, and *AtMYB122*, which were identified as important regulators of the indolic GSL biosynthetic pathway, significantly reduced the transcript levels of *AtCYP79B2*, *AtCYP79B3*, *AtCYP83B1*, *AtUTG74B1*, *AtSOT16,* and *3'-phosphoadenosine 5'-phosphosulphate transporter* (*PAPST1*) genes, which are involved in the indolic GSL biosynthetic pathway (Frerigmann and Gigolashvili 2014; Guo et al. 2013; Sønderby et al. 2010). The three *MYB* transcription factors exhibit both additive and epistatic interactions in the regulation of indolic GSL biosynthesis (Frerigmann and Gigolashvili 2014). Lines lacking the two main regulators of indolic GSL biosynthesis, *MYB34* and *MYB51*, exhibit a significant reduction in total indolic GSLs, demonstrating the importance of these two genes for indolic GSL biosynthesis. Previous research also showed that *MYB34* and *MYB51* have distinct roles in indolic GSL production, functioning in different tissues or under different environmental conditions. *MYB51* is the central regulator of indolic GSL biosynthesis in shoots and is activated by salicylic acid (SA) and ethylene (ET) treatments. By contrast, *MYB34* regulates indolic GSL biosynthesis mainly in the roots and functions in abscisic acid (ABA) and methyl jasmonate (MeJA) signaling. Interestingly, *MYB51* appears to regulate indolic GSL biosynthesis in roots in the *myb34* mutant. *MYB122* only plays an accessory role in indolic GSL biosynthesis and in JA/ET-induced GSL biosynthesis (Frerigmann and Gigolashvili 2014).

In addition to the MYB transcription factors, some other regulators of GSL biosynthesis have also been characterized. *Arabidopsis* CaM-binding protein IQ-DOMAIN1 (IQD1) binds calmodulin in a Ca^{2+}-dependent manner and is a positive regulator of total GSL accumulation during biotic stress responses, with a gain-of-function *IQD1* mutation resulting in elevated levels of both indole and aliphatic GSLs and a reduction in insect herbivory and infestation (Laluk et al. 2012; Levy et al. 2005). Another CaM-binding transcription factor SIGNAL RESPONSIVE1 (AtSR1) also proved to be a key regulator of GSL levels through transcriptional regulation of several genes involved in GSL metabolism, including *AtIQD1*, *AtMYB51*, and *AtSOT16*, and is a negative regulator for herbivory tolerance in *Arabidopsis* (Laluk et al. 2012). *AtSLIM1* was identified as a central transcription factor that negatively regulates both aliphatic and indolic biosynthesis under sulfur-limiting conditions and downregulates *AtMBY34* transcription (Maruyama-Nakashita et al. 2006). Another characterized regulator of GLS biosynthesis is DNA-binding-with-one-finger (DOF) transcription factor *AtDof1.1* (also known as *AtOBP2*), which is induced by wounding and herbivore attack and MeJA treatment, and specifically upregulates *CYP83B1* expression and promotes indolic GSL accumulation (Skirycz et al. 2006). Although *AtDof1.1* does not seem to regulate the expression of *CYP79F1* and *CYP79F2*, the aliphatic GSL content was altered in *AtDof1.1* overexpression lines (Skirycz et al. 2006). Loss-of-function mutations of *Arabidopsis TERMINAL FLOWER2* (*TFL2*, also known as *LHP1* or *TU8*) significantly increased the abundance of four long-chain aliphatic GSLs in the seeds, whereas indolyl-3-methyl GSL levels were significantly reduced relative to the wild type, leading to a reduction in symptoms resulting from infection by the obligate biotrophic fungus *Plasmodiophora brassicae*, which causes clubroot disease, a damaging disease in Brassicaceae (Kim et al. 2004; Le Roux et al. 2014). In addition, TFL2 regulates heterochromatin formation and represses the expression of genes involved in flowering time, floral organ identity, meiosis, and seed maturation (Nakahigashi et al. 2005).

12.3.3 GSL Transport Genes

The GSLs are believed to be synthesized mainly in rosette leaves and silique walls and then to be relocated to embryos through phloem by specific transporters (Lu et al. 2014). In *Arabidopsis*, GSLs have successfully been eliminated from the seeds by silencing two recently identified nitrate/peptide transporter family members, *GTR1* and *GTR2*, which suggests that manipulation of these two transporters may increase the nutritional value of crops and be used in biotechnological approaches to control the allocation of GSLs to seeds in *Brassica* crops (Nour-Eldin et al. 2012). The *gtr2* single mutant exhibited a significant reduction in total GSL levels in seeds and a threefold increase in aliphatic GSLs in source tissues (i.e., senescent leaves and silique walls), but no significant changes in GSL content in the seeds (Jorgensen et al. 2015; Nour-Eldin et al. 2012). In the *gtr1 gtr2* double mutant, aliphatic and indolic GSLs were absent in the seeds, but exhibited a more than tenfold increase in source tissues, demonstrating that both plasma membrane-localized transporters are essential for long-distance GSL transport to the seeds and are responsible for loading GSLs from the apoplasm into the phloem, and finally for determining the tissue-specific distribution of GSLs in plants (Nour-Eldin et al. 2012). Identifying these two GSL transporters provides a strategy for breeding *Brassica* varieties that contain extremely low levels of total GSLs in the seeds but high levels in the green tissues by reducing functional GTR activity and blocking the translocation of GSLs.

12.3.4 GSL Breakdown Genes

Numerous studies to date have focused on the beneficial effects of GSLs and their breakdown products on human health and plant defense, and on their negative effects on animal nutrition. In the well-studied GSL–myrosinase-specifier protein system, myrosinases hydrolyse GSLs in the presence of water, producing a series of degradation products (Wittstock and Burow 2010). The types of products of myrosinase hydrolysis depend on the structure of the parent GSLs, reaction conditions, and availability of epithiospecifier proteins (ESPs) and nitrile-specifier proteins (NSPs) (Kissen and Bones 2009).

In *Arabidopsis*, six genes (*TGG1-TGG6*) encoding classical myrosinases have been identified on two chromosomes (Xu et al. 2004). Among these genes, *TGG1* and *TGG2* were tandem duplicates of *TGG3*, while *TGG5* and *TGG6* were tandem duplicates of *TGG4*. These duplicated genes share the same gene structure as their parent genes. Although *TGG3* and *TGG6* are predominantly expressed in specific tissues (Xu et al. 2004), both are probably pseudogenes that encode non-functional proteins due to multiple frameshift mutations (Wang et al. 2009). *TGG1* is expressed in myrosin cells, stomatal guard cells, and phloem cells of all the aboveground organs except the seeds (Barth and Jander 2006; Xue et al. 1995). Similar to *TGG1*, *TGG2* is also highly expressed in the aboveground tissues (Xu et al. 2004), but is much less abundant in the rosette leaves than is *TGG1*, and was not detected in guard cells (Zhao et al. 2008). *TGG4* and *TGG5* are primarily expressed in the roots. Despite the distinct expression patterns and the difference in vitro myrosinase activities of TGG1 and TGG2, GSL breakdown in the crushed leaves of *TGG1* or *TGG2* single mutants is basically unchanged, indicating that the two myrosinases may have redundant functions (Barth and Jander 2006). Leaf extracts of *TGG1 TGG2* double mutants had no detectable in vitro myrosinase activity on exogenously applied aliphatic GSLs, and endogenous aliphatic GSLs were no longer broken down in disrupted leaf material of the double mutant (Barth and Jander 2006). However, myrosinase-independent breakdown of indolic GSLs still slowly proceeds, indicating the presence of a breakdown pathway for these GSLs that is independent of TGG1 and TGG2.

Several specifier proteins, such as ESPs and NSPs, myrosinase-associated proteins (MyAPs), such as EPITHIOSPECIFIER-MODIFIER1 (ESM1), MODIFIED VACUOLE PHENOTYPE1 (MVP1), and enzymes involved in further metabolism, such as nitrilases, have been shown to be involved in the generation of diversified GSL metabolic products in *Arabidopsis* (Wittstock and Burow 2010). Specifier proteins do not exhibit hydrolytic activity on GSLs, but affect the outcome of GSL hydrolysis products. In the absence of specifier proteins, ITCs are typically formed at neutral pH (Bones and Rossiter 2006). ESPs and the related thiocyanate-forming proteins (TFPs) catalyze the formation of epithionitrile, in the presence of GSLs with terminal double bonds in the side chain and ferrous ions, while the formation of thiocyanate purely depends on TFPs (Wittstock and Burow 2010). NSPs are involved in simple nitrile formation at acidic pH values, but do not catalyze epithionitrile or thiocyanate formation. The simple nitrile can be further converted by nitrilases (NITs) to a carboxylic acid in the presence of a specifier protein (Vorwerk et al. 2001; Wittstock and Burow 2010). ESP function is inhibited by ESM1, leading to decreased simple nitrile formation and increased ITC production for benzyl and alkyl GSLs, but not for alkenyl GSLs (Zhang et al. 2006). Cloning and sequence analysis of *ESM1* revealed that it encodes a putative endoplasmic reticulum (ER) binding protein and that allelic variation in this gene contributes to the variation in GSL breakdown among different *Arabidopsis* accessions (Zhang et al. 2006). *MVP1* is expressed ubiquitously and encodes another MyAP-like protein that is closely related to ESM1. The *mvp1* mutant is impaired in endomembrane protein trafficking and shows a significant increase in simple nitrile production from allyl GSLs (Agee et al. 2010). Interestingly, MVP1 interacts with TGG2 and the PYK10 complex, but not with TGG1 in vitro, suggesting that MVP1 functions in the quality control of GSL hydrolysis by contributing to the proper tonoplast localization of TGG2 and in ER body-related defense systems by regulating the PYK10 complex (Agee et al. 2010; Nakano et al. 2012). An atypical myrosinase gene, *PEN2*, which may be limited to indole GSL hydrolysis and is required for pathogen resistance, was recently identified in *Arabidopsis* (Bednarek and Osbourn 2009).

12.4 Evolution of GSL-Related Genes in *B. Napus* and Its Parental Species

12.4.1 Identification of GSL-Related Genes from *B. Napus* and Its Parental Species

To identify GSL-related genes from *B. napus* and its parental species *B. rapa* and *B. oleracea*, we used the sequences of 58 GSL biosynthesis, 3 GSL transport, and 17 GSL breakdown genes characterized in *A. thaliana* as queries against the four publicly available genomes of *Brassica* crops based on a combination of syntenic and nonsyntenic homology analyses (Table 12.3). We identified 119, 119, 134, and 240 GSL biosynthetic genes in *B. rapa*, *B. oleracea* var. *capitata*, *B. oleracea* var. *italica*, and *B. napus* (both 120 genes in A and C subgenomes), respectively (Fig. 12.1). The fact that more GSL biosynthetic genes were identified in *B. oleracea* var. *italica* than in the other three *Brassica* crops is mainly a consequence of the expansion of genes responsible for core structure formation and side-chain modification. For three *Arabidopsis* GSL transporters, there are 8 orthologs in both *B. rapa* and two subgenomes of *B. napus*, while only 7 and 6 orthologs exist in *B. oleracea* var. *capitata* and *B. oleracea* var. *italica*, respectively. The number of GSL breakdown genes is almost identical among *B. rapa*, *B. oleracea* var. *capitata*, and two subgenomes of *B. napus*, while *B. oleracea* var. *italica* contains many more.

After the split with *Arabidopsis*, the *Brassica* progenitor species experienced a whole-genome triplication (WGT) and subsequently diverged into three diploid *Brassica* species, *B. rapa*, *B. oleracea*, and *B. nigra*. As a young allopolyploid species, *B. napus* was formed from multiple independent hybridization events between ancestors of the diploids *B. rapa* (A genome donor) and *B. oleracea* (C genome donor) (Nagaharu 1935). Hence, we found that most multi-copy genes might have originated from WGT events and that several gene families involved in GSL metabolism

Table 12.3 GSL-related genes in *Arabidopsis* and in *B. napus* and its parental species

Pathway	*Arabidopsis*	*B. rapa*	*B. oleracea* var. *capitata*	*B. oleracea* var. *italica*	A subgenome of *B. napus*	C subgenome of *B. napus*
GSL biosynthesis	58	119	119	134	120	120
Transcription factors	9	21	21	23	20	20
Side-chain elongation	10	18	20	20	20	19
Core structure formation	18	37	38	43	39	37
Side-chain modification	15	28	23	33	26	30
Co-substrate pathways	6	15	17	15	15	14
GSL transport	3	8	7	6	8	8
GSL breakdown	17	39	41	51	38	40
Total	78	166	167	191	165	168

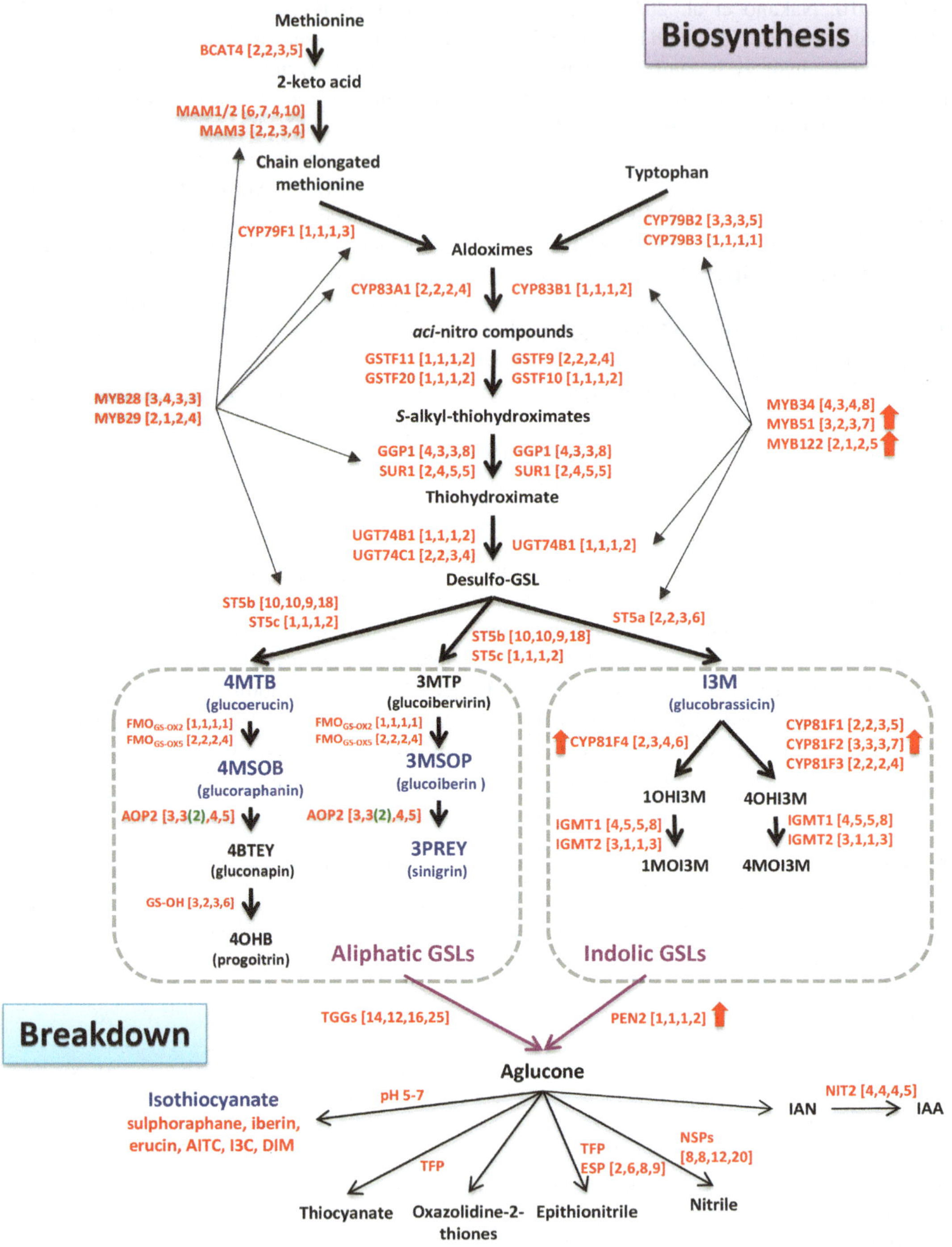
Biosynthesis
Methionine
BCAT4 [2,2,3,5]
2-keto acid
MAM1/2 [6,7,4,10]
MAM3 [2,2,3,4]
Chain elongated methionine
Typtophan
CYP79F1 [1,1,1,3]
CYP79B2 [3,3,3,5]
CYP79B3 [1,1,1,1]
Aldoximes
CYP83A1 [2,2,2,4]
CYP83B1 [1,1,1,2]
aci-nitro compounds
GSTF11 [1,1,1,2]
GSTF20 [1,1,1,2]
GSTF9 [2,2,2,4]
GSTF10 [1,1,1,2]
MYB28 [3,4,3,3]
MYB29 [2,1,2,4]
S-alkyl-thiohydroximates
MYB34 [4,3,4,8]
MYB51 [3,2,3,7]
MYB122 [2,1,2,5
GGP1 [4,3,3,8]
SUR1 [2,4,5,5]
GGP1 [4,3,3,8]
SUR1 [2,4,5,5]
Thiohydroximate
UGT74B1 [1,1,1,2]
UGT74C1 [2,2,3,4]
UGT74B1 [1,1,1,2]
Desulfo-GSL
ST5b [10,10,9,18]
ST5c [1,1,1,2]
ST5a [2,2,3,6]
ST5b [10,10,9,18]
ST5c [1,1,1,2]
4MTB
(glucoerucin)
3MTP
(glucoibervirin)
I3M
(glucobrassicin)
FMOGS-OX2 [1,1,1,1]
FMOGS-OX5 [2,2,2,4]
FMOGS-OX2 [1,1,1,1]
FMOGS-OX5 [2,2,2,4]
CYP81F4 [2,3,4,6]
CYP81F1 [2,2,3,5]
CYP81F2 [3,3,3,7]
CYP81F3 [2,2,2,4]
4MSOB
(glucoraphanin)
3MSOP
(glucoiberin)
1OHI3M
4OHI3M
AOP2 [3,3(2),4,5]
AOP2 [3,3(2),4,5]
IGMT1 [4,5,5,8]
IGMT2 [3,1,1,3]
IGMT1 [4,5,5,8]
IGMT2 [3,1,1,3]
4BTEY
(gluconapin)
3PREY
(sinigrin)
1MOI3M
4MOI3M
GS-OH [3,2,3,6]
4OHB
(progoitrin)
Aliphatic GSLs
Indolic GSLs
Breakdown
TGGs [14,12,16,25]
PEN2 [1,1,1,2]
Aglucone
Isothiocyanate
sulphoraphane, iberin, erucin, AITC, I3C, DIM
pH 5-7
IAN
NIT2 [4,4,4,5]
IAA
TFP
TFP
ESP [2,6,8,9]
NSPs
[8,8,12,20]
Thiocyanate
Oxazolidine-2-thiones
Epithionitrile
Nitrile

◂ **Fig. 12.1** Comparison of aliphatic and indolic glucosinolate biosynthetic and breakdown genes in *A. thaliana*, *B. rapa*, *B. oleracea* var. *capitata*, and *B. napus*. The copy number of GSL biosynthetic genes in *A. thaliana*, *B. rapa*, *B. oleracea* var. *capitata* and *B. napus* is listed in square brackets. Potential anticancer substances/precursors are highlighted in blue bold. The most important transcription factors, amino acid chain elongation and side-chain modification loci MYB28 (HAG1), MAMs, and AOP2, are highlighted in red bold, with the number in parentheses (green) representing the number of non-functional genes. 1MOI3 M: 1-methoxyindol-3-ylmethyl GSL; 1OHI3 M: 1-hydroxyindol-3-ylmethyl GSL; 3 MSOP: 3-methylsulfinylpropyl GSL; 3 MTP: 3-methylthiopropyl GSL; 3PREY: 2-Propenyl GSL; 4BTEY: 3-butenyl GSL; 4BzOB: 4-benzoyloxybutyl GSL; 4MOI3 M: 4-methoxyindol-3-ylmethyl GSL; 4OHB, 4-hydroxybutyl GSL; 4OHI3M: 4-hydroxyindol-3-ylmethyl GSL; 4MSOB: 4-methylsulfinylbutyl GSL; 4MTB, 4-methylthiobutyl GSL; AITC: allyl isothiocyanate; DIM: 3,3'-diindolymethane; ESP: epithiospecifier protein; I3C: indole-3-carbinol; IAA: indole-3-acetaldehyde; IAN: indole-3-acetonitile; I3M: indolyl-3-methyl GSL; NSP: nitrile-specifier protein; TFP: thiocyanate-forming protein; and TGG: thioglucoside glucohydrolase (Figure reprinted, with modifications, from Liu et al. (2014) under a CC BY license (Creative Commons Attribution 4.0 International License))

or transport also experienced homeologous gene loss events after the WGT, leading to the formation of 13 conserved single-copy GSL biosynthesis genes and single copies of GSL transport (*PEN3*) and breakdown (*PEN2*) genes in *B. rapa*, *B. oleracea*, and two subgenomes of *B. napus*. The 78 GSL-related genes present in *Arabidopsis* represent 0.28% of all *Arabidopsis* genes, while the GSL-related genes in *B. rapa*, *B. oleracea* var. *capitata*, *B. oleracea* var. *italica*, and *B. napus* represent 0.40, 0.36, 0.33, and 0.33% of all predicted genes in the corresponding species, indicating that the expansion levels and total numbers of GSL-related genes in *Brassica* crops are similar to the whole-genome gene expansion levels of the corresponding crops (*P*-value > 0.05).

To reveal the retention status of the GSL-related genes after the WGT, we determined the ratio of single- to multi-copy paralogous genes involved in various steps of GSL metabolism (Table 12.4). The proportion of total paralogous sets with different copy numbers over the whole genome was used as background, and we found that the expansion levels of transcription factors, side-chain modification, and breakdown genes in *B. rapa* were significantly higher than those of their backgrounds ($P < 0.05$). The same trends were observed for GSL breakdown genes in two *B. oleracea* genomes and for transcription factors in *B. oleracea* var. *italica*, indicating that a specific subset of GSL-related genes was retained in *B. oleracea*. Over-retention of GSL transcription factors occurred in the C subgenome of *B. napus*, while those associated with side-chain modification and breakdown were only over-retained in the A subgenome of *B. napus*. It seems that the GSL-related genes responsible for chain elongation, core structure formation, co-substrate pathways, and transport did not experience significant expansion, since they showed no significant difference from the background (Table 12.4). However, the GSL-related genes were significantly retained in all four studied *Brassica* crops, since the ratio of single- to multi-copy paralogous genes was significantly smaller than the background (*P*-value < 0.05), suggesting that GSL-related genes expanded in *B. rapa* and *B. oleracea* and were retained in the two subgenomes of *B. napus*. Tandem duplication (TD) also contributed greatly to the evolution of GSL-related genes in both *Arabidopsis* and *Brassica* species. We identified 11 TD events in *Arabidopsis* GSL-related genes, including 8 and 3 events associated with GSL biosynthesis and breakdown, respectively. We found that 21 pairs of paralogous genes had undergone more recent TD events after WGT in two *B. oleracea* crops and two subgenomes of *B. napus*. For example, *SOT18* consists of 10 copies in *B. rapa*, *B. oleracea* var. *capitata*, and the C subgenome of *B. napus*, and 9 and 8 copies in *B. oleracea* var. *italica* and A subgenome of *B. napus*, respectively. At least six *SOT18* genes originated from three TD events in all of these *Brassica* species, implying that these ancient TD events might have occurred after the *Arabidopsis–Brassica* split and before divergence of *B. rapa* and *B. oleracea*.

Similar to the findings of a previous study in *B. rapa* (Wang et al. 2011a), we found that a total

Table 12.4 Number of single- and multi-copy paralogs of GSL-related genes and their ratios among *Brassica* crops

Role in GSL metabolism	*B. rapa*			*B. oleracea* var. *capitata*			*B. oleracea* var. *italica*			A subgenome of *B. napus*			C subgenome of *B. napus*		
	Single[a]	Multi[b]	*P*-value[c]	Single[a]	Multi[b]	*P*-value[c]	Single[a]	Multi[b]	*P*-value[c]	Single[a]	Multi[b]	*P*-value[c]	Single[a]	Multi[b]	*P*-value[c]
Transcription factor	0	8	0.0021**	2	6	0.1572	0	8	0.0098**	1	7	0.0811	0	8	0.0102**
Chain elongation	2	6	0.1553	2	6	0.1572	3	5	0.7359	2	6	0.3083	2	6	0.3097
Core structure formation	8	9	0.6317	8	9	0.6342	8	9	1.0000	7	10	0.8116	7	9	1.0000
Side-chain modification	1	10	0.0039**	2	8	0.0529	2	9	0.1261	1	8	0.0475*	2	8	0.2014
Co-substrate pathways	2	4	0.4245	2	4	0.4270	2	4	0.6959	2	4	0.6966	2	4	0.6973
Transport	1	2	0.5987	1	2	0.6015	1	2	1.0000	1	2	1.0000	1	2	1.0000
Breakdown	2	10	0.0165*	3	10	0.0474*	2	12	0.0282*	2	11	0.0472*	4	9	0.4070
Total	16	49	0.0000**	20	45	0.0003**	18	49	0.0029**	16	48	0.0015**	18	46	0.0078**
Background [d]	9172	7868		8977	7841		8641	10,420		8520	10,398		8460	10,401	

[a]The number of single-copy paralogs. [b]The number of multi-copy paralogs. [c] The proportion of total paralogous sets with different copy numbers over the whole genome was used as background to calculate the *P*-value using Fisher's test. The "*" and "**" indicate *P*-values of less than 0.05 and 0.01, respectively, which means the ratio of single to multiple copies of these kinds of GSL-related genes shows a significant difference from the background. [d] The background data were derived from Liu et al. (2014) and Chalhoub et al. (2014)

of 11 GSL-related genes in *Arabidopsis* have no orthologs in the studied *Brassica* genomes, including a transcription factor (*MYB76*), two amino acid side-chain elongation genes (*IPMDH3* and *IPMI SSU3*), one core structure formation gene (*CYP79F2*) for long-chain aliphatic GSL, four side-chain modification genes ($FMO_{GS\text{-}OX1}$, $FMO_{GS\text{-}OX3}$, $FMO_{GS\text{-}OX4}$ and *AOP3*), and three GSL breakdown genes (*NSP3*, *NIT1*, and *NIT3*). It seems that the loss of these genes is not indispensable for GSL biosynthesis and breakdown, as paralogs with similar functions are present in the *Brassica* species.

12.4.2 Evolution of GSL Biosynthesis Genes Influencing Variation in GSL Profiles in *B. napus* and Its Parental Species

To date, more than 20 kinds of GSLs have been identified in commercial *Brassica* crops. The diversity of GSL types and variation in GSL profiles in these *Brassica* species are largely due to the evolution of GSL-related genes. In our study, we mainly focused on the evolution of *MAM* and *AOP* gene families in the four *Brassica* crops.

The *MAM* genes encode methylthioalkylmalate synthase, which is involved in amino acid chain elongation, and gave rise to GSLs with diverse chain lengths during the biosynthesis of methionine-derived GSLs (Zhang et al. 2015a, b, b). The phylogenetic and synteny relationships of *MAM* genes from 13 sequenced Brassicaceae species indicated that the *MAM* genes taken two independent lineage-specific evolution routes after the divergence from *Aethionema arabicum*. In the lineage I species such as *A. thaliana*, the *MAM* loci evolved three tandem genes encoding enzymes responsible for the biosynthesis of aliphatic GSLs with different carbon chain lengths, while in lineage II species such as *Brassica* crops, the *MAM* loci encode enzymes responsible for the biosynthesis of short-chain aliphatic GSLs (Zhang et al. 2015). In *Arabidopsis*, the *MAM* family contains three tandemly duplicated and functionally diverse members, *MAM1*, *MAM2*, and *MAM3* (*MAM-L*). Functional analysis demonstrated that MAM2 and MAM1 catalyze the condensation of the first and the first two elongation cycles for the synthesis of short-chain Met-derived aliphatic GSLs (3C and 4C), respectively, while MAM3 catalyzes the formation of all aliphatic GSLs, especially long-chain GSLs (6C, 7C, and 8C) (Textor et al. 2007).

In *B. rapa*, *B. oleracea* var. *capitata*, and *B. oleracea* var. *italica*, *MAM1/MAM2* genes experienced independent TD after WGT to produce 6, 7, and 6 orthologs, respectively (Fig. 12.1). Due to gene loss that occurred after the formation of *B. napus* from the fusion of two parental species, only 5 and 3 orthologs were retained in the A and C subgenomes of *B. napus*. The greatest diversity of GSL side-chain structures in *Brassica* is observed within *B. oleracea*. The main GSLs in this species (i.e., PRO, NAP, GRA, and SIN) are restricted to either 3C or 4C side-chain lengths (Liu et al. 2014). In contrast to the diversity observed in *B. oleracea*, *B. nigra* and the amphidiploid *B. carinata* only have the 3C GSL and SIN, and *B. juncea* mainly has 3C and 4C GSLs (SIN and NAP). *B. rapa* and *B. napus* lack 3C GSLs and predominately possess a mixture of 4C GSLs (NAP and PRO and their hydroxylated homologs), with small amounts 5C GSL GBS. Thus, all of these *Brassica* species can be considered to have functional alleles at the *MAM1/MAM2* loci, while some variation occurred at the *MAM3* locus, which led to the existence of 5C GSL in *B. rapa* and *B. napus*. Based on our analyses of expression patterns and phylogenetic and syntenic relationships, we identified a pair of genes, *Bol017070* and *Bra013007*, which are the only orthologs with high expression in *B. oleracea* var. *capitata*, but are silenced in *B. rapa* (Liu et al. 2014). Their two descendant orthologs in *B. napus*, *BnaA03g39720D* and *BnaCnng21190D*, both showed weak expression in roots and silenced in siliques simultaneously, implying that *Bol017070* might greatly promote the accumulation of the 3C GSL anticancer precursor SIN in *B. oleracea*. At the *MAM3* locus, one orthologous group of genes, *Bra008532*, *Bol040636*, *BnaA02g36350D*, and *BnaC02g27590D*, showed no expression due to pseudogenization. In another *MAM3* orthologous group, expression of *Bra018524* is much higher than that of *Bol016496*, *BnaA02g20830D*, and *BnaC02g26810D*. Expression differences of *MAM3* genes among *Brassica* crops most likely resulted in the increased biosynthesis of the 5C GSLs GBN and GNL in *B. rapa*.

In addition to *MAM* genes, *AOPs* are other crucial regulators of variation in aliphatic GSL profiles in Brassicaceae species (Hasan et al. 2008). Previous phylogenetic analyses showed that the core Brassicaceae species have retained AOP1, while AOP2 is retained by most of the lineage II species (excluding *Sisymbrium irio* and *Raphanus sativus*), and AOP3 by lineage I species. The variation in AOP2/AOP3 has led to different aliphatic GSL profiles in each lineage (Al-Shehbaz and Al-Shammary 1987). While the function of GSL-AOP1 is currently unknown, AOP2 catalyzes the conversion of methylsulfinylalkyl GSLs (GRA and GIB) to alkenyl GSLs (NAP and SIN), and the GS-OH locus can further convert NAP to PRO. AOP3 is associated with the production of hydroxyalkyl GSL, a compound not found in *Brassica* crops. When both AOPs are non-functional, the plant accumulates the methylsulfinylalkyl GSL precursor (Liu et al. 2014). Genetic variation at *AOP2* is also linked to increased GSL accumulation, since its expression promotes the transcription of most GSL biosynthetic genes and two R2R3 domain MYB transcription factors (*MYB28* and *MYB29*) of the pathway, suggesting that *AOP2* plays a role in the positive feedback loop controlling aliphatic GSL biosynthesis (Burow et al. 2015).

Phylogenetic and BLASTN analysis indicated that the genomes of *B. rapa*, *B. oleracea* var. *capitata,* and *B. napus* possess 3, 3, and 5 orthologs of *AOP2* and contain 3, 2, and 7 orthologs of *AOP1,* respectively (Fig. 12.2). Not all *Brassica* species have an ortholog of *AtAOP3*, and such species are unable to produce hydroxyalkyl GSLs. Similar to our results, a natural frameshift mutation resulting from a 2-bp deletion was identified in broccoli, which accumulates GRA by ceasing downstream biosynthesis of other 4C aliphatic GSLs (Li and Quiros 2003). In our previous study, we found that 2 non-functional *AOP2* genes contributing to the accumulation of GRA due to the presence of premature stop codons (Liu et al. 2014). Hence, it would be a useful strategy to enhance the GRA concentrations in *Brassica* crops by blocking the side-chain modification pathway downstream of GRA through silencing of all orthologs of *AOP2*. Recently, this strategy has been successfully applied in the metabolic engineering for increasing the anticancer compound GRA by suppressing *AOP2* gene family in both *B. juncea* and *B. napus* (Liu et al. 2012; Augustine and Bisht. 2015). In *B. rapa*, all three *BrAOP2* paralogs have been proved to be active but functionally diverged (Zhang et al. 2015). Expression patterns of five *AOP2* genes in *B. napus* are quite different, *BnaA09g01260D* and *BnaC09g004 10D* showed the highest expression in siliques, while the rest *AOP2* paralogs showed higher expression in flower and stem (Fig. 12.2), implying that these *Bna.AOP2* genes might be functional. These results provide insight into the relationship between observed GSL profiles and the evolution of GSL biosynthesis genes and explain why anticancer compound GRA is abundant in *B. oleracea*, but not in *B. rapa* and *B. napus*. The *AOP2* genes in *B. rapa* and *B. napus* are functional, reflecting the fact that the dominant GSLs are NAP and PRO in both *B. rapa* and *B. napus*.

12.4.3 Evolution of Major Genes Controlling the Seed GSL Content in *B. napus*

Quantitative trait locus (QTL) mapping and association mapping (AM) are powerful methods for analyzing the genetic structure of quantitative traits and have been widely used to characterize the total seed GSL contents and profiles in different populations of *B. napus* (Fu et al. 2015; Hasan et al. 2008; Li et al. 2014; Uzunova et al. 1995). Recently, the orthologs of *HAG1* (*MYB28*), which controls aliphatic GSL biosynthesis in *Arabidopsis*, were suggested as candidates for major QTLs on A09, C02, C07, and C09 of *B. napus*. These QTLs were detected

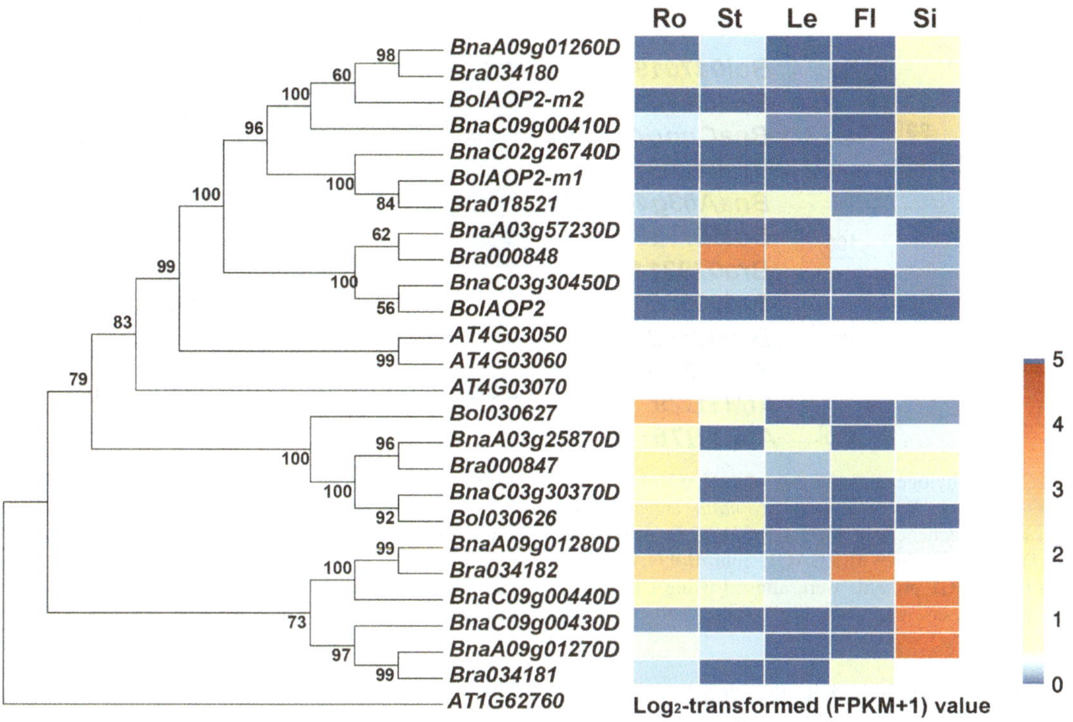

Fig. 12.2 Phylogenetic analysis of three AtAOP genes and orthologs in *B. rapa*, *B. oleracea* var. *capitata*, and *B. napus*. Full-length sequences of AOP proteins from *Arabidopsis*, *B. rapa*, *B. oleracea* var. *capitata*, and *B. napus* were aligned using ClustalW2. The phylogenetic tree (left panel) was constructed using MEGA 6.0 and the neighbor-joining method (1000 bootstrap replicates). Expression levels of *Brassica AOP* genes were derived from Tong et al. (2013) and Liu et al. (2014) and are presented as the $\log_2$-transformed (FPKM + 1) values

independently in different studies using different methods, including conventional QTL mapping, AM, and associative transcriptomic analysis (Li et al. 2014; Lu et al. 2014; Zhao and Meng 2003). Howell et al. (2003) detected four QTLs that together accounted for at least 76% of the phenotypic variation in the accumulation of GSLs in *B. napus* seeds and revealed that the QTLs on A09, C02, and C09 were homoeologous loci (Howell et al. 2003). Harper et al. (2012) revealed that the *HAG1* transcription factor gene family was a candidate in the quantitative control of GSL content of *B. napus* and that the orthologous genes on C02 and A09 had been lost from the low-GSL accessions (Harper et al. 2012). In our study, we identified three copies of *HAG1* genes (*BnaA03g40190D*, *BnaCnng43220D*, and *BnaC09g05300D*) from the genome sequence of the French homozygous *B. napus* winter line "Darmor-*bzh*," which is a double-low *B. napus* cultivar lacking detectable levels of erucic acid in the seed oil and with a low seed GSL content (Chalhoub et al. 2014). We found that the *AtHAG1* orthologs on A09 and C02 were deleted from the double-low *B. napus* cultivar "Darmor-*bzh*," leading to a reduction in seed GSL accumulation. The expression patterns of the three *Bna.HAG1* genes were investigated in an elite semi-winter double-low *B. napus* cultivar "Zhongsuang No. 11," which is widely cultivated in the Yangtze River region of China (Fig. 12.3). Among the three retained *Bna.HAG1* genes, neither *BnaA03g40190D* nor *BnaCnng 43220D* was expressed in siliques, indicating that the proteins encoded by these two genes probably lost DNA-binding activity for seed GSL accumulation. *BnaC09g05300D* exhibited the highest transcription levels in the root, followed by the stem and flower, and was expressed at very low levels in the leaf and siliques. Sequence

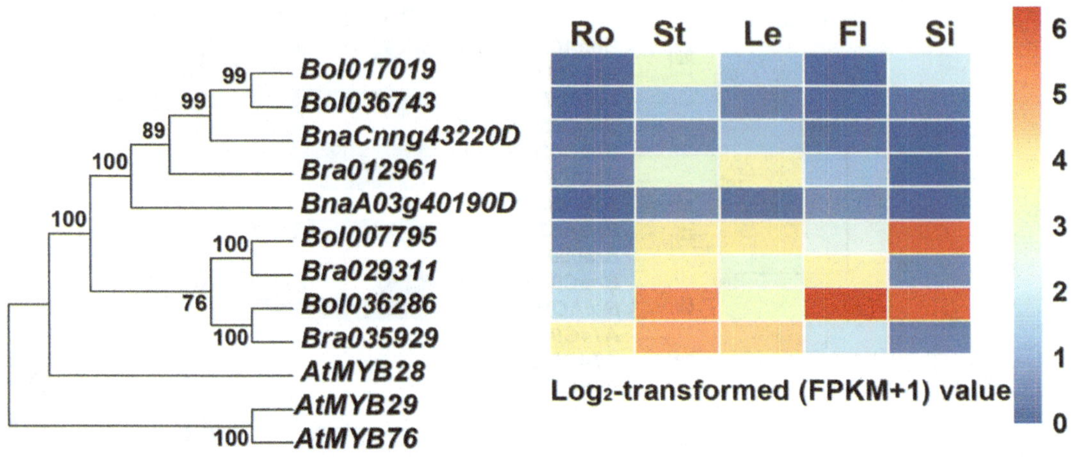

Fig. 12.3 Phylogenetic analysis of AtHAG1 and orthologs in *B. rapa*, *B. oleracea* var. *capitata*, and *B. napus*. Full-length sequences of AtHAG1 (MYB28), AtMYB29, AtMYB76, and three Bra.HAG1, four Bol.HAG1, and two Bna.HAG1 proteins were aligned using ClustalW2. The phylogenetic tree (left panel) was constructed using MEGA 6.0 and the neighbor-joining method (1000 bootstrap replicates). The BnaC09g05300D protein sequence was too short to be excluded in the phylogenetic analysis. Expression levels of *Brassica HAG1* genes were derived from Tong et al. (2013) and Liu et al. (2014) and are presented as the log_2-transformed (FPKM + 1) values

alignment revealed that the *BnaC09g05300D* coding sequence is only 420 bp long, much shorter than that of *AtHAG1* and other members of the *HAG1* gene family in *Brassica* crops, but the intact MYB DNA-binding domain (PF00249) was still predicted to exist in the BnaC09g 05300D protein sequence. These data suggest that the *Bna.HAG1* gene family experienced not only gene loss due to segment deletion, but also loss of most function in the seeds during the breeding of low-GSL *B. napus*. In current low-GSL *B. napus* accessions, *BnaC09g05300D*, which controls the biosynthesis of aliphatic GSLs, might be the only functional *Bna.HAG1* gene. Therefore, it is possible to further reduce the seed GSL content in low-GSL *B. napus* lines by silencing *BnaC09g05300D* expression.

12.4.4 Evolution of GSL Transport Genes in *B. napus*

The GSLs are believed to be synthesized mainly in the roots and vegetative tissues and accumulate abundantly in the embryos, where no de novo synthesis occurs (Nour-Eldin and Halkier 2013). Therefore, there must be specific transporters that are responsible for the relocation of GSLs from source tissues to embryos through the phloem. Recently, two members of the nitrate/peptide transporter family in *Arabidopsis*, GTR1 and GTR2, were identified as high-affinity plasma membrane-localized, GSL-specific proton symporters in a screen of an in vitro library of *Arabidopsis* transporters (Nour-Eldin et al. 2012). Previous studies suggested that GTR2 is essential for loading GSLs into the phloem, while GTR1 additionally may be involved in distributing GSLs within the leaf. Importantly, GTR1 and GTR2 are essential for the long-distance transport of both aliphatic and indole GSLs to seeds, because the *gtr1 gtr2* double mutant had only trace levels of GSLs in seeds and a concomitant increase in rosettes and silique walls (Nour-Eldin et al. 2012). However, it is notable that indole GSLs are transported between rosettes and roots in the absence of GTRs, suggesting the existence of an indole glucosinolate-specific transporter besides GTR1 and GTR2 (Jorgensen et al. 2015).

We identified 32 orthologs of *AtGTR* in the four *Brassica* crops we investigated, including

15 *GTR1* and 17 *GTR2* genes. Phylogenetic analysis and tissue-specific expression detection showed that the transcription levels of most *Bna. GTR* genes are lower than those of orthologs in the parental species *B. rapa* or *B. oleracea* var. *capitata* (Fig. 12.4). For example, *Bra029248* and *Bol020699* showed higher expression than *BnaA02g33530D* and *BnaC02g42260D*. The expression of GSL-related genes was determined in the Chinese double-low *B. napus* cultivar "Zhongsuang No. 11." This analysis indicated that the expression of *Bna.GTR* genes is reduced in this cultivar, suggesting that the reduced transport of GSLs from source tissues to seeds accounts for the hypo-accumulation of GSLs in the seeds of this low-GSL content variety. For each *AtGTR* gene, we identified at least one *Bna. GTR* ortholog with high expression (Fig. 12.4). For instance, *BnaA09g06190D*, *BnaC09g 05810D*, *BnaC03g51560D,* and *BnaC03g 75950D* which might be the major GTR members responsible for the long-distance transport of GSL in the *B. napus* cultivar "Zhongsuang No. 11," were expressed at higher levels than

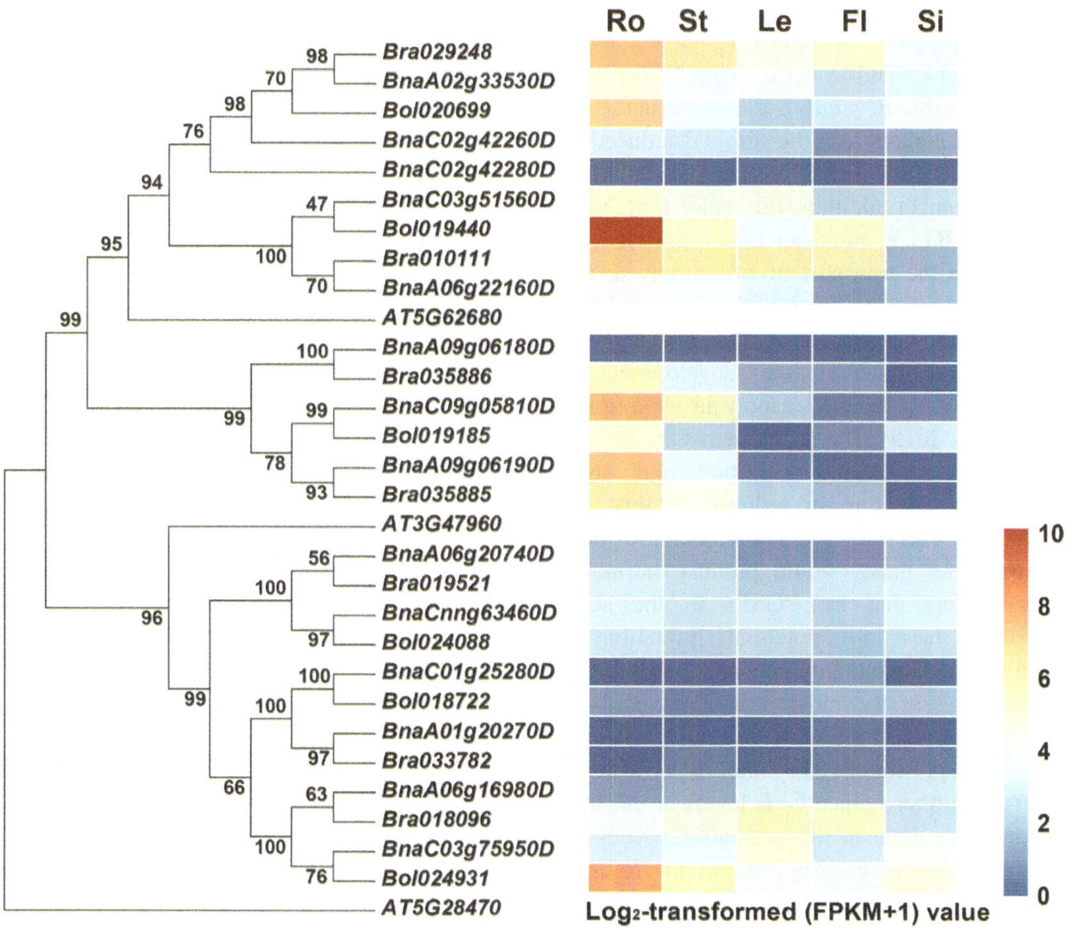

Fig. 12.4 Phylogenetic analysis of two AtGTR genes and orthologs in *B. rapa*, *B. oleracea* var. *capitata,* and *B. napus*. Full-length sequences of GTR proteins from *Arabidopsis*, *B. rapa*, *B. oleracea* var. *capitata,* and *B. napus* were aligned using ClustalW2. The phylogenetic tree (left panel) was constructed using MEGA 6.0 and the neighbor-joining method (1000 bootstrap replicates). Expression levels of *Brassica GTR* genes were derived from Tong et al. (2013) and Liu et al. (2014) and are presented as the log_2-transformed (FPKM + 1) values

other members. Lu et al. (2014) reported that the transcript abundance in the leaves of the candidate gene involved in GSL transport, *BnaA. GTR2a,* located on chromosome A02, was correlated with seed GSL content, accounting for 18.8% of the phenotypic variation in seed GSL content between *B. napus* cultivars (Lu et al. 2014). Recently, we also found that *Bna.GRT2* on chromosome A09 is a candidate GSL transporter and is associated with seed GSL content based on AM analysis of seed GSL content using the 60K *Brassica* Infinium SNP array in 520 *B. napus* accessions. These results strongly suggest that transport engineering can be used to eliminate antinutritional GSLs in seeds by silencing GTR transporters in *B. napus*.

Indole GSL 4HGBS is the major GSL present in the low-GSL *B. napus* varieties. Whether the total GSL content can be further reduced by silencing all of the *Bna.GTR* genes merits further investigation. In addition, the major GSL transporter, GTR1, is multifunctional and may be involved in the transport of structurally distinct compounds, including GSLs, jasmonoyl-isoleucine, and gibberellin, and may positively regulate stamen development by mediating gibberellin transport in *Arabidopsis* (Saito et al. 2015). The *gtr1* mutants are severely impaired in filament elongation and anther dehiscence, resulting in reduced fertility, and hence, it is uncertain whether silencing of all of the *Bna.GTR* genes would produce normal *B. napus* plants that lack GSLs in the seeds. Although there are potential limitations in genetic engineering applications, the *Bna.GTR* genes represent the most promising regulation loci among the GSL-related genes and have potential applications in molecular breeding efforts to further reduce GSL levels in the seeds and increase them in the vegetative tissues and roots, where they play important roles in enhancing biotic and/or abiotic resistance in *B. napus*.

Acknowledgements This work was supported by grants from the National Science Foundation of China (31571701, 31171619, U1302266, and 31401412), the 973 Project (2015CB250100), the 863 Project (2013AA102602), the 111 Project (B12006) and the Fundamental Research Funds for the Central Universities (XDJK2012A009 and XDJK2013C031).

References

Agee AE, Surpin M, Sohn EJ, Girke T, Rosado A, Kram BW et al (2010) MODIFIED VACUOLE PHENOTYPE1 is an *Arabidopsis* myrosinase-associated protein involved in endomembrane protein trafficking. Plant Physiol 152:120–132

Agerbirk N, Olsen CE (2012) Glucosinolate structures in evolution. Phytochemistry 77:16–45

Al-Shehbaz IA, Al-Shammary KI (1987) Distribution and chemotaxonomic significance of glucosinolates in certain Middle-Eastern *cruciferae*. Biochem Syst Ecol 15:559–569

Augustine R, Bisht NC (2015) Biofortification of oilseed *Brassica juncea* with the anti-cancer compound glucoraphanin by suppressing *GSL-ALK* gene family. Sci Rep 5:18005

Bak S, Feyereisen R (2001) The involvement of two P450 enzymes, CYP83B1 and CYP83A1, in auxin homeostasis and glucosinolate biosynthesis. Plant Physiol 127:108–118

Barth C, Jander G (2006) *Arabidopsis* myrosinases TGG1 and TGG2 have redundant function in glucosinolate breakdown and insect defense. Plant J 46:549–562

Bednarek P, Osbourn A (2009) Plant-microbe interactions: chemical diversity in plant defense. Science 324:746–748

Bednarek P, Pislewska-Bednarek M, Svatos A, Schneider B, Doubsky J, Mansurova M et al (2009) A glucosinolate metabolism pathway in living plant cells mediates broad-spectrum antifungal defense. Science 323:101–106

Bellostas N, Sørensen JC, Sørensen H (2007) Profiling glucosinolates in vegetative and reproductive tissues of four *Brassica* species of the U-triangle for their biofumigation potential. J Sci Food Agr 87:1586–1594

Benderoth M, Textor S, Windsor AJ, Mitchell-Olds T, Gershenzon J, Kroymann J (2006) Positive selection driving diversification in plant secondary metabolism. Proc Nat Acad Sci USA 103:9118–9123

Bones AM, Rossiter JT (2006) The enzymic and chemically induced decomposition of glucosinolates. Phytochemistry 67:1053–1067

Burow M, Atwell S, Francisco M, Kerwin RE, Halkier BA, Kliebenstein DJ (2015) The Glucosinolate biosynthetic gene *AOP2* mediates feed-back regulation of jasmonic acid signaling in *Arabidopsis*. Mol Plant 8:1201–1212

Burow M, Halkier BA, Kliebenstein DJ (2010) Regulatory networks of glucosinolates shape *Arabidopsis thaliana* fitness. Curr Opin Plant Biol 13:348–353

Cartea ME, Velasco P (2008) Glucosinolates in *Brassica* foods: bioavailability in food and significance for human health. Phytochem Rev 7:213–229

Cartea ME, Velasco P, Obregón S, Padilla G, de Haro A (2008) Seasonal variation in glucosinolate content in *Brassica oleracea* crops grown in northwestern Spain. Phytochemistry 69:403–410

Chalhoub B, Denoeud F, Liu S, Parkin IA, Tang H, Wang X et al (2014) Plant genetics. Early allopolyploid evolution in the post-neolithic *Brassica napus* oilseed genome. Science 345:950–953

Chen S, Glawischnig E, Jørgensen K, Naur P, Jørgensen B, Olsen CE, Hansen CH, Rasmussen H, Pickett JA, Halkier BA (2003) CYP79F1 and CYP79F2 have distinct functions in the biosynthesis of aliphatic glucosinolates in *Arabidopsis*. Plant J 33:923–937

Daxenbichler ME, Spencer GF, Carlson DG, Rose GB, Brinker AM, Powell RG (1991) Glucosinolate composition of seeds from 297 species of wild plants. Phytochemistry 30:2623–2638

Dubos C, Stracke R, Grotewold E, Weisshaar B, Martin C, Lepiniec L (2010) MYB transcription factors in *Arabidopsis*. Trends Plant Sci 15:573–581

Fahey JW, Zalcmann AT, Talalay P (2001) The chemical diversity and distribution of glucosinolates and isothiocyanates among plants. Phytochemistry 56:5–51

Font R, del Rio-Celestino M, Cartea E, de Haro-Bailon A (2005) Quantification of glucosinolates in leaves of leaf rape (*Brassica napus* ssp. *pabularia*) by near-infrared spectroscopy. Phytochemistry 66:175–185

Frerigmann H, Gigolashvili T (2014) MYB34, MYB51, and MYB122 distinctly regulate indolic glucosinolate biosynthesis in *Arabidopsis thaliana*. Mol Plant 7:814–828

Fu Y, Lu K, Qian L, Mei J, Wei D, Peng X et al (2015) Development of genic cleavage markers in association with seed glucosinolate content in canola. Theor Appl Genet 128:1029–1037

Gigolashvili T, Berger B, Flügge U-I (2008a) Specific and coordinated control of indolic and aliphatic glucosinolate biosynthesis by R2R3-MYB transcription factors in *Arabidopsis thaliana*. Phytochem Rev 8:3–13

Gigolashvili T, Engqvist M, Yatusevich R, Muller C, Flugge UI (2008b) HAG2/MYB76 and HAG3/MYB29 exert a specific and coordinated control on the regulation of aliphatic glucosinolate biosynthesis in *Arabidopsis thaliana*. New Phytol 177:627–642

Grubb CD, Abel S (2006) Glucosinolate metabolism and its control. Trends Plant Sci 11:89–100

Grubb CD, Zipp BJ, Ludwig-Müller J, Masuno MN, Molinski TF, Abel S (2004) *Arabidopsis* glucosyltransferase UGT74B1 functions in glucosinolate biosynthesis and auxin homeostasis. Plant J 40:893–908

Grubb CD, Zipp BJ, Kopycki J, Schubert M, Quint M, Lim EK, Bowles DJ, Pedras MS, Abel S (2014) Comparative analysis of *Arabidopsis* UGT74 glucosyltransferases reveals a special role of UGT74C1 in glucosinolate biosynthesis. Plant J 79:92–105

Gu Z-X, Guo Q-H, Gu Y-J (2012) Factors influencing glucoraphanin and sulforaphane formation in *Brassica* plants: a review. J Integr Agr 11:1804–1816

Guo R, Qian H, Shen W, Liu L, Zhang M, Cai C, Zhao Y, Qiao J, Wang Q (2013) BZR1 and BES1 participate in regulation of glucosinolate biosynthesis by brassinosteroids in *Arabidopsis*. J Exp Bot 64:2401–2412

Halkier BA, Du L (1997) The biosynthesis of glucosinolates. Trends Plant Sci 2:425–431

Halkier BA, Gershenzon J (2006) Biology and biochemistry of glucosinolates. Annu Rev Plant Biol 57:303–333

Hansen BG, Kerwin RE, Ober JA, Lambrix VM, Mitchell-Olds T, Gershenzon J et al (2008) A novel 2-oxoacid-dependent dioxygenase involved in the formation of the goiterogenic 2-hydroxybut-3-enyl glucosinolate and generalist insect resistance in *Arabidopsis*. Plant Physiol 148:2096–2108

Harper AL, Trick M, Higgins J, Fraser F, Clissold L, Wells R et al (2012) Associative transcriptomics of traits in the polyploid crop species *Brassica napus*. Nat Biotechnol 30:798–802

Hasan M, Friedt W, Pons-Kuhnemann J, Freitag NM, Link K, Snowdon RJ (2008) Association of gene-linked SSR markers to seed glucosinolate content in oilseed rape (*Brassica napus* ssp. *napus*). Theor Appl Genet 116:1035–1049

Hirai MY, Sugiyama K, Sawada Y, Tohge T, Obayashi T, Suzuki A et al (2007) Omics-based identification of *Arabidopsis* Myb transcription factors regulating aliphatic glucosinolate biosynthesis. Proc Natl Acad Sci USA 104:6478–6483

Howell PM, Sharpe AG, Lydiate DJ (2003) Homoeologous loci control the accumulation of seed glucosinolates in oilseed rape (*Brassica napus*). Genome 46:454–460

Ishida M, Hara M, Fukino N, Kakizaki T, Morimitsu Y (2014) Glucosinolate metabolism, functionality and breeding for the improvement of Brassicaceae vegetables. Breed Sci 64:48–59

Jeffery EH, Brown AF, Kurilich AC, Keck AS, Matusheski N, Klein BP et al (2003) Variation in content of bioactive components in broccoli. J Food Compos Anal 16:323–330

Jorgensen ME, Nour-Eldin HH, Halkier BA (2015) Transport of defense compounds from source to sink: lessons learned from glucosinolates. Trends Plant Sci 20:508–514

Khajali F, Slominski B (2012) Factors that affect the nutritive value of canola meal for poultry. Poultry Sci 91:2564–2575

Kim JH, Durrett TP, Last RL, Jander G (2004) Characterization of the *Arabidopsis* TU8 glucosinolate mutation, an allele of *TERMINAL FLOWER2*. Plant Mol Biol 54:671–682

Kissen R, Bones AM (2009) Nitrile-specifier proteins involved in glucosinolate hydrolysis in *Arabidopsis thaliana*. J Biol Chem 284:12057–12070

Kliebenstein DJ, Lambrix VM, Reichelt M, Gershenzon J, Mitchell-Olds T (2001) Gene duplication in the diversification of secondary metabolism: tandem 2-oxoglutarate–dependent dioxygenases control glucosinolate biosynthesis in *Arabidopsis*. Plant Cell 13:681–693

Kroymann J, Donnerhacke S, Schnabelrauch D, Mitchell-Olds T (2003) Evolutionary dynamics of an *Arabidopsis* insect resistance quantitative trait locus. Proc Nat Acad Sci USA 100:14587–14592

Kushad MM, Brown AF, Kurilich AC, Juvik JA, Klein BP, Wallig MA et al (1999) Variation of glucosinolates in vegetable crops of *Brassica oleracea*. J Agri Food Chem 47:1541–1548

Laluk K, Prasad KV, Savchenko T, Celesnik H, Dehesh K, Levy M et al (2012) The calmodulin-binding transcription factor SIGNAL RESPONSIVE1 is a novel regulator of glucosinolate metabolism and herbivory tolerance in *Arabidopsis*. Plant Cell Physiol 53:2008–2015

Le Roux C, Del Prete S, Boutet-Mercey S, Perreau F, Balagué C, Roby D et al (2014) The hnRNP-Q protein LIF2 participates in the plant immune response. PLoS ONE 9:e99343

Leeson S, Atteh J, Summers J (1987) The replacement value of canola meal for soybean meal in poultry diets. Can J Anim Sci 67:151–158

Levy M, Wang Q, Kaspi R, Parrella MP, Abel S (2005) Arabidopsis IQD1, a novel calmodulin-binding nuclear protein, stimulates glucosinolate accumulation and plant defense. Plant J 43:79–96

Li G, Quiros CF (2003) *In planta* side-chain glucosinolate modification in *Arabidopsis* by introduction of dioxygenase *Brassica* homolog BoGSL-ALK. Theor Appl Genet 106:1116–1121

Li P, Zhao Y, Zhang W, Ding X, Yang M, Wang X et al (2005) Analysis of glucosinolate components and profiles in *Brassica napus*. Sci Agri Sin 38:1346–1352

Li J, Hansen BG, Ober JA, Kliebenstein DJ, Halkier BA (2008) Subclade of flavin-monooxygenases involved in aliphatic glucosinolate biosynthesis. Plant Physiol 148:1721–1733

Li F, Chen B, Xu K, Wu J, Song W, Bancroft I et al (2014) Genome-wide association study dissects the genetic architecture of seed weight and seed quality in rapeseed (*Brassica napus* L.). DNA Res 21:355–367

Liu Z, Hirani AH, McVetty PB, Daayf F, Quiros CF, Li G (2012) Reducing progoitrin and enriching glucoraphanin in *Braasica napus* seeds through silencing of the GSL-ALK gene family. Plant Mol Biol 79:179–189

Liu S, Liu Y, Yang X, Tong C, Edwards D, Parkin IA et al (2014) The *Brassica oleracea* genome reveals the asymmetrical evolution of polyploid genomes. Nat Commun 5:3930

Lu G, Harper AL, Trick M, Morgan C, Fraser F, O'Neill C et al (2014) Associative transcriptomics study dissects the genetic architecture of seed glucosinolate content in *Brassica napus*. DNA Res 21:613–625

Maruyama-Nakashita A, Nakamura Y, Tohge T, Saito K, Takahashi H (2006) *Arabidopsis* SLIM1 is a central transcriptional regulator of plant sulfur response and metabolism. Plant Cell 18:3235–3251

Mikkelsen MD, Petersen B, Olsen C, Halkier B (2002) Biosynthesis and metabolic engineering of glucosinolates. Amino Acids 22:279–295

Mikkelsen MD, Naur P, Halkier BA (2004) *Arabidopsis* mutants in the C-S lyase of glucosinolate biosynthesis establish a critical role for indole-3-acetaldoxime in auxin homeostasis. Plant J 37:770–777

Nagaharu U (1935) Genome-analysis in *Brassica* with special reference to the experimental formation of *B. napus* and peculiar mode of fertilization. Japan J Bot 7:389–452

Nakahigashi K, Jasencakova Z, Schubert I, Goto K (2005) The *Arabidopsis* heterochromatin protein1 homolog (TERMINAL FLOWER2) silences genes within the euchromatic region but not genes positioned in heterochromatin. Plant Cell Physiol 46:1747–1756

Nakano RT, Matsushima R, Nagano AJ, Fukao Y, Fujiwara M, Kondo M et al (2012) ERMO3/MVP1/GOLD36 is involved in a cell type-specific mechanism for maintaining er morphology in *Arabidopsis thaliana*. PLoS ONE 7:e49103

Naur P, Petersen BL, Mikkelsen MD, Bak S, Rasmussen H, Olsen CE et al (2003) CYP83A1 and CYP83B1, two nonredundant cytochrome P450 enzymes metabolizing oximes in the biosynthesis of glucosinolates in *Arabidopsis*. Plant Physiol 133:63–72

Neal CS, Fredericks DP, Griffiths CA, Neale AD (2010) The characterisation of AOP2: a gene associated with the biosynthesis of aliphatic alkenyl glucosinolates in *Arabidopsis thaliana*. BMC Plant Biol 10:170

Nour-Eldin HH, Halkier BA (2013) The emerging field of transport engineering of plant specialized metabolites. Curr Opin Biotechnol 24:263–270

Nour-Eldin HH, Andersen TG, Burow M, Madsen SR, Jorgensen ME, Olsen CE et al (2012) NRT/PTR transporters are essential for translocation of glucosinolate defence compounds to seeds. Nature 488:531–534

Padilla G, Cartea ME, Velasco P, de Haro A, Ordas A (2007) Variation of glucosinolates in vegetable crops of *Brassica rapa*. Phytochemistry 68:536–545

Parkin IAP, Koh C, Tang H, Robinson SJ, Kagale S, Clarke WE et al (2014) Transcriptome and methylome profiling reveals relics of genome dominance in the mesopolyploid *Brassica oleracea*. Genome Biol 15:R77

Pfalz M, Vogel H, Kroymann J (2009) The gene controlling the indole glucosinolate modifier1 quantitative trait locus alters indole glucosinolate structures and aphid resistance in *Arabidopsis*. Plant Cell 21:985–999

Piotrowski M, Schemenewitz A, Lopukhina A, Müller A, Janowitz T, Weiler EW et al (2004) Desulfoglucosinolate sulfotransferases from *Arabidopsis thaliana* catalyze the final step in the biosynthesis of the glucosinolate core structure. J Biol Chem 279:50717–50725

Radojčić Redovniković I, Glivetić T, Delonga K, Vorkapić-Furač J (2008) Glucosinolates and their potential role in plant. Period Biol 110:297–309

Rosa EA (1997) Glucosinolates from flower buds of Portuguese *Brassica* crops. Phytochemistry 44:1415–1419

Saito H, Oikawa T, Hamamoto S, Ishimaru Y, Kanamori-Sato M, Sasaki-Sekimoto Y et al (2015) The jasmonate-responsive GTR1 transporter is required for gibberellin-mediated stamen development in *Arabidopsis*. Nat Commun 6:6095

Sang J, Minchinton I, Johnstone P, Truscott R (1984) Glucosinolate profiles in the seed, root and leaf tissue of cabbage, mustard, rapeseed, radish and swede. Can J Plant Sci 64:77–93

Skirycz A, Reichelt M, Burow M, Birkemeyer C, Rolcik J, Kopka J et al (2006) DOF transcription factor AtDof1.1 (OBP2) is part of a regulatory network controlling glucosinolate biosynthesis in *Arabidopsis*. Plant J 47:10–24

Sønderby IE, Geu-Flores F, Halkier BA (2010) Biosynthesis of glucosinolates–gene discovery and beyond. Trends Plant Sci 15:283–290

Sun B, Liu N, Zhao Y, Yan H, Wang Q (2011) Variation of glucosinolates in three edible parts of Chinese kale (*Brassica alboglabra* Bailey) varieties. Food Chem 124:941–947

Textor S, Bartram S, Kroymann J, Falk KL, Hick A, Pickett JA et al (2004) Biosynthesis of methionine-derived glucosinolates in *Arabidopsis thaliana*: recombinant expression and characterization of methylthioalkylmalate synthase, the condensing enzyme of the chain-elongation cycle. Planta 218:1026–1035

Textor S, De Kraker J-W, Hause B, Gershenzon J, Tokuhisa JG (2007) MAM3 catalyzes the formation of all aliphatic glucosinolate chain lengths in *Arabidopsis*. Plant Physiol 144:60–71

Tian Q, Rosselot RA, Schwartz SJ (2005) Quantitative determination of intact glucosinolates in broccoli, broccoli sprouts, brussels sprouts, and cauliflower by high-performance liquid chromatography–electrospray ionization–tandem mass spectrometry. Anal Biochem 343:93–99

Tong C, Wang X, Yu J, Wu J, Li W, Huang J et al (2013) Comprehensive analysis of RNA-seq data reveals the complexity of the transcriptome in *Brassica rapa*. BMC Genomics 14:689

Uzunova M, Ecke W, Weissleder K, Röbbelen G (1995) Mapping the genome of rapeseed (*Brassica napus* L.). I. construction of an RFLP linkage map and localization of QTLs for seed glucosinolate content. Theor Appl Genet 90:194–204

Vorwerk S, Biernacki S, Hillebrand H, Janzik I, Müller A, Weiler EW et al (2001) Enzymatic characterization of the recombinant *Arabidopsis thaliana* nitrilase subfamily encoded by the *NIT2*/*NIT1*/*NIT3*-gene cluster. Planta 212:508–516

Wang H, Wu J, Sun S, Liu B, Cheng F, Sun R et al (2011a) Glucosinolate biosynthetic genes in *Brassica rapa*. Gene 487:135–142

Wang X, Wang H, Wang J, Sun R, Wu J, Liu S et al (2011b) The genome of the mesopolyploid crop species *Brassica rapa*. Nat Genet 43:1035–1039

Wang M, Sun X, Tan D, Gong S, Meijer J, Zhang J (2009) The two non-functional myrosinase genes TGG3 and TGG6 in *Arabidopsis* are expressed predominantly in pollen. Plant Sci 177:371–375

Wang J, Gu H, Yu H, Zhao Z, Sheng X, Zhang X (2012) Genotypic variation of glucosinolates in broccoli (*Brassica oleracea* var. *italica*) florets from China. Food Chem 133:735–741

Warwick SI (2011) Brassicaceae in agriculture. In: Schmidt R, Bancroft I (eds) Genetics and genomics of the Brassicaceae. Springer, New York, pp 33–65

West L, Tsui I, Haas G (2002) Single column approach for the liquid chromatographic separation of polar and non-polar glucosinolates from broccoli sprouts and seeds. J Chromatogr A 966:227–232

Wittstock U, Halkier BA (2002) Glucosinolate research in the *Arabidopsis* era. Trends Plant Sci 7:263–270

Wittstock U, Burow M (2010) Glucosinolate breakdown in *Arabidopsis*: mechanism, regulation and biological significance. Arabidopsis Book 8:e0134

Xu Z, Escamilla-Treviño L, Zeng L, Lalgondar M, Bevan D, Winkel B, Mohamed A, Cheng C-L, Shih M-C, Poulton J, Esen A (2004) Functional genomic analysis of *Arabidopsis thaliana* glycoside hydrolase family 1. Plant Mol Biol 55:343–367

Xue J, Jørgensen M, Pihlgren U, Rask L (1995) The myrosinase gene family in *Arabidopsis thaliana*: gene organization, expression and evolution. Plant Mol Biol 27:911–922

Yang B, Quiros CF (2010) Survey of glucosinolate variation in leaves of *Brassica rapa* crops. Genet Resour Crop Ev 57:1079–1089

Zhang Z, Ober JA, Kliebenstein DJ (2006) The gene controlling the quantitative trait locus EPITHIOSPECIFIER MODIFIER1 alters glucosinolate hydrolysis and insect resistance in *Arabidopsis*. Plant Cell 18:1524–1536

Zhang J, Liu Z, Liang J, Wu J, Cheng F, Wang X (2015) Three genes encoding AOP2, a protein involved in aliphatic glucosinolate biosynthesis, are differentially expressed in *Brassica rapa*. J Exp Bot 66:6205–6218

Zhang J, Liu Z, Liang J, Wu J, Cheng F, Wang X (2015a) Three genes encoding AOP2, a protein involved in aliphatic glucosinolate biosynthesis, are differentially expressed in *Brassica rapa*. J Exp Bot 66:6205–6218

Zhang J, Wang X, Cheng F, Wu J, Liang J, Yang W et al (2015b) Lineage-specific evolution of Methylthioalkylmalate synthases (MAMs) involved in glucosinolates biosynthesis. Front Plant Sci 6:18

Zhao J, Meng J (2003) Detection of loci controlling seed glucosinolate content and their association with Sclerotinia resistance in *Brassica napus*. Plant Breed 122:19–23

Zhao Z, Zhang W, Stanley BA, Assmann SM (2008) Functional proteomics of *Arabidopsis thaliana* guard cells uncovers new stomatal signaling pathways. Plant Cell 20:3210–3226

Case Study for Trait-Related Gene Evolution: Disease Resistance Genes in *Brassica napus*

13

Aria Dolatabadian, Hua Yang and Jacqueline Batley

Abstract

Brassica species are among the most important crop species worldwide. Amongst these, *Brassica napus* L. (canola/rapeseed/oilseed rape) is the most economically important species providing a healthy oil and a highly nutritious livestock feed. *Brassica napus* is severely affected by several diseases, such as downy mildew and blackleg, causing a reduction in productivity and quality. For decades farmers have been using agronomic practices and pesticides to control the diseases. Today, advances in genetics and genomics accelerate the process towards developing disease resistant canola cultivars to enable growers to manage diseases economically and safely. To date, many reports have appeared about disease resistance in *Brassica* species. This chapter provides information about disease resistance in *B. napus* emphasising defence response, resistance genes and classification. The chapter also provides an account of genomics studies for identifying disease resistance genes in *Brassica* species.

A. Dolatabadian
School of Biological Sciences, University of Western Australia, Crawley, WA 6009, Australia

H. Yang
School of Agriculture and Food Science, The University of Queensland, St Lucia, QLD 4072, Australia

J. Batley (✉)
School of Biological Sciences and Institute of Agriculture, University of Western Australia, Crawley, WA 6009, Australia
e-mail: Jacqueline.batley@uwa.edu.au

13.1 Introduction

Brassica napus (canola/rapeseed/oilseed rape) has an amphidiploid genome (AACC, $n = 19$) and originated from interspecific hybridisation between the diploids *Brassica rapa* (Asian cabbage or turnip, AA genome) and *Brassica oleracea* (cabbage, cauliflower, Brussels sprouts, etc., CC genome). It is one of the most economically important oilseed crops in the world and is an essential source of edible vegetable oil and proteins for human and animal consumption. Globally, canola has been cultivated over about 36.4 Mha with total production approximately 72.5 M tonnes (FAO 2014). *Brassica* species yield strongly depends on the crop-breeding programs (cultivar), agronomic practices, soil fertility, environmental conditions and pest and disease management. Plant diseases are responsible for dramatic *Brassica* species yield loss, and the impact of disease outbreaks poses a threat to global food security worldwide. Consequently, development of effective strategies to control the diseases should be taken into account where appropriate. One of these strategies is to

S. Liu et al. (eds.), *The Brassica napus Genome*, Compendium of Plant Genomes,
https://doi.org/10.1007/978-3-319-43694-4_13

identify resistance genes (*R*-genes) in different *Brassica* species and understand the genetic interaction between plant *R*-genes and the pathogen avirulence (*Avr*) genes, as plant disease resistance genes play a critical role in providing resistance against pathogens. It has been reported that *R*-genes provide an economical and environmentally responsible solution to control plant diseases, such that cloning of these genes would enable durable *R*-gene deployment strategies (Steuernagel et al. 2016). Therefore, an understanding of the relationship between race-specific *R*-genes and their corresponding pathogen *Avr* genes (Flor 1971) is required for the effective deployment of resistance genetics in *Brassica* species.

13.2 Defence Response Activation

Plants have developed defence mechanisms (non-specific and specific barriers) to protect themselves from disease. The non-specific barriers consist of the plant's external structures, for example rigid cell walls, a waxy cuticle on the leaf surface and epidermal hairs on the surface of plant (Fu and Dong 2013), and preformed chemicals (Freeman and Beattie 2008). Whilst these non-specific barriers prevent many pathogens invading before they are able to cause extensive damage, a small amount of pathogens manage to evade these initial barriers, activating the plant innate immune system (specific barriers). This is divided into two immune system branches based on the pathogen molecules that trigger the responses; pattern-triggered immunity (PTI) where slowly evolving pathogen-associated molecular patterns (PAMPs) trigger basal defence responses, or effector-triggered immunity (ETI) (Cui et al. 2015; Wu et al. 2014), in which specific pathogen effectors, targeted to disrupt PTI, either directly or indirectly trigger specific *R*-genes (Jones and Dangl 2006; Katagiri and Tsuda 2010). PAMP occurs at the plant cell surface with the recognition of conserved microbial groups such as lipopolysaccharides and peptidoglycans (Peele et al. 2014). The PAMPs are recognized by cognate pattern-recognition receptors (PRRs) (Chisholm et al. 2006).

13.3 Plant Disease Resistance (*R*) Genes

Plant disease resistance (*R*) genes play an important role in triggering the genetic resistance defence mechanisms in plants. The *Hm1* gene from maize was the first plant *R*-gene isolated in 1992 (Johal and Briggs 1992). Since then, a large number of *R* genes conferring resistance against different pathogens (viruses, bacteria, fungi and nematodes) have been cloned from a wide range of plant species (Pandolfi et al. 2016; Sanseverino et al. 2013; Yu et al. 2014).

Plant disease *R*-genes interact with corresponding pathogen *Avr* genes. This gene for gene interaction (or genes-for-genes) activates the signal transduction cascades that may involve protein phosphorylation, ion fluxes, reactive oxygen species and other signalling events (Dangl and Jones 2001) that finally turn on complex defence responses against pathogen attack. This is termed an incompatible interaction (Dangl and Jones 2001). In most cases, a single *R*-gene can offer complete resistance to one or more strains of certain pathogen, when conveyed to a previously susceptible plant of the same species (McDowell and Woffenden 2003); thereupon, *R*-genes have been widely used in plant molecular biology and resistance breeding programs for decades (Pink 2002). *R*-genes encode putative receptors that respond to the products of *Avr* genes expressed by the pathogen during infection (McDowell and Woffenden 2003). In Brassicas, this qualitative resistance is seedling resistance, single-gene race specific and expressed during the cotyledon stage (Delourme et al. 2006). This resistance typically depends on the presence of an *R*-gene in the plant and a corresponding *Avr* gene in the pathogen, where if the *Avr* gene does not correspond to the *R*-gene in the plant, the plant is susceptible to disease. This is a very effective resistance and operates through *R*-gene activity when a pathogen infects the cotyledons of the seedling, subsequently preventing infection spread to the whole plant (Johnson and Lewis 1994).

13.4 Resistance Genes Classification

Generally, *R*-genes can be divided into eight groups based on their amino acid motif organization and their membrane-spanning domains (Gururani et al. 2012): (1) the genes encoding for cytoplasm proteins with a NBS (Nucleotide Binding Site), a C-terminal LRR (Leucine Rich Repeat) and a putative CC (Coiled–Coil) at the N-terminus, (2) cytoplasmic proteins which possess LRR and NBS motifs and an N-terminal domain with homology to the mammalian TIR (Toll/Interleukin Receptor) domain, (3) an extra cytoplasmic LRR, attached to a trans-membrane domain (TrD), (4) an extracellular LRR domain, a TrD and an intracellular serine–threonine kinase (KIN) domain, (5) the putative extracellular LRRs, along with a protein degradation domain (proline–glycine–serine–threonine) (PEST), and endocytosis cell signalling domain (ECS) that might target the protein for receptor-mediated endocytos, (6) a TrD, fused to a putative C–C, (7) a new member of the TIR–NBS–LRR *R*-protein class having a C-terminal extension with a putative nuclear localization signal (NLS) and an amino acid domain (WRKY) domain, (8) the enzymatic *R*-genes, which contain neither LRR nor NBS groups. The maize *Hm1* gene, which provides protection against *Cochliobolus carbonum,* is an example of an enzymatic *R*-gene (Johal and Briggs 1992). The *Hm1* encodes the enzyme HC toxin reductase, which detoxifies a specific cyclic tetra-peptide toxin produced by the fungus (HC toxin) that is essential for pathogenicity. However, the majority of these genes encode proteins containing a nucleotide-binding site (NBS) and leucine-reach repeats (LRR) (Dangl et al. 2013).

NBS-encoding resistance genes have been annotated in many monocot and dicot species, pioneered by research in *Arabidopsis thaliana* (TAIR10). In the reference genome of *A. thaliana*, 149 R-proteins harbour a LRR motif, of which 83 are composed of TIR-NB-LRR and 51 have CC-NB-LRR domains (Meyers et al. 2003). Genome-wide analysis of NBS-LRR genes in many plant species including rice (Monosi et al. 2004), sorghum (Paterson et al. 2009), *Arabidopsis* (Meyers et al. 2003; Tan et al. 2007) and papaya (Porter et al. 2009) indicated that they are widely distributed throughout the genome with approximately 0.6–1.8% of genes encoding NBS-LRRs at a density of 0.3–1.6 genes per megabase. Furthermore, a common feature of most plant *R*-genes is the presence of LRR motifs that play a key role in recognition of pathogen effectors by facilitating protein–protein interactions (McDowell and Woffenden 2003). A typical plant genome contains hundreds of NLR-encoding genes, many of which reside in complex clusters of linked paralogs (Hulbert et al. 2001). The clustered arrangement of these genes may be a critical attribute for the generation of novel resistance specificities (Meyers et al. 2003).

13.5 Genome-Wide Analysis of Brassica NBS-LRR Genes

Sequencing of *Brassica* genomes over the last decade has resulted in reference genome sequences for *B. napus* (Chalhoub et al. 2014), *B. rapa* (Wang et al. 2011) and *B. oleracea* (Liu et al. 2014; Parkin et al. 2014) permitting a comprehensive study of *R*-genes in these *Brassica* species. In an initial study on the *B. rapa* genome, a lower number of NBS-LRR genes were predicted than found in other sequenced crops (Mun et al. 2009). However, they estimated the number of NBS-LRR genes in the *B. rapa* genome should be higher than in *Arabidopsis* (Mun et al. 2009). Moreover, almost 50% of NBS family members were detected as tandem arrays within homogenous clusters suggesting tandem duplication in combination with polyploidy played an important role in the expansion of NBS-LRR genes in the *Brassica* genome (Fourmann et al. 2001; Mun et al. 2009; Vicente and King 2001). Alamery (2015) has recently found 641, 443 and 249 in *B. napus*, *B.*

Table 13.1 Statistics of predicted NBS-encoding genes in sequenced *Brassica* species

Categories	Total NBS genes	Method	References
B. napus	641	MEME/MAST and CNL and TNL BLAST	Alamery (2015)
B. oleracea	443	MEME/MAST and CNL and TNL BLAST	Alamery (2015)
B. oleracea	157	HMM	Yu et al. (2014)
B. rapa	249	MEME/MAST and CNL and TNL BLAST	Alamery (2015)
B. rapa	206	HMM	Yu et al. (2014)

MEME/MAST Multiple Em for motif elicitation/Motif alignment search tool
HMM Hidden markov model profile corresponding to the Pfam NBS (NB-ARC) family PF00931 domain

oleracea and *B. rapa*, respectively (Table 13.1), whereas Yu and co-workers (2014) had previously identified 157, 206 and 167 NBS-LRR genes in *B. oleracea*, *B. rapa* and *A. thaliana* (Table 13.1). The difference in gene number resulted from different methods used (given in Table 13.1) to predict the genes. They found that the number of NBS-LRR genes in these three species was close despite the differences in genome size and complexity. This is surprising as *B. rapa* and *B. oleracea* would be expected to host a greater number of NBS-LRR genes than *Arabidopsis* because the *Brassica* genomes have undergone triplication following divergence from their common ancestor with *Arabidopsis*.

Genetic and genomics studies have demonstrated that the majority of NSB-LRR genes are present in gene clusters (as tightly linked genes) in plant genomes (Hulbert et al. 2001) conferring different resistance specificities (Leister 2004). The clustered arrangement of these genes may be a critical attribute of the generation of novel resistance specificities via gene duplication or recombination (Meyers et al. 2003). Thus, race-specific resistance might have evolved and be present in clusters resulting from tandem duplications of paralogous sequences (Kaur et al. 2009; McDowell and Simon 2006; Meyers et al. 2005). Indeed, previous studies revealed that *R*-genes are often present as tightly linked genes with high homology and are prone to gene duplication and recombination, and thus evolve more rapidly than the rest of genome (Grant et al. 1998). For this reason, functional polymorphism at *R*-loci could maintain multiple alleles or linked genes with different recognition capabilities that would recognize novel pathogen variants (Grant et al. 1998).

Alamery (2015) identified 641 candidate nucleotide-binding site–leucine-rich repeat (NBS-LRR) genes in *B. napus* using MEME/MAST analysis and CNL and TNL consensus sequences. The complexity of the *B. napus* genome duplication and recombination were attributed to causing the high number NBS-LRR genes in *B. napus*. The TNL genes comprised the majority (about 70%) of the NBS-LRR genes and showed more domain diversity compared to CNL genes (Alamery 2015). This is similar to the proportions reported in *B. rapa* (Mun et al. 2009; Yu et al. 2014), *Arabidopsis* (Meyers et al. 2003; Yu et al. 2014), *Medicago truncatula* (Ameline-Torregrosa et al. 2008), poplar (Kohler et al. 2008) and linseed (Kale et al. 2012). All 641 NBS-LRR genes were found to be distributed randomly and unevenly on all 19 chromosomes in *B. napus cv.* Darmor (Table 13.2). About 59% NBS-LRRs were located in clusters in the *B. napus* genome (Alamery 2015), compared to 61% of all NBS-LRR genes in clusters in *Arabidopsis* (Meyers et al. 2003), 50% in *M. truncatula* (Ameline-Torregrosa et al. 2008) and 58% in potato (Jupe et al. 2012). Large numbers of TNL genes were found to occur in clusters and tended to form large clusters, whereas most of the CNLs were not in clusters (singletons) (Alamery 2015).

Alamery (2015) found 641 candidate NBS-LRR genes in *B. napus*, of which 366 (57%) were classified as typical or regular NBS-LRR genes, with 124 CNLs and 242 TNLs. These 366 genes show highly conserved NBS

Table 13.2 Distribution of candidate NBS-LRR genes in *B. napus* (Alamery 2015)

Genome	Chromosome	CNL	TNL	Total
A genome *B. napus* cv. Darmor	A1	11	17	28
	A2	11	33	44
	A3	4	20	24
	A4	1	12	13
	A5	6	4	10
	A6	14	7	21
	A7	3	16	19
	A8	7	17	24
	A9	20	38	58
	A10	4	7	11
	Total	81	171	252
C Genome *B. napus* cv. Darmor	C1	13	24	37
	C2	4	47	51
	C3	10	46	56
	C4	6	12	18
	C5	8	11	19
	C6	11	33	44
	C7	8	30	38
	C8	18	21	39
	C9	17	62	79
	Total	95	286	381
	Unassigned	4	4	8
	Total	180	461	641

regions and complete open reading frames. The remaining 188 genes (29%) were classified as non-regular genes because of the lack of specific domains. These genes were classified into three distinct groups for TNLs and CNLs. Non-regular TNLs were classified as TN (74), N (9) and NL (43), whereas non-regular CNLs were classified as CN (24), N (8) and NL (23). These non-regular genes were described as partial or truncated within the N-terminal domains and/or have an absence of LRR domains. In addition, 87 TIR genes which lack both NBS and LRR domains were also identified in *B. napus*. Furthermore, 75 NBS-LRR genes (44 CNL and 31 TNL (21 TIR + 10 TNL) were encoded by a single reading frame without introns (single exon). Chalhoub et al. (2014) identified 425 nucleotide-binding site–leucine-rich repeat (NBS-LRR) sequences encoding resistance gene homologs (245 on C_n and 180 on A_n). Of these, 75% (153 A_n and 224 C_n) were syntenic to A_r and C_o progenitors. They confirmed the absence of five NBS-LRR genes from the A_n subgenome, three from the C_n subgenome and three from *B. rapa* (A_r), with none absent from *B. oleracea* C_o. This variation may reflect differential selection for resistance to diseases (Chalhoub et al. 2014).

13.6 Resistance Genes and Important Brassica Pathogens

In the three most important cultivated *Brassica* species, *B. napus, B. rapa* and *B. oleracea*, clubroot (caused by *Plasmodiophora brassicae*), downy mildew (caused by *Hyaloperonospora*

Parasitica) and blackleg (caused by *Leptosphaeria maculans*) diseases have become an increasingly serious threat to *Brassica* crops around the world and have been a major focus of *Brassica* disease resistance research in recent years (Delourme et al. 2011).

Clubroot disease, caused by the soil-borne obligate biotroph pathogen *P. brassicae*, occurs worldwide in all cruciferous vegetable and oil crops. The characteristic symptom of clubroot is the development of large, disorganized growths (clubs) on the roots of susceptible plants, leading to wilting, stunted growth and premature ripening (Hirani and Li 2015). Although, different sources of resistance against clubroot have been found in *B. oleracea*, *B. rapa* and *B. napus* (Diederichsen et al. 2009; Hirai 2006; Piao et al. 2009; Suwabe et al. 2006; Verma et al. 2014; Yu et al. 2016), only the genes *Crr1a* and *CRa*, encoding TIR-NBS-LRR proteins, have been identified and cloned from *B. rapa* (Hatakeyama et al. 2013; Ueno et al. 2012) (Table 13.3).

Blackleg is the most serious fungal disease of Brassicas worldwide; it not only causes significant (average 15–48%) yield losses, but also affects the quality of oilseed worldwide. *L. maculans* can infect any part of the plant. The fungus not only kills seedlings and young plants, but also grows systemically within the host and leads to stem canker development (Hammond et al. 1985; Howlett et al. 2001; Raman et al. 2013). To date, twenty-one major *R*-genes, controlling resistance to blackleg disease have been genetically mapped in the *Brassica* species; *B. rapa*, *B. napus*, *B. juncea* and *B. nigra*, and several of them are probably allelic variants (Balesdent et al. 2002, 2013; Delourme et al. 2006; Long et al. 2011; Rimmer 2006; Rimmer and van den Berg 1992; Tollenaere et al. 2012; Van de Wouw et al. 2009; Yu et al. 2005, 2008). Only two genes, *LepR3* and *Rlm2*, have been cloned (Larkan et al. 2013, 2015)(Table 13.3).

Downy mildew (caused by *H. parasitica*) causes damage to production of *Brassica* species worldwide. It is severely destructive to young seedlings, although the disease still causes yield and quality reduction at adult-plant stages. Several major qualitative resistance loci to downy mildew have been identified in *B. rapa* and *B. oleracea* (Carlier et al. 2012; Farinhó et al. 2007; Kim et al. 2011; Yu et al. 2009). In addition, multiple sources of resistance have been reported in *B. napus* by Ge et al. (2008) who showed that the resistance is more likely to be controlled by a major resistance gene.

There is increasing interest in the advanced use of genomic approaches for identifying new resistance genes in plants. Genome-wide studies can enhance the identification of genes that encode for resistance traits. Genome-wide motif searches identified 1134 genes in the lettuce, which were potentially involved in pathogen recognition (Christopoulou et al. 2015). In addition, Mun and co-workers (2009) used genome-wide studies to identify NBS-encoding genes in the *B. rapa* genome and identified 92 non-redundant NBS-encoding genes [30 CC-NBS-LRR (CNL) and 62 TIR-NBS-LRR (TNL) genes] in approximately 100 Mbp of *B. rapa* euchromatic genome sequence. Similarly, Chen et al. (2015) found numerous resistance gene analogues genes in cotton genome, so that almost half of them were found to be located in clusters, which evolved by sequence exchanges,

Table 13.3 Resistance genes cloned from *Brassicas*

Species	Resistance gene	Pathogen	Gene location	Type	References
B. rapa	*Crr1a*	*Plasmodiophora brassicae*	A 08	TIR-NBS-LRR	Hatakeyama et al. (2013)
	CRa		A 03	TIR-NBS-LRR	Ueno et al. (2012)
B. napus	*Rlm2*	*Leptosphaeria maculans*	A 10	Cf-9	Larkan et al. (2015)
	LepR3		A 10	Cf-9	Larkan et al. (2013), (2015)

tandem duplications and segmental duplications. Although many outstanding studies have been conducted on plants molecular resistance mechanisms, our overall knowledge about this is still very limited. For example, we still know too little about the structural basis of pathogen recognition (McDowell and Woffenden 2003). It is expected that more *R* genes will be identified and cloned using functional genomic tools, which greatly accelerate the speed of discovery and provide new insights into interactions between plants and pathogens. Accordingly, genome-wide identification of *R*-genes in plant genomes would give insights into the evolution of disease resistance genes and help in functional validation of these genes and also to understand molecular mechanism of disease resistance and their evolution.

References

Alamery SF (2015) Genome wide identification of NBS-LRR genes in *Brassica* and their association with disease resistance in *Brassica napus*. PhD Thesis, The University of Queensland. School of Agriculture and Food Sciences

Ameline-Torregrosa C, Wang BB, O'Bleness MS, Deshpande S, Zhu H, Roe B, Young ND, Cannon SB (2008) Identification and characterization of nucleotide-binding site-leucine-rich repeat genes in the model plant *Medicago truncatula*. Plant Physiol 146:5–21

Balesdent MH, Attard A, Kühn ML, Rouxel T (2002) New avirulence genes in the phytopathogenic fungus *Leptosphaeria maculans*. Phytopathology 92:1122–1133

Balesdent MH, Fudal I, Ollivier B, Bally P, Grandaubert J, Eber F, Chevre AM, Leflon M, Rouxel T (2013) The dispensable chromosome of *Leptosphaeria maculans* shelters an effector gene conferring avirulence towards *Brassica rapa*. The New Phytologist 198:887–898

Carlier JD, Alabaça CA, Coelho PS, Monteiro AA, Leitão JM (2012) The downy mildew resistance locus *Pp523* is located on chromosome C8 of *Brassica oleracea* L. Plant Breeding 131:170–175

Chalhoub B, Denoeud F, Liu S, Parkin IAP, Tang H, Wang X, Chiquet J, Belcram H, Tong C, Samans B, Corréa M, Da Silva C, Just J, Falentin C, Koh CS, Le Clainche I, Bernard M, Bento P, Noel B, Labadie K, Alberti A, Charles M, Arnaud D, Guo H, Daviaud C, Alamery S, Jabbari K, Zhao M, Edger PP, Chelaifa H, Tack D, Lassalle G, Mestiri I, Schnel N, Le Paslier M-C, Fan G, Renault V, Bayer PE, Golicz AA, Manoli S, Lee T-H, Thi VHD, Chalabi S, Hu Q, Fan C, Tollenaere R, Lu Y, Battail C, Shen J, Sidebottom CHD, Wang X, Canaguier A, Chauveau A, Bérard A, Deniot G, Guan M, Liu Z, Sun F, Lim YP, Lyons E, Town CD, Bancroft I, Wang X, Meng J, Ma J, Pires JC, King GJ, Brunel D, Delourme R, Renard M, Aury J-M, Adams KL, Batley J, Snowdon RJ, Tost J, Edwards D, Zhou Y, Hua W, Sharpe AG, Paterson AH, Guan C, Wincker P (2014) Early allopolyploid evolution in the post-Neolithic *Brassica napus* oilseed genome. Science 345:950–953

Chen J-Y, Huang J-Q, Li N-Y, Ma X-F, Wang J-L, Liu C, Liu Y-F, Liang Y, Bao Y-M, Dai X-F (2015) Genome-wide analysis of the gene families of resistance gene analogues in cotton and their response to verticillium wilt. BMC Plant Biol 15:148

Chisholm ST, Coaker G, Day B, Staskawicz BJ (2006) Host-microbe interactions: shaping the evolution of the plant immune response. Cell 124:803–814

Christopoulou M, Wo SR-C, Kozik A, McHale LK, Truco M-J, Wroblewski T, Michelmore RW (2015) Genome-wide architecture of disease resistance genes in lettuce. G3: Genes, Genomes, Genetics 5:2655–2669

Cui H, Tsuda K, Parker JE (2015) Effector-triggered immunity: from pathogen perception to robust defense. Annu Rev Plant Biol 66:487–511

Dangl JL, Jones JD (2001) Plant pathogens and integrated defence responses to infection. Nature 411:826–833

Dangl JL, Horvath DM, Staskawicz BJ (2013) Pivoting the plant immune system from dissection to deployment. Science 341:746–751

Delourme R, Chèvre AM, Brun H, Rouxel T, Balesdent MH, Dias JS, Salisbury P, Renard M, Rimmer SR (2006) Major gene and polygenic resistance to *Leptosphaeria maculans* in oilseed rape (*Brassica napus*). Eur J Plant Pathol 114:41–52

Delourme R, Barbetti MJ, Snowdon R, Zhao J, Manazanares-Dauleux M (2011) Genetics and genomics of disease resistance. Genetics, genomics and breeding of oilseed brassicas. Science Publishers, Boca Raton, FL, PP. 276–318

Diederichsen E, Frauen M, Linders EG, Hatakeyama K, Hirai M (2009) Status and perspectives of clubroot resistance breeding in crucifer crops. J Plant Growth Regul 28:265–281

Farinhó M, Coelho P, Monteiro A, Leitão J (2007) SCAR and CAPS markers flanking the *Brassica oleracea* L. *Pp523* downy mildew resistance locus demarcate a genomic region syntenic to the top arm end of *Arabidopsis thaliana* L. chromosome 1. Euphytica 157:215–221

Flor HH (1971) Current status of the gene-for-gene concept. Annu Rev Phytopathol 9:275–296

Food and agriculture organization of the United Nations. (2014). http://www.fao.org/home/en/. Accessed 15 Dec 2014

Fourmann M, Charlot F, Froger N, Delourme R, Brunel D (2001) Expression, mapping, and genetic variability of *Brassica napus* disease resistance gene analogues. Genome 44:1083–1099

Freeman BC, Beattie GA (2008) An overview of plant defenses against pathogens and herbivores. The Plant Health Instructor

Fu ZQ, Dong X (2013) Systemic acquired resistance: turning local infection into global defense. Annu Rev Plant Biol 64:839–863

Ge X, Li H, Han S, Sivasithamparam K, Barbetti M (2008) Evaluation of Australian *Brassica napus* genotypes for resistance to the downy mildew pathogen, *Hyaloperonospora parasitica*. Crop Pasture Science 59:1030–1034

Grant MR, McDowell JM, Sharpe AG, de Torres Zabala M, Lydiate DJ, Dangl JL (1998) Independent deletions of a pathogen-resistance gene in *Brassica* and *Arabidopsis*. Proc Natl Acad Sci 95:15843–15848

Gururani MA, Venkatesh J, Upadhyaya CP, Nookaraju A, Pandey SK, Park SW (2012) Plant disease resistance genes: current status and future directions. Physiol Mol Plant Pathol 78:51–65

Hammond KE, Lewis BG, Musa TM (1985) A systemic pathway in the infection of oilseed rape plants by *Leptosphaeria maculans*

Hatakeyama K, Suwabe K, Tomita RN, Kato T, Nunome T, Fukuoka H, Matsumoto S (2013) Identification and characterization of *Crr1a*, a gene for resistance to clubroot disease (*Plasmodiophora brassicae Woronin*) in *Brassica rapa* L. PLoS ONE 8:e54745

Hirai M (2006) Genetic analysis of clubroot resistance in *Brassica* crops. Breed Sci 56:223–229

Hirani AH, Li G (2015) Understanding the genetics of clubroot resistance for effectively controlling this disease in *Brassica* species

Howlett BJ, Idnurm A, Pedras MS (2001) *Leptosphaeria maculans*, the causal agent of blackleg disease of *Brassicas*. Fungal Genet Biol FG & B 33:1–14

Hulbert SH, Webb CA, Smith SM, Sun Q (2001) Resistance gene complexes: evolution and utilization. Annu Rev Phytopathol 39:285–312

Johal G, Briggs S (1992) Reductase activity encoded by the *HM1* disease resistance gene in maize. Science 258:985–987

Johnson RD, Lewis BG (1994) Variation in host range, systemic infection and epidemiology of *Leptosphaeria maculans*. Plant Pathol 43:269–277

Jones JD, Dangl JL (2006) The plant immune system. Nature 444:323–329

Jupe F, Pritchard L, Etherington GJ, Mackenzie K, Cock PJ, Wright F, Sharma SK, Bolser D, Bryan GJ, Jones JD, Hein I (2012) Identification and localisation of the NB-LRR gene family within the potato genome. BMC Genom 13:75

Kale SM, Pardeshi VC, Barvkar VT, Gupta VS, Kadoo NY (2012) Genome-wide identification and characterization of nucleotide binding site leucine-rich repeat genes in linseed reveal distinct patterns of gene structure. Genome 56:91–99

Katagiri F, Tsuda K (2010) Understanding the plant immune system. Mol Plant Microbe Interact 23:1531–1536

Kaur S, Cogan N, Ye G, Baillie R, Hand M, Ling A, Mcgearey A, Kaur J, Hopkins C, Todorovic M (2009) Genetic map construction and QTL mapping of resistance to blackleg (*Leptosphaeria maculans*) disease in Australian canola (*Brassica napus* L.) cultivars. Theor Appl Genet 120:71–83

Kim S, Song YH, Lee J-Y, Choi SR, Dhandapani V, Jang CS, Lim YP, Han T (2011) Identification of the *BrRHP1* locus that confers resistance to downy mildew in Chinese cabbage (*Brassica rapa* ssp. *pekinensis*) and development of linked molecular markers. Theor Appl Genet 123:1183–1192

Kohler A, Rinaldi C, Duplessis S, Baucher M, Geelen D, Duchaussoy F, Meyers BC, Boerjan W, Martin F (2008) Genome-wide identification of NBS resistance genes in *Populus trichocarpa*. Plant Mol Biol 66:619–636

Larkan NJ, Lydiate DJ, Parkin IAP, Nelson MN, Epp DJ, Cowling WA, Rimmer SR, Borhan MH (2013) The *Brassica napus* blackleg resistance gene *LepR3* encodes a receptor-like protein triggered by the *Leptosphaeria maculans* effector AVRLM1. New Phytol 197:595–605

Larkan NJ, Ma L, Borhan MH (2015) The Brassica napus receptor-like protein RLM2 is encoded by a second allele of the *LepR3/Rlm2* blackleg resistance locus. Plant Biotechnol J 13:983–992

Leister D (2004) Tandem and segmental gene duplication and recombination in the evolution of plant disease resistance genes. Trends Genet 20:116–122

Liu S, Liu Y, Yang X, Tong C, Edwards D, Parkin IAP, Zhao M, Ma J, Yu J, Huang S, Wang X, Wang J, Lu K, Fang Z, Bancroft I, Yang T-J, Hu Q, Wang X, Yue Z, Li H, Yang L, Wu J, Zhou Q, Wang W, King GJ, Pires JC, Lu C, Wu Z, Sampath P, Wang Z, Guo H, Pan S, Yang L, Min J, Zhang D, Jin D, Li W, Belcram H, Tu J, Guan M, Qi C, Du D, Li J, Jiang L, Batley J, Sharpe AG, Park B-S, Ruperao P, Cheng F, Waminal NE, Huang Y, Dong C, Wang L, Li J, Hu Z, Zhuang M, Huang Y, Huang J, Shi J, Mei D, Liu J, Lee T-H, Wang J, Jin H, Li Z, Li X, Zhang J, Xiao L, Zhou Y, Liu Z, Liu X, Qin R, Tang X, Liu W, Wang Y, Zhang Y, Lee J, Kim HH, Denoeud F, Xu X, Liang X, Hua W, Wang X, Wang J, Chalhoub B, Paterson AH (2014) The *Brassica oleracea* genome reveals the asymmetrical evolution of polyploid genomes. Nat Commun 5

Long Y, Wang Z, Sun Z, Fernando DW, McVetty PB, Li G (2011) Identification of two blackleg resistance genes and fine mapping of one of these two genes in a *Brassica napus* canola cultivar 'Surpass 400'. Theor Appl Genet 122:1223–1231

McDowell JM, Woffenden BJ (2003) Plant disease resistance genes: recent insights and potential applications. Trends Biotechnol 21:178–183
McDowell JM, Simon SA (2006) Recent insights into R gene evolution. Molecular Plant Pathology 7:437–448
Meyers BC, Kozik A, Griego A, Kuang HH, Michelmore RW (2003) Genome-wide analysiso f NBS-LRR-encoding genes in *Arabidopsis*. Plant Cell 15:809–834
Meyers BC, Kaushik S, Nandety RS (2005) Evolving disease resistance genes. Curr Opin Plant Biol 8:129–134
Monosi B, Wisser R, Pennill L, Hulbert S (2004) Full-genome analysis of resistance gene homologues in rice. Theor Appl Genet 109:1434–1447
Mun JH, Yu HJ, Park S, Park BS (2009) Genome-wide identification of NBS-encoding resistance genes in *Brassica rapa*. Mol Genet Genomics 282:617–631
Pandolfi V, Neto JR, Silva MD, Amorim LL, Wanderley-Nogueira AC, de Oliveira Silva RL, Kido EA, Crovella S, Iseppon AM (2016) Resistance (R) genes: applications and prospects for plant biotechnology and breeding. Curr Protein pept Sci
Parkin I, Koh C, Tang H, Robinson S, Kagale S, Clarke W, Town C, Nixon J, Krishnakumar V, Bidwell S, Denoeud F, Belcram H, Links M, Just J, Clarke C, Bender T, Huebert T, Mason A, Pires C, Barker G, Moore J, Walley P, Manoli S, Batley J, Edwards D, Nelson M, Wang X, Paterson A, King G, Bancroft I, Chalhoub B, Sharpe A (2014) Transcriptome and methylome profiling reveals relics of genome dominance in the mesopolyploid *Brassica oleracea*. Genome Biol 15:R77
Paterson AH, Bowers JE, Bruggmann R, Dubchak I, Grimwood J, Gundlach H, Haberer G, Hellsten U, Mitros T, Poliakov A (2009) The *Sorghum bicolor* genome and the diversification of grasses. Nature 457:551–556
Peele HM, Guan N, Fogelqvist J, Dixelius C (2014) Loss and retention of resistance genes in five species of the *Brassicaceae* family. BMC Plant Biol 14:298
Piao Z, Ramchiary N, Lim YP (2009) Genetics of clubroot resistance in *Brassica* species. J Plant Growth Regul 28:252–264
Pink DA (2002) Strategies using genes for non-durable disease resistance. Euphytica 124:227–236
Porter BW, Paidi M, Ming R, Alam M, Nishijima WT, Zhu YJ (2009) Genome-wide analysis of *Carica papaya* reveals a small NBS resistance gene family. Mol Genet Genomics 281:609–626
Raman H, Raman R, Larkan N (2013) Genetic dissection of blackleg resistance loci in rapeseed (*Brassica napus* L.). Plant Breed Lab Fields 4:86–120
Rimmer SR (2006) Resistance genes to *Leptosphaeria maculans* in *Brassica napus*. Can J Plant Path 28: S288–S297
Rimmer SR, van den Berg CGJ (1992) Resistance of oilseed *Brassica* spp. to blackleg caused by *Leptosphaeria maculans*. Can J Plant Path 14:56–66
Sanseverino W, Hermoso A, D'Alessandro R, Vlasova A, Andolfo G, Frusciante L, Lowy E, Roma G, Ercolano MR (2013) PRGdb 2.0: towards a community-based database model for the analysis of R-genes in plants. Nucleic Acids Res 41:D1167–D1171
Steuernagel B, Periyannan SK, Hernández-Pinzón I, Witek K, Rouse MN, Yu G, Hatta A, Ayliffe M, Bariana H, Jones JD (2016) Rapid cloning of disease-resistance genes in plants using mutagenesis and sequence capture. Nat Biotechnol 34:652–655
Suwabe K, Tsukazaki H, Iketani H, Hatakeyama K, Kondo M, Fujimura M, Nunome T, Fukuoka H, Hirai M, Matsumoto S (2006) Simple sequence repeat-based comparative genomics between *Brassica rapa* and *Arabidopsis thaliana*: the genetic origin of clubroot resistance. Genetics 173:309–319
Tan X, Meyers BC, Kozik A, West MA, Morgante M, St Clair DA, Bent AF, Michelmore RW (2007) Global expression analysis of nucleotide binding site-leucine rich repeat-encoding and related genes in *Arabidopsis*. BMC Plant Biol 7:1
Tollenaere R, Hayward A, Dalton-Morgan J, Campbell E, Lee JR, Lorenc MT, Manoli S, Stiller J, Raman R, Raman H, Edwards D, Batley J (2012) Identification and characterization of candidate *Rlm4* blackleg resistance genes in *Brassica napus* using next-generation sequencing. Plant Biotechnol J 10:709–715
Ueno H, Matsumoto E, Aruga D, Kitagawa S, Matsumura H, Hayashida N (2012) Molecular characterization of the *CRa* gene conferring clubroot resistance in *Brassica rapa*. Plant Mol Biol 80:621–629
Van de Wouw AP, Marcroft SJ, Barbetti MJ, Hua L, Salisbury PA, Gout L, Rouxel T, Howlett BJ, Balesdent MH (2009) Dual control of avirulence in *Leptosphaeria maculans* towards a *Brassica napus* cultivar with '*sylvestris-derived*' resistance suggests involvement of two resistance genes. Plant Pathol 58:305–313
Verma SS, Rahman MH, Deyholos MK, Basu U, Kav NN (2014) Differential expression of miRNAs in *Brassica napus* root following infection with *Plasmodiophora* brassicae. PLoS ONE 9:e86648
Vicente JG, King GJ (2001) Characterisation of disease resistance gene-like sequences in *Brassica oleracea* L. Theor Appl Genet 102:555–563
Wang X, Wang H, Wang J, Sun R, Wu J, Liu S, Bai Y, Mun J-H, Bancroft I, Cheng F, Huang S, Li X, Hua W, Wang J, Wang X, Freeling M, Pires JC, Paterson AH, Chalhoub B, Wang B, Hayward A, Sharpe AG, Park B-S, Weisshaar B, Liu B, Li B, Liu B, Tong C, Song C, Duran C, Peng C, Geng C, Koh C, Lin C, Edwards D, Mu D, Shen D, Soumpourou E, Li F, Fraser F, Conant G, Lassalle G, King GJ, Bonnema G, Tang H, Wang H, Belcram H, Zhou H, Hirakawa H, Abe H, Guo H, Wang H, Jin H, Parkin IAP, Batley J, Kim J-S, Just J, Li J, Xu J, Deng J, Kim JA, Li J, Yu J, Meng J, Wang J, Min J,

Poulain J, Wang J, Hatakeyama K, Wu K, Wang L, Fang L, Trick M, Links MG, Zhao M, Jin M, Ramchiary N, Drou N, Berkman PJ, Cai Q, Huang Q, Li R, Tabata S, Cheng S, Zhang S, Zhang S, Huang S, Sato S, Sun S, Kwon S-J, Choi S-R, Lee T-H, Fan W, Zhao X, Tan X, Xu X, Wang Y, Qiu Y, Yin Y, Li Y, Du Y, Liao Y, Lim Y, Narusaka Y, Wang Y, Wang Z, Li Z, Wang Z, Xiong Z, Zhang Z (2011) The genome of the mesopolyploid crop species *Brassica rapa*. Nat Genet 43:1035–1039

Wu L, Chen H, Curtis C, Fu ZQ (2014) Go in for the kill: how plants deploy effector-triggered immunity to combat pathogens. Virulence 5:710–721

Yu F, Lydiate DJ, Rimmer SR (2005) Identification of two novel genes for blackleg resistance in *Brassica napus*. Theor Appl Genet 110:969–979

Yu F, Lydiate DJ, Rimmer SR (2008) Identification and mapping of a third blackleg resistance locus in *Brassica napus* derived from *B. rapa* subsp. *sylvestris*. Genome 51:64–72

Yu S, Zhang F, Yu R, Zou Y, Qi J, Zhao X, Yu Y, Zhang D, Li L (2009) Genetic mapping and localization of a major QTL for seedling resistance to downy mildew in Chinese cabbage (*Brassica rapa* ssp. *pekinensis*). Mol Breeding 23:573–590

Yu J, Tehrim S, Zhang FQ, Tong CB, Huang JY, Cheng XH, Dong CH, Zhou YQ, Qin R, Hua W (2014) Genome-wide comparative analysis of NBS-encoding genes between *Brassica* species and *Arabidopsis thaliana*. BMC Genom 15:3

Yu F, Zhang X, Huang Z, Chu M, Song T, Falk KC, Deora A, Chen Q, Zhang Y, McGregor L (2016) Identification of genome-wide variants and discovery of variants associated with *brassica rapa* clubroot resistance gene *Rcr1* through bulked segregant rna sequencing. PLoS ONE 11:e0153218

Brassica napus Genomic Resources 14

Graham J. King and Abdul Baten

Abstract

This chapter provides an overview of the various physical experimental and reference data resources available in the public domain to support evolutionary, comparative and functional analysis of *Brassica napus* genomes, and that underpin pre-breeding R&D and breeding activities for canola and related oilseed crops. Many of these resources have become available as a result of international cooperation through the Multinational *Brassica* Genome Project (MBGP). The role and establishment of nomenclature and other standards for *Brassica* species is outlined. Specific plant resources and their use in mutant screening, mapping and genome-wide association studies are described, along with genomic tools such as genetic marker and transcriptome platforms. The range of current genomic data, information resources, bioinformatics tools and analysis pipelines for *B. napus* are introduced, together with the available dedicated *B. napus*-specific genome browsers and related online sites. In addition, databases dedicated to managing phenotypic trait, trial and related data are described, and future requirements are identified for enabling greater integration of phenotypic and genotypic information and tools to collate and navigate increasingly complex data sets. Due to rapidly changing genomic technologies and funding support, the coverage of resources described here is unlikely to be comprehensive. However, many additional details are to be found either hosted at or linked from the http://www.brassica.info website on behalf of the MBGP. The *Brassica* database (BRAD) maintained at http://Brassicadb.org/brad/ also continues to provide a valuable set of reference information.

14.1 Introduction

Brassica napus (AC genome, $n = 19$) has undergone at least two rounds of domestication and appears to have arisen during the last ~ 7000 years as an allodiploid (amphidiploid) from the domesticated diploids *Brassica rapa* (A genome, $n = 10$) and *Brassica oleracea* (C genome, $n = 9$). The second round of domestication commenced around 40 years ago with the genetic selection and introduction of genotypes carrying rare alleles contributing to 'double-low' seed erucic acid and glucosinolate content. The subsequent adoption of these canola-type rapeseeds worldwide has led to *B. napus* becoming the third largest oilseed crop. *B. napus* also contribute to minor vegetable and fodder crops, including swede turnip forms. This unique

G. J. King (✉) · A. Baten
Southern Cross Plant Science, Southern Cross University, Lismore PO Box 157 NSW 2480, Australia
e-mail: graham.king@scu.edu.au

S. Liu et al. (eds.), *The Brassica napus Genome*, Compendium of Plant Genomes,
https://doi.org/10.1007/978-3-319-43694-4_14

heritage means that relevant genomic resources arise not only from *B. napus* itself, but also from the progenitor species.

The Multinational *Brassica* Genome Project was established as a collective of *Brassica* researchers who had a common interest in developing public domain and commercially pre-competitive experimental resources and analysis platforms. The A genome of *B. rapa* was nominated as the first *Brassica* to be sequenced (Wang et al. 2011a, b) followed by the C genome of *B. oleracea* (Parkin et al. 2014; Liu et al. 2014) and the AC genome of *B. napus* (Chalhoub et al. 2014). Resources for both *B. rapa* and *B. oleracea* are therefore pertinent to *B. napus*. It is apparent from recent re-sequencing studies that the *Brassica* pan-genome (Bayer et al. 2017; Hurgobin et al. 2017) consists of conserved chromosomal regions interspersed with more variable sections, some of which appear to be hot spots of variation. Moreover, hybridization during modern breeding is likely to result in rapid remodelling of the genome (Szadkowski et al. 2010).

Brassica researchers have benefited from the relatively close evolutionary relationship with the model plant species *Arabidopsis thaliana*. The sequencing of Arabidopsis (The *Arabidopsis* Genome Initiative 2000) and ongoing functional annotation have provided a rich source of candidate genes that predominantly have been able to predict corresponding function within crop brassicas. There are exceptions and subtleties to these orthologous relationships, where ancestral whole-genome duplications (WGDs) have inflated paralogous copy numbers so that single-copy genes in Arabidopsis are represented by up to six or more copies in *B. napus*, arising from the combined A and C diploid triplicated loci. Although these segmental duplication events have provided the opportunity for neo- and sub-functionalization (Schiessl et al. 2017), there is overwhelming evidence for a process of gene loss following WGD (Mun et al. 2009; Wang et al. 2011a, b; Chalhoub et al. 2014).

In outlining the availability of any existing experimental resources, it is important to recognize their often transitory nature, due to turnover of genomic technologies, as well as requirements for ongoing funding support. Whilst 'soft' genomic resources such as genome sequences, associated annotation and bioinformatics pipelines may appear to be relatively easy to coordinate and maintain with distributed effort, they do require well-described and comprehensive meta-data in order to have lasting value. For physical genetic and genomic resources, ongoing maintenance is often more challenging. For example, whilst there has been ongoing research community support for maintaining seed and other resources in repositories such as the Arabidopsis Biological Resource Center (ABRC; https://abrc.osu.edu/) and Tomato Genetic Resources Center (TGRC; http://tgrc.ucdavis.edu/), such a capability has not been coordinated for *Brassica* crop researchers, beyond the FastPlants (https://fastplants.org/) initiative originally established by Paul Williams in Wisconsin, along with the international network of ex situ crop genetic resources. The latter are increasingly being mined for genome association studies, and yet the maintenance of specific plant lineages relating to reported allelic sequences is typically not within the remit of such seed bank institutions. However, there are some exceptions to this situation, and where these exist for *Brassica* genomic resources, they are described in later sections.

In contrast, for many data-based genomic resources, the future is more promising with availability of cloud storage, and tools for the establishment of global standards within research communities expected to contribute to data persistence. Recognition of the need for establishment of data standards and adoption of FAIR (Findable, Accessible, Interoperable and Re-usable) data principles (Wilkinson et al. 2016) are driving efforts to establish a more cohesive and cumulative set of common resources. In the current era, the Multinational *Brassica* Genome Project (MBGP) steering group will continue to play a key role in communicating and coordinating efforts worldwide.

In this chapter, an overview of various genomic experimental and data resources available for *B. napus* is provided. Whilst not comprehensive, many additional details are to be found either hosted at or linked from the http://www.brassica.

info website on behalf of the MBGP. The *Brassica* database (BRAD) maintained at http://Brassicadb.org/brad/ also provides a valuable set of reference information.

14.2 Nomenclature and Other Standards

The full benefit from the accumulated data that continue to be generated for *Brassica* genomics will be greatly enhanced where it is possible to describe, annotate and locate existing experimental resources and data sets unambiguously. To maximize the benefits from sharing physical and data resources, it is important to avoid ambiguity and be able to provide quality assurance and clarity. This places a premium on adopting explicit data definitions, and recognition and resolution of, for example, context-specific synonyms and homonyms. Although common nomenclature systems have been established for, e.g. *Brassica* chromosomes and functional genes, for other key entities such as plant populations, cultivars and derived lines and phenotypic trait descriptors, the widespread adoption of standards has yet to be achieved. For plant materials, this is of increasing importance where identification and tracking of specific allelic sequence variants require attribution to explicit lineages. More generally, the lack of consistent naming systems and look-up registries currently limits the ability to carry out comparative analysis in silico, particularly in relating trait genetics for multiple studies with annotated genomes. It is hoped that in the near future, the MBGP oversee the establishment of universally accessible data registries for a wide range of *B. napus* bio-samples (Barrett et al. 2012), along with encouraging the use of data description approaches based on crop ontologies and development of a comprehensive *B. napus* trait dictionary.

14.2.1 Plant Materials and Populations

Advances in genomics research are based on access to fixed (e.g. homozygous) reference biological samples. For *B. napus*, the establishment of correctly ordered and oriented reference genomic sequences has been dependent on anchoring to dense genetic linkage maps developed from segregating populations derived from biparental crosses. Relevant plant materials are discussed below (Sect. 14.3).

14.2.2 Chromosomes and Linkage Groups

The agreed standard for naming *Brassica* A and C chromosomes was ratified by the MBGP steering committee in 2007. This assigned consistent chromosome/linkage group nomenclature to the canonical diploid *Brassica* genomes in the 'triangle of U' as follows: 'A', 'B' and 'C', with the A and C assignments of A1-A10 and C1-C9 based on the founding evidence presented by Parkin et al. (1995) and Sharpe et al. (1995), and subsequently adopted in published linkage maps and genome sequences.

14.2.3 Gene Models

The adoption of fixed standards for annotated *Brassica* genomes has to date not been consistent and is currently in the process of harmonization. There is a strong requirement to ensure that ambiguity is reduced so that homonyms do not arise—i.e. similar or identical string identifiers adopted for distinct gene loci. The establishment of a universally agreed standard needs to take into account not only the observed inflation/deflation of chromosome gene orders which typify the pan-genome (He et al. 2015; Bayer et al. 2017), but also the relatively frequent phenomenon of reciprocal and non-reciprocal exchange of chromosomal segments (Sharpe et al. 1995; Hurgobin et al. 2017). *B. napus* was the first genome (Chalhoub et al. 2014) for which the gene nomenclature standards proposed by Multinational *Brassica* Genome Project (MBGP) were adopted.

The current MBGP gene model standard is of the form: BnaC01g010030.1D, comprised formally of the genus/species and genome designations, which follow the convention of Østergaard and King (2008) (http://www.Brassica.info/info/genome_annotation.php), with chromosome numbers assigned with leading zero (thus BnaC01 for chromosome C01 in *B. napus).* Gene models are assigned numbers decatonically (e.g. 10, 20, 30) with 5-digit leading zero integers from top to bottom of correctly orientated pseudochromosome sequence. This allows for additional or alternative gene models to be inserted. A default version number of 1 (e.g. for different splicing models) is assigned after a '.'. Following this, in order to distinguish between reference sequences from different plant genotypes (e.g. Darmor-bzh and Tapidor for *B. napus*), a single capital letter is allocated (e.g. 'D' or 'T').

14.2.4 Gene Functional Names

A standardized nomenclature was proposed (Østergaard and King 2008) for genes described within the *Brassica* genus. This enables a distinction to be made between copies associated with the different haploid genomes, as well as at paralogous loci. This nomenclature system is valuable where a body of experimental evidence has validated gene function and can supplement the annotation above, as is the case for many other species. For example, the *FRUITFULL* functional gene in Arabidopsis is designated as *FUL* and associated with gene model AT5G60910.1, whereas in *B. napus,* the orthologues have the functional gene names *BnaA.FUL.a* (gene model = BnaA03g39830D) *BnaA.FUL.b, BnaC.FUL.a,* etc.

14.2.5 Trials, Phenotypic Traits and Other Standards

At the present time, the *Brassica* Trait Ontology (BRaTO; https://github.com/Brassica-Trait-Ontology/brato) is being constructed to host trait information to describe *Brassica* crop data, and is based on the Crop Ontology Trait Dictionary (Shrestha et al. 2012). Terms are being collected from various projects, with development currently supported by researchers in the UK, Australia and France. A compilation of identifiers are defined for Variables, Trait, Methods and Scales, comprised of an ontology code given by CropOntology.org (CO_348), followed by 7 digits, where digits and ontology code are separated by a colon in the form: CO_348:XXXXXXX.

14.3 Plant Resources

The availability of complete genome sequence data has enabled an increasing number of studies to unravel allelic variation across the relevant gene pools. For *B. napus,* this has provided opportunity to explore not only breeding pedigrees, historical cultivars and landraces, but also the secondary gene pools of *B. rapa* and *B. oleracea.* Genome-wide association studies (GWAS) are now capitalizing on efforts over preceding decades in collecting and conserving genetic resources in gene seed banks and in breeders' collections. From these ex situ resource collections, a series of representative core collections, diversity sets and GWAS panels have been derived. These include those for which extensive multi-environment phenotyping has taken place, as well as SNP screening using technologies of increasing resolution that are uncovering the pattern of pan-genome variation.

14.3.1 *B. napus* Diversity Collections

The ERANET-ASSYST collection comprises a diversity set of over 500 cultivar lines and continues to be used for a range of studies (Körber et al. 2015). The OREGIN *B. napus* core diversity collection was assembled in the UK and underwent a process of inbreeding and generation of doubled haploids in order to reduce the level of heterozygosity. The resulting *B. napus* diversity fixed foundation set has been assessed in replicated trials for a wide range of traits (Bennett et al. 2017), with phenotypic data collated in the *Brassica* Information Portal (https://bip.earlham.ac.uk/). Other sets based on a range of different germplasm collections have been used, although it is unclear whether the materials are generally available (Lu et al. 2016; Wang et al. 2018).

14.3.2 Genetic Mapping Populations

The establishment of reference biparental segregating populations for *B. napus* was essential for the establishment of initial genetic linkage maps, and their subsequent expansion and increase in density in order to anchor genome sequences. They have also been valuable for a very wide range of quantitative trait loci (QTL) studies. In *B. napus,* such populations have often made use of anther culture and microspore technology to generate doubled haploid lines from F1 hybrids, and so establish unselected 'immortal' populations. The value of such populations lies in their ability to establish a deeper understanding of genotype × environment (G × E) interactions, as well as pleiotropic effects of single loci and epistatic interactions between loci from replicated trials. Summaries of these and other populations are outlined at http://www.brassica.info/resource/plant.php.

14.3.3 Mutant Populations

It is unclear how many EMS *B. napus* mutant populations are currently available, although a number have been developed in the past (Wang et al. 2008; Gilchrist et al. 2013). However, for studies representing the A genome, the EMS-TILLING population of *B. rapa* var *trilocularis* line R-o-18 (Stephenson et al. 2010), now managed by RevGenUK, has been particularly successful and enabled a wide range of functional gene characterization studies to progress. A related chemically induced hypomethylated population of R-o-18 has also been generated and initial characterization carried out to demonstrate value of this approach for studying epigenetic variation (Amoah et al. 2012).

14.4 Genomic Tools

Completed genome sequences provide an ongoing resource for many experimental studies, although their value is only fully realized when combined with tools such as DNA markers that are able to provide associations between specific loci, alleles and transcripts along with phenotypic trait variation. As for other organisms, there has been a progressive development of technology platforms for *B. napus* to facilitate whole-genome analysis, whilst reducing per-allele call cost.

14.4.1 Marker Platforms

14.4.1.1 SSRs

Microsatellites (simple sequence repeats, SSRs) and other locus-specific marker platforms have been developed over the past 20–25 years, with increasing density and specificity for *B. napus*. Early efforts generated progressively larger sets of SSR markers and generated multiplexed assays (e.g. Tommasini et al. 2003; Piquemal et al. 2005; Iniguez-Luy et al. 2009; Radoev et al. 2008). A systematic re-evaluation of existing SSR markers was carried out (Li et al. 2013) and identified 2,701 putative single-locus monotypic amplicons in *B. napus*, from which a set of 230 high-quality SSR markers was established and mapped onto the 19 chromosomes. Commercial platforms of accumulated *B. napus* SSRs are also available from companies such as TraitGenetics (www.traitgenetics.com).

14.4.1.2 Single Nucleotide Polymorphisms (SNPs)

A number of dedicated SNP platforms have been established for *B. napus*, whilst other technologies make use of genotype-by-sequencing (GBS) or associated re-sequencing approaches (Schmutzer et al. 2015). A pioneering effort to detect SNPs systematically in the *B. napus* genome involved transcriptome sequencing (Trick et al. 2009). This was followed by increasingly refined re-sequencing of transcriptome (Harper et al. 2012). A NimbleGen Inc. platform has been described (Delourme et al. 2013), as well as a DArT (Diversity Arrays Technology) platform with a 1,359 anchored array which allowed construction of a consensus map from six *B. napus* segregating populations (Raman et al. 2013). More recently, a high-density single nucleotide polymorphism (SNP) Illumina

Infinium® array has been developed for *B. napus* based on a consortium approach to generating sufficient representative SNP discovery data sets from a range of germplasm. This contains 52,157 markers (Clarke et al. 2016; Mason et al. 2017) and is being used for a wide range of association and breeding studies (Qu et al. 2017; Wang et al. 2017; Wei et al. 2016). A specific-locus amplified fragment sequencing (SLAF)-based platform has been developed that identified a subset of 201,817 SNPs with minor allele frequency > 0.05 and has successfully been used in linkage disequilibrium (LD) studies (Zhou et al. 2017).

14.4.1.3 Transcriptome

In the early stages of genome sequencing and accumulation of EST transcript sequences, the Affymetrix GeneChip® *Brassica* Exon 1.0 ST Array (Love et al. 2010) based on 135,000 ESTs was generated by pooling resources from consortia within the MBGP. This has successfully been used in a number of studies and commercially for *B. napus* and the constituent diploid A genome (Graham et al. 2014). Due to decreasing costs, RNA-seq is effectively replacing such platforms. Published data for RNA-seq are not managed in a dedicated system for *B. napus*, but are searchable via SRA (https://www.ncbi.nlm.nih.gov/sra) or ENA (https://www.ebi.ac.uk/ena).

14.4.1.4 BAC Libraries

Prior to the widespread adoption of NGS sequencing platforms for whole-genome sequencing, the development of bacterial artificial chromosome (BAC) insert libraries of ~60–180 kbp was effective in facilitating genome sequencing of 'golden paths' of inserts anchored to specific linkage groups/chromosomes of *Brassica*. Whilst this approach was effective in establishing an early whole pseudochromosome sequence for *Brassicarapa* A3 (Mun et al. 2010), it was then abandoned. However, there are situations in which BAC clones can still provide a useful tool for unravelling the physical arrangement of tandemly duplicated or complex loci (Ryder et al. 2001). BAC clones have been successfully used to anchor *Brassica* genome sequence directly to chromosomes using fluorescent in situ hybridization (FISH) (Howell et al. 2002, 2005; Xiong and Pires 2011), which has been a valuable tool in the development of *B. napus* genomics, especially in understanding and verifying the physical arrangement and relationships between ancestral chromosome segments and rearrangements (Deng et al. 2016; Howell et al. 2008).

14.5 Data, Information Resources, Bioinformatics Tools and Analysis Pipelines

As outlined in the Introduction, the relevant databases and information resources lack a formal description in an associated publication and are subject to volatility. The '*Brassica*.info' website (http://www.Brassica.info) was established on behalf of the MBGP in 2002 to provide a portal to collate information on a wide range of information and physical resources relevant to *Brassica* genomics. The site is currently undergoing an update to reflect recent availability of various resources and will soon provide links to an inventory to facilitate the first stages of establishing a MBGP *Brassica* Information System.

14.6 Genome Browsers and Related Sites

Reference genome sequences for *B. napus* have now been established for more than one genotype (Chalhoub et al. 2014; Bayer et al. 2017). As well as the reference sequences of pseudochromosomes appearing in the NCBI, EMBL and DDBJ repositories, advanced genome browsers with multiple functionality provide a tool for researchers to navigate annotation, associated transcript and other features as these are established (Table 14.1). Each of these repositories provides not only online browser/navigation tools but also a variety of download capabilities in GFF3, FASTA and other formats. There are currently some limitations in terms of providing a comprehensive platform for users to navigate and align data sets, including links to and from

Table 14.1 *B. napus* genome databases, browsers and related informatics resources

Site	Content	FTP	URL
Brassica.info	MBGP reference information, links, online publication	–	a
Genoscope	Genome browser	–	b
Brassica database (BRAD)	Browser, downloads, analysis tools and reference information	Yes	c
Ensembl	Genome browsers, analysis pipelines	Yes	d
CoGe	Comparative genomics resource	–	e
Brassica genome gateway	Legacy genomics data sets, links	–	f
NCBI genome	Genome browser, genetic maps	–	g
EMBL/ENA	Generic DNA, RNA sequence archives	–	h, i
Brassica information portal	Plant populations, lines, phenotypic trait, trial, DNA marker, linkage maps	–	j
Oil crops genomics database	Browser, analysis tools, reference information, downloads, linkage maps, DNA marker, and population genomic, transcriptomic and metabolomics data	Yes	k

URL: a: http://Brassica.info/; b: http://www.genoscope.cns.fr/Brassicanapus/; c: http://Brassicadb.org/brad/; d: http://plants.ensembl.org/Brassica_napus/Info/Index; e: https://genomevolution.org/CoGe/GenomeInfo.pl?gid=20192; f: http://Brassica.nbi.ac.uk; g: https://www.ncbi.nlm.nih.gov/genome/annotation_euk/Brassica_napus/100/; h: https://www.ebi.ac.uk/ena; i: https://www.ncbi.nlm.nih.gov/sra; j: https://bip.earlham.ac.uk/; k: www.OCRI-genomics.org/

established genetic linkage maps. The latter had previously been demonstrated within the InterStoreDB framework between genetic markers from the integrated map of *B. napus* (Wang et al. 2011a, b) displayed in CMap, and anchored to the draft *B. rapa* A genome implementation in Ensembl (Love et al. 2012).

In addition, there is increasing requirement to represent the *B. napus* pan-genome and associated re-sequencing and SNP data, along with GWAS data sets.

14.6.1 Genoscope Browser

The canonical *B. napus* genome is hosted within the Genoscope website (http://www.genoscope.cns.fr/Brassicanapus/) and provides an integrated database for genomic sequence data from multiple genome sequencing projects across taxa. Four functional modules include a browser, BLAT sequence alignment server, synteny analysis and a download feature. The generic UCSC genome browser has a chromosome-centric representation of the *B. napus* genome and allows navigation based on the pseudochromosome assembly. Specific genes or genomic regions may be located. The basic local alignment tool (BLAT) is employed to facilitate fast search of sequence (DNA, RNA or protein) against the genome. Users may also explore syntenic regions between multiple *Brassica* species and may download data sets including genome, predicted genes and proteins, gene annotations and other annotated genomic elements.

14.6.2 Ensembl Plants

The generic Ensembl Plants site (http://plants.ensembl.org) curated by the European Bioinformatics Institute (EBI) manages and provides access for a range of plant genomes which meet minimum criteria of completeness and levels of annotation. It provides persistent visualization of annotated genomes, along with a wide range of associated tracks, comparative genomic analysis and flexible download of genome sequence segments, annotated genes, transcripts or proteins. One of the key benefits of navigating a genome via Ensembl is the ability to draw on the standard set of data analyses that include the Compara gene tree and Variant pipelines. The later may include a variant table, image and representation of structural variants.

Ensembl Plants currently includes the *B. napus* Darmor-bzh reference genome (http://plants.ensembl.org/Brassica_napus/Info/Index), as well as the original published versions of the *B. rapa* Chiifu-401 'A' and *B. oleracea* TO1000 'C' genomes. The implementations currently include navigable outputs from the Compara analysis pipeline that for any gene model includes genomic alignments, gene tree, gene gain/loss tree, orthologues and paralogues.

14.6.3 *Brassica* Database (BRAD)

The *Brassica* Database (BRAD; http://Brassicadb.org/brad/) contains a valuable compilation of source and analysed data sets, along with searchable online tools to enable a wide range of comparative or evolutionary studies to be undertaken. The initial data sets curated focused on the 'A' genome of *B. rapa,* but has been extended to include genome annotation, transcript and translated protein data for selected members of the *Brassicaceae*, including *Brassica* species and Arabidopsis thaliana. Specific subsets of data have also been curated including *Brassica*-specific gene families relating to classes such as glucosinolates, parasite resistance, flowering and transcription factors.

The user interface allows search based on annotated gene name and extraction of detailed annotation including gene ontology terms. As with Ensembl implementation, queried genes or proteins can be searched against the annotated genome using BLAST, and flanking regions of a gene, RNA, transposon and genetic markers may be extracted based on genomic coordinates.

14.6.4 Specialized Sequence Databases

The *B. napus* genome is also available in the Comparative Genomics online platform CoGE (https://genomevolution.org/coge/), which is powerful in its ability to integrate reference and user-supplied genome data and comparative genomics tools from a single and secure access point. It facilitates uploading and sharing of genomes by individual researchers and research consortia, as well as comparative genomics and evolutionary study of closely related species.

The *Brassica* Genome Gateway (http://Brassica.nbi.ac.uk/) contains historical genomic data of *Brassica* species and information from a range of *Brassica* genome sequencing projects. A number of other generic and specific resources have been developed that allow search and comparison of particular genomic sequence subclasses, in particular those relating to transposable elements and other repeat sequences. *Brassica*TED (Murukarthick et al. 2014) is a public domain database for utilization of miniature transposable elements in *Brassica* species.

14.7 Genetic and Trait Data

14.7.1 *Brassica* Information Portal

The *Brassica* Information Portal (BIP; https://bip.earlham.ac.uk/) was launched in 2016 and is managed by the Earlham Institute (formerly The Genome Analysis Centre) in the UK. The portal (Eckes et al. 2017) is based on the CropStoreDB relational database schema (Love et al. 2012; Eckes et al. 2017; Leibovici et al. 2017) and provides a repository for *Brassica* population and trait scoring information related to pre-breeding genetics and genomics studies. Descriptions of a range of *B. napus* and other *Brassica* linkage maps, along with marker assays and related plant populations and lines, are curated, together with detailed meta-data outlining provenance, status and ownership of data. The web interface allows users to browse information about quantitative trait loci, with links to curated *Brassica* phenotype experimental data, along with genotype information stored in external data sources. Advanced data submission capabilities and APIs have been developed to enable users to store and publish their own study results in the portal.

14.7.2 CropStoreDB

The CropStoreDB data curation pipeline (http://www.cropstoredb.org/) was developed as a generic platform for collating a wide range of genetics and related data and implemented to facilitate navigation between genome and traits. Its first use-case for curation was for *Brassica* species. *B. napus* genetic map and sequence-tagged marker data (Wang et al. 2011a, b) have been managed within CropStoreDB, along with trait and QTL data. The database schema and original curated *Brassica* data formed the basis of BIP (Eckes et al. 2017), and extensibility to include GIS (geolocation) data was demonstrated in the GRASP initiative (Lebovici et al. 2017).

14.7.3 Proteomic and Metabolomics Data

Whilst there have been a few studies published in recent years (Desclos et al. 2009; Zhu et al. 2009; Girondé et al. 2016; Geng et al. 2017), at present, there are no searchable resources dedicated to navigating proteomics data sets for *B. napus*. As with proteomics, only a few studies have been published in recent years describing metabolomics analysis of *B. napus* (Farag et al. 2012; Kortesniemi et al. 2015); at present, there are no searchable resources outlining metabolomics data sets for *B. napus*. Initial information about biochemical pathways is available for *B. rapa* (https://www.plantcyc.org/databases/brapafpsccyc/1.0).

14.8 Conclusion

As outlined in this chapter, there are rich and varied experimental and data resources available in the public domain to support evolutionary, comparative and functional analysis of *B. napus* genomes and underpin pre-breeding R&D and breeding activities for oilseed and related crops. These resources have been accumulated and made available primarily by public-sector researchers, often in productive partnership with private sector breeding companies, and facilitated through the long-standing cooperative endeavours of the Multinational *Brassica* Genome Project. Commercial development and release of new cultivars will increasingly be dependent upon the availability of updated reference genomic resources to guide the selection and recombination of beneficial alleles. Future requirements include greater integration of phenotype to genotype and tools to collate and navigate increasingly complex data sets.

References

Amoah S, Kurup S, Rodriguez Lopez C, Welham S, Powers S, Hopkins C, Wilkinson M, King GJ (2012) A hypomethylated population of *Brassica* rapa for forward and reverse epi-genetics. BMC Plant Biol 12:193. https://doi.org/10.1186/1471-2229-12-193

Barrett T, Clark K, Gevorgyan R, et al (2012) BioProject and BioSample databases at NCBI: facilitating capture and organization of metadata. Nucl Acids Res 40 (Database issue):D57–D63. http://doi.org/10.1093/nar/gkr1163

Bayer PE, Hurgobin B, Golicz AA, Chan CK, Yuan Y, Lee H, Renton M, Meng J, Li R, Long Y, Zou J, Bancroft I, Chalhoub B, King GJ, Batley J, Edwards D (2017) Assembly and comparison of two closely related *Brassica napus* genomes. Plant Biotechnol J. https://doi.org/10.1111/pbi.12742

Bennett EJ, Brignell CJ, Carion PWC, Cook SM, Eastmond PJ, Teakle G, Hammond JP, Love C, King GJ, Roberts JA, Wagstaff C (2017) Development of a statistical crop model to explain the relationship between seed yield and phenotypic diversity within the *Brassica napus* genepool. Agronomy 7:31. https://doi.org/10.3390/agronomy7020031

Chalhoub B, Denoued F, Liu S, Parking IAP, King GJ et al (2014) Early allopolyploid evolution in the post-neolithic *Brassica napus* oilseed genome. Science 345:950–953

Clarke WE, Higgins EE, Plieske J et al (2016) A high-density SNP genotyping array for *Brassica napus* and its ancestral diploid species based on optimised selection of single-locus markers in the allotetraploid genome. TAG Theor Appl Genet 129(10):1887–1899. https://doi.org/10.1007/s00122-016-2746-7

Data Brief (Jul 2 2015) 4:357–362. https://doi.org/10.1016/j.dib.2015.06.016 (eCollection Sept 2015)

Delourme R, Falentin C, Fomeju BF, Boillot M, Lassalle G, André I, Duarte J, Gauthier V, Lucante N, Marty A, Pauchon M, Pichon JP, Ribière N, Trotoux G, Blanchard P, Rivière N, Martinant JP, Pauquet J (2013) High-density SNP-based genetic map development and linkage disequilibrium assessment in

Brassica napus L. BMC Genom 22(14):120. https://doi.org/10.1186/1471-2164-14-120

Deng Z, Li X, Wang Z, Jiang Y, Wan L, Dong F, Chen F, Hong D, Yang G (2016) Map-based cloning reveals the complex organization of the BnRf locus and leads to the identification of BnRf(b), a male sterility gene, in *Brassica napus*. Theor Appl Genet 129(1):53–64. https://doi.org/10.1007/s00122-015-2608-8 (Epub 3 Oct 2015)

Desclos M, Etienne P, Coquet L, Jouenne T, Bonnefoy J, Segura R, Reze S, Ourry A, Avice JC (2009) A combined 15 N tracing/proteomics study in *Brassica napus* reveals the chronology of proteomics events associated with N remobilisation during leaf senescence induced by nitrate limitation or starvation. Proteomics 9(13):3580–3608. https://doi.org/10.1002/pmic.200800984

Eckes AH, Gubala T, Nowakowski P, Szymczyszyn T, Wells R, Irwin JA, Horro C, Hancock JM, King GJ, Dyer SC, Jurkowski W (2017) Introducing the *Brassica* Information Portal: towards integrating genotypic and phenotypic *Brassica* crop data. F1000Research 6:465. https://doi.org/10.12688/f1000research.11301.2)

Farag MA, Sharaf Eldin MG, Kassem H, Abou el Fetouh M (2012) Metabolome classification of *Brassica napus* L. organs via UPLC-QTOF-PDA-MS and their anti-oxidant potential. Phytochem Anal 24 (3):277–287. https://doi.org/10.1002/pca.2408 (Epub 11 Oct 2012)

Geng S, Yu B, Zhu N, Dufresne C, Chen S (2017) Metabolomics and Proteomics of *Brassica napus* guard cells in response to low CO_2. Front Mol Biosci 25(4):51. https://doi.org/10.3389/fmolb.2017.00051 (eCollection 2017)

Gilchrist EJ, Sidebottom CHD, Koh CS, MacInnes T, Sharpe AG, Haughn GW (2013) A mutant *Brassica napus* (Canola) population for the identification of new genetic diversity via TILLING and next generation sequencing. PLoS ONE. 8(12):e84303. https://doi.org/10.1371/journal.pone.0084303 (Virolle M-J ed)

Girondé A, Poret M, Etienne P, et al (2016) A comparative study of proteolytic mechanisms during leaf senescence of four genotypes of winter oilseed rape highlighted relevant physiological and molecular traits for NRE improvement. Plants 5(1):1. https://doi.org/10.3390/plants5010001 (Balazadeh S ed.)

Graham NS, Hammond JP, Lysenko A, Mayes S, King GJ et al (2014) Ge netical and comparative genomics of *Brassica* under altered Ca supply identifies Arabidopsis Ca-transporter orthologues. Plant Cell 26:2818–2830

Harper AL, Trick M, Higgins J, Fraser F, Clissold L, Wells R, Hattori C, Werner P, Bancroft I (2012) Associative transcriptomics of traits in the polyploid crop species *Brassica napus*. Nat Biotechnol 30 (8):798–802

He Z, Cheng F, Li Y, Wang X, Parkin IA, Chalhoub B, Liu S, Bancroft I (2015) Construction of *Brassica* A and C genome-based ordered pan-transcriptomes for use in rapeseed genomic research

Howell EC, Barker GC, Jones GH, Kearsey MJ, King GJ, Kop EP, Ryder CD, Teakle GR, Vicente JG, Armstrong SJ (2002) Integration of a cytogenetic map with a genetic linkage map of *Brassica oleracea*. Genetics 161:1225–1234

Howell EC, Armstrong SJ, Barker GC, Jones GH, King GJ, Ryder CD, Kearsey MJ (2005) Physical organisation of the major duplication on *Brassica oleracea* chromosome O6 revealed through fluorescence in situ hybridisation with Arabidopsis and *Brassica* BACs. Genome. 48(6):1093–1103

Howell EC, Kearsey MJ, Jones GH, King GJ, Armstrong SJ (2008) A and C genome distinction and chromosome identification in *Brassica napus* by sequential FISH and GISH. Genetics 180:1849–1857

Hurgobin B, Golicz A, Bayer P, Chan K, Tirnaz S, Dolatabadian A, Schiessl S, Samans B, Montenegro J, Parkin I, Pires C, Chalhoub B, King G, Snowdon R, Batley J, Edwards D (2017) Homoeologous exchange is a major cause of gene presence/absence variation in the amphidiploid *Brassica napus. Plant Biotech J.* https://doi.org/10.1111/pbi.12867

Iniguez-Luy FL, Lukens L, Farnham MW, Amasino RM, Osborn TC (2009) Development of public immortal mapping populations, molecular markers and linkage maps for rapid cycling *Brassica rapa* and *B. oleracea*. Theor Appl Genet 120(1):31–43. https://doi.org/10.1007/s00122-009-1157-4 (Epub 26 Sep 2009)

Körber N, Bus A, Li J, Higgins J, Bancroft I, Higgins EE, Parkin IA, Salazar-Colqui B, Snowdon RJ, Stich B (2015) Seedling development traits in *Brassica napus* examined by gene expression analysis and association mapping. BMC Plant Biol 9(15):136. https://doi.org/10.1186/s12870-015-0496-3

Kortesniemi M, Vuorinen AL, Sinkkonen J, Yang B, Rajala A, Kallio H (2015) NMR metabolomics of ripened and developing oilseed rape (*Brassica napus*) and turnip rape (*Brassica rapa*. Food Chem 1 (172):63–70. https://doi.org/10.1016/j.foodchem.2014.09.040 (Epub 16 Sep 2014)

Leibovici DG, Meek S, Anand S, Santos R, Morley J, Jackson MJ, Mayes S, Ray R, Hodgman C, Baten A, King GJ, Brailsford T, Vu T, Karunaratne AS, Azman R, Walker S, Azam-Ali S (2017) Geospatial binding for transdisciplinary research in crop science: the GRASP-GFS initiative. In: Mavridis A, Ifadis I (eds) Advances in geospatial technology series. IGI Global. https://doi.org/10.1186/s40965-017-0034-3

Li H, Younas M, Wang X, Li X, Chen L, Zhao B, Chen X, Xu J, Hou F, Hong B, Liu G, Zhao H, Wu X, Du H, Wu J, Liu K (2013) Development of a core set of single-locus SSR markers for allotetraploid rapeseed (*Brassica napus* L.). Theor Appl Genet 126 (4):937–947. https://doi.org/10.1007/s00122-012-2027-z (Epub 13 Dec 2012)

Liu S, Liu Y, Yang X, King GJ et al (2014) The *Brassica oleracea* genome reveals the asymmetrical evolution of polyploid genomes. Nat Commun 5:3930

Love CG, Graham NS, Lochlainn S, Bowen HC, May ST, White PJ, Broadley MR, Hammond JP, King GJ

(2010) A *Brassica* exon array for whole-transcript gene expression profiling. PLoS ONE 5(9):e12812
Love C, Andongabo A, Wang J, Carion P, Rawlings C, King GJ (2012) InterstoreDB: a generic integration resource for genetic and genomic data. J Intergrative Plant Biol 54(5):345–355
Lu K, Xiao Z, Jian H et al (2016) A combination of genome-wide association and transcriptome analysis reveals candidate genes controlling harvest index-related traits in *Brassica napus*. Sci Rep 6:36452. https://doi.org/10.1038/srep36452
Mason AS, Higgins EE, Snowdon RJ, Batley J, Stein A, Werner C, Parkin IA (2017) A user guide to the Brassica 60 K Illumina Infinium™ SNP genotyping array. Theor Appl Genet 130(4):621–633. https://doi.org/10.1007/s00122-016-2849-1 (Epub 20 Feb 2017)
Mun J-H, Kwon S-J, Seol Y-J et al (2010) Sequence and structure of *Brassica rapa* chromosome A3. Genome Biol 11(9):R94. https://doi.org/10.1186/gb-2010-11-9-r94
Murukarthick J, Sampath P, Lee SC et al (2014) *Brassica* TED—a public database for utilization of miniature transposable elements in *Brassica* species. BMC Res Notes 7:379. https://doi.org/10.1186/1756-0500-7-379
Parkin IA, Sharpe AG, Keith DJ, Lydiate DJ (1995) Identification of the A and C genomes of amphidiploid *Brassica napus* (oilseed rape). Genome 38(6):1122–1131
Parkin IAP, Koh C, Tang H, Robinson SJ, King GJ et al (2014) Transcriptome and methylome profiling reveals relics of genome dominance in the mesopolyploid *Brassica oleracea*. Genome Biol 15:R77. https://doi.org/10.1186/gb-2014-15-6-r77
Piquemal J, Cinquin E, Couton F, Rondeau C, Seignoret E, Doucet I, Perret D, Villeger MJ, Vincourt P, Blanchard (2005) Construction of an oilseed rape (*Brassica napus* L.) genetic map with SSR markers. Theor Appl Genet 111(8):1514–1523 (Epub 10 Nov 2005)
Qu C, Jia L, Fu F et al (2017) Genome-wide association mapping and Identification of candidate genes for fatty acid composition in *Brassica napus* L. using SNP markers. BMC Genom 18:232. https://doi.org/10.1186/s12864-017-3607-8
Radoev M, Becker HC, Ecke W (2008) Genetic analysis of heterosis for yield and yield components in rapeseed (*Brassica napus* L.) by quantitative trait locus mapping. Genetics 179(3):1547–1558. https://doi.org/10.1534/genetics.108.089680
Raman H, Raman R, Kilian A, Detering F, Long Y, Edwards D, Parkin IA, Sharpe AG, Nelson MN, Larkan N, Zou J, Meng J, Aslam MN, Batley J, Cowling WA, Lydiate D (2013) A consensus map of rapeseed (*Brassica napus* L.) based on diversity array technology markers: applications in genetic dissection of qualitative and quantitative traits. BMC Genom 23 (14):277. https://doi.org/10.1186/1471-2164-14-277
Schiessl S, Huettel B, Kuehn D, Reinhardt R, Snowdon R (2017) Post-polyploidisation morphotype diversification associates with gene copy number variation. Sci Rep 6(7):41845. https://doi.org/10.1038/srep41845
Schmutzer T, Samans B, Dyrszka E, Ulpinnis C, Weise S, Stengel D, Colmsee C, Lespinasse D, Micic Z, Abel S, Duchscherer P, Breuer F, Abbadi A, Leckband G, Snowdon R, Scholz U (2015) Species-wide genome sequence and nucleotide polymorphisms from the model allopolyploid plant *Brassica napus*. Sci Data 8 (2):150072. https://doi.org/10.1038/sdata.2015.72
Sharpe AG, Parkin IA, Keith DJ, Lydiate DJ (1995) Frequent nonreciprocal translocations in the amphidiploid genome of oilseed rape (*Brassica napus*. Genome 38(6):1112–1121
Shrestha R, Matteis L, Skofic M et al (2012) Bridging the phenotypic and genetic data useful for integrated breeding through a data annotation using the crop ontology developed by the crop communities of practice. Front Physiol 3:326. https://doi.org/10.3389/fphys.2012.00326
Stephenson P, Baker D, Girin T, Perez A, Amoah S, King GJ, Ostergaard L (2010) A rich TILLING resource for studying gene function in *Brassica rapa*. BMC Plant Biol 10:62
Szadkowski E, Eber F, Huteau V, Lod M, Huneau C, Belcram H, Coriton O, Manzanares-Dauleux MJ, Delourme R, King GJ, Chalhoub B, Jenczewski E, Chvre AM (2010) The first meiosis of resynthesized *Brassica napus*, a genome blender. New Phytol 186:102–112
The Arabidopsis Genome Initiative (2000) Analysis of the genome sequence of the flowering plant *Arabidopsis thaliana*. Nature 408:796–881. https://doi.org/10.1038/35048692
Tommasini L, Batley J, Arnold GM, Cooke RJ, Donini P, Lee D, Law JR, Lowe C, Moule C, Trick M, Edwards KJ (2003) The development of multiplex simple sequence repeat (SSR) markers to complement distinctness, uniformity and stability testing of rape (*Brassica napus* L.) varieties. Theor Appl Genet 106 (6):1091–1101 (Epub 22 Oct 2002)
Trick M, Long Y, Meng J, Bancroft I (2009) Single nucleotide polymorphism (SNP) discovery in the polyploid *Brassica napus* using Solexa transcriptome sequencing. Plant Biotechnol J 7(4):334–346. https://doi.org/10.1111/j.1467-7652.2008.00396.x (Epub 21 Jan 2009)
Wang N, Wang Y, Tian F, King GJ, Zhang C, Long Y, Shi L, Meng J (2008) A functional genomics resource for *Brassica napus*: development of an EMS mutagenized population and discovery of FAE1 point mutations by TILLING. New Phytol 180:751–765

Wang J, Lydiate DJ, Parkin IAP, Falentin C, Delourme R, Carion P, King GJ (2011a) Integration of linkage maps for the amphidiploid *Brassica napus*, and comparative mapping with *Arabidopsis* and *Brassica rapa*. BMC Genom 12:101

Wang X1, Wang H, Wang J, Sun R, Wu J, Liu S, Bai Y, Mun JH, Bancroft I, Cheng F, Huang S, Li X, Hua W, Wang J, Wang X, Freeling M, Pires JC, Paterson AH, Chalhoub B, Wang B, Hayward A, Sharpe AG, Park BS, Weisshaar B, Liu B, Li B, Liu B, Tong C, Song C, Duran C, Peng C, Geng C, Koh C, Lin C, Edwards D, Mu D, Shen D, Soumpourou E, Li F, Fraser F, Conant G, Lassalle G, King GJ, Bonnema G, Tang H, Wang H, Belcram H, Zhou H, Hirakawa H, Abe H, Guo H, Wang H, Jin H, Parkin IA, Batley J, Kim JS, Just J, Li J, Xu J, Deng J, Kim JA, Li J, Yu J, Meng J, Wang J, Min J, Poulain J, Wang J, Hatakeyama K, Wu K, Wang L, Fang L, Trick M, Links MG, Zhao M, Jin M, Ramchiary N, Drou N, Berkman PJ, Cai Q, Huang Q, Li R, Tabata S, Cheng S, Zhang S, Zhang S, Huang S, Sato S, Sun S, Kwon SJ, Choi SR, Lee TH, Fan W, Zhao X, Tan X, Xu X, Wang Y, Qiu Y, Yin Y, Li Y, Du Y, Liao Y, Lim Y, Narusaka Y, Wang Y, Wang Z, Li Z, Wang Z, Xiong Z, Zhang Z (2011b) *Brassica rapa* Genome Sequencing Project Consortium. The genome of the mesopolyploid crop species *Brassica rapa*. Nat Genet 28, 43(10):1035–9. https://doi.org/10.1038/ng.919

Wang X, Long Y, Wang N, Zou J, Ding G, Broadley MR, White PJ, Yuan P, Zhang Q, Luo Z, Liu P, Zhao H, Zhang Y, Cai H, King GJ, Xu F, Meng J, Shi L (2017) Breeding histories and selection criteria for oilseed rape in Europe and China identified by genome wide pedigree dissection. Sci Rep 7(1):1916. https://doi.org/10.1038/s41598-017-02188-z

Wang B, Wu Z, Li Z, Zhang Q, Hu J, Xiao Y, Cai D, Wu J, King GJ, Li H, Liu K (2018) Dissection of the genetic architecture of three seed-quality traits and consequences for breeding in *Brassica napus*. Plant Biotech J. https://doi.org/10.1111/pbi.12873

Wei L, Jian H, Lu K, Filardo F, Yin N, Liu L, Li J et al (2016) Genome-wide association analysis and differential expression analysis of resistance to Sclerotinia stem rot in *Brassica napus*. Plant Biotechnol J 14 (6):1368–1380

Wilkinson MD, Dumontier M, IJ Aalbersberg et al (2016) The FAIR guiding principles for scientific data management and stewardship. Sci Data 3:160018. https://doi.org/10.1038/sdata.2016.18

Xiong Z, Pires JC (2011) Karyotype and identification of all homoeologous chromosomes of allopolyploid *Brassica napus* and its Diploid Progenitors. Genetics 187(1):37–49. https://doi.org/10.1534/genetics.110.122473

Zhou Q, Zhou C, Zheng W et al (2017) Genome-wide SNP markers based on SLAF-seq uncover breeding traces in rapeseed (*Brassica napus* L.). Front Plant Sci 8:648. https://doi.org/10.3389/fpls.2017.00648

Zhu M, Dai S, McClung S, Yan X, Chen S (2009) Functional differentiation of *Brassica napus* guard cells and mesophyll cells revealed by comparative proteomics. Mol Cell Proteomics 8(4):752–766. https://doi.org/10.1074/mcp.M800343-MCP200 (Epub 22 Dec 2008)

Mun J-H, Kwon S-J, Yang T-J, Seol Y-J, Jin M, Kim J-A, Lim M-H, Kim JS, Baek S, Choi B-S, Yu H-J, Kim D-S, Kim N, Lim K-B, Lee S-I, Hahn J-H, Lim YP, Bancroft I, Park B-S (2009) Genome-wide comparative analysis of the Brassica rapa gene space reveals genome shrinkage and differential loss of duplicated genes after whole genome triplication. Genome Biology 10(10):R111

Østergaard L, King GJ (2008) Standardized gene nomenclature for the Brassica genus. Plant Methods 4(1):10

Ryder CD, Smith LB, Teakle GR, King GJ (2001) Contrasting genome organisation: two regions of the genome compared with collinear regions of the genome. Genome 44(5):808–817

Genome-Facilitated Breeding of Oilseed Rape

15

Christian Werner and Rod Snowdon

Abstract

Brassica napus ssp. *napus* (rapeseed, oilseed rape, canola) became a major global oilseed crop through intensive breeding during the last five decades. The implementation of large-scale metabolic screening of seeds to identify mutants carrying low seed erucic acid and glucosinolate content, respectively, facilitated the use of these variants in backcrossing programmes which established the species as a "new" global oilseed crop with exceptional oil quality and high meal quality. Divergent ecogeographical forms were adapted to agricultural systems in North America (spring-type canola), Europe (predominantly winter-type oilseed rape) and Asia/Australia (semi-winter rapeseed/canola forms), and the establishment of hybrid breeding systems during the 1990s increased seed yield and yield stability and established *B. napus* as an important cash crop for farmers and breeders. Also around this time, cytogenetic studies and the first genetic maps for *B. napus*, developed with restriction fragment length polymorphism (RFLP) markers, revealed first evidence of unusual rearrangements among homoeologous chromosomes, and evidence grew for complex allohexaploidization among the diploid subgenomes. However, it was not until the *B. napus* genome sequence was deciphered two decades later that the broad extent and consequences of large-scale and small-scale rearrangements in the *B. napus* genome, and the unexpected impact of complex genome structural rearrangements on traits of importance for breeding, became apparent. As more and more *B. napus* genome sequences become available, and new methods for high-throughput screening enable detailed associations of genome features with simple and quantitative traits, breeders are beginning to appreciate the importance of post-polyploidization genome structural variation for a multitude of important traits in this recent allopolyploid crop. The use of genome data, high-throughput genotyping techniques, genomic selection, and genome-based hybrid performance prediction is already changing the way that *B. napus* breeders identify useful diversity and implement it in their breeding programmes. Access to high-quality genome assemblies, vast genomic datasets, and large-scale digital phenomic datasets will play a key role in future implementation of omics-assisted breeding in *B. napus* breeding. This chapter provides an overview of the impact of the *B. napus* genome on breeding progress and the opportunities provided by genomics technologies for future breeding.

C. Werner · R. Snowdon (✉)
Department of Plant Breeding, Justus Liebig University, Giessen, Germany
e-mail: Rod.Snowdon@agrar.uni-giessen.de

S. Liu et al. (eds.), *The Brassica napus Genome*, Compendium of Plant Genomes,
https://doi.org/10.1007/978-3-319-43694-4_15

15.1 History and Bottlenecks in Oilseed Rape/Canola Breeding

Oilseed rape and canola (*Brassica napus* ssp. *napus*; genome AACC, $2n = 38$) deliver one of the world's most important sources of healthy vegetable oil, along with a high-quality extraction meal for animal nutrition. The huge global success of this crop was achieved during just the past four decades as an explicit result of breeding. In particularly, this involved intensive selection to dramatically improve the seed oil composition (through replacement of C22:1 erucic acid with C18:1 oleic acid) and to reduce antinutritive components in the seed meal (especially a dramatic reduction of seed glucosinolate content). A detailed description of *B. napus* breeding history and achievements, breeding aims and methods is provided by (Friedt and Snowdon 2010a, b).

As a facultative out-crosser conducive to tissue culture, *B. napus* presents breeders with diverse opportunities for implementation of different breeding methods. Prior to the successful implementation of pollination control mechanisms, early generations of so-called double-low (00) canola-quality cultivars, with low seed erucic acid and glucosinolate content, were generally developed as inbred line varieties, taking advantage of classical pedigree selection techniques. Application of doubled-haploid techniques (Weber et al. 2005) later played an important role in reducing breeding cycle intervals for line varieties and is today an important asset in generating homozygous parental lines for hybrid breeding. The latter has today become the predominant breeding method in the major oilseed rape production areas of the world, based on the discovery and development of stable cytoplasmic and genic male-sterility systems for systematic generation of pure F1-hybrid seed (reviewed by Friedt and Snowdon 2010a, b).

A recent allotetraploid species (Chalhoub et al. 2014), *B. napus* carries the full chromosome complements of its two progenitors, believed to be Asian cabbage/turnip (*Brassica rapa*; AA, $2n = 20$) and Mediterranean cabbage (*Brassica oleracea*; CC, $2n = 18$) (Chalhoub et al. 2014; Gomez-Campo 1999; Olsson 1960; Song and Osborn 1992). Because no wild *B. napus* forms are known, it is assumed that the species appeared relatively recently, when the parental species began being cultivated in geographical proximity (Friedt and Snowdon 2010a, b). Only a few independent interspecific hybridisation events are believed to have contributed as *B. napus* species founders (Allender and King 2010). The initial allopolyploidisation events are thought to have contributed significantly to the establishment of novel trait diversity, via genome restructuring and gene conversion (Chalhoub et al. 2014; Szadkowski et al. 2010). Some evidence suggests that the origin of adaptive variation—which enabled the de novo allopolyploid to become a highly successful crop species in highly diverse ecogeographical regions—was due to the influence of homoeologous chromosome restructuring affecting flowering-time regulatory genes and other fundamental adaptive processes (Chalhoub et al. 2014; Schiessl et al. 2014). The different ecotypes of *B. napus* today represent the most important oilseed crop cultivated in Europe (where predominantly winter oilseed rape is grown), Canada (spring-type canola), China and Australia (both predominantly semi-winter rapeseed). The differentiation into winter, semi-winter and spring forms is governed by central flowering-time pathway genes that control winter hardiness and the requirement for vernalisation, a prerequisite for winter-annual forms to promote the onset of flowering. This classification represents the most significant genetic differentiation among today's primary gene pools for oilseed rape and canola breeding (Fig. 15.1; reviewed by Friedt and Snowdon 2010a, b).

Within each of the three oilseed-type *B. napus* genepools, the genetic diversity available for breeding was significantly eroded by the strict breeding bottlenecks imposed by selection for

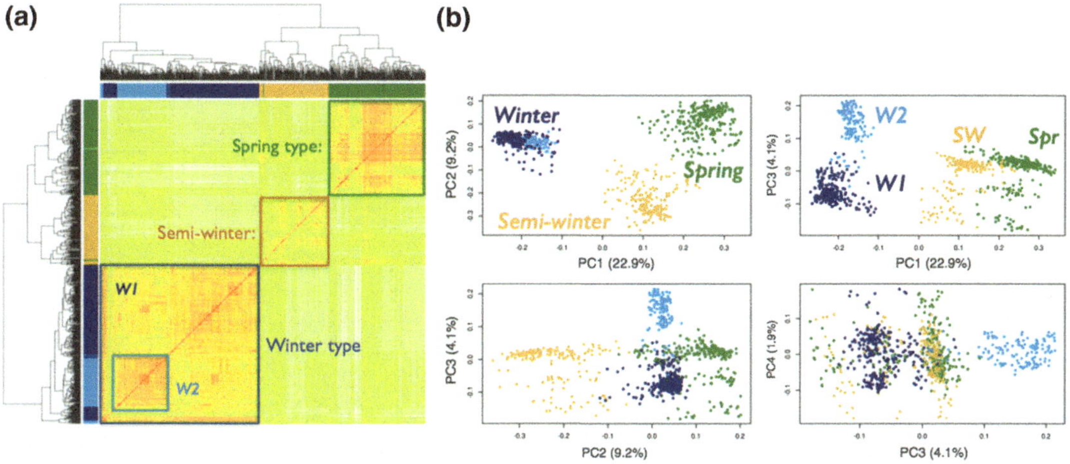

Fig. 15.1 Principle coordinate analysis describing genetic variation among a panel of 850 *B. napus* breeding lines representing the primary gene pools of European winter oilseed rape, Asian semi-winter rapeseed and North American spring canola, respectively. The first two principle coordinates, which distinguish these three gene pools based on major genes controlling flowering behaviour and vernalisation requirement, account for the vast majority of the between-pool genetic diversity, whereas intense selection during breeding has strongly eroded within-pool diversity in all three pools

elimination of erucic acid and for low glucosinolate oilseed forms. Most modern double-low "canola" varieties, with zero erucic acid and low glucosinolate contents, share a common pedigree that traces back to the respective donor cultivars for these two traits, "Liho" and "Bronowski" (Downey and Rakow 1987), and genomic regions inherited from both ancestors still can be identified in contemporary varieties (Bancroft et al. 2011). As a result of these severe genetic bottlenecks, major oilseed rape breeding pools in Europe, North America, Australia and Asia are extremely narrow (Becker 2011; Bus et al. 2011; Cowling 2007; Diers et al. 1996; Hasan et al. 2006). Chromosomes carrying the respective mutations discriminating flowering behaviour and seed 00-quality are associated with strong signatures of selection and exhibit particularly low levels of diversity among modern cultivars (Fig. 15.2; Qian et al. 2014). Disruption of extended stretches of linkage disequilibrium (LD) associated with the responsible genes or quantitative trait loci (QTL) is made more difficult, particularly on C-subgenome chromosomes, by low levels of recombination (Hatzig et al. 2015a; Qian et al. 2014; Schiessl et al. 2015).

15.2 Principles of Breeding Theory and Their Implications in Rapeseed Breeding

The ultimate aim of plant breeding is to modify the characteristics of crop plants in order to adapt them to the needs and wishes of humanity. Plant breeding, one of the oldest and most important technologies developed by humans, provided the basis for civilisations to rise, and its continuous progress is still crucial to maintaining various aspects of our contemporary way of life. Originally established out of the sole necessity to provide food for survival and feed for livestock, today's crop breeding programmes need to focus on a vast number of properties besides high yield, including many different quality traits as well as tolerances and resistances to numerous biotic and abiotic stress factors. Additionally, alternative uses of crop products in non-food,

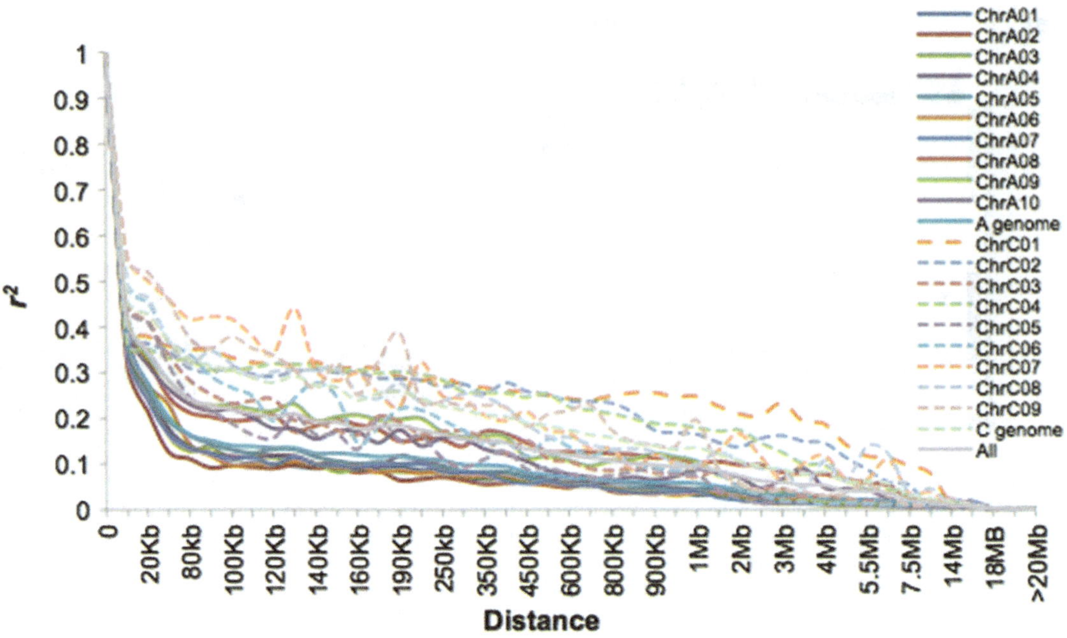

Fig. 15.2 Patterns of linkage disequilibrium (LD, $r^2 = 0.1$) across the 19 haploid chromosomes of semi-winter-type *B. napus*, measured with 24,994 single-copy SNP markers. The solid lines represent LD decay in A-subgenome chromosomes, while the dashed lines represent LD decay in C-subgenome chromosomes. Figure reprinted, without modifications, from Qian et al. (2014) under the Creative Commons Attribution License CC BY 4.0

non-feed sectors, for example, as resources for bioenergy or industrial feedstocks, increase the complexity of requirements that plant varieties have to meet when released to the market.

Rapeseed is no exception in this regard, with very stringent seed quality criteria imposing a severe bottleneck on breeders that needs to be addressed while still maintaining yield, oil content and resistance to diseases, pests, weeds, or lodging, along with a high level of environmental resilience. There is also increasing demand for oilseed rape cultivars with variable oil quality characteristics, for example, high-oleic, low-linolenic (HOLL) forms, based on mutations in fatty acid biosynthesis genes, that are bred as a source of heat-stable oil for the frying industry (Wittkop et al. 2009). Breeding of hybrid cultivars further increases the challenge involved in combining all of these diverse genetic prerequisites in new cultivars, as all traits that do not exhibit a dominant mode of inheritance must be complemented in both maternal and paternal components of every new hybrid.

Plant scientists benefit today from a magnitude of extremely beneficial modern technologies to help enhance breeding efficacy. These include access to greenhouses and winter nurseries, which enable increased numbers of breeding cycles per year, along with tissue culture and genetic marker technologies that can accelerate trait fixation and selection, respectively. Nevertheless, the general processes of plant breeding remain fundamentally simple and have not changed since their inception. This basic scheme of a breeding programme can be illustrated by the methodology of recurrent selection, which consists of three elementary steps:

Identification and introgression of new, favourable genetic diversity;
Recombination among crossing partners in a genetically diverse base population;

Selection and fixation of superior phenotypes.

Depending on the mating system of a plant species, new varieties can be produced and successively optimised by this simple alternation between expansion and restriction of genetic diversity. However, although the latter of these three steps is obviously an intrinsic part of continuous improvement of elite varieties, the maintenance and introgression of new allelic diversity into narrow genepools is an equally essential prerequisite for long-term breeding success. In this chapter, we describe how the availability of the *B. napus* reference genome sequence can contribute to identification, implementation and selection of genetic diversity for efficient and sustainable plant breeding. This has particular relevance with regard to identifying and recapturing lost allelic diversity in breeding pools, to counter the negative effects of drift and linkage drag imposed by selection bottlenecks. Special emphasis will be placed on the obstacles that rapeseed and canola breeders have to deal with when trying to combine desirable traits from different genetic backgrounds in one superior cultivar, and how knowledge about the genome, and powerful genomic technologies and tools, can help to overcome these hurdles.

15.3 Sources of Diversity for Rapeseed Breeding

The low level of diversity in modern breeding pools constitutes a difficult problem for oilseed rape and canola breeders, since genetic diversity is the ultimate underlying prerequisite for breeding success. The absence of non-cultivated wild forms of *B. napus* represents a severe constraint for breeders that sets rapeseed apart from most other crops, where rich sources of diversity are available in the primary gene pool of the species. Additionally, the diversity represented in international genebanks exhibits relatively high uniformity throughout each of the major eco-geographical groups described above (Fig. 15.1). Overcoming this diversity, bottleneck is a major ongoing necessity for breeders, and there have been frequent examples for use of interspecific hybridisations involving the diploid progenitor species, in order to enrich traits where selection has eliminated the required breadth of allelic variation from modern cultivars. Fortunately, the propensity of diploid Brassica species to form interspecific hybrids, which gave rise to the allopolyploid species of U's "Brassica triangle" (Nagahuru 1935), facilitates the relatively straightforward reconstitution of synthetic *B. napus* via embryo rescue techniques (Abel et al. 2005; Girke et al. 2012). Other interspecific hybrids are also possible, so that the A- and C-subgenomes can potentially be enriched using the diversity present throughout the entire triangle of U (Chen et al. 2011; Fu et al. 2012; Li et al. 2006; Mason et al. 2010, 2014; Qian et al. 2005).

Examples for the use of synthetic *B. napus*, from *B. rapa* × *B. oleracea* crosses, for transfer of specific traits to breeding programmes include attempts to improve disease or pest resistance (Diederichsen et al. 2009; Juergens et al. 2010; Mei et al. 2011; Rygulla et al. 2007a, b; Snowdon et al. 2000), creation of novel diversity to improve seed quality (Badani et al. 2006; Liu et al. 2012; Stein et al. 2013), and utilisation of synthetic *B. napus* to diversify hybrid breeding pools (Chen et al. 2011; Girke et al. 2012; Li et al. 2006, 2007; Seyis et al. 2003; Snowdon et al. 2015).

In almost all cases, synthetic *B. napus* forms show very poor fertility, seed viability and vigour, which create great challenge for breeders attempting to implement novel diversity via interspecific crosses. It is now known that the basis for this fitness tradeoff is a breakdown in control of homoeologous chromosome pairing during the first meiotic generations, under control of the PrBn locus (Cifuentes et al. 2010; Nicolas et al. 2009; Szadkowski et al. 2011). On the other hand, in rare cases normal chromosome pairing —and subsequent recovery of fertility and vigour —is reinstated after a number of self-fertilisation generations, making it possible to recover stable resynthesised rapeseed from synthetic *B. napus* accessions. Such materials represent a completely novel gene pool for improvement of heterotic potential, for example, (Jesske et al. 2013; Snowdon et al. 2015). On the other hand, synthetic *B. napus* accessions rarely possess the required adaptation attributes in terms of flowering behaviour, seldom show desirable oilseed characters like high seed yield or oil content, and they almost never exhibit the double-low seed quality criteria required for modern oilseed rape and canola production. Nevertheless, synthetic *B. napus* can be readily backcrossed to natural *B. napus*, enabling potential transfer of new genetic diversity for enrichment of breeding genepools. As we describe later, genomic information and tools can accelerate the targeted transfer of useful diversity while preserving essential trait loci for adaptation and seed quality.

The genome restructuring accompanying the breakdown of homoeologous chromosome pairing in synthetic *B. napus* certainly constitutes a challenge for breeders. On the other hand, however, since the availability of a reference genome template for genome resequencing it is now becoming clear that the same process also represents a significant and valuable source of novel genetic variation for evolutionary and breeding selection (Chalhoub et al. 2014; Schiessl et al. 2014; Samans et al. 2017). Small homoeologous exchanges appear to be a common remnant of allopolyploidisation in natural *B. napus* (Chalhoub et al. 2014). Like large-scale homoeologous exchanges, small exchanges giving rise to gene conversions at the single gene and single nucleotide level show a directional bias towards exchanges in which larger segments from the C-subgenome are replaced by their smaller homoeologous counterparts from the A-subgenome (Fig. 15.3). Although the mechanisms underlying this bias are still unclear, this may drive selection for genome size reduction (Samans et al. 2017). On the other hand, it also may cause reduction in C-subgenome diversity, a common observation in all major *B. napus* germplasm pools (Qian et al. 2014; Schiessl et al. 2015; Voss-Fels and Snowdon 2015). Implementation of novel *B. napus* resyntheses can potentially induce de novo genetic variation by homoeologous exchanges and gene conversion. In some cases novel traits, like yellow seed colour or reduced antinutritive seed coat components, can be derived from synthetic *B. napus*, or crosses of *B. napus* to other closely related species (Badani et al. 2006; Li et al. 2012; Lipsa et al. 2012). Dissection of the underlying QTL with the help of genomic reference sequences has provided indications that some of this variation may be associated with homoeologous gene conversions (Liu et al. 2012; Stein et al. 2013). Further examples have been identified for homoeologous gene conversions influencing many processes and pathways of fundamental importance for oilseed rape breeding, for example, the oil biosynthesis and glucosinolate pathways, or within clusters of nucleotide binding site–leucine-rich repeat (NBS–LRR) resistance genes (Chalhoub et al. 2014). These examples provide important clues about how newly arising allopolyploid species have overcome the severe diversity bottleneck associated with rare allopolyploidisation events, to rapidly generate the necessary genetic variation for species survival (Samans et al. 2017).

15.4 Identifying and Accessing Novel Diversity Using Genomic Tools

High-throughput genome analysis techniques represent a powerful technology to investigate and describe genetic diversity in crop species. In *B. napus*, a species with relatively low overall diversity, there have been numerous studies in recent years which used genome-wide molecular marker collections to characterise genetic diversity, thereby facilitating the application of such information to assemble diversity collections for use in large-scale population genetic analyses and genome-wide allele-trait association studies.

The primary prerequisite for identification of diversity that is of interest to breeders is to collect and fix genetic diversity in representative germplasm sets. Such collections enable easy access to the allelic variation present within the species. Great progress has been made in assembly and characterisation of representative germplasm collections for *B. napus* during the past two decades. One of the largest collections of diversity,

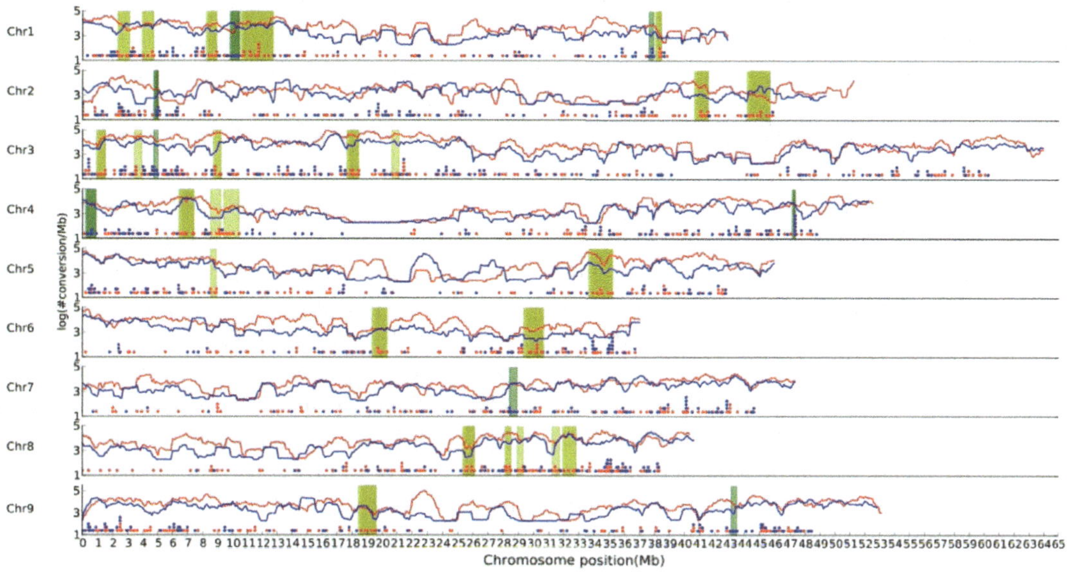

Fig. 15.3 Influence of homoeologous exchanges on quantitative trait variation in *B. napus*, with examples from exchanges between chromosomes A_n2 and C_n2. (A, D) Coverage depth obtained along chromosome A_n2 after mapping Illumina sequence reads from seven natural and one resynthesized *B. napus* genotypes to the "Darmor-bzh" reference genome (B and C) coverage depth obtained for A_r2 and C_o2 chromosomes, respectively, after mapping > 21 genome-equivalents of Illumina sequence reads from *B. napus* "Darmor-bzh" on concatenated genome assemblies from *B. rapa* and *B. oleracea*, respectively. (D) Similar to (A), where the C_n2 chromosome of Darmor-bzh is displayed, egmental homoeologous exchanges are revealed based on sequence read coverage analysis, where a duplication (red) is revealed by significantly greater coverage for a given segment than the rest of the genome (black) and a deletion (blue) by little or no coverage for the corresponding homoeologous segment. Sizes of chromosomes are indicated in Mb. Black arrows in (A) indicate exchanges involving glucosinolate (GSL) and *FLOWERING LOCUS C* (FLC) genes. Figure reprinted with permission from Chalhoub et al. (2014); Copyright 2014 by the American Association for the Advancement of Science

particularly for European and North American oilseed, fodder rape and kale forms, along with rutabaga (swede) forms, was assembled in a European consortium that collected, genotyped and phenotyped various fixed diversity sets from different sources (Bus et al. 2011) and has made seed samples and genotype data available to international users for population genetic analyses and association studies (Bus et al. 2014; Gajardo et al. 2015; Hatzig et al. 2015a; Korber et al. 2015, 2012; Nagel et al. 2011; Schiessl et al. 2015). Other diversity collections with more focus on Asian gene pools have been assembled and used for diversity analysis, LD mapping and association studies by (Wang et al. 2014) and (Qian et al. 2014). Recently, different national research programmes in Germany, Canada, France and China have initiated the generation, genotyping and phenotypic characterisation of large, nested association mapping (NAM; McMullen et al. 2009; Yu et al. 2008) populations for *B. napus*. The intention of NAM populations is to combine the high power of QTL detection provided by structured mapping populations with the considerably greater resolution offered by genetically diverse, non-related populations. NAM populations are generated by crossing a single, adapted accession with a large number of genetically diverse founder accessions, and subsequently creating fixed recombinant sub-families from each of these crosses. The result is a vast, immortal, interrelated population of fixed, half-sibling individuals, which after genotyping with high-density SNP markers can be used for high-resolution, high-power QTL detection.

The most advanced *B. napus* NAM population resource to date consists of over 2500 doubled-haploid (DH) or recombinant inbred line (RIL) progenies, generated by a public–private German research consortium, from crosses of 50 highly divergent founder accessions with a common elite winter-type oilseed rape parent (Snowdon et al. 2013). The founders comprise 20 synthetic *B. napus* accessions with extremely diverse parental origins, including additional C-subgenome donors beyond *B. oleracea*. Furthermore, 30 natural *B. napus* accessions were selected using genome-wide SNP marker data to span the entire species-wide diversity present in the *B. napus* genepool. All of the NAM founder accessions were subjected to genomic resequencing for identification of sequence variation (Schmutzer et al. 2015), and the entire population was genotyped with the Illumina Brassica 60k SNP consortium genotyping array. The resulting dataset is one of the most comprehensive generated to date for a structured *B. napus* mapping population of this large size. In combination with the founder genome sequences, the high-density SNP marker data of each progeny enables high-resolution recombination breakpoint analysis, facilitating accurate genome sequence reconstitution for each of the 2500 NAM lines. The combination of high-density genotype data and structured population is ideal for high-resolution, high-power association studies and access to the underlying sequence variants for each detected QTL constitutes a powerful platform for identification of trait-related candidate genes along with simultaneous allele mining for breeding purposes. This unique resource provides breeders with unprecedented scope for detailed investigations into the genetic mechanisms underlying complex traits, to identify and implement novel variation and to discover the behaviour of interesting genetic variants in highly diverse genetic backgrounds. In combination with state-of-the-art genome data and techniques, highly diverse but adapted NAM populations (and similar multi-parent mapping populations) represent a major prebreeding resource in breeders' aims to recover and implement lost diversity in eroded breeding pools.

The availability of the first reference genome assemblies for *B. napus* (Chalhoub et al. 2014), *B. rapa* (Wang et al. 2011) and *B. oleracea* (Liu et al. 2014; Parkin et al. 2014) provides a template for resequencing activities to capture and characterise species-wide diversity that can be mined and used for *B. napus* improvement. This can involve various strategies to target different genomic targets. For example, (Clarke et al. 2013) described an array-based, reduced-representation sequence-capture approach to survey diversity and trait associations spanning chromosome regions that were identified as meta-QTL for numerous resistances, seed quality and yield-related traits. A bead-based sequence-capture strategy was used by (Schiessl et al. 2014) to capture all homologous copies of over 30 flowering-time regulatory genes, uncovering broad diversity including presence–absence and copy-number variants associated? with adaptive and evolutionary traits. Smaller target regions can be effectively assayed for sequence diversity in large populations using long-read next-generation sequencing protocols that specifically target PCR amplicons for genes of interest (e.g. Gholami et al. 2012; Stein et al. 2013). Alternatively, transcriptome sequencing using mRNAseq, or other reduced-representation RNA sequencing approaches, can give deep insight into variation at a pathway or development-related level. For example, transcriptome sequencing in a genetically diverse *B. napus* collection was used to unravel genome structural variants (Bancroft et al. 2011; Higgins et al. 2012) and facilitate associations of important breeding traits with global gene expression patterns caused by structural variants (Harper et al. 2012).

Ultimately, whole-genome resequencing provides the most powerful technique available for discovery of sequence variations, either for direct analysis of genome-scale variation (e.g. Schmutzer et al. 2015; Snowdon et al. 2015) or for identification of variants for implementation in high-throughput, array-based genotyping platforms. The Illumina Brassica 60k SNP Infinium consortium genotyping array, released in 2013 (Edwards et al. 2013; Mason et al. 2017; Clarke et al. 2016), was largely based on SNPs derived from genomic resequencing and transcriptome sequencing in diverse *B. napus* collections (e.g. Trick et al. 2009). As shown by an ever-increasing number of publications in recent years, high-density *B. napus* SNP array platforms, genomic skim sequencing or reduced-representation genotyping-by-sequencing procedures represent extremely powerful platforms for high-resolution genetic mapping in *B. napus* (Cai et al. 2014, 2015; Delourme et al. 2013; Liu et al. 2013; Raman et al. 2013, 2014), genetic diversity analysis (Qian et al. 2014; Wang et al. 2014), and LD-based genome-wide association studies (Bus et al. 2014; Gajardo et al. 2015; Hatzig et al. 2015a; Korber et al. 2012, 2015; Schiessl et al. 2015). For breeders, these methods create the opportunity to more accurately decipher major-effect loci associated with key traits of interest, and identify tightly linked markers for deployment in breeding programmes. On the other hand, such methods also provide unprecedented insight into the genetic complexity of quantitative traits and their regulation in the context of epistasis, pleiotropy and genetic background (Edwards et al. 2013; Voss-Fels and Snowdon 2015).

The availability of high-resolution genomics platforms for trait dissection and gene discovery in *B. napus* and its close relatives has enormous potential for breeding in association with so-called new breeding technologies based on genome editing procedures (Belhaj et al. 2013; Bortesi and Fischer 2015; Maiti et al. 2015). The essential prerequisite for effective application of genome editing is a detailed knowledge of the genes underlying expression of desired traits, along with their genotype × environment (G * E) and pleiotropic (trait-trait) interactions. After more than 30 years of genetic marker applications in oilseed rape (Snowdon and Friedt 2004; Snowdon and Luy 2012), map-based cloning methods remain a laborious and time-consuming technique that have failed to discover more than a handful of major-effect genes underlying traits of real commercial importance. Unfortunately, in *B. napus*, recombination-based fine-mapping techniques are complicated by the intrinsic difficulties of genetic mapping in an allopolyploid genome, in which a considerable level of segmental homoeologous duplication and rearrangement is present. This can lead to skewed segregation of markers in vast portions of the genome, particularly when synthetic *B. napus* accessions (which are common donors of novel diversity for trait and gene discovery) are used as mapping parents. Analysis of genome structural rearrangements, using bioinformatic techniques associated with high-coverage sequencing data (Samans et al. 2017) or genome-wide SNP calls (Grandke et al. 2017), promise considerable new progress in mapping of genes for important traits. Access to large, well-characterised and densely genotyped mapping and association populations provides additional impulse for discovery of genes underlying even complex traits (e.g. Qian et al. 2016 accepted). Sequencing-based forward-genetic approaches in populations carrying induced or natural quantitative variation also accelerate the discovery of novel mutants causal for traits of interest (Schneeberger et al. 2009; Takagi et al. 2013). Associative expression analyses based on messenger RNA sequencing (mRNAseq) can provide similarly high-resolution access to allelic variants underlying quantitative trait variations (Harper et al. 2012). With access to the *B. napus* reference genome as a template, the implementation of such techniques for gene discovery in oilseed rape can be expected to gain considerable momentum in coming years. Breeders will inevitably profit from the ability to more rapidly and efficiently identify interesting gene variants, and effectively deploy them directly in elite breeding lines using new breeding technologies. Indeed, canola and oilseed rape were among the earliest test cases for implementation of genome editing technologies in cultivar development. Using an oligonucleotide-directed mutagenesis approach, the North American company CIBUS (San Diego, CA, USA) has recently developed spring canola and winter oilseed rape cultivars carrying an engineered herbicide tolerance based on a modified acetohydroxyacid synthase (AHAS) gene. Herbicide tolerance, imparted by genetic modification or through natural mutations, is today a trait with enormous importance for production of spring-sown canola in North America. On the other hand, the stronger vigour of winter-type rapeseed reduces the importance of this trait in European production, and lack of public acceptance for genetically modified crops has prevented implementation of genetically engineered herbicide tolerance in European oilseed rape. Products of genome editing can be indistinguishable from natural point mutations, and are therefore not encompassed by present legislation that strictly governs the use of GM crops in most European countries. Until the implementation of regulatory changes that accommodate the new possibilities enabled by genome editing methods (Huang et al. 2016), the newly registered herbicide tolerant cultivars from CIBUS provide a provocative but highly interesting test case that demonstrates the enormous power of new breeding technologies in oilseed rape and canola variety development.

15.5 The Breeders Equation: Optimising Rapeseed Breeding Progress by Genomics

Under natural conditions, the process of selection can basically be defined as the differential survival and reproduction of individuals due to differences in phenotype (Zimmer and Emlen 2015). As a consequence, this evolutionary force determines the further development of a population and the constitution of subsequent generations. Interestingly, the effect of natural selection on a population can be characterised by only two fundamental parameters: (I) selection intensity and (II) proportion of phenotypic variance that can be attributed to genetic effects. Plant breeders have adapted and exploited this relatively simple principle for centuries, to shape the characteristics of their breeding populations. However, in order to utilise the best genotypes and create superior varieties, breeders have to a great extent assumed nature's role as a selecting force. In contrast to its natural equivalent, artificial selection tries to act with high specificity on a target trait (or a combination of traits), with the aim of directionally changing the phenotypic mean value of a population according to the demands of mankind (Becker 2011).

An observable shift in a population, regardless of whether caused by natural or artificial factors, can be easily quantified by comparing the mean value before and after selection. The interval describing the change of the mean value is defined as the "selection response" (R), which can be described mathematically by the following term, commonly referred to as the "breeder´s equation":

$$R = S * h^2$$

whereby S represents the directional selection differential, and h^2 illustrates the heritability of a trait (Lynch and Walsh 1998). The selection differential is the difference between the phenotypic mean value of the whole population and the phenotypic mean value of a selected fraction of this population within one generation. The effectiveness of transferring the underlying trait to the next generation by selection is then determined by the heritability, which describes the fraction of phenotypic variance that can be attributed to genetic effects.

15.6 Components of the Breeder's Equation and Their Interactions

The quantities of the directional selection interval (S) and the heritability (h^2) already allow drawing a couple of basic conclusions about how selection affects a population. However, the intention of a plant breeder is not to simply measure the selection response, but rather to predict it and consequently design a breeding programme that aims at combining specific desirable characteristics in an elite variety. For this purpose, the breeder´s equation can also be expressed in a slightly modified form that defines the expected selection response per year:

$$R = \frac{i\sigma_G h}{L}$$

According to this expansion of the formula for breeding progress per unit of time, the expected selection response depends on three main factors: (I) The square root of the genetic variance, representing the amount of trait variation that is caused by genotype in a breeding population, (II) the square root of the heritability, equivalent to the selection accuracy (r), which is the correlation between observed and predicted phenotype, and (III) the selection intensity (i), a standardised coefficient that describes the difference between the mean of the whole population and the mean of the selected subpopulation in standard deviations. Finally, by dividing the product of these three parameters by the generation interval (L), the complete selection response per generation is subdivided into selection response over time.

This simple equation is one of the most powerful tools available to breeders, since it not only allows predicting the selection response, but also offers the opportunity to direct selection response by adjusting the respective factors. However, the three variables in the numerator are associated in a close interdependency. Thus, in order to achieve a sustainable, effective long-term response to selection, it is impossible to modify single factors without taking the other, interacting factors into account, meaning that genetic gain is ultimately dependent on a balanced improvement of all contributing factors. In particular, maintenance of high selection intensity (i) over multiple generations permanently necessitates a sufficient amount of genetic variation in the breeding pool. This underscores the overriding importance for breeders of continuous genetic enrichment of their breeding pools, to counteract the potential erosion of diversity by selection and drift. Rapeseed and canola breeding pools can have notoriously low effective population sizes, illustrated dramatically by the extreme example of Australian canola breeding, which was estimated to have been established from as few as 11 founder accessions (Cowling 2007). Implementation of specific traits requiring introgression of multiple gene loci and QTL from a limited donor gene pool can lead to eroded diversity due to strong selection: For example, cultivars combining low seed glucosinolate content with HOLL oil quality require selection of at least three major QTL for low glucosinolates, combined with selection for the two fatty acid elongase (*FAE1*) loci reducing erucic acid content, along with combined mutations at different fatty acid desaturate loci to reduce linolenic acid content.

Additionally, the selection intensity applied in a breeding programme must be defined under consideration of heritability, to ensure that the desired genetic determinants of the traits under selection are transferred to the next generation. Many key traits in canola, such as fatty acid composition, oil content, seed glucosinolate content and flowering behaviour exhibit high or moderately high heritability, which favours high selection intensity. On the other hand, traits with strong environmental interaction, like germination and field emergence (Hatzig et al. 2015a), have extremely low heritability and a consequently poor response to high selection intensity.

In the following sections, we elucidate how modern genomic techniques and methodologies can be exploited to fine-tune and coordinate these three parameters in oilseed rape breeding, enabling simultaneous preservation and enrichment of genetic variation along with maximisation of selection gain per generation interval. Furthermore, we explain how simultaneous reduction of the generation time will ultimately optimise breeding programmes for effectiveness, sustainability and long-term success.

15.7 Increasing Selection Intensity Without Reducing Genetic Variance

Selection is the strongest and most fundamental instrument with which breeders can shape the genetic constitution of their breeding pools. Essentially, the procedure of generating a high performance variety is based on utilising the best genotypes of a population while discarding poor ones. In a stepwise manner, the number of tested genotypes is reduced, while at the same time the most promising candidates are assessed more intensively within various environments and over multiple years. In the short term, this strategy gives rise to efficient elite varieties that are optimally adapted to present nutritional, environmental and sociopolitical requirements. Conditions change over time, however, and breeding programmes inevitably lead to an elimination of allelic diversity which in future might be vital to deal for instance with newly emerging biotic and abiotic factors. For example, dispersal of clubroot disease from Europe to intensive canola and rapeseed production areas in Canada and China necessitates the development of new resistant varieties in these regions (Chai et al. 2014; Rempel et al. 2014). Similarly, modifications of fertilisation legislation in Europe entail breeding of winter oilseed rape cultivars with improved yields under reduced nitrogen input (Stahl 2017), and climate change aspects may mitigate the necessity for more drought-tolerant cultivars in many major growing areas (Hatzig et al. 2014, 2015b; Hohmann et al. 2016, in press). To address such emerging changes in long-term breeding programmes, breeders must permanently have access to comprehensive allelic diversity which encompasses both present and future trait requirements. Hence, in order to ensure a permanent and sustainable breeding success, it is of absolute importance to counter this narrowing of gene pools with a constant, sufficient introgression of novel diversity. Traditionally, the challenge of re-expanding restricted allelic variation had to be met by the mere assessment of phenotypic features and genealogical records, making it difficult to choose suitable genetic resources. In contrast, modern genome-based methods offer exceptionally powerful technology for targeted identification, introgression and conservation of allelic diversity within breeding populations.

Among the most valuable tools presently being used by commercial rapeseed and canola breeders is the Illumina Brassica 60k SNP array (Mason et al. 2017, Clarke et al. 2016). With minimal financial expenditure, the enormous amount of genome-wide SNP markers allows a rapid genotyping of complete populations. As a result, diversity within as well as between populations can be easily assessed on a whole genome or chromosome-wise scale, or even targeted to specific chromosomal regions. Furthermore, due to its high resolution, reproducibility and accuracy, the SNP array technology facilitates an extremely reliable characterisation of plant genotypes and calculation of population structural parameters of significance to breeders, including different measurements of genetic distance, LD, heterozygosity and fixation indices. Such statistics provide extremely important information regarding genome or chromosome regions showing undesirable lack of diversity due to selective sweeps (Voss-Fels and Snowdon 2015). Marker-assisted background selection facilitated by genome-wide SNP arrays can help overcome the negative effects of such genome features in eroded breeding populations. Such strategies are particularly important when implemented in recurrent selection programmes to specifically replenish reduced allelic variation, for example based on crosses between ecogeographical gene pools with synthetic or new-type *B. napus*. Nevertheless, it must be remembered that practical breeders are ultimately not only interested in genetic diversity per se (for example to increase heterotic potential), but specifically in "useful" variation related to optimisation of specific target traits. Therefore, an essential part of genomic data mining for identification of desirable variation is related to the detection of causal associations between genotypic and phenotypic variation. In this context, genome-wide association studies, based on high-density SNP marker analyses in genetically diverse populations, represent a powerful method not only for quantitative trait dissection, but also for identification of useful phenotypic variation for traits of interest (Bus et al. 2014; Gajardo et al. 2015; Hatzig et al. 2015a; Korber et al. 2012, 2015; Nagel et al. 2011; Schiessl et al. 2015).

Despite their inarguable power in terms of cost-efficient genotyping for diversity analyses and genetic dissection of complex traits, SNP array platforms carry the intrinsic disadvantage of only being able to detect already known variation. Because only SNPs that are relatively common in *B. napus* breeding pools are represented on the most commonly used arrays, a general ascertainment bias is inevitable, meaning that very rare alleles and detailed sequence differences (including copy-number and presence–absence variants) cannot generally be detected. Reduced-representation sequencing approaches overcome this ascertainment bias and can potentially increase the potential to identify and introgress completely novel diversity, an important consideration for development of heterotic pools.

15.8 Improving Selection Accuracy for Low-Heritability Traits—Accurate Phenotyping as an Essential Basis for Training of GS Models

As illustrated above, sufficient genetic diversity is an absolutely fundamental prerequisite to identify and select superior genotypes in a breeding population. The breeder's equation makes it clear that the amount of variation in a trait is in direct proportional interaction with selection intensity, so that an increase in genetic diversity generally results in a higher selection response (*R*). Expressed another way, increased diversity facilitates sustained selection response while simultaneously narrowing the fraction of selected individuals.

The third key determining factor in the numerator of the breeder's equation is the heritability, which in general terms can be defined as the variance in the phenotype that is attributable to genotypic variance among individuals in a population. This partitioning of the total phenotypic variance into a genetic and a non-genetic component allows estimation of the relative importance of the determinants of phenotype, in particular the role of heredity. Strictly speaking, a distinction must be made between heritability in the broad sense, measuring the relative importance of the total genotypic variance (V_G/V_P), and heritability in the narrow sense, which refers to the relative importance of the additive variance (V_A/V_P) (see Falconer and MacKay 1996). However, this distinction is irrelevant for a general consideration about the importance of heritability in the breeder's equation and will thus be ignored in the following section.

Being implemented as the square root of the heritability, the meaning of "*h*" can also be interpreted as the transferability of trait expression from the parental population to their offspring. Therefore, if the heritability of a trait is known, "*h*" serves as a measure of how accurately the performance of the next generation can be predicted from the performance of its progenitors.

It is intuitive that the smaller the heritability of a trait, the more breeding material has to be comprehensively tested to reliably evaluate the value of a genotype for a population, since non-genetic factors can extremely distort reliable estimates. This implies, in terms of the breeder′s equation, that selection intensity has to be lowered to maintain selection response, taking the available genetic diversity for granted. Consequently, resources for extensive field trials have to be allocated to test a greater fraction of the breeding population, including candidates that carry less useful alleles. Complex polygenic traits with low heritability compound the accurate estimation of the true genotypic value of an individual, increase G * E interactions and necessitate a high number of phenotypic measurements over numerous environments and years in order to exclude non-genetic error.

In rapeseed breeding, there are relatively few important traits with rather simple genetic inheritance and high heritability that can be dealt with effectively by phenotypic selection in early generations by testing in just one or few environments. A notable exception is selection for low erucic acid, which can be achieved already in F_2 generations by half-seed analysis of fatty acid composition via gas chromatography (Wittkop et al. 2009). Besides major-gene resistances against blackleg and clubroot disease, which in some cases are amenable to seedling cotyledon tests or marker-assisted selection, respectively, most other important disease resistances in rapeseed are characterised by quantitative inheritance, low heritability and ineffective or poorly reproducible field screening procedures (e.g. resistance against *Sclerotinia sclerotiorum* or

Verticillium disease). The same is true for traits like emergence, abiotic stress tolerance, lodging and nutrient use efficiency, necessitating large-scale multi-location field evaluations of advanced breeding lines over multiple years to achieve adequate selection accuracy. This greatly increases the cost associated with complex trait selection and restricts the size of progenies which breeders are able to test effectively under field conditions.

In contrast to the phenotype, the genotype of an inbred breeding line remains stable over environments and years. Thus, with the advent of molecular markers in the 1980s the idea arose to directly identify the genetic factors responsible for traits, in order to improve selection accuracy. Although marker identification still necessitated initial phenotyping of experimental populations, once a marker-trait association was established, a simple marker screening could theoretically replace phenotyping efforts. Early hybridisation-based an PCR-based markers enabled the identification of various genes and quantitative trait loci (QTL) in bi-parental studies (Delourme et al. 1994; Ecke et al. 1995; Foisset et al. 1995, 1996; Uzunova et al. 1995), representing the first step from traditional phenotypic selection to marker-assisted selection (MAS) and marker-assisted recurrent selection (MARS). However, after initial euphoria it became clear that these assumptions worked quite well for monogenic traits, and sometimes also for major-effect QTL, but not for quantitative traits that are influenced by numerous genes, the environment and G * E interaction. Unfortunately, the latter are the high-value traits, ultimately determining the seed and oil yield components, for which rapeseed breeders require more effective selection techniques. Hence, a major advance in breeding progress could be achieved by selection methods that are able to more effectively implement genome information associated with complex traits.

In the classical model of the genetics of quantitative or complex traits, the phenotypic value of an individual is controlled by an infinite number of genes, each with an infinitesimal effect, as well as by non-genetic factors (Goddard 2009). With regard to traits falling into this category, MAS/MARS is severely limited by the proportion of the genetic variance that can be explained by detected QTL (Meuwissen 2009; Meuwissen and Goddard 2001). The inability of classical QTL approaches to deal with highly polygenic traits led to the development of a statistical approach that enabled a simultaneous estimation of effects for all available marker alleles without setting an arbitrary significance threshold. Also, this method no longer focused on finding specific marker-trait associations, but rather used all available marker information to calculate a "genomic estimated breeding value" (GEBV). The term "genomic selection" (GS; Meuwissen 2007) refers to the use of dense markers covering the whole genome so that all genetic variance can be captured by these markers. An elementary assumption underlying GS is that the markers are dense enough to be in LD with all involved major-effect and minor-effect QTL in order to calculate allele effects (Goddard 2009).

The general procedure of GS is relatively simple. One part of the breeding population is used as a reference for the statistical model, called the "training set". The individuals in the training set are both genotyped and phenotyped, representing the base material for the calculation of allelic effects at all loci. All alleles that contribute to genetic variation are captured in this way, even if the effects of the individual loci are very small. Further lines under selection then only have to be genotyped for the markers. The GEBV of the non-phenotyped lines is the sum of all the marker effects predicted from the training population (Hayes et al. 2009a).

Classical linear approaches that try to estimate fixed effects for every marker allele are unable to deal with very large marker numbers that exorbitantly exceed the number of observations, since not enough degrees of freedom are available to simultaneously fit all effects using ordinary least squares regression (Lande and Thompson 1990). However, modern statistical approaches that treat allele effects as random are able to circumvent this problem, facilitated by high-speed computing platforms able to process extremely large data volumes. The most commonly used GS models can basically be categorised, depending on their statistical background, into marker-based linear mixed model approaches using best linear unbiased prediction (RR-BLUP or G-BLUP) and Bayesian approaches. The two concepts differ primarily in their assumptions on the genetic variance that is used to predict allele effects. The detailed theory of GS and statistical models would go far beyond the scope of this chapter, but an excellent and comprehensive review is provided by (Lorenz et al. 2011a).

Because of its assumption of an infinitesimal model, GS is highly reliant on a dense marker covering of the genome. In *B. napus*, high-density marker genotyping is readily achieved using the Illumina Infinium Brassica 60k SNP array, which lays the foundation for fast and cheap, straightforward GS in rapeseed breeding populations. The conserved structure of LD decay on many *B. napus* chromosomes suggests that lower numbers of SNPs may also be sufficient; hence, subsets of selected SNPs that account for variation across LD blocks may give equally accurate predictions. Although to date few empirical studies have been performed describing GS in rapeseed, the effectiveness of GS techniques has been demonstrated in several other crop and animal species in recent years (reviewed by Heslot et al. 2015; Meuwissen et al. 2013), with high prediction accuracies providing an opportunity to revolutionise complex trait selection.

According to Hayes et al. (2009b), the following four parameters essentially determine the accuracy of GEBV predictions:

I. The level of LD between markers and QTL (which is adjustable by marker density) whereby single markers must be in sufficient LD to allow accurate effect prediction;
II. The size of the training population from which SNP effects are estimated;
III. The heritability of the trait: With greater heritability, fewer records are necessary to achieve equal accuracy;
IV. The distribution of QTL effects (which depends on the degree of quantitative inheritance): If many QTL with small effects contribute to trait variation, then a large number of phenotypic records will be required to accurately estimate these effects.

The first two of these factors are under the control of the experimenters, whereas the last two are not. However, selection and introgression of new diversity will alter the constitution of breeding populations over time, thereby changing the pattern of LD between SNPs and QTL (Muir 2007). Therefore, constant updating of the reference population under consideration of these factors is absolutely necessary to ensure a continuously high accuracy for maximisation of long-term response from GS (Muir 2007; Goddard 2009). This aspect is also important for rapeseed breeders who wish to reduce genotyping costs by implementing reduced numbers of SNPs, for example on smaller arrays which select markers based on LD: Changes in population structure, recombination and LD during the course of a breeding programme through introduction of novel diversity may render selected marker sets inefficient for prediction of effects imparted by the novel diversity. This consideration is especially important in rapeseed, where some chromosomes have very long stretches of conserved LD and low recombination in breeding populations, so that they may presently be predictable using small numbers of markers. However, the aim of breeders must actually be to disrupt these conserved LD blocks and introduce new diversity, which will require denser SNP placement to account for new recombinants.

Another very important parameter that still represents a major bottleneck in GS, particularly in plant breeding, is phenotyping accuracy in the training population. Rapeseed has a notoriously flexible plant architecture that reacts with extremely high phenotypic plasticity to environmental disruptions or planting density. For example, plants grown singly in pots or containers, or at the edges of rows, exhibit a completely different branching behaviour and yield parameters to plants growing in the middle of dense plots. Border effects are therefore extreme, particularly where neighbouring plots in field trials show different plant height, lodging resistance, flowering-time, disease resistance or maturation characters. Efficient field evaluations of yield and yield-related characters in rapeseed therefore rely on sufficiently replicated trials, in multiple locations and years, which are best performed using plot-in-plot techniques that deal appropriately with neighbour effects. Breeders aiming to capture and describe genetic variance for complex traits like seed yield, nitrogen use efficiency, drought tolerance or highly quantitative disease resistances (e.g. Verticillium or Sclerotinia) are extremely dependent on large-scale, well-designed field studies. Because accurate yield assessments in rapeseed require particularly large plots, it can be difficult to assess very large populations under uniform field conditions. Genome-based selection strategies would therefore offer an opportunity to prescreen breeding lines for accessions with high GEBV predictions. On the other hand, the accuracy of the predictions is equally dependent on the availability of accurate phenotype data from training populations. The experimental design for field evaluations of training and validation populations is thus of utmost importance for the success of GS.

Compared to classical selection based on phenotypic records, genomic selection exploiting dense genome-wide SNPs, combined with highly accurate phenotyping, offers unprecedented opportunities to improve selection accuracy in traits with even low heritability. Selection intensity can be potentially increased, selecting only those genotypes which are promising, and resources in the field can be allocated to test only the most promising candidates ("selection at the top").

15.9 Speeding Up Breeding—Genomic Selection in Double-Haploid Progenies

The last parameter in the breeder's equation, which we have not addressed thus far, is the generation interval (L), determining the response to selection per time unit. This is a particularly important factor in European winter rapeseed, which has a long annual growing season of around 11 months and a challenging turnaround time between field seasons of only around one month. A number of technical approaches are already implemented to speed up oilseed rape breeding, for example, production of recombinant inbred lines by single-seed descent in the greenhouse, using stress-induced flowering to accelerate the transition from the vegetative to generative phase. Marker-assisted backcrossing (with both foreground and background markers) is particularly useful for single-gene traits that are expressed late in the growing season, like male-sterility and-fertility restoration, or for transfer of major resistance genes and QTL. In rapeseed hybrid breeding programmes, an essential part of the breeding process is to produce highly homozygous lines that serve as potential hybrid parents. By conventional methods, this involves several generations of selfing, adding considerably to the time needed to generate a new hybrid cultivar.

The production of induced doubled-haploid (DH) lines from haploid gametes via cell cultures enables completely homozygous lines to be produced in a single generation, and most hybrid breeding programmes today implement DH lines as paternal hybrid parents (whereas mother lines are generally derived from marker-assisted backcrossing to introgress male-sterility mutations). In the first year of a classical DH line breeding programme, however, considerable effort is still required to generate sufficient seed quantities for multi-environment field selection of per se performance, and until seed bulking is completed in the first year after DH production it is only possible to grow single micro-plots at one or two locations. A strict selection is nevertheless necessary to reduce costs, and this selection process is a key area where breeders may benefit from the application of genomic selection to predict the performance of a DH line based only on its genotype. The combination of GS with DH line production creates a revolutionary, highly efficient tool for rapeseed hybrid breeding. Production and direct assessment of completely homozygous rapeseed-lines can potentially be performed on haploid platelets prior to seed production, enabling selection at least one generation earlier and consequently leading to an enormous increase in the response to selection per year, while simultaneously shortening the generation interval. Firstly, hybrid parent candidates with high predicted per se performance can be made available within a single generation, and secondly their genome-based assessment enables a targeted and stringent selection, resulting in an increased prediction accuracy of trait expression (*h*) and an enhanced selection intensity (*i*). The use of GS to predict general combining ability (GCA) in DH progenies (Jan et al. 2016) can potentially reallocate resources more efficiently from greenhouse expenditure to genotyping expenditure. The potential for preselection of test hybrid combinations based on predicted GCA, rather than per se performance, potentially enables field selection to be performed by breeders at a higher performance level than was previously achievable.

15.10 Further Opportunities for Applying and Expanding Genomic Selection in Oilseed Rape

Since genomic selection is based on flexible statistical models (Lorenz et al. 2011b), different fixed and random factors easily can be included and taken into account during performance prediction and selection. Presently, in rapeseed there are virtually no empirical values available to evaluate GS-assisted breeding. However, studies in other plant species like maize (Riedelsheimer et al. 2012, 2013), rice (Spindel et al. 2015), wheat (Poland et al. 2012; Rutkoski et al. 2011, 2014, 2015), barley (Iwata and Jannink 2011; Zhong et al. 2009) and sugar beet (Hofheinz et al. 2012) allow inferences on how basic GS models could be adopted and further advanced to enhance prediction accuracy in *B. napus*.

One opportunity to extend the prediction model is by integration of the environment as a factor (Heslot et al. 2014, 2013). The key question in this regard is how to design a model to optimally analyse large, unbalanced, multi-environment trials, as are commonly performed in rapeseed breeding programmes. By including allele effect variation across several environments, under consideration of G * E interactions in a linear mixed model, the ultimate goal would be to enable prediction of the performance of a non-phenotyped individual in an untested environment, by incorporation of the effects of alleles, the environment and the influence of G * E interaction.

This, in turn, leads to another important question: What is the optimal target population of environments (TPE) to train a GS model? The TPE is the combination of environments representative for the regions where the cultivars will be grown, defined by all abiotic and biotic factors (Heslot et al. 2015). When accounting for the environment and G * E interaction for prediction, this is a critical question, since non-representative environments will negatively affect the prediction of allele effects and hence reduce prediction accuracy of the GEBV of an individual (Heslot et al. 2015).

Another desirable modification to better adapt GS models to the demands of plant breeders would be creation and application of a genomic selection index (GSI) (Ceron-Rojas et al. 2015). A GSI is a linear combination of GEBVs for different traits, weighted by their economic relevance, with the aim being to analyse several traits simultaneously in a multi-trait model. The use of indices in plant breeding was traditionally relatively uncommon in comparison to animal breeding, because in crops selection normally takes place for different traits at different times, in different environments (Heslot et al. 2015). In rapeseed, for example, phenotypic selection for traits such as yield is usually not available from the preliminary trials used to select on traits like seed oil quality, lodging, flowering-time or monogenic blackleg resistances.

Furthermore, the lack of balanced experimental data, which is a common problem in rapeseed breeding programmes, makes it difficult to accurately analyse field trials. Implementation of more flexible statistical models in GS may help to solve these problems, because (I) for many traits missing phenotypic information can potentially be estimated based only on genotypic data, and (II) linear mixed models offer the possibility to deal with unbalanced data in which not all individuals are (equally) phenotyped for all traits (Heslot et al. 2015).

On the other hand, several further issues need to be considered before GS becomes fully established in rapeseed breeding. From a breeder's point of view, the perhaps most important point is how, where and when to include GS in practical breeding processes. A fundamental question is the optimal allocation of resources with regards to an ideal combination of different selection approaches, i.e. how to efficiently combine classical phenotypic selection, MAS and GS at different breeding stages. Several questions still remain unanswered in regard to GS, for example how to design a representative training population, how and when this should be updated, and how to deal with loss and reconstitution of diversity in GS programmes. An elementary shift associated with GS is that the unit of evaluation is not the phenotype of the individual anymore, but the allele (Knapp and Bridges 1990), although the unit of selection remains the individual. This might necessitate a rethinking of contemporary experimental strategies and necessitate new experimental and statistical approaches to accurately estimate allele effects. An up-to-date overview discussing all of these topics, along with additional considerations regarding adoption on GS as a standard method in plant breeding, is provided by (Heslot et al. 2015).

15.11 Optimising Heterosis in Oilseed Rape

In most of the world's important rapeseed and canola production areas, hybrids are rapidly overtaking inbred line cultivars as the major source of seed for growers, and the benefits of hybrids in terms of yield stability suggest that this trend will continue. In comparison to other major hybrid crops, like maize, however, the level of heterosis in rapeseed is relatively poor. This lack of heterotic potential can be attributed to a number of factors specific to the origin and history or the species: Firstly, the lack of genetic diversity in the primary gene pool limits heterotic potential per se. Secondly, the strong selection for essential quality traits, and for specific flowering variants in different ecogeographical pools, has severely eroded diversity in all breeding pools. Thirdly, for many decades most breeders treated rapeseed as a classical inbreeding crop, generating inbred line varieties for which exploitation of novel diversity across the entire available gene pool was more important than differentiation of crossing partners into clear heterotic pools.

The result of this history is visible today as poor differentiation among heterotic pools. Only over the past two decades have breeders made a conscious effort to actively separate their breeding materials into distinct pools in an effort to increase heterosis. Making this switch from decades of line breeding to a complete focus on hybrid breeding is a major challenge for breeders which can be potentially accelerated by genomic information. In particular, genome-enabled discovery and introgression of novel diversity from the secondary gene pools, represented by related Brassica oilseeds, is a first step to gene pool enrichment without loss of seed quality and adaptation. The use of genome-wide SNP profiles for detailed characterisation of haplotype structures between potential crossing partners and in their offspring represents another means to clearly differentiate heterotic pools on a genome sequence level. Finally, detailed characterisation and differentiation of structural genome variation, taking advantage of additive effects from the Brassica pan-genome, will inevitably maximise additive heterosis effects, as has been inadvertently achieved by many decades of strict pool separation in maize, for example.

15.12 Performance Prediction of Oilseed Rape Hybrids

Simple linear genomic selection models use a mixed model approach that treats the effects of marker alleles as random and only the phenotypic population mean as fixed effect. The prediction of the average values for all available SNP marker alleles is described by the following equation:

$$y = 1\mu + Zu$$

where y represents the predicted GEBV, μ is the phenotypic population mean (as a fixed effect), Z is the genotype * marker matrix and u is the vector of marker effects, represented by average values of the parental alleles (as random effects).

Through the addition of μ as a fixed effect and the random effects in u, it is possible to predict the GEBV of a hybrid based on the genotype data from its parental lines via in silico generation of the hybrid marker profile. Publications of hybrid performance prediction in rapeseed based on genome-wide SNP markers are not yet available, however many commercial breeders are beginning to test prediction models for potential applications, for example to preselect potential hybrid combinations using parameters other than per se performance, which requires multi-environment field trials. As an example, we generated in silico genotype data from a diverse population of 476 F1-hybrids using genome-wide SNP profiles of their respective parental lines. A performance prediction model based on RR-BLUP was run in 100 iterations, using a randomly chosen training population (TP: 70% of the total population) to calibrate the model and the remaining 30% for validation (validation population; VP). This rather rudimentary prediction already gave promising results that demonstrate the potential of GS for hybrid prediction: Prediction accuracies (r) ranged from ~35–42% for low-heritability traits like seed yield and emergence, up to 75–82% for the more heritable traits seed oil and glucosinolate content, respectively. These promising accuracies, which reflect the selection accuracies that can be achieved for these traits by field phenotyping in multiple years and environments, were achieved although linear mixed models like RR-BLUP rely solely on the sum of the average effects of an allele and neglect interactions between alleles or loci. In other words, this form of prediction corresponds to the GEBV for the hybrid component determined solely by GCA, whereas intra-locus dominance effects and inter-locus epistatic effects are not considered in particular. This method performs satisfactorily in populations where GCA is the major contributor to heterosis, for example, in the highly differentiated heterotic pools of maize. As described above, however, rapeseed breeders are far from achieving clear heterotic pool differentiation, and dominance and epistatic effects contributing to specific combining ability (SCA) can play a key role in heterotic expression in *B. napus* (Basunanda et al. 2010). Therefore, hybrid prediction models that additionally account for SCA effects are expected to greatly improve performance of rapeseed hybrid prediction in the immediate future.

Although allelic interaction effects are to some extent already included in the average effect of an allele, dominance and epistasis are not explicitly considered in calculations of the breeding value (Falconer and MacKay 1996). However, these factors are crucial determinants of heterosis and hybrid performance. To include these two components into a hybrid prediction model, the basic GS linear mixed model can be expanded by a dominance matrix, implemented to calculate effects for heterozygous gene loci (Piepho 2009; Zhao et al. 2013a). The application of such a statistical model necessitates the decomposition of the complete genotypic variance into GCA and SCA variance components. The relative importance of the marker allele effects (GCA) versus the allelic interaction effects (SCA) thereby results from the ratio between the two variance components. Although the design of the dominance matrix does not account for complete SCA, since it fails to consider epistasis, the ability of such computationally simple and robust model extensions to improve hybrid prediction accuracy was already demonstrated in other crops (Reif et al. 2013; Zhao et al. 2013b).

Ultimately, successful implementation of hybrid prediction for improved heterotic performance will rely on novel diversity to enrich heterotic pools. Early results from test hybrids using genetically diverse BnNAM lines suggests that information from the rapeseed genome can play a key role in creating, characterising and implementing new hybrid breeding pools carrying "heterotic haplotypes" with elevated hybrid performance (Snowdon et al. 2015). Improved genome assemblies, providing more detailed information on structural variants contributing to intra-locus dominance effects, may prove a key to this implementation. Effective, breeder-friendly data management, access and analysis tools represent a final, essential component for effective exploitation of *B. napus* genome data in rapeseed and canola breeding.

References

Abel S, Mollers C, Becker HC (2005) Development of synthetic *Brassica napus* lines for the analysis of "fixed heterosis" in allopolyploid plants. Euphytica 146:157–163

Allender CJ, King GJ (2010) Origins of the amphiploid species *Brassica napus* L. investigated by chloroplast and nuclear molecular markers. Bmc Plant Biology 10

Badani AG, Snowdon RJ, Wittkop B, Lipsa FD, Baetzel R, Horn R, De Haro A, Font R, Luhs W, Friedt W (2006) Colocalization of a partially dominant gene for yellow seed colour with a major QTL influencing acid detergent fibre (ADF) content in different crosses of oilseed rape (*Brassica napus*). Genome 49:1499–1509

Bancroft I, Morgan C, Fraser F, Higgins J, Wells R, Clissold L, Baker D, Long Y, Meng J, Wang X, Liu S, Trick M (2011) Dissecting the genome of the polyploid crop oilseed rape by transcriptome sequencing. Nat Biotechnol 29:762–766

Basunanda P, Radoev M, Ecke W, Friedt W, Becker HC, Snowdon RJ (2010) Comparative mapping of quantitative trait loci involved in heterosis for seedling and yield traits in oilseed rape (*Brassica napus* L.). Theor Appl Genet 120:271–281

Becker H (2011) Pflanzenzüchtung. Ulmer, Stuttgart

Belhaj K, Chaparro-Garcia A, Kamoun S, Nekrasov V (2013) Plant genome editing made easy: targeted mutagenesis in model and crop plants using the CRISPR/Cas system. Plant Methods 9

Bortesi L, Fischer R (2015) The CRISPR/Cas9 system for plant genome editing and beyond. Biotechnol Adv 33:41–52

Bus A, Korber N, Snowdon RJ, Stich B (2011) Patterns of molecular variation in a species-wide germplasm set of *Brassica napus*. Theor Appl Genet 123:1413–1423

Bus A, Korber N, Parkin IAP, Samans B, Snowdon RJ, Li JQ, Stich B (2014) Species- and genome-wide dissection of the shoot ionome in *Brassica napus* and its relationship to seedling development. Front Plant Sci 5

Cai G, Yang Q, Yi B, Fan C, Edwards D, Batley J, Zhou Y (2014) A complex recombination pattern in the genome of allotetraploid *Brassica napus* as revealed by a high-density genetic map. PLoS ONE 9:e109910

Cai G, Yang Q, Yi B, Fan C, Zhang C, Edwards D, Batley J, Zhou Y (2015) A bi-filtering method for processing single nucleotide polymorphism array data improves the quality of genetic map and accuracy of quantitative trait locus mapping in doubled haploid populations of polyploid *Brassica napus*. BMC Genom 16:409

Ceron-Rojas JJ, Crossa J, Arief VN, Basford K, Rutkoski J, Jarquin D, Alvarado G, Beyene Y, Semagn K, DeLacy I (2015) A genomic selection index applied to simulated and real data. G3-Genes Genom Genet 5:2155–2164

Chai AL, Xie XW, Shi YX, Li BJ (2014) Special issue: research status of clubroot (*Plasmodiophora brassicae*) on cruciferous crops in China. Can J Plant Pathol 36:142–153

Chalhoub B, Denoeud F, Liu SY, Parkin IAP, Tang HB, Wang XY, Chiquet J, Belcram H, Tong CB, Samans B, Correa M, Da Silva C, Just J, Falentin C, Koh CS, Le Clainche I, Bernard M, Bento P, Noel B, Labadie K, Alberti A, Charles M, Arnaud D, Guo H, Daviaud C, Alamery S, Jabbari K, Zhao MX, Edger PP, Chelaifa H, Tack D, Lassalle G, Mestiri I, Schnel N, Le Paslier MC, Fan GY, Renault V, Bayer PE, Golicz AA, Manoli S, Lee TH, Thi VHD, Chalabi S, Hu Q, Fan CC, Tollenaere R, Lu YH, Battail C, Shen JX, Sidebottom CHD, Wang XF, Canaguier A, Chauveau A, Berard A, Deniot G, Guan M, Liu ZS, Sun FM, Lim YP, Lyons E, Town CD, Bancroft I, Wang XW, Meng JL, Ma JX, Pires JC, King GJ, Brunel D, Delourme R, Renard M, Aury JM, Adams KL, Batley J, Snowdon RJ, Tost J, Edwards D, Zhou YM, Hua W, Sharpe AG, Paterson AH, Guan CY, Wincker P (2014) Early allopolyploid evolution in the post-neolithic *Brassica napus* oilseed genome. Science 345:950–953

Chen S, Nelson MN, Chevre AM, Jenczewski E, Li ZY, Mason AS, Meng JL, Plummer JA, Pradhan A, Siddique KHM, Snowdon RJ, Yan GJ, Zhou WJ, Cowling WA (2011) Trigenomic bridges for Brassica improvement. Crit Rev Plant Sci 30:524–547

Cifuentes M, Eber F, Lucas MO, Lode M, Chevre AM, Jenczewski E (2010) Repeated polyploidy drove different levels of crossover suppression between homoeologous chromosomes in *Brassica napus* allohaploids. Plant Cell 22:2265–2276

Clarke WE, Parkin IA, Gajardo HA, Gerhardt DJ, Higgins E, Sidebottom C, Sharpe AG, Snowdon RJ, Federico ML, Iniguez-Luy FL (2013) Genomic DNA enrichment using sequence capture microarrays: a novel approach to discover sequence nucleotide polymorphisms (SNP) in *Brassica napus* L. Plos One 8

Clarke WE, Higgins EE, Plieske J, Wieseke R, Sidebottom C, Khedikar Y, Batley J, Edwards D, Meng J, Li R, Taylor Lawley C, Pauquet J, Laga B, Cheung W, Iniguez-Luy F, Dyrszka E, Rae S, Stich B, Snowdon RJ, Sharpe AG, Ganal MW, Parkin IAP (2016) A

high-density SNP genotyping array for *Brassica napus* and its ancestral diploid species based on optimised selection of single-locus markers in the allotetraploid genome. Theor Appl Genet 129:1887–1899
Cowling WA (2007) Genetic diversity in Australian canola and implications for crop breeding for changing future environments. Field Crop Res 104:103–111
Delourme R, Bouchereau A, Hubert N, Renard M, Landry BS (1994) Identification of RAPD markers linked to a fertility restorer gene for the ogura radish cytoplasmic male-sterility of rapeseed (*Brassica napus* L). Theor Appl Genet 88:741–748
Delourme R, Falentin C, Fomeju BF, Boillot M, Lassalle G, Andre I, Duarte J, Gauthier V, Lucante N, Marty A, Pauchon M, Pichon JP, Ribiere N, Trotoux G, Blanchard P, Riviere N, Martinant JP, Pauquet J (2013) High-density SNP-based genetic map development and linkage disequilibrium assessment in *Brassica napus* L. Bmc Genomics 14
Diederichsen E, Frauen M, Linders EGA, Hatakeyama K, Hirai M (2009) Status and perspectives of clubroot resistance breeding in crucifer crops. J Plant Growth Regul 28:265–281
Diers BW, McVetty PBE, Osborn TC (1996) Relationship between heterosis and genetic distance based on restriction fragment length polymorphism markers in oilseed rape (*Brassica napus* L). Crop Sci 36:79–83
Downey RK, Rakow G (1987) Rapeseed and mustard. In: Fehr WR (ed) Principles of cultivar development. Macmillan, New York, pp 437–486
Ecke W, Uzunova M, Weissleder K (1995) Mapping the genome of rapeseed (*Brassica-Napus* L). 2. Localization of genes-controlling erucic-acid synthesis and seed oil content. Theor Appl Genet 91:972–977
Edwards D, Batley J, Snowdon RJ (2013) Accessing complex crop genomes with next-generation sequencing. Theor Appl Genet 126:1–11
Falconer DS, MacKay TFC (1996) Introduction to quantitative genetics, 4th edn. Longmans Green, Harlow, Essex, UK
Foisset N, Delourme R, Barret P, Renard M (1995) Molecular tagging of the dwarf breizh (Bzh) gene in *Brassica-napus*. Theor Appl Genet 91:756–761
Foisset N, Delourme R, Barret P, Hubert N, Landry BS, Renard M (1996) Molecular-mapping analysis in *Brassica napus* using isozyme, RAPD and RFLP markers on a doubled haploid progeny. Theor Appl Genet 93:1017–1025
Friedt W, Snowdon RJ (2010a) Oil crops breeding. In: Vollmann J, Istvan R (eds) Handbook of plant breeding. Springer Verlag, Berlin/Heidelberg, pp 91–126
Friedt W, Snowdon RJ (2010b) Oilseed rape. In: Vollmann J, Rajcan I (eds) Handbook of plant breeding—oil crops breeding. Springer, Berlin/Heidelberg, pp 91–126
Fu DH, Qian W, Zou J, Meng JL (2012) Genetic dissection of intersubgenomic heterosis in *Brassica napus* carrying genomic components of B. rapa. Euphytica 184:151–164
Gajardo HA, Wittkop B, Soto-Cerda B, Higgins EE, Parkin IAP, Snowdon RJ, Federico ML, Iniguez-Luy FL (2015) Association mapping of seed quality traits in *Brassica napus* L. using GWAS and candidate QTL approaches. Mol Breed 35
Gholami M, Bekele WA, Schondelmaier J, Snowdon RJ (2012) A tailed PCR procedure for cost-effective, two-order multiplex sequencing of candidate genes in polyploid plants. Plant Biotechnol J 10:635–645
Girke A, Schierholt A, Becker HC (2012) Extending the rapeseed gene pool with resynthesized *Brassica napus* II: heterosis. Theor Appl Genet 124:1017–1026
Goddard M (2009) Genomic selection: prediction of accuracy and maximisation of long term response. Genetica 136:245–257
Gomez-Campo C (1999) Biology of Brassica Coenospecies. Elsevier Science
Grandke F, Snowdon R, Samans B (2017) gsrc: an R package for genome structure rearrangement calling. Bioinformatics 33:545–546
Harper AL, Trick M, Higgins J, Fraser F, Clissold L, Wells R, Hattori C, Werner P, Bancroft I (2012) Associative transcriptomics of traits in the polyploid crop species *Brassica napus*. Nat Biotechnol 30:798–802
Hasan M, Seyis F, Badani AG, Pons-Kuhnemann J, Friedt W, Luhs W, Snowdon RJ (2006) Analysis of genetic diversity in the *Brassica napus* L. gene pool using SSR markers. Genet Resour Crop Ev 53:793–802
Hatzig S, Zaharia LI, Abrams S, Hohmann M, Legoahec L, Bouchereau A, Nesi N, Snowdon RJ (2014) Early osmotic adjustment responses in drought-resistant and drought-sensitive oilseed rape. J Integrative Plant Biol 56:797–809
Hatzig SV, Frisch M, Breuer F, Nesi N, Ducoumau S, Wagner MH, Leckband G, Abbadi A, Snowdon RJ (2015a) Genome-wide association mapping unravels the genetic control of seed germination and vigor in *Brassica napus*. Front Plant Sci 6
Hatzig SV, Schiessl S, Stahl A, Snowdon RJ (2015b) Characterizing root response phenotypes by neural network analysis. J Exp Bot 66:5617–5624
Hayes BJ, Visscher PM, Goddard ME (2009) Increased accuracy of artificial selection by using the realized relationship matrix. Genet Res 91:47–60
Hayes BJ, Bowman PJ, Chamberlain AJ, Goddard ME (2009) Genomic selection in dairy cattle: progress and challenges. J Dairy Sci 92:433–443
Heslot N, Jannink JL, Sorrells ME (2013) Using genomic prediction to characterize environments and optimize prediction accuracy in applied breeding data. Crop Sci 53:921–933
Heslot N, Akdemir D, Sorrells ME, Jannink JL (2014) Integrating environmental covariates and crop modeling into the genomic selection framework to predict genotype by environment interactions. Theor Appl Genet 127:463–480
Heslot N, Jannink JL, Sorrells ME (2015) Perspectives for genomic selection applications and research in plants. Crop Sci 55:1–12

Higgins J, Magusin A, Trick M, Fraser F, Bancroft I (2012) Use of mRNA-seq to discriminate contributions to the transcriptome from the constituent genomes of the polyploid crop species *Brassica napus*. BMC Genom 13:247

Hofheinz N, Borchardt D, Weissleder K, Frisch M (2012) Genome-based prediction of test cross performance in two subsequent breeding cycles. Theor Appl Genet 125:1639–1645

Hohmann M, Stahl A, Rudloff J, Wittkop B, Snowdon RJ (2016) Not a load of rubbish: simulated field trials in large-scale containers. Plant Cell Environ 39:2064–2073

Huang SW, Weigel D, Beachy RN, Li JY (2016) A proposed regulatory framework for genome-edited crops. Nat Genet 48:109–111

Iwata H, Jannink JL (2011) Accuracy of genomic selection prediction in barley breeding programs: A simulation study based on the real single nucleotide polymorphism data of barley breeding lines. Crop Sci 51:1915–1927

Jan HU, Abbadi A, Lucke S, Nichols RA, Snowdon RJ (2016) Genomic prediction of testcross performance in canola (*Brassica napus*). Plos One 11

Jesske T, Olberg B, Schierholt A, Becker HC (2013) Resynthesized lines from domesticated and wild *Brassica taxa* and their hybrids with *B-napus* L.: genetic diversity and hybrid yield. Theor Appl Genet 126:1053–1065

Juergens M, Paetsch C, Kramer I, Zahn M, Rabenstein F, Schondelmaier J, Schliephake E, Snowdon R, Friedt W, Ordon F (2010) Genetic analyses of the host-pathogen system Turnip yellows virus (TuYV)-rapeseed (Brassica napus L.) and development of molecular markers for TuYV-resistance. Theor Applied Genet 120:735–744

Knapp SJ, Bridges WC (1990) Using molecular markers to estimate quantitative trait locus parameters—power and genetic variances for unreplicated and replicated progeny. Genetics 126:769–777

Korber N, Wittkop B, Bus A, Friedt W, Snowdon RJ, Stich B (2012) Seedling development in a *Brassica napus* diversity set and its relationship to agronomic performance. Theor Appl Genet 125:1275–1287

Korber N, Bus A, Li J, Higgins J, Bancroft I, Higgins EE, Parkin IAP, Salazar-Colqui B, Snowdon RJ, Stich B (2015) Seedling development traits in *Brassica napus* examined by gene expression analysis and association mapping. Bmc Plant Biol 15

Lande R, Thompson R (1990) Efficiency of marker-assisted selection in the improvement of quantitative traits. Genetics 124:743–756

Li MT, Chen X, Meng JL (2006) Intersubgenomic heterosis in rapeseed production with a partial new-typed *Brassica napus* containing subgenome A (r) firom *B. rapa* and C-c from *Brassica carinata*. Crop Sci 46:234–242

Li M, Liu J, Wang Y, Yu L, Meng J (2007) Production of partial new-typed *Brassica napus* by introgression of genomic components from *B. rapa* and *B. carinata*. J Genet Genomics 34:460–468

Li AM, Jiang JJ, Zhang YT, Snowdon RJ, Liang GH, Wang YP (2012) Molecular and cytological characterization of introgression lines in yellow seed derived from somatic hybrids between *Brassica napus* and *Sinapis alba*. Mol Breed 29:209–219

Lipsa FD, Snowdon R, Friedt W (2012) Quantitative genetic analysis of condensed tannins in oilseed rape meal. Euphytica 184:195–205

Liu LZ, Stein A, Wittkop B, Sarvari P, Li JN, Yan XY, Dreyer F, Frauen M, Friedt W, Snowdon RJ (2012) A knockout mutation in the lignin biosynthesis gene CCR1 explains a major QTL for acid detergent lignin content in *Brassica napus* seeds. Theor Appl Genet 124:1573–1586

Liu LZ, Qu CM, Wittkop B, Yi B, Xiao Y, He YJ, Snowdon RJ, Li JN (2013) A high-density SNP map for accurate mapping of seed fibre QTL in *Brassica napus* L. Plos One 8

Liu S, Liu Y, Yang X, Tong C, Edwards D, Parkin IA, Zhao M, Ma J, Yu J, Huang S, Wang X, Wang J, Lu K, Fang Z, Bancroft I, Yang TJ, Hu Q, Wang X, Yue Z, Li H, Yang L, Wu J, Zhou Q, Wang W, King GJ, Pires JC, Lu C, Wu Z, Sampath P, Wang Z, Guo H, Pan S, Yang L, Min J, Zhang D, Jin D, Li W, Belcram H, Tu J, Guan M, Qi C, Du D, Li J, Jiang L, Batley J, Sharpe AG, Park BS, Ruperao P, Cheng F, Waminal NE, Huang Y, Dong C, Wang L, Li J, Hu Z, Zhuang M, Huang Y, Huang J, Shi J, Mei D, Liu J, Lee TH, Wang J, Jin H, Li Z, Li X, Zhang J, Xiao L, Zhou Y, Liu Z, Liu X, Qin R, Tang X, Liu W, Wang Y, Zhang Y, Lee J, Kim HH, Denoeud F, Xu X, Liang X, Hua W, Wang X, Wang J, Chalhoub B, Paterson AH (2014) The *Brassica oleracea* genome reveals the asymmetrical evolution of polyploid genomes. Nat Commun 5:3930

Lorenz AJ, Chao S, Asoro FG, Heffner EL, Hayashi T, Iwata H, Smith KP, Sorrells ME, Jannink J-L (2011a) Chapter two—genomic selection in plant breeding: knowledge and prospects. In: Donald LS (ed) Advances in agronomy. Academic Press, pp 77–123

Lorenz AJ, Chao SM, Asoro FG, Heffner EL, Hayashi T, Iwata H, Smith KP, Sorrells ME, Jannink JL (2011b) Genomic selection in plant breeding: knowledge and prospects. Adv Agron 110:77–123

Lynch M, Walsh B (1998) Genetics and analysis of quantitative traits. Sinauer

Maiti IB, Sahoo DK, Shreshta A, Patro S, Dey N (2015) Vectors to Modify plant genome using CRISPR-Cas9-sgRNA genome editing system, application to reduce harmful products in tobacco, and in developing broad Spectrum virus resistance in crop plants like Brassica. In Vitro Cell Dev-An 51:S39–S39

Mason AS, Huteau V, Eber F, Coriton O, Yan G, Nelson MN, Cowling WA, Chevre AM (2010) Genome structure affects the rate of autosyndesis

and allosyndesis in AABC, BBAC and CCAB Brassica interspecific hybrids. Chromosome Res 18:655–666
Mason AS, Nelson MN, Takahira J, Cowling WA, Alves GM, Chaudhuri A, Chen N, Ragu ME, Dalton-Morgan J, Coriton O, Huteau V, Eber F, Chevre AM, Batley J (2014) The fate of chromosomes and alleles in an allohexaploid Brassica population. Genetics 197:273–283
Mason AS, Higgins EE, Snowdon RJ, Batley J, Stein A, Werner C, Parkin IA (2017) A user guide to the Brassica 60K Illumina Infinium SNP genotyping array. Theor Appl Genet 130:621–633
McMullen MD, Kresovich S, Villeda HS, Bradbury P, Li HH, Sun Q, Flint-Garcia S, Thornsberry J, Acharya C, Bottoms C, Brown P, Browne C, Eller M, Guill K, Harjes C, Kroon D, Lepak N, Mitchell SE, Peterson B, Pressoir G, Romero S, Rosas MO, Salvo S, Yates H, Hanson M, Jones E, Smith S, Glaubitz JC, Goodman M, Ware D, Holland JB, Buckler ES (2009) Genetic properties of the maize nested association mapping population. Science 325:737–740
Mei J, Qian L, Disi JO, Yang X, Li Q, Li J, Frauen M, Cai D, Qian W (2011) Identification of resistant sources against Sclerotinia sclerotiorum in Brassica species with emphasis on *B. oleracea*. Euphytica 177:393–399
Meuwissen T (2007) Genomic selection: marker assisted selection on a genome wide scale. J Anim Breed Genet 124:321–322
Meuwissen T (2009) Towards consensus on how to measure neutral genetic diversity? J Anim Breed Genet 126:333–334
Meuwissen THE, Goddard ME (2001) Prediction of identity by descent probabilities from marker-haplotypes. Genet Sel Evol 33:605–634
Meuwissen T, Hayes B, Goddard M (2013) Accelerating improvement of livestock with genomic selection. Annu Rev Anim Biosci 1:221–237
Muir WM (2007) Comparison of genomic and traditional BLUP-estimated breeding value accuracy and selection response under alternative trait and genomic parameters. J Anim Breed Genet 124:342–355
Nagel M, Rosenhauer M, Willner E, Snowdon RJ, Friedt W, Borner A (2011) Seed longevity in oilseed rape (*Brassica napus* L.)—genetic variation and QTL mapping. Plant Genet Resour-C 9:260–263
Nicolas SD, Leflon M, Monod H, Eber F, Coriton O, Huteau V, Chevre AM, Jenczewski E (2009) Genetic regulation of meiotic cross-overs between related genomes in Brassica napus haploids and hybrids. Plant Cell 21:373–385
Olsson G (1960) Studies on some plant breeding problems in Brassica and Sinapis. Berlingska Boktr, Ystad, Sweden
Parkin IA, Koh C, Tang H, Robinson SJ, Kagale S, Clarke WE, Town CD, Nixon J, Krishnakumar V, Bidwell SL, Denoeud F, Belcram H, Links MG, Just J, Clarke C, Bender T, Huebert T, Mason AS, Pires JC, Barker G, Moore J, Walley PG, Manoli S, Batley J, Edwards D, Nelson MN, Wang X, Paterson AH, King G, Bancroft I, Chalhoub B, Sharpe AG (2014) Transcriptome and methylome profiling reveals relics of genome dominance in the mesopolyploid *Brassica oleracea*. Genome Biol 15:R77
Piepho HP (2009) Ridge regression and extensions for genomewide selection in maize. Crop Sci 49:1165–1176
Poland J, Endelman J, Dawson J, Rutkoski J, Wu SY, Manes Y, Dreisigacker S, Crossa J, Sanchez-Villeda H, Sorrells M, Jannink JL (2012) Genomic selection in wheat breeding using genotyping-by-sequencing. Plant Genome-Us 5:103–113
Qian W, Chen X, Fu D, Zou J, Meng J (2005) Intersubgenomic heterosis in seed yield potential observed in a new type of *Brassica napus* introgressed with partial *Brassica rapa* genome. Theor Appl Genet 110:1187–1194
Qian L, Qian W, Snowdon RJ (2014) Sub-genomic selection patterns as a signature of breeding in the allopolyploid *Brassica napus* genome. BMC Genom 15:1170
Qian L, Voss-Fels K, Cui Y, Jan HU, Samans B, Obermeier C, Qian W, Snowdon RJ (2016) Deletion of a staygreen gene associates with adaptive selection in Brassica napus. Mol Plant 9:1559–1569
Raman H, Raman R, Kilian A, Detering F, Long Y, Edwards D, Parkin IAP, Sharpe AG, Nelson MN, Larkan N, Zou J, Meng JL, Aslam MN, Batley J, Cowling WA, Lydiate D (2013) A consensus map of rapeseed (*Brassica napus* L.) based on diversity array technology markers: applications in genetic dissection of qualitative and quantitative traits. Bmc Genomics 14
Raman H, Dalton-Morgan J, Diffey S, Raman R, Alamery S, Edwards D, Batley J (2014) SNP markers-based map construction and genome-wide linkage analysis in *Brassica napus*. Plant Biotech J 12:851–860
Reif JC, Zhao YS, Wurschum T, Gowda M, Hahn V (2013) Genomic prediction of sunflower hybrid performance. Plant Breed 132:107–114
Rempel CB, Hutton SN, Jurke CJ (2014) Clubroot and the importance of canola in Canada. Can J Plant Pathol 36:19–26
Riedelsheimer C, Czedik-Eysenberg A, Grieder C, Lisec J, Technow F, Sulpice R, Altmann T, Stitt M, Willmitzer L, Melchinger AE (2012) Genomic and metabolic prediction of complex heterotic traits in hybrid maize. Nat Genet 44:217–220
Riedelsheimer C, Endelman JB, Stange M, Sorrells ME, Jannink JL, Melchinger AE (2013) Genomic predictability of interconnected biparental maize populations. Genetics 194:493–503
Rutkoski JE, Heffner EL, Sorrells ME (2011) Genomic selection for durable stem rust resistance in wheat. Euphytica 179:161–173
Rutkoski JE, Poland JA, Singh RP, Huerta-Espino J, Bhavani S, Barbier H, Rouse MN, Jannink JL,

Sorrells ME (2014) Genomic Selection for quantitative adult plant stem rust resistance in wheat. Plant Genome-Us 7
Rutkoski J, Singh RP, Huerta-Espino J, Bhavani S, Poland J, Jannink JL, Sorrells ME (2015) Genetic gain from phenotypic and genomic selection for quantitative resistance to stem rust of wheat. Plant Genome-Us 8
Rygulla W, Friedt W, Seyis F, Luhs W, Eynck C, von Tiedemann A, Snowdon RJ (2007a) Combination of resistance to *Verticillium longisporum* from zero erucic acid *Brassica oleracea* and oilseed Brassica rapa genotypes in resynthesized rapeseed (*Brassica napus*) lines. Plant Breeding 126:596–602
Rygulla W, Snowdon RJ, Eynck C, Koopmann B, von Tiedemann A, Luhs W, Friedt W (2007b) Broadening the genetic basis of *Verticillium longisporum* resistance in *Brassica napus* by interspecific hybridization. Phytopathology 97:1391–1396
Samans B, Chalhoub B, Snowdon RJ (2017) Surviving a genome collision: genomic signatures of allopolyploidization in the recent crop species Brassica napus. Plant Genome 10:3835
Schiessl S, Samans B, Huttel B, Reinhard R, Snowdon RJ (2014) Capturing sequence variation among flowering-time regulatory gene homologs in the allopolyploid crop species *Brassica napus*. Front Plant Sci 5
Schiessl S, Iniguez-Luy F, Qian W, Snowdon RJ (2015) Diverse regulatory factors associate with flowering time and yield responses in winter-type *Brassica napus*. BMC Genomics 16
Schmutzer T, Samans B, Dyrszka E, Ulpinnis C, Weise S, Stengel D, Colmsee C, Lespinasse D, Micic Z, Abel S, Duchscherer P, Breuer F, Abbadi A, Leckband G, Snowdon R, Scholz U (2015) Species-wide genome sequence and nucleotide polymorphisms from the model allopolyploid plant *Brassica napus*. Sci Data 2:150072
Schneeberger K, Ossowski S, Lanz C, Juul T, Petersen AH, Nielsen KL, Jorgensen JE, Weigel D, Andersen SU (2009) SHOREmap: simultaneous mapping and mutation identification by deep sequencing. Nat Methods 6:550–551
Seyis F, Snowdon RJ, Luhs W, Friedt W (2003) Molecular characterization of novel resynthesized rapeseed (*Brassica napus*) lines and analysis of their genetic diversity in comparison with spring rapeseed cultivars. Plant Breed 122:473–478
Snowdon RJ, Friedt W (2004) Molecular markers in Brassica oilseed breeding: current status and future possibilities. Plant Breed 123:1–8
Snowdon RJ, Winter H, Diestel A, Sacristan MD (2000) Development and characterisation of Brassica napus-Sinapis arvensis addition lines exhibiting resistance to Leptosphaeria maculans. Theor Applied Genet 101:1008–1014
Snowdon RJ, Luy FLI (2012) Potential to improve oilseed rape and canola breeding in the genomics era. Plant Breed 131:351–360
Snowdon RJ, Abbadi A, Kox T, Schmutzer T, Leckband G (2015) Heterotic haplotype capture: precision breeding for hybrid performance. Trends Plant Sci 20:410–413
Song K, Osborn TC (1992) Polyphyletic origins of *Brassica-napus*—new evidence based on organelle and nuclear RFLP analyses. Genome 35:992–1001
Spindel J, Begum H, Akdemir D, Virk P, Collard B, Redona E, Atlin G, Jannink JL, McCouch SR (2015) Genomic selection and association mapping in rice (Oryza sativa): effect of trait genetic architecture, training population composition, marker number and statistical model on accuracy of rice genomic selection in elite. Trop Rice Breed Lines, Plos Genetics, p 11
Stahl A, Pfeifer M, Frisch M, Wittkop B, Snowdon RJ (2017) Recent genetic gains in nitrogen use efficiency in oilseed rape. Front Plant Sci 8:963
Stein A, Wittkop B, Liu LZ, Obermeier C, Friedt W, Snowdon RJ (2013) Dissection of a major QTL for seed colour and fibre content in *Brassica napus* reveals colocalization with candidate genes for phenylpropanoid biosynthesis and flavonoid deposition. Plant Breed 132:382–389
Snowdon R, Samans B, Léon J, Becker H, Ordon F, Scholz U, Weise S, von Virén N, Nagel K, Fiorani F, Schurr U, Li J, Stich B, MH, Brummermann K, Micic Z, Ahlemeyer J, Stelling D, Duchscherer P, Kaiser J, Breuer F, SA, Wolf M, Abbadi A, Leckband G (2013) PreBreed-yield: population genomics resources for nested association mapping and precision breeding in winter oilseed rape. In: International Plant and Animal Genome Conference XXI, San Diego, CA, USA, 2013
Szadkowski E, Eber F, Huteau V, Lode M, Huneau C, Belcram H, Coriton O, Manzanares-Dauleux MJ, Delourme R, King GJ, Chalhoub B, Jenczewski E, Chevre AM (2010) The first meiosis of resynthesized *Brassica napus*, a genome blender. New Phytol 186:102–112
Szadkowski E, Eber F, Huteau V, Lode M, Coriton O, Jenczewski E, Chevre AM (2011) Polyploid formation pathways have an impact on genetic rearrangements in resynthesized *Brassica napus*. New Phytol 191:884–894
Takagi H, Abe A, Yoshida K, Kosugi S, Natsume S, Mitsuoka C, Uemura A, Utsushi H, Tamiru M, Takuno S, Innan H, Cano LM, Kamoun S, Terauchi R (2013) QTL-seq: rapid mapping of quantitative trait loci in rice by whole genome resequencing of DNA from two bulked populations. Plant J 74:174–183
Trick M, Long Y, Meng J, Bancroft I (2009) Single nucleotide polymorphism (SNP) discovery in the polyploid *Brassica napus* using Solexa transcriptome sequencing. Plant Biotechnol J 7:334–346
Nagahuru U (1935) Genome analysis in Brassica with special reference to the experimental formation of *B. napus* and peculiar mode of fertilization. Jap J Bot 7:389–453

Uzunova M, Ecke W, Weissleder K, Robbelen G (1995) Mapping the genome of rapeseed (*Brassica-Napus* L). 1. Construction of an RFLP linkage map and localization of QTLS for seed glucosinolate content. Theor Appl Genet 90:194–204

Voss-Fels K, Snowdon RJ (2015) Understanding and utilizing crop genome diversity via high-resolution genotyping. Plant Biotech J 1–9

Wang X, Wang H, Wang J, Sun R, Wu J, Liu S, Bai Y, Mun JH, Bancroft I, Cheng F, Huang S, Li X, Hua W, Wang J, Wang X, Freeling M, Pires JC, Paterson AH, Chalhoub B, Wang B, Hayward A, Sharpe AG, Park BS, Weisshaar B, Liu B, Li B, Liu B, Tong C, Song C, Duran C, Peng C, Geng C, Koh C, Lin C, Edwards D, Mu D, Shen D, Soumpourou E, Li F, Fraser F, Conant G, Lassalle G, King GJ, Bonnema G, Tang H, Wang H, Belcram H, Zhou H, Hirakawa H, Abe H, Guo H, Wang H, Jin H, Parkin IA, Batley J, Kim JS, Just J, Li J, Xu J, Deng J, Kim JA, Li J, Yu J, Meng J, Wang J, Min J, Poulain J, Wang J, Hatakeyama K, Wu K, Wang L, Fang L, Trick M, Links MG, Zhao M, Jin M, Ramchiary N, Drou N, Berkman PJ, Cai Q, Huang Q, Li R, Tabata S, Cheng S, Zhang S, Zhang S, Huang S, Sato S, Sun S, Kwon SJ, Choi SR, Lee TH, Fan W, Zhao X, Tan X, Xu X, Wang Y, Qiu Y, Yin Y, Li Y, Du Y, Liao Y, Lim Y, Narusaka Y, Wang Y, Wang Z, Li Z, Wang Z, Xiong Z, Zhang Z, Brassica rapa Genome Sequencing Project C (2011) The genome of the mesopolyploid crop species *Brassica rapa*. Nat Genet 43:1035–1039

Wang N, Li F, Chen BY, Xu K, Yan GX, Qiao JW, Li J, Gao GZ, Bancroft I, Meng JL, King GJ, Wu XM (2014) Genome-wide investigation of genetic changes during modern breeding of *Brassica napus*. Theor Appl Genet 127:1817–1829

Weber S, Unker F, Friedt W (2005) Improved doubled haploid production protocol for *Brassica napus* using microspore colchicine treatment in vitro and ploidy determination by flow cytometry. Plant Breed 124:511–513

Wittkop B, Snowdon RJ, Friedt W (2009) Status and perspectives of breeding for enhanced yield and quality of oilseed crops for Europe. Euphytica 170:131–140

Yu JM, Holland JB, McMullen MD, Buckler ES (2008) Genetic design and statistical power of nested association mapping in maize. Genetics 178:539–551

Zhao YS, Gowda M, Liu WX, Wurschum T, Maurer HP, Longin FH, Ranc N, Piepho HP, Reif JC (2013a) Choice of shrinkage parameter and prediction of genomic breeding values in elite maize breeding populations. Plant Breed 132:99–106

Zhao YS, Zeng J, Fernando R, Reif JC (2013b) Genomic prediction of hybrid wheat performance. Crop Sci 53:802–810

Zhong SQ, Dekkers JCM, Fernando RL, Jannink JL (2009) Factors affecting accuracy from genomic selection in populations derived from multiple inbred lines: a barley case study. Genetics 182:355–364

Zimmer C, Emlen DJ (2015) Evolution: making sense of life. Roberts and Company, Greenwood Village, Colorado, USA

Future Prospects for Structural, Functional, and Evolutionary Genomics

16

Shengyi Liu and Rod Snowdon

Abstract

Completion of the genome assemblies of the first three *Brassica napus* genotypes provided a reference for genome evolution research, gene discovery, and breeding of Brassica crops—in particular the availability of a reference genome has greatly facilitated mapping of trait loci. After the previous chapters, this chapter provides future prospects on three aspects: (1) Structural genomics—The current *B. napus* reference genomes remain a great space for improvement. This is urgent need for at least one chromosome-level assembly should be achieved which corrects collapsed genomic regions such as highly repeated sequences from the previous versions. Highly structural variation also necessitated construction of a *B. napus* pan-genome which describes species-level structural variation in as much detail as possible. Meanwhile, substantial improvement should be undertaken to the annotations of genomic composition and gene models, in order to provide a set of comprehensive annotations including non-coding RNA and alternative splicing transcripts. (2) Functional genomics—There are many ways for making use of the *B. napus* genome resources to assist genetics-related research. With genome resources and related technologies, one can further speed up discovery of molecular markers and functional genes by linkage mapping, association mapping, syntenic comparison approaches, and their combination (among themselves and with other omics data). (3) Genome evolution—This gives insight into allopolyploid *B. napus* genome origin, its dynamic genome structure variation, genetic diversity, and selection patterns. Insight into the processes of multiple cycles of "whole-genome duplication and subsequent diploidization", such as structural variation and its underlying mechanism, patterns and origins of duplicate gene expression changes, the relative contributions of duplicate genes to trait expression, and asymmetrical recombination and selection between *B. napus* subgenomes and regional "hot" and "cold" spots, will broaden our understanding of *B. napus* polyploid genetic diversity and benefit breeding method innovation.

S. Liu (✉)
Key Laboratory of Biology and Genetic Improvement of Oil Crops, the Ministry of Agriculture, Oil Crops Research Institute of the Chinese Academy of Agricultural Sciences, Wuhan 430062, China
e-mail: liusy@oilcrops.cn

R. Snowdon (✉)
Plant Breeding Department, IFZ Research Centre for Biosystems, Land Use and Nutrition, Justus Liebig University, Heinrich-Buff-Ring 26-32, 35392 Giessen, Germany
e-mail: rod.snowdon@agrar.uni-giessen.de

S. Liu et al. (eds.), *The Brassica napus Genome*, Compendium of Plant Genomes,
https://doi.org/10.1007/978-3-319-43694-4_16

16.1 Introduction

As with other important plant species, successful sequencing, assembling into pseudomolecules or chromosomes, and genome annotation of the major crop species *Brassica napus* not only opened a door of understanding the characteristics of a recent allopolyploid plant genome, but also substantially promoted research in all aspects of genetics and breeding along with numerous aspects of polyploid evolution and related plant science areas. The previous chapters of the book are rich in examples for applied genomics in genetics and breeding research. The development of trait markers, mapping of genes controlling traits, and cloning of target genes have grown at an exponential pace in *B. napus* after its annotated genome became available (Fig. 16.1 and also see Chap. 3 by Delourme et al., this book). Consequently, molecular marker-aided breeding has been considerably enhanced, and research in genome-based design breeding has been nurturing with accumulative trait-associated genes or markers and allelic variants. Importantly, the *B. napus* genome assembly enabled elucidation of the syntenic relationships with the *Arabidopsis thaliana* genome greatly facilitating transfer of Arabidopsis gene function and pathway information to this important oilseed crop.

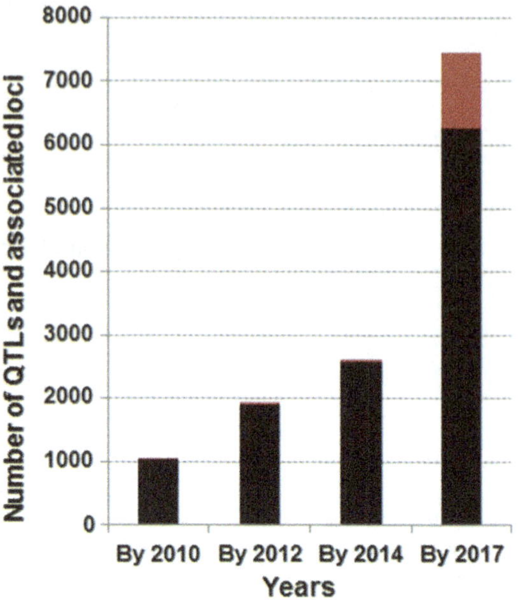

Fig. 16.1 *Brassica napus* QTLs published (black) and associated loci (brown; about one-third unpublished) at the different stages of available genome resources. The *B. rapa* genome sequence became available in 2011, the Brassica genome-wide Illumina 60K SNP array available in 2012, and the *B. napus* and *B. oleracea* genome sequences available in 2014, respectively. Before *B. napus* genome released, the *B. rapa* genome could be used for molecular marker development due to high synteny of the *B. rapa* and *B. napus* A subgenome. Design of the 60K SNP array was based on the three genome sequences of *B. rapa*, *B. oleracea*, and *B. napus*

These upsurges will continue. With the development of more mapping populations, functional genome studies will continuously increase, and evolution and many other related research areas will step into a new stage. New technologies already open opportunities for "gold standard" Brassica genome assemblies by updating the current reference genome or creating new de novo assemblies along with construction of a pan-genome containing species-level structural variation. These developments will further substantially facilitate all these researches mentioned above.

After comprehensive reviews on the topics of from the economic and academic importance of *B. napus* (Chap. 1 by Friedt, Tu and Fu, this book), cytogenetics (Chap. 2 by Chevre et al., this book), trait locus mapping (Chap. 3 by Delourme et al., this book), and genome-facilitated breeding (Chap. 15 by Werner and Snowdon, this book) in the context of the genome as a resource and tool, to characteristics of the nuclear and cytoplasmic genomes, and further case studies on important trait-related gene families, this chapter provides forward-looking prospects on structural genomics, functional genomics, and genome evolution.

16.2 Structural Genomics

At the time of writing, there are three *B. napus* genome sequences available already and de novo assembled from the genotypes Darmor-*bzh* (Chalhoub et al. 2014), Zhongshuang11 (Sun et al. 2017), and Tapidor (Bayer et al. 2017). These reference genomes, particularly the first published Darmor-*bzh* genome, have provided unprecedented rich resources and been greatly advancing the aforementioned genetic and related studies. However, owing to the limitations of sequencing technology and the lack of a more powerful assembly strategy and pipeline, the present versions of the three *B. napus* genomes are only draft assemblies, and a large portion of the genome has not yet been assembled. Based on the theoretical estimations, the *B. napus* genome size is found to be 1130–1345 Mb (Bayer et al. 2017; Chalhoub et al. 2014; Johnston et al. 2005). The published Darmor-*bzh* genome was sequenced with a combination of 454 GS-FLX + Titanium, Sanger and Illumina HiSeq technologies, and assembled with SOAP (8) and Newbler (Roche). In this strategy, Illumina HiSeq reads were used for correcting and gap filling. The final assembly was 849.7 Mb in size, covering ~79% of the genome size. Of the 849.7 Mb, 84% (712.3 Mb) was genetically anchored with a high-resolution SNP genetic map, yielding pseudomolecules for the 19 chromosomes. For the Zhongshuang11 genome, adoption of a BAC-to-BAC strategy complemented with the assembling of reads from Illumina whole-genome libraries resulted in a size of ~976 Mb, with a scaffold N50 of 602.22 kb and a contig N50 of 39.57 kb, all slightly bigger than those of Darmor-*bzh*. The third assembled genome is about 634 Mb in size with an N50 of 197 kb, all smaller than those of Zhongshuang11. Subsequently, the Darmor-*bzh* genome was improved by adding 153 Mb in size into the pseudomolecules (Bayer et al. 2017). Therefore, compared to the estimated sizes of the *B. napus* genomes, no more than 86.4% was assembled for any of genotype, no more than 73.4% anchored onto genetic maps, and no more than 57.2% is placed into the pseudomolecules. Furthermore, recent studies have showed that transposable element-derived small RNAs (mainly 24-nt RNA) play pivotal roles in the regulation of gene expression, e.g., via RNA-directed DNA methylation, and more importantly, there are a huge number of different small RNAs, but there are massive quantities of transposable elements missing in the present reference genomes. There are therefore an urgent need to update genome assembly quality (assembly size, percent anchoring, and percent placement) up to a gold standard for a single genotype in order to further solve remaining difficulties in gene fine mapping and cloning.

To assemble gold standard genomes is becoming increasingly feasible as new sequencing and assembling technologies emerge. Since the second-generation sequencing technologies and corresponding assembling pipelines were proved to be successful in assembling complex plant genomes, the later assembling pipelines, such as ALLPATHS-LG (Gnerre et al. 2011), MaS-65 uRCA (Zimin et al. 2013), and SOAPdenovo2 (Luo et al. 2012), can produce contigs by using longer k-mers over 100 bp or even dynamic k-mers (Platanus) (Kajitani et al. 2014), which can greatly increase the quality of the assembly. While sequence reads from the Illumina sequencing platform now achieves over 250 bp long reads, many projects have already moved on to third-generation sequencing technologies, such as Pacbio SMRT (single-molecule real-time sequencing), Bionano DLS (Direct Label and Stain high-resolution optical mapping) and Oxford Nanopore semi-conductor sequencing. Combinations of these new technologies, particularly with high-through chromosome conformation capture, Hi-C), can produce a chromosomal arm-level or even whole chromosome-level assembly (Badouin et al. 2017; Jarvis et al. 2017; Jiao et al. 2017). Such assembling into pseudomolecules (chromosomes) can be completed without a genetic map for scaffold alignment, although a highly dense genetic map can more accurately orchestrate the assembling procedure. Illumina sequence reads can be used to fill gaps and correct base errors that are still prevalent in the third-generation technologies. These chromosome-level assemblies through these latest technology combinations can be regarded as today's gold standard reference genomes covering nearly all collapsed genomic regions such as highly repeated segments, and thus greatly increasing the usefulness of a reference genome. Generation of such gold standard *B. napus* reference genomes is now possible and urgently required for research in genomics and genetic improvement.

Improvement in the annotation of a reference genome is however a continuous and ongoing task. There are several shortcomings in the present three assembled *B. napus* genomes: (1) Many gene models (about 30%) have not yet had expression data support; (2) alternative splicing transcripts (AST; also called variants or isoforms) from different tissues have not been identified and annotated, and some of the already identified ASTs are not correct, because Illumina short read mapping to the genome often causes artifacts coming from highly similar duplicate sequences; (3) small RNA and lncRNA have not been annotated, although they are known to play broad and important roles in the regulation of gene expression; and (4) there remains no gene information in unassembled genomic regions. Third-generation sequencing technologies enable RNA sequencing with very long reads spanning whole bodies of almost all individual genes and thus allow us to distinguish duplicated genes accurately and different ASTs from individual genes. As long as we sequence representative tissues/organs at different development stages and with various stress treatments, these techniques can capture a majority of expressed genes along with their ASTs. Comprehensive, high quality gene annotations for new, gold standard reference genomes represent a greatly needed future resource to facilitate further progress in Brassica research.

Furthermore, a majority feature of *B. napus* species is the vast amount of genome structural variation among different genotypes and there is growing evidence for links to many important traits (see below for details). To adequately assess, understand and utilise the genome structural variation, multiple gold standard de novo genome assemblies are required for representative genotypes across the species, which are also required for construction of a high-quality pan-genome as an essential resource for the Brassica research community.

A number of genomes of other Brassicaceae species and relatives have been sequenced after the model species *A. thaliana*, e.g., *A. lyrata* (Hu et al. 2011), *Capsella rubella* (Slotte et al. 2013), *Schrenkiella parvula* (syn. *Thellungiella parvula*) (Dassanayake et al. 2013), *Leavenworthia alabamica* (Haudry et al. 2013), *Sisymbrium irio* (Haudry et al. 2013), *Aethionema arabicum* (Haudry et al. 2013), *Brassica nigra,* and *Brassica juncea* (Yang et al. 2016), and more and more will be sequenced in coming years. It is very important to analyze and establish syntenic and orthologous relationship or syntenic/orthologous blocks among these species, because the structural relationships are potentially extremely useful for the comparison of interspecific genomic structures and function and transfer of information, and for genome evolution studies.

An integrated genomics resource is a key tool for accessing and using genome data (see Chap. 14 by King and Baten, this book). There are several publicly accessible Brassica genomics databases which can be used to conduct search and comparative analyses, for example http://www.genoscope.cns.fr/brassicanapus/, www.OCRI-genomics.org, along with the general information site www.braasica.info which serves as a community resource to link users to relevant information and data sources. There are several broader databases which contain Brassica genome data, for example, http://plants.ensembl.org, http://www.plantgdb.org/, https://genomevolution.org/cog, but these still lack intra- and interspecific comparative genomic information. Specific genomic databases urgently need to improve genome browser functions that compare and display information on synteny blocks and genome composition annotation and provide links to high-density genetic map information and QTL locations in public germplasm collections. Visual identification of structure variation is also a very useful tool of high relevance to *B. napus* (He et al. 2017). Future databases should integrate genetic materials, genomics, transcriptomics, metabolomics, proteomics, and phenomics and provide easy search and comparison tools with visual outcomes. This is not only because these comparative tools can be used for transferring knowledge from species to species, but also because these tools can help investigators who are not working in the field of genomics to apply genomics in their own fields.

16.3 Functional Genomics

The public availability of the assembled *B. napus* genomes has greatly facilitated studies on molecular marker development for important traits (Fig. 16.1), cloning, and functional regulation of genes controlling important traits. These have accelerated progress in the molecular marker-aided selection and further genome-based design breeding. Such facilitation will continue and particularly will be significantly enhanced by completion of a pan-genome construction. There are many ways for use of the *B. napus* genome resources to assist genetics-related research, some of which are outlined below.

Genome-wide association studies (GWAS) represent a powerful, high-throughput method for the detection of trait loci (association mapping). Coupling high-throughput methods of cost-effective genotyping by whole-genome re-sequencing with automated phenotyping under controlled conditions or with drones under field conditions provides new opportunities to exploid genome data to uncover responsible QTL and genes. The *B. napus* reference genome enables whole-genome re-sequencing to detect a variety of genetic variations such as SNPs, InDels, large segmental deletion, copy number variants , and homoeologous exchange (HE) in large populations comprising diverse core sets or representative germplasm collections of a crop species. These variations provide a chance to understand genetic structure and recombination landscape of the species. When the variations are associated with phenotypic data, we can not only map important traits, but also detect selection signals such as selective sweeps and fixation indices (F_{ST}). Although morphological variation in *B. napus* is not so large as its progenitors *Brassica rapa* and *Brassica oleracea*, genomic structural variation is very large (Chalhoub et al. 2014; He et al. 2017; Stein et al. 2017). GWAS is a powerful tool to detect these variations linked to traits—many hundreds of trait loci can potentially be assayed and compared in a single population.

Mapping trait loci via biparental segregation populations has been most widely used to date (linkage mapping). The typical primary populations are F2, RILs, and DH, and from them, some derived secondary populations for fine mapping can be generated. Such populations are generally useful to detect only a limited number of traits that segregate between the parent pairs, and few identified QTLs are economically relevant. On the other hand, individual research groups generally lack the resources to deal with multiple biparental segregation populations or multi-parental populations for QTL mapping. Traditional electrophoresis-based genotyping methods are time-consuming and laborious and thus expensive. Also, costs of genotyping by genome re-sequencing of large population are still not affordable for all users, even at a very low price, particularly for crops like *B. napus* with a large, complex genome. Bulked segregant analysis (BSA) has therefore become a popular option for QTL delineation because of the vast reduction in sample numbers for whole-genome re-sequencing. For a large number of samples to be genotyped, an alternative method is the high-throughput SNP arrays designed from whole-genome re-sequencing of a representative germplasm collection. This may be a reason that a leap increase has occurred in the *B. napus* QTL number (Fig. 16.1). The array technology is especially cost-effective for a crop with a large genome size. As more mapping population are generated and genotyped with new genotyping technologies, links between genetic mapping studies and (pan) genome sequences will become continually stronger and provide enormous added value.

A combination of linkage and association methodologies should provide the most robust and powerful approach to fine-map target loci underlying traits (Liu et al. 2015; Ott et al. 2011). Their combination takes their complementary advantages together. For example, QTL regions are usually large in linkage mapping of a biparental population, but associated loci are relatively much smaller in association mapping in a diverse natural population. Therefore, targeted association analysis of a previously identified QTL region and/or its genes can be implemented in a association mapping population if the whole genome-wide association analysis has no signal detected in the physical region, also providing allelic variation information that cannot be obtained from a biparental population.

Syntenic or orthologous relationships represent a useful approach for functional gene studies as well as evolution studies. In particular, syntenic relationships provide a highly valuable starting point to transfer information on *A. thaliana* gene function and related pathways to *B. napus*. Many previous studies used homologous Arabidopsis genes with known functions to infer or predict *B. napus* genes. With the availability of the *B. napus* genome, these homologous comparisons can be limited to syntenic orthologs and paralogs, in order to better ensure their evolutionary origin, as most of the orthologous and paralogous genes have large changes in DNA sequences, and thus, sequence comparison makes it hard to distinguish their homologous relationship. In the publications on the genomes of *B. napus* and its progenitors *B. rapa* and *B. oleracea*, syntenic or subgenomic relationships have been established between the three species and of these species with *A. thaliana*,the basal eudicot *Vitis vinifera* (The French–Italian Public Consortium for Grapevine Genome Characterization 2007) and even the basal angiosperm *Amborella trichopoda* (Amborella Genome Project 2013). Syntenic genes are a group of genes or a genomic block, usually more than five required for construction of a syntenic block, that structurally retained from an ancestral genome . Thus, interspecific comparison of syntenic blocks provide strong evidence of orthologous gene relationships and strongly facilitate functional inference. There are about 17,000 *A. thaliana* protein-coding genes (about 70% of the total) syntenic to 32,700 genes (around 75% of the total) of A or C genome in *B. rapa, B. oleracea, and B. napus*. More than 21,000 genes have two or more copies (duplicated genes) in A or C genomes (see details in Chap. 5 by Tang et al., this book). With analyses of gene and regulatory sequence similarity/changes, domains, and gene expression in different tissues or organs in comparison with functionally known Arabidopsis orthologous genes, one may be able to get concrete evidence on *B. napus* gene functions, e.g., candidate genes in a mapped genomic region, before commencing time-consuming fine mapping or gene function experiments. Furthermore, more than 50% of the total A or C genes have an orthologous relationship, and thus, we can also compare the orthologous genes of *B. napus* A and C genomes to help identification of gene functions and selection roles (like asymmetrical selection).

In addition, the above methods can be flexibly integrated with other approaches or data such as transcriptomes and mutant information available to narrow down target regions or exclude non-target genes. With the reference genome as a physical axis, published and unpublished genetic maps and QTL information can be integrated into one consensus map, which will greatly facilitate mapping work.

Numerous recent studies described SNP discovery for QTL mapping and gene cloning based on time-saving and high-throughput second-generation sequencing (e.g. Bancroft et al. 2011; Bus et al. 2012; Clarke et al. 2013; Huang et al. 2013; Trick et al. 2009). Bancroft et al. (2011) developed genetic maps comprising 21,323 SNPs (arranged in 887 bins, a cluster of SNPs in a genomic region in which every two SNPs are with zero recombination) derived from transcriptome sequence analyses. A genetic map with 14,675 SNPs corresponding to 895 genetic bins was constructed using a modified double-digested restriction-site associated DNA (ddRAD) sequencing technique (Chen et al. 2013). Furthermore, development of SNP arrays designed from whole-genome re-sequencing of a small number of representative germplasm accessions made genotyping even more convenient and time-saving. There were several different Illumina Infinium arrays developed and used for genetic map construction and QTL mapping. An 8K array including 7322 SNP markers was used to build an integrated genetic map comprising 5764 SNPs and 1603 PCR markers from four biparental segregating populations (Delourme et al. 2013). A 6K array (Dalton-Morgan et al. 2014) allowed 631 and 1667 SNPs to be incorporated into two genetic maps built by Raman et al. (2014) and Cai et al. (2014), respectively. In 2012, an International Brassica SNP Consortium designed a *B. napus* 60K SNP Infinium array with 52,157 SNPs, which was manufactured by Illumina Inc., San Diego, CA, USA (Edwards et al. 2013; Snowdon and Iniguez Luy 2012; Clarke et al. 2016; Mason et al. 2017). Using the array, the maps with ~9K or 11K SNPs were built (Liu et al. 2013; Zhang et al. 2014) and many more thereafter. The combination of SNPs from the various high-density genotyping systems led to the construction of an integrated genetic map comprising 41,001 markers mapped into 7287 genetic bins, which was the most dense single genetic map (Chalhoub et al. 2014).

With the availability of the reference genome, more than 7300 QTLs and associated loci have meanwhile mapped (Fig. 16.1; see details in Chap. 3 by Delourme et al., this book), and the efficiency of map-based cloning has improved considerably. For gene cloning, fine mapping of underlying genes for important traits is essential. Using the results of fine mapping and targeted regional association, Liu et al. (2015) identified seven putative ORFs for a QTL, explaining ca. 30% of the phenotypic variation for seed weight and silique length. Based on the genome sequence of *B. napus*, the seven genes including the upstream regulatory and coding regions were cloned and their sequences were compared with the two parental lines to exclude non-target genes. Finally, an auxin response factor 18 (ARF18) with a 165-bp deletion was selected as the target casual gene and subsequently was further validated through gene expression and over expression analyses (Liu et al. 2015). Other trait loci such as those controlling plant height ("dwarf" trait), disease resistance, Polima cytoplasmic male sterility (cms), genic male sterility along with restoration loci were fine-mapped and cloned for their target genes (Larkan et al. 2013; Liu et al. 2010; Xia et al. 2016; Yi et al. 2010). Although QTL map-based cloning and fine mapping are often laborious with a low success rate, combinations of high-throughput methods are accelerating the process and will lead to cloning of more and more genes.

For *B. napus*, genotyping with new high-throughput SNP arrays is still cost-effective compared to whole-genome re-sequencing. The previous and present Illumina Infinium arrays mentioned above were designed from whole-genome re-sequencing of a small collection of representative germplasm, and the SNPs called were based on incomplete genome assembly, and thus, some genomic regions are with no SNPs. These arrays may not meet research requirement that needs higher resolution and more reasonable SNP distribution throughout the genome. A new array should be designed based on SNPs called with an updated and more complete genome assembly and from a large and diverse *B. napus* population and provide higher resolution of genotyping, e.g., 120K or more SNPs, and more rational SNP distribution throughout the genome. Because many trait loci have been mapped, a new array should incorporate these trait markers for use in breeding selection. Smaller, custom breeding arrays which incorporate trait markers for use in molecular marker-aided selection and genome-based design breeding are already in use by commercial breeders.

16.4 Genome Evolution

Numerous studies have clearly indicated that angiosperm plants experienced several cycles of polyploidization events (whole-genome duplications, WGD, or triplication, WGT). For example, the Brassica lineage includes the following: (1) the *Zeta* event before the divergence between gymnosperms and angiosperms, (2) the *Epsilon* event occurred in the angiosperm plant ancestor, (3) the *Gamma* event occurred in the eudicot plant ancestor (The French–Italian Public Consortium for Grapevine Genome Characterization 2007; Jiao et al. 2011; Amborella Genome Project 2013), (4) the *Beta* and *Alpha* events occurred in the Brassicaceae ancestor (Bowers et al. 2003; Lyons et al. 2008; Vision et al. 2000), (5) the triplication event shared by Brassicaceae species (Liu et al. 2014; Lysak et al. 2005; Parkin et al. 2005; Wang et al. 2011), and (6) Brassica allopolyploids (Chalhoub et al. 2014). Why have so many cycles of polyploidization events occurred in flowering plants? It is commonly observed that polyploids have many advantages such as vigor (heterosis or bigger biomass), adaptation, and many others (Comai 2005; Dubcovsky and Dvorak 2007; Jackson and Chen 2010; Masterson 1994), and it is considered that polyploidization events enabled species radiation/burst which consequently have had a profound impact on the earth ecological system (Ainouche and Jenczewski 2010; Soltis and Soltis 2003; Van de Peer 2011). The timing of many ancient polyploidization events was estimated to be after dramatic climate changes in the earth surface (Beilstein et al. 2010; Kagale et al. 2014). Therefore, the study on the mechanisms of plant polyploid genome evolution is a core part for understanding of the mechanisms of plant genome evolution, a frontier area in plant science.

The *B. napus* genome contains three clearly detectable events of whole-genome duplication or triplication (polyploidization events): Alpha (around 5500 gene pairs shared in *A. thaliana, B. rapa, B. oleracea,* and *B. napus* A or C genome), triplication (about 16,300 gene pairs or triplets shared in *B. rapa, B. oleracea,* and *B. napus* A or C genome), and duplication (hybridization of *B. rapa* and *B. oleracea* to form *B. napus*). Genes that originated from ancient shared WGD events can also be traced, from the *Beta* duplication event prior to *Alpha* to the *Gamma* triplication event represented by the basal eudicot *V. vinifera* (The French–Italian Public Consortium for Grapevine Genome Characterization 2007), and some of the genes can even be traced for syntenic relationships with the basal angiosperm genome of *Amborella trichopoda*. Together with rich *A. thaliana* information on genome structure, evolution, and gene functions, this knowledge makes *B. napus* an excellent and unique model to address questions of crop allopolyploid genome evolution, particularly for duplicated genes when compared to *A. thaliana, B. rapa,* and *B. oleracea*. Besides the traceable shared WGD events and their syntenic genes, this unique model is represented by a reasonable evolutionary time frame: paleo- (*A. thaliana*), meso- (*B. rapa* and *B. oleracea*), and neo- (*B. napus*) polyploidization events occurring ~35 million years, ~15 million years, and ~7500 years ago, respectively, along with a clear lineage relation including the two direct progenitors to the allopolyploid *B. napus*.

Comparative studies on the A and C genomes of *B. rapa*, *B. oleracea,* and *B. napus* led to a hypothesis that evolution of the subgenomes is symmetrical, and such asymmetry is a molecular mechanism driving Brassica allopolyploid genome evolution. Here asymmetry means divergence or differentiation, imbalance, disproportion, and unparallel genome or duplicated gene evolution, all related to structure determinants. The symmetrical evolution of the A and C genomes can be observed in several layers: (1) Asymmetrical transposable element (TE) amplification between least fractionation (LF), mid-fractionation (MF1), and most fractionation (MF2) within A or C, and between A and C, where there are more TEs in LF than MF1 and MF2, and more in C than A (Chalhoub et al. 2014; Liu et al. 2014; Lysak et al. 2005; Parkin et al. 2005; Wang et al. 2011); (2) asymmetrical tandem duplication amplification between the genomes A and C (Liu et al. 2014); (3) asymmetry in gene loss/deletion and HE, PAV, and CNV between LF, MF1, and MF2, and between A and C (Chalhoub et al. 2014; Liu et al. 2014; Lysak et al. 2005; Parkin et al. 2005; Snowdon et al. 2015; Stein et al. 2017; Wang et al. 2011); (4) divergence of duplicates' DNA sequences between LF, MF1, and MF2, and between A and C (Chalhoub et al. 2014; Liu et al. 2014; Lysak et al. 2005; Parkin et al. 2005; Wang et al. 2011; unpublished data); (5) asymmetrical divergence of duplicates' expression and alternative splicing variants (Chalhoub et al. 2014; Liu et al. 2014; Lysak et al. 2005; Parkin et al. 2005; Wang et al. 2011); (6) asymmetry in epigenetics–methylation, ncRNA (unpublished data); (7) recombination dominance of LF over MFs and An over Cn (unpublished data); and (8) asymmetry in trait QTLs and selection (unpublished data).

Of the structure variations, the present studies have highlighted important roles of HE in variations of genome structure and phenotypes (see Chap. 2 by Chevre et al., Chap. 4 by Sun et al. and Chap. 7 by Samans et al., this book). Segmental exchanges between homoeologous chromosomes are frequent and widespread throughout the *B. napus* genome. Earlier mapping studies with RFLP markers revealed both the loss and the duplication of homoeologous loci in mapping populations of natural accessions and re-synthesized polyploids (Gaeta et al. 2007; Osborn et al. 2003; Parkin et al. 1995; Pires et al. 2004; Sharpe et al. 1995; Udall et al. 2005), but their extent was showed to be much lower in natural *B. napus* than in re-synthesized lines (Chalhoub et al. 2014; Gaeta et al. 2007; Rousseau-Gueutin et al. 2016; Song et al. 1995). It was considered that re-synthesized lines are specifically prone to homoeologous rearrangements, including deletions, duplications, and translocations (Gaeta et al. 2007; Szadkowski et al. 2010; Xiong et al. 2011). Similarly, comparative genome analyses among extant *B. oleracea, B. rapa,* and *B. napus* have found little evidence for extensive rearrangement in genome microstructure (Rana et al. 2004). In large-scale genome sequencing projects, however, these HEs and caused genetic and phenotypic variation have had stronger evidence at the DNA sequence and expression levels (Chalhoub et al. 2014; Delourme et al. 2013; He et al. 2017; also see Chap. 4 by Sun et al. and Chap. 7 by Samans et al., this book). Chalhoub et al. (2014) detected a large number of HEs in eight genotypes including one re-synthetic line; of them, 17 HEs are large. Sequences from these genotypes revealed both shared and specific segmental HEs. HE sizes varied, and most frequent occurrence is between chromosomes An1-Cn1, An2-Cn2, and An9-Cn9. In total, there are more than 1300 genes involved in these HEs, particularly those carrying a replacement of Cn2 genomic fragment with FLC homologs by An2 in the Asian semi-winter oilseed forms Yudal and Aburamasari and a replacement of Cn9 by An10 in late-flowering swedes. These loci correspond to important QTLs for vernalization requirement and flowering time. By combining genetic maps and QTLs with parental re-sequencing data, multiple trait loci have involved in HEs where genetic positions may be different from their actual physical positions when compared to the Darmor-*bzh* reference genome (Chalhoub et al. 2014; Stein et al. 2017). The first clear example of HE in *B. napus* directly impacting a trait was that a translocation between A9 and C8 resulted in two copies of an A9 allele, affecting seed fiber content (Liu et al. 2014; Stein et al. 2017). The native *bnaA.ccr1.A9* alleles in some accessions are a functional mutant possibly affecting amino acids near the enzyme active sites expected to impair the enzyme activity. The fragment carrying the *bnaA. ccr1.A9* alleles replaced the C08 homoeolog accompanied with a frameshift mutation that resulted in complete loss of function of the gene on C08 due to a stop codon in exon 1 (Liu et al. 2012). For expression divergence of duplicated genes, the published (Chalhoub et al. 2014) and unpublished data from the whole genome-wide studies all showed that the majority of An and Cn homoeologous pairs have similar expression levels, and different tissues showed a portion (around 20%) of total genes with higher expression for An homoeologs than Cn homoeologs or vice versa. There is obvious subgenome dominance in gene expression as observed in many fractionated subgenomes including triplicated subgenomes in An and Cn and other old subgenomes in other plants (Liu et al. 2014; Schnable et al. 2011; Wang et al. 2011). Gene expression is generally inversely related to CpG, CHG, and CHH cytosine DNA methylation. Methyl bisulfite sequencing in Darmor-*bzh* showed 4–8% higher methylation in Cn genes than in their homoeologous An genes, and other several genotypes have similar trends (unpublished data), possibly because of greater transposon density in the Cn subgenome. Of the ~3100 gene pairs with differential gene body and/or untranslated region methylation between An and Cn homoeologs in both roots and leaves, 51% were equally expressed. Only ~34% showed higher expression for the less methylated homoeologs, and the remaining ~15% showed the opposite pattern. Effects of DNA methylation within or near genes on gene expression has been highlighted for its importance because emerging 24-nt RNA-directed DNA methylation mechanism might be of significance when considering that Cn carries more TE and produces much more 24-nt RNA which may act as a trans-factor to affect An gene expression (unpublished data).

Extremely narrow genetic diversity is expected in modern rapeseed breeding pools due to genetic bottlenecks from the small number of founder allopolyploidization events during the origin of *B. napus* (Allender and King 2010), strong adaptive selection in strict eco-geographic gene pools, and intensive agronomic selection during recent breeding for essential seed quality traits (Snowdon et al. 2015). Therefore, use of the diploid progenitor species harboring important variation particularly for disease resistance (Rygulla et al. 2007a, b; Werner et al. 2007) or to enhance heterotic pools, has become a focus in which large numbers of re-synthetised allotetraploid *B. napus* accessions were produced. However, very few of these re-synthetics were bred into elite commercial varieties due to generally weak performance including low fertility and vigour. Cytogenetic and sequencing analysis revealed that re-synthesized oilseed rape can have extreme structural variation, particularly HE, causing chromosome instable and less adaptive. However, the parents *B. rapa* and *B. oleracea* for re-synthetics should have more divergence in their genomes than those of extant natural oilseed rape, and thus, the former genomes/chromosomes should be less stable than the latter in terms of meiosis behavior. Therefore, it is reasoned that natural selection plays an important role in the establishment and maintenance of fertile natural allopolyploids that have stabilized chromosome inheritance and a few advantageous chromosomal rearrangements (Gaeta and Pires 2010). In fact, there were differences in structural variation in re-synthetic lines, e.g., 16.2 and 41.5% of the genome affected by the genomic rearrangement events in the two synthetic lines 1012-98 and R53, respectively, while 8% in the natural accession Express 617 (Stein et al. 2017). This suggests that selection may be a key to determine which structural variants are retained in successful *B. napus* allopolyploids. On the other hand, however, natural germplasm of oilseed rape presents rich genetic variation such as SNPs, small Indel, large segmental deletions, and duplication, and many present alleles are linked to traits among different varieties (Snowdon et al. 2015; unpublished data). Furthermore, polyploids have great plasticity that allows high-pressure selection to create new phenotypes. In other words, it is hypothesised that oilseed rape genetic diversity is harbored by genome structural plasticity waiting for selection. In summary, based on genome data and knowledge, taking advantage of polyploid plasticity, re-synthetics and introgression with a focus on breaking of "silencing" Cn will open a door to broaden genetic diversity.

As mentioned above, allopolyploid *B. napus* is an excellent model to study polyploid genome evolution, particularly in relation to structural variation such as asymmetry and HE generation. Structural variation is a genetic/genomic foundation of polyploid evolution and biodiversity formation. However, detailed structural variation study needs a "gold standard" pan-genome to enable exact detection of large-scale and small-scale structural variation and both reciprocal and non-reciprocal HE events. Given that structural variations such as HE, intra-genomic conversion, and deletion, are mediated by recombination, a detailed landscape of recombination rate along whole genome needs to be defined, in which heterozygosity existing in genomes should be resolved in calculating recombination rate and other related parameters such as LD block and haplotype structures. To explore reasons or molecular mechanisms of structural variation, 3D genomics or spatial structure studies are promising emerging approaches to uncover how or why structural variation is generated, for example, whether and how sequence similarity can promote genome fragmentation or HE, and what selection force are involved. Allopolyploids like *B. napus* carry two relative but distinct subgenomes, and thus, their different genetic components may comprise trans-factors functioning in subgenome interaction, such as 24-nt small RNA-directed DNA methylation, which may affect chromatin status and gene expression.

References

Ainouche ML, Jenczewski E (2010) Focus on polyploidy. New Phytol 186:1–4

Allender CJ, King GJ (2010) Origins of the amphiploid species *Brassica napus* L. investigated by chloroplast and nuclear molecular markers. BMC Plant Biol 10:54

Amborella Genome Project, A.G (2013) The Amborella genome and the evolution of flowering plants. Science 342:1241089

Badouin H, Gouzy J, Grassa CJ, Murat F, Staton SE, Cottret L, Lelandais-Brière C, Owens GL, Carrère S, Mayjonade B et al (2017) The sunflower genome provides insights into oil metabolism, flowering and Asterid evolution. Nature 546:148–152

Bancroft I, Morgan C, Fraser F, Higgins J, Wells R, Clissold L, Baker D, Long Y, Meng J, Wang X et al (2011) Dissecting the genome of the polyploid crop oilseed rape by transcriptome sequencing. Nat Biotechnol 29:762–766

Bayer PE, Hurgobin B, Golicz AA, Chan C-KK, Yuan Y, Lee H, Renton M, Meng J, Li R, Long Y et al (2017) Assembly and comparison of two closely related *Brassica napus* genomes. Plant Biotechnol J 15:1602–1610

Beilstein MA, Nagalingum NS, Clements MD, Manchester SR, Mathews S (2010) Dated molecular phylogenies indicate a Miocene origin for *Arabidopsis thaliana*. Proc Nat Acad Sci 107:18724–18728

Bowers JE, Chapman BA, Rong J, Paterson AH (2003) Unravelling angiosperm genome evolution by phylogenetic analysis of chromosomal duplication events. Nature 422:433–438

Bus A, Hecht J, Huettel B, Reinhardt R, Stich B (2012) High-throughput polymorphism detection and genotyping in *Brassica napus* using next-generation RAD sequencing. BMC Genom 13:281

Cai G, Yang Q, Yi B, Fan C, Edwards D, Batley J, Zhou Y (2014) A complex recombination pattern in the genome of allotetraploid *Brassica napus* as revealed by a high-density genetic map. PLoS ONE 9:e109910

Chalhoub B, Denoeud F, Liu S, Parkin IA, Tang H, Wang X, Chiquet J, Belcram H, Tong C, Samans B et al (2014) Early allopolyploid evolution in the post-neolithic *Brassica napus* oilseed genome. Science 345:950–953

Chen X, Li X, Zhang B, Xu J, Wu Z, Wang B, Li H, Younas M, Huang L, Luo Y et al (2013) Detection and genotyping of restriction fragment associated polymorphisms in polyploid crops with a pseudo-reference sequence: a case study in allotetraploid *Brassica napus*. BMC Genom 14:346

Clarke WE, Parkin IA, Gajardo HA, Gerhardt DJ, Higgins E, Sidebottom C, Sharpe AG, Snowdon RJ, Federico ML, Iniguez-Luy FL (2013) Genomic DNA enrichment using sequence capture microarrays: a novel approach to discover sequence nucleotide polymorphisms (SNP) in *Brassica napus* L. PLoS ONE 8:e81992

Clarke WE, Higgins EE, Plieske J, Wieseke R, Sidebottom C, Khedikar Y, Batley J, Edwards D, Meng J, Li R, Lawler CT, Bouquet J, Laga B, Chung W, Iniguez Luy F, Dyrszka E, Rae S, Stich B, Snowdon RJ, Sharpe AG, Ganal MW, Parkin IAP (2016) A high-density SNP genotyping array for *Brassica napus* and its ancestral diploid species based on optimised selection of single-locus markers in the allotetraploid genome. Theor Appl Genet 129:1887–1899.

Comai L (2005) The advantages and disadvantages of being polyploid. Nat Rev Genet 6:836–846

Dalton-Morgan J, Hayward A, Alamery S, Tollenaere R, Mason AS, Campbell E, Patel D, Lorenc MT, Yi B, Long Y et al (2014) A high-throughput SNP array in the amphidiploid species *Brassica napus* shows diversity in resistance genes. Funct Integrative Genomics 14:643–655

Dassanayake M, Oh DH, Haas JS, Hernandez A, Hong H, Ali S (2013) The genome of the extremophile crucifer *Thellungiella parvula*. Nat Genet 43:913–918

Delourme R, Falentin C, Fomeju B, Boillot M, Lassalle G, André I, Duarte J, Gauthier V, Lucante N, Marty A et al (2013) High-density SNP-based genetic map development and linkage disequilibrium assessment in *Brassica napus* L. BMC Genom 14:120

Dubcovsky J, Dvorak J (2007) Genome plasticity a key factor in the success of polyploid wheat under domestication. Science 316:1862–1866

Edwards D, Batley J, Snowdon RJ (2013) Accessing complex crop genomes with next-generation sequencing. Theor Appl Genet 126:1–11

Gaeta RT, Pires JC (2010) Homoeologous recombination in allopolyploids: the polyploid ratchet. New Phytol 186:18–28

Gaeta R, Pires J, Iniguez-Luy F, Leon E, Osborn T (2007) Genomic changes in resynthesized *Brassica napus* and their effect on gene expression and phenotype. Plant Cell 19:3403–3417

Gnerre S, MacCallum I, Przybylski D, Ribeiro FJ, Burton JN, Walker BJ, Sharpe T, Hall G, Shea TP, Sykes S et al (2011) High-quality draft assemblies of mammalian genomes from massively parallel sequence data. Proc Nat Acad Sci 108:1513–1518

Haudry A, Platts AE, Vello E, Hoen DR, Leclercq M, Williamson RJ (2013) An atlas of over 90,000 conserved noncoding sequences provides insight into crucifer regulatory regions. Nat Genet 45:891–898

He Z, Wang L, Harper AL, Havlickova L, Pradhan AK, Parkin IAP, Bancroft I (2017) Extensive homoeologous genome exchanges in allopolyploid crops revealed by mRNAseq-based visualization. Plant Biotechnol J 15:594–604

Hu TT, Pattyn P, Bakker EG, Cao J, Cheng JF, Clark RM (2011) The *Arabidopsis lyrata* genome sequence and the basis of rapid genome size change. Nat Genet 43:476–481

Huang S, Deng L, Guan M, Li J, Lu K, Wang H, Fu D, Mason AS, Liu S, Hua W (2013) Identification of genome-wide single nucleotide polymorphisms in allopolyploid crop *Brassica napus*. BMC Genom 14:717

Jackson S, Chen ZJ (2010) Genomic and expression plasticity of polyploidy. Current Opin Plant Biol 13:153–159

Jarvis DE, Ho YS, Lightfoot DJ, Schmockel SM, Li B, Borm TJA, Ohyanagi H, Mineta K, Michell CT, Saber N et al (2017) The genome of *Chenopodium quinoa*. Nature 542:307–312

Jiao Y, Wickett NJ, Ayyampalayam S, Chanderbali AS, Landherr L, Ralph PE, Tomsho LP, Hu Y, Liang H, Soltis PS (2011) Ancestral polyploidy in seed plants and angiosperms. Nature 473:97–100

Jiao Y, Peluso P, Shi J, Liang T, Stitzer MC, Wang B, Campbell MS, Stein JC, Wei X, Chin CS et al (2017) Improved maize reference genome with single-molecule technologies. Nature 546:524–527

Johnston JS, Pepper AE, Hall AE, Chen ZJ, Hodnett G, Drabek J, Lopez R, Price HJ (2005) Evolution of genome size in Brassicaceae. Ann Bot 95:229–235

Kagale S, Robinson SJ, Nixon J, Xiao R, Huebert T, Condie J, Kessler D, Clarke WE, Edger PP, Links MG et al (2014) Polyploid evolution of the Brassicaceae during the Cenozoic Era. Plant Cell 26:2777–2791

Kajitani R, Toshimoto K, Noguchi H, Toyoda A, Ogura Y, Okuno M, Yabana M, Harada M, Nagayasu E, Maruyama H et al (2014) Efficient de novo assembly of highly heterozygous genomes from whole-genome shotgun short reads. Genome Res 24:1384–1395

Larkan NJ, Lydiate DJ, Parkin IAP, Nelson MN, Epp DJ, Cowling WA, Rimmer SR, Borhan MH (2013) The *Brassica napus* blackleg resistance gene *LepR3* encodes a receptor-like protein triggered by the *Leptosphaeria maculans* effector AVRLM1. New Phytol 197:595–605

Liu C, Wang J, Huang T, Wang F, Yuan F, Cheng X, Zhang Y, Shi S, Wu J, Liu K (2010) A missense mutation in the VHYNP motif of a DELLA protein causes a semi-dwarf mutant phenotype in *Brassica napus*. Theor Appl Genet 121:249–258

Liu L, Stein A, Wittkop B, Sarvari P, Li J, Yan X, Dreyer F, Frauen M, Friedt W, Snowdon RJ (2012) A knockout mutation in the lignin biosynthesis gene CCR1 explains a major QTL for acid detergent lignin content in *Brassica napus* seeds. Theor Appl Genet 124:1573–1586

Liu L, Qu C, Wittkop B, Yi B, Xiao Y, He Y, Snowdon RJ, Li J (2013) A high-density SNP map for accurate mapping of seed fibre QTL in *Brassica napus* L. PLoS ONE 8:e83052

Liu S, Liu Y, Yang X, Tong C, Edwards D, Parkin IA, Zhao M, Ma J, Yu J, Huang S et al (2014) The *Brassica oleracea* genome reveals the asymmetrical evolution of polyploid genomes. Nat Commun 5:3930

Liu J, Hua W, Hu Z, Yang H, Zhang L, Li R, Deng L, Sun X, Wang X, Wang H (2015) Natural variation in ARF18 gene simultaneously affects seed weight and silique length in polyploid rapeseed. Proc Natl Acad Sci USA 112:5123–5132

Luo R, Liu B, Xie Y, Li Z, Huang W, Yuan J, He G, Chen Y, Pan Q, Liu Y et al (2012) SOAPdenovo2: an empirically improved memory-efficient short-read *de novo* assembler. GigaScience 1:18

Lyons E, Pedersen B, Kane J, Alam M, Ming R, Tang H, Wang X, Bowers J, Paterson A, Lisch D et al (2008) Finding and comparing syntenic regions among Arabidopsis and the outgroups papaya, poplar, and grape: CoGe with rosids. Plant Physiol 148:1772–1781

Lysak MA, Koch MA, Pecinka A, Schubert I (2005) Chromosome triplication found across the tribe Brassiceae. Genome Res 15:516–525

Mason AS, Higgins EE, Stein A, Werner C, Batley J, Parkin IAP, Snowdon RJ (2017) A user guide to the Brassica 60K Illumina Infinium™ SNP genotyping array. Theor Appl Genet 130:621–633.

Masterson J (1994) Stomatal size in fossil plants: evidence for polyploidy in majority of angiosperms. Science 264:421–424

Osborn TC, Butrulle DV, Sharpe AG, Pickering KJ, Parkin IAP, Parker JS, Lydiate DJ (2003) Detection and effects of a homeologous reciprocal transposition in *Brassica napus*. Genetics 165:1569–1577

Ott J, Kamatani Y, Lathrop M (2011) Family-based designs for genome-wide association studies. Nat Rev Genet 12:465–474

Parkin IAP, Sharpe AG, Keith DJ, Lydiate DJ (1995) Identification of the A and C genomes of amphidiploid *Brassica napus* (oilseed rape). Genome 38: 1122–1131

Parkin IA, Gulden SM, Sharpe AG, Lukens L, Trick M, Osborn TC, Lydiate DJ (2005) Segmental structure of the *Brassica napus* genome based on comparative analysis with *Arabidopsis thaliana*. Genetics 171:765–781

Pires JC, Zhao J, Schranz ME, Leon EJ, Quijada PA, Lukens LN, Osborn TC (2004) Flowering time divergence and genomic rearrangements in resynthesized Brassica polyploids (Brassicaceae). Biol J Linn Soc 82:675–688

Raman H, Dalton-Morgan J, Diffey S, Raman R, Alamery S, Edwards D, Batley J (2014) SNP markers-based map construction and genome-wide linkage analysis in *Brassica napus*. Plant Biotechnol J 12:851–860

Rana D, Boogaart T, O'Neill CM, Hynes L, Bent E, Macpherson L, Park JY, Lim YP, Bancroft I (2004) Conservation of the microstructure of genome segments in *Brassica napus* and its diploid relatives. Plant J 40:725–733

Rousseau-Gueutin M, Morice J, Coriton O, Huteau V, Trotoux G, Nègre S, Falentin C, Deniot G, Gilet M, Eber F et al (2016) The impact of open pollination on the structural evolutionary dynamics, meiotic behavior and fertility of resynthesized allotetraploid *Brassica napus* L. G3: genes. Genomes Genet 7:705–717

Rygulla W, Friedt W, Seyis F, Lühs W, Eynck C, Tiedemann AV, Snowdon RJ (2007a) Combination of resistance to *Verticillium longisporum* from zero erucic acid *Brassica oleracea* and oilseed *Brassica rapa* genotypes in resynthesized rapeseed (*Brassica napus*) lines. Plant Breed 126:596–602

Rygulla W, Snowdon RJ, Eynck C, Koopmann B, von Tiedemann A, Lühs W, Friedt W (2007b) Broadening the genetic basis of *Verticillium longisporum* resistance in *Brassica napus* by interspecific hybridization. Phytopathology 97:1391–1396

Schnable JC, Springer NM, Freeling M (2011) Differentiation of the maize subgenomes by genome dominance and both ancient and ongoing gene loss. Proc Nat Acad Sci 108:4069–4074

Sharpe AG, Parkin IAP, Keith DJ, Lydiate DJ (1995) Frequent nonreciprocal translocations in the amphidiploid genome of oilseed rape (*Brassica napus*). Genome 38:1112–1121

Slotte T, Hazzouri KM, Ågren JA, Koenig D, Maumus F, Guo YL (2013) The *Capsella rubella* genome and the genomic consequences of rapid mating system evolution. Nat Genet 45:831–835

Snowdon RJ, Iniguez Luy FL (2012) Potential to improve oilseed rape and canola breeding in the genomics era. Plant Breed 131:351–360

Snowdon RJ, Abbadi A, Kox T, Schmutzer T, Leckband G (2015) Heterotic haplotype capture: precision breeding for hybrid performance. Trends Plant Sci 20:410–413

Soltis D, Soltis P (2003) The role of phylogenetics in comparative genetics. Plant Physiol 132:1790–1800

Song K, Lu P, Tang K, Osborn TC (1995) Rapid genome change in synthetic polyploids of Brassica and its implications for polyploid evolution. Proc Nat Acad Sci 92:7719–7723

Stein A, Coriton O, Rousseau-Gueutin M, Samans B, Schiessl SV, Obermeier C, Parkin IA, Chèvre AM, Snowdon RJ (2017) Mapping of homoeologous chromosome exchanges influencing quantitative trait variation in *Brassica napus*. Plant Biotechnol J 15:1478–1489

Sun F, Fan G, Hu Q, Zhou Y, Guan M, Tong C, Li J, Du D, Qi C, Jiang L et al (2017) The high-quality genome of *Brassica napus* cultivar 'ZS11' reveals the introgression history in semi-winter morphotype. Plant J 92:452–468

Szadkowski E, Eber F, Huteau V, Lodé M, Huneau C, Belcram H, Coriton O, Manzanares-Dauleux MJ, Delourme R, King GJ et al (2010) The first meiosis of resynthesized *Brassica napus*, a genome blender. New Phytol 186:102–112

The French-Italian Public Consortium for Grapevine Genome Characterization (2007) The grapevine genome sequence suggests ancestral hexaploidization in major angiosperm phyla. Nature 449:463–467

Trick M, Long Y, Meng J, Bancroft I (2009) Single nucleotide polymorphism (SNP) discovery in the polyploid *Brassica napus* using Solexa transcriptome sequencing. Plant Biotechnol J 7:334–346

Udall JA, Quijada PA, Osborn TC (2005) Detection of chromosomal rearrangements derived from homeologous recombination in four mapping populations of *Brassica napus* L. Genetics 169:967–979

Van de Peer Y (2011) A mystery unveiled. Genome Biol 12:113

Vision TJ, Brown DG, Tanksley SD (2000) The origins of genomic duplications in Arabidopsis. Science 290:2114–2117

Wang X, Wang H, Wang J, Sun R, Wu J, Liu S (2011) The genome of the mesopolyploid crop species *Brassica rapa*. Nat Genet 43:1035–1039

Werner S, Diederichsen E, Frauen M, Schondelmaier J, Jung C (2007) Genetic mapping of clubroot resistance genes in oilseed rape. Theor Appl Genet 116:363

Xia S, Wang Z, Zhang H, Hu K, Zhang Z, Qin M, Dun X, Yi B, Wen J, Ma C et al (2016) Altered transcription and neofunctionalization of duplicated genes rescue the harmful effects of a chimeric gene in *Brassica napus*. Plant Cell 28:2060–2078

Xiong Z, Gaeta RT, Pires JC (2011) Homoeologous shuffling and chromosome compensation maintain genome balance in resynthesized allopolyploid *Brassica napus*. Proc Nat Acad Sci 108:7908–7913

Yang J, Liu D, Wang X, Ji C, Cheng F, Liu B, Hu Z, Chen S, Pental D, Ju Y et al (2016) The genome sequence of allopolyploid *Brassica juncea* and analysis of differential homoeolog gene expression influencing selection. Nat Genet 48:1225–1232

Yi B, Zeng F, Lei S, Chen Y, Yao X, Zhu Y, Wen J, Shen J, Ma C, Tu J et al (2010) Two duplicate CYP704B1-homologous genes *BnMs1* and *BnMs2* are required for pollen exine formation and tapetal development in *Brassica napus*. The Plant J 63:925–938

Zhang D, Hua Y, Wang X, Zhao H, Shi L, Xu F (2014) A high-density genetic map identifies a novel major QTL for boron efficiency in oilseed rape (*Brassica napus* L.). PLoS ONE 9:e112089

Zimin AV, Marçais G, Puiu D, Roberts M, Salzberg SL, Yorke JA (2013) The MaSuRCA genome assembler. Bioinformatics 29:2669–2677

The manufacturer's authorised representative in the EU is Springer
Nature Customer Service Centre GmbH, Europaplatz 3, 69115 Heidelberg,
Germany. If you have any concerns regarding our products, please
contact ProductSafety@springernature.com

Printed and bound by CPI Group (UK) Ltd, Croydon, CR0 4YY
30/09/2026
02225658-0020